Springer Series on Polymer and Composite Materials

Series editor

Susheel Kalia, Dehradun, India

More information about this series at http://www.springer.com/series/13173

Pradip Kumar Dutta
Editor

Chitin and Chitosan for Regenerative Medicine

Springer

Editor
Pradip Kumar Dutta
Department of Chemistry
MN National Institute of Technology
Allahabad, Uttar Pradesh
India

ISSN 2364-1878 ISSN 2364-1886 (electronic)
Springer Series on Polymer and Composite Materials
ISBN 978-81-322-2510-2 ISBN 978-81-322-2511-9 (eBook)
DOI 10.1007/978-81-322-2511-9

Library of Congress Control Number: 2015944502

Springer New Delhi Heidelberg New York Dordrecht London

Printed on acid-free paper

Springer (India) Pvt. Ltd. is part of Springer Science+Business Media (www.springer.com)

Preface

Polymers, the spectacular world of macromolecules have delivered their services to our society in various fields of life. In most cases, biopolymers are vulnerable and their identity to specific functions, especially for medicinal applications are very important. Owing to their biocompatible, nontoxic, and biodegradable nature, biopolymers (rather to say biomaterials in a broad sense) give a better option to be used in advancement of the biomedical field. Regenerative medicine is an area where tissue engineering, stem cell research, gene therapy, and therapeutic cloning are the collective work toward rebuilding or replacement of missing or injured body parts. Regenerative medicine is a blessing for our society where around 10 % of the entire world's populations suffering from a disability. Biomaterials, especially chitin and chitosan match up all the characteristics required in the field of regenerative medicine.

The present two volumes entitled "Chitin and Chitosan for Regenerative Medicine: Part I—Focus on Tissue Engineering and Part II—Focus on Therapeutics, Functionalization and Computer-aided Techniques" were conceived to provide broad and innovative information not only related to tissue engineering but also on other therapeutic and biomedical applications based on chitin/chitosan and their various derivatives in the field of regenerative medicine like quantum dots, nano-medicines, drug delivery, hydrogels, and scaffolds. The book consists of 13 chapters written in such a manner that will surely meet the expectations of scientists as well as researchers from various disciplines.

Part I will mainly focus on tissue engineering and its applications for regenerative medicine. "Chitosan Hydrogels for Regenerative Engineering" reviews the various methods used for preparing chitosan-based hydrogels and their applications as cell, protein and drug delivery vehicles in pharmaceutical, biomedical sciences, and tissue engineering. "Prospects of Bioactive Chitosan-Based Scaffolds in Tissue Engineering and Regenerative Medicine" will focus on the synthesis of several biologically active chitin and chitosan-based scaffolds for tissue engineering and other related strategies to enhance the activity of prepared scaffolds. "Chitosan-Based Scaffolds for Cartilage Regeneration" deals with issues related to cartilage damage which causes osteoarthritis and how different types of chitosan-based scaffolds are synthesized that can be utilized for regeneration of damaged tissues.

"Composite Chitosan Calcium Phosphate Scaffolds for Cartilage Tissue Engineering" explains the processes for the fabrication of chitosan-calcium phosphate (CHI-CaP) composite scaffolds for the enhancement in the field of cartilage tissue engineering along with its physical characteristics and possible aspects of the scaffold's degradation. "Chitosan-Gelatin Composite Scaffolds in Bone Tissue Engineering" highlights the importance of chitosan-gelatin-based composite scaffolds in bone tissue engineering along with its preparation techniques and its physical and biological characteristics. "Chitin and Chitosan Nanocomposites for Tissue Engineering" provides novel approaches at the juncture between biology and nanotechnology to develop encouraging ecofriendly biopolymer nanocomposites based on chitin and chitosan. "Chitin, Chitosan and Silk Fibroin Electrospun Nanofibrous Scaffolds: A Prospective Approach for Regenerative Medicine" discusses the current advancements in the field of electrospun nanofibrous scaffolds-based chitin, chitosan, and silk fibroin highlighting tissue engineering for regenerative medicine. Part II will focus on various therapeutics and computer aided techniques for regenerative medicine. Here, "Chitosan: A Potential Therapeutic Dressing Material for Wound Healing" will mainly focuses on how to develop dressing material for wound healing by combining natural biopolymers (chitin and chitosan), synthetic polymer and nanoparticles which can be available as a biomaterial for regenerative medicine. "Recent Advances in Chitosan Based Nanomedicines for Cancer Chemotherapy" highlights the fabrication process and the possible function of chitosan-based derivatives in cancer chemotherapy. "Chitosan: A Promising Substrate for Regenerative Medicine in Drug Formulation" reviews the chitosan-based formulation with potential medicinal uses to deliver an enhanced knowledge of utilization of chitosan in regenerative medicine. "D-Glucosamine and *N*-Acetyl D-Glucosamine: Their Potential Use as Regenerative Medicine" explains the importance of chitin and chitosan oligosaccharides, N-acetyglucosamine, and D-glucosamine, as drug carriers for molecular therapeutics like in the drug and gene delivery systems and also its role in imaging for tumor and cancer detection. "Functionalized Chitosan: A Quantum Dot Based Approach for Regenerative Medicine" explains how chitosan-quantum dots are utilized in regenerative medicine and also discusses their potential barriers of using techniques. "Development and Selection of Porous Scaffolds using Computer Aided Tissue Engineering" describes the selection of biomaterials, its facilitated properties, experimental methods, knowledge of computer-based biomodeling to synthesize scaffolds using computer aided tissue engineering (CATE) which acts as an important tool for the fabrication of scaffolds especially for regenerative medicine.

Last but not the least, I would like to thank all contributors for their generous support, the publisher for accepting our book, all research scholars and staff members of Polymer Research Laboratory, Chemistry Department, and the administrative head of MNNIT Allahabad, India for their encouragement and cooperation, without which it would have been extremely difficult to complete this task on time.

Allahabad, India, April 2015 Pradip Kumar Dutta

Contents

About the Editor

Dr. Pradip Kumar Dutta obtained his M.Sc. (Chemistry) and Ph.D. (Polymer Chemistry) from Indian Institute of Technology, Kharagpur in 1987 and 1993, respectively. He started his career as a research scientist in Birla Research Institute, Nagda, Madhya Pradesh, India in 1992, before receiving his Ph.D. degree. His strong interest in academics drew him to one of the best engineering institutes, Shri G.S. Institute of Technology and Science (SGSITS) in Indore, Madhya Pradesh in 1993. He served there for about 10 years in different posts such as lecturer, senior lecturer in chemistry and coordinator for Continuing Education Program (All India Council for Technical Education, Government of India). In 2002, Dr. Dutta joined as Reader in Chemistry in Motilal Nehru National Institute of Technology (Deemed University), Allahabad, India. His progress in academic activities through teaching, research projects, guiding doctoral and postdoctoral students, research collaboration with similar researchers in India and abroad is continuing. His research interests include synthesis and modification of polymers, nanomaterials/composites, functional polymers, drug delivery, wound management, tissue engineering, food preservation, etc. He has about 200 papers published in national and international journals, 18 book articles/chapters, delivered 20 h video-lecture programs, and prepared 18 course modules. He has already supervised 16 M.Sc., 18 M.Tech./M.Phil./M.Pharma and nine Ph.D. theses. Presently, five Ph.D. students are working under him. He was awarded Commonwealth Academic Staff Fellowship-2007 and visited York University, York, UK. He was also awarded the Chinese Academy of Sciences & Third World Academy of Sciences (CAS-TWAS) visiting scholar fellowship 2004, 2006 and 2009 and visited Changchun Institute of Applied Chemistry, Changchun, China for collaborative research work. Besides these, Dr. Dutta has extensively visited countries like South Korea, Japan, Turkey, Switzerland, USA for academic purposes. He is the founder editor of International Journal of Asian Chitin Journal since 2005, founder member of Indian Chitin and Chitosan Society, reviewer of various international journals, and Fellow of Royal Society of Chemistry (UK).

Part I
Focus on Tissue Engineering

Chitosan Hydrogels for Regenerative Engineering

Aiswaria Padmanabhan and Lakshmi S. Nair

Abstract Research in the field of hydrogels has been actively growing for the past couple of decades. Hydrogels are crosslinked polymers with high water content. They can be prepared from natural, synthetic, and composite polymers using different chemical and physical crosslinking methods. Hydrogels have been widely explored for the delivery of bioactive molecules, drugs, and for other therapeutic applications. Chitosan-based hydrogels have unique advantages owing to their biocompatibility, biodegradability, antimicrobial activity, mucoadhesivity, and low toxicity. This chapter reviews the different methods used for preparing chitosan-based hydrogels and their applications as cell, protein, and drug delivery vehicles to support tissue regeneration.

Keywords Chitosan · Hydrogels · Therapeutic applications · Regenerative engineering

1 Introduction

1.1 Chitosan

Chitosan is derived from chitin, which is a naturally occurring linear polysaccharide. Chitin is composed of repeating units of *N*-acetyl-D-glucosamine, as shown in Fig. 1.

A. Padmanabhan · L.S. Nair
Department of Materials Science and Engineering, University of Connecticut, Storrs, CT 06269, USA

L.S. Nair
Department of Biomedical Engineering, University of Connecticut, Storrs, CT 06269, USA

L.S. Nair (✉)
Department of Orthopaedic Surgery, Institute for Regenerative Engineering, Raymond and Beverly Sackler Center for Biomedical, Biological, Physical and Engineering Sciences, University of Connecticut Health Center, E-7041, MC-3711 263 Farmington Avenue, Farmington, CT 06030, USA
e-mail: nair@uchc.edu

P.K. Dutta (ed.), *Chitin and Chitosan for Regenerative Medicine*, Springer Series on Polymer and Composite Materials, DOI 10.1007/978-81-322-2511-9_1

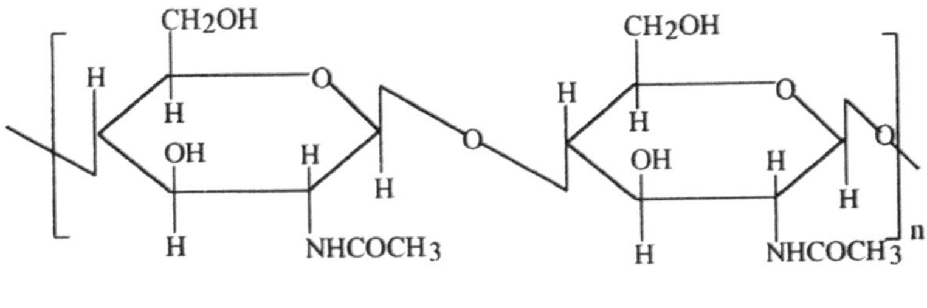

Chitin

Fig. 1 Structure of chitin [1]

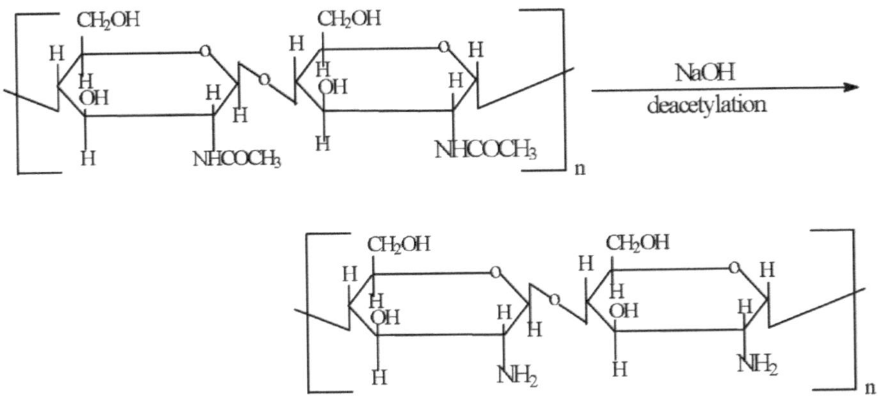

Fig. 2 Preparation of chitosan by the deacetylation of chitin [1]

Crab and shrimp shells currently serve as the common source of chitin. Despite its favorable structural and biological properties, the insolubility of chitin in water and common organic solvents is a limitation that prevents the extensive use of chitin for biomedical applications [1, 2]. So several approaches have been investigated to increase the aqueous solubility of chitin, which include deacetylation [2], carboxymethylation [3], and sulfation [4]. Among these, deacetylation of chitin by alkaline treatment is the most commonly used approach. The derivative of chitin with degree of deacetylation of approximately 50 % is known as "chitosan" and it is soluble in aqueous acidic solutions [2]. Chitosan is therefore a copolymer that comprises of *N*-acetyl-D-glucosamine and deacetylated D-glucosamine units [1, 2, 4–7]. The structure of chitosan obtained by complete deacetylation of chitin is shown in Fig. 2.

The most commonly used methods to determine the degree of deacetylation are ^{1}H (liquid state), ^{13}C (solid state), and ^{15}N (solid state) nuclear magnetic resonance (NMR) spectroscopy. Among these, ^{1}H is extensively used to determine acetyl groups in soluble samples. Other methods include infra-red (IR) and ultraviolet (UV) spectrometry, elemental analysis, potentiometric titration, and enzymatic reaction [2]. Degree of deacetylation can significantly affect the biological as well as physiochemical properties of chitosan [8]. Chitosan in solid state is reported to be a semicrystalline polymer [2]. Cartier et al. determined the crystallinity of chitosan using X-ray and electron diffraction methods, which allow the identification of the unit cell parameters. Electron diffraction of fully deacetylated chitosan single crystal indicated an orthorhombic unit cell with lattice parameters $a = 0.807$, $b = 0.844$, and $c = 1.034$ nm [9]. Molecular weight is another important parameter that determines the physicochemical and biological properties of chitosan. It varies with the chitin source from which chitosan is obtained and decreases with increase in degree of deacetylation [8]. Several methods can be used to determine the molecular weight of chitosan. The selection of appropriate solvent system that does not lead to significant aggregation of chitosan is necessary while determining the molecular weight. A 0.3 M acetic acid/0.2 M sodium acetate (pH = 4.5) solution has been reported to be a suitable solvent system [10]. Mark–Houwink equation given below Eq. (1) is commonly used to determine the viscosity average molecular weight of chitosan.

$$[\eta] = \mathrm{KM}^{\mathrm{a}} \tag{1}$$

Where, η is the intrinsic viscosity, M is the molecular weight, and K and a are the experimentally determined parameters for a given solvent system. If K and a are known, molecular weight can then be obtained by intrinsic viscosity measurements [11].

As mentioned earlier, the main motivation behind modifying chitin to obtain chitosan is to take advantage of the improved solubility. In solutions of pH less than ~6.0, the amine groups in chitosan become protonated, thereby allowing chitosan to be soluble in aqueous acidic solutions. The protonation does not occur in basic solutions, leading to a solubility–insolubility transition at pH value of ~6.5 [5]. The exact pH above which chitosan becomes insoluble, however, depends on the degree of deacetylation, the distribution of the acetyl groups on the linear chain, and the molecular weight. Another advantage of chitosan is that it can be chemically modified using the reactive amine group at the C-2 position or hydroxyl groups at the C-3 and C-6 positions to alter its functionality. Particularly, this allows the functionalization of chitosan for different biological applications [2]. Table 1 lists some of the chemically modified chitosan derivatives that are water soluble along with some of their biomedical applications.

1.2 Hydrogels

Hydrogels are 3D networks of hydrophilic polymers that have the ability to imbibe aqueous fluids and swell while remaining mechanically stable [23–25]. Hydrogels

Table 1 Functionalized derivatives of chitosan [2, 12]

Derivatives of chitosan	Solubility	Example of applications	References
Carboxymethylchitosan	Water soluble	Drug delivery, tissue engineering	[13, 14]
Glycol chitosan	Water soluble	Drug delivery	[15, 16]
PEG-grafted chitosan	Water soluble	Drug delivery, tissue engineering	[17]
Sulfated chitosan	Water soluble	Tissue engineering, anticoagulant	[18, 19]
N-Methylene phosphonic chitosan	Water soluble	Gene delivery	[20, 21]
Cyclodextrin grafted chitosan	Water soluble	Drug delivery	[22]

are attractive candidates for regenerative engineering applications mainly due to their biocompatibility, biodegradability, ability to mimic natural tissue, and promote cell attachment and proliferation [25, 26]. Due to the excellent permeability of hydrogels, they are used as cell and drug delivery vehicles. Hydrogels used for regenerative engineering applications can be either preformed or injectable. Preformed hydrogels are fabricated in vitro followed by transplantation in vivo [27]. Preformed hydrogels can be prepared in different forms—membranes, coatings, pressed powder matrices, molded solids, encapsulated solids, and microparticles—depending on the intended application [23–25, 28]. On the other hand, injectable hydrogels are formed by injecting polymer solutions into the body that can then form gels in situ [27]. Injectable hydrogels have shown potential for minimally invasive delivery of cells as well as biomolecules due to their easily tailorable physical properties. The starting polymers are free-flowing liquid solutions that are easy to handle, can form gels according to the size and shape of the wound/defect area [29], and also provide homogenous distribution of cells in gels [28]. Hydrogels are also useful for the targeted, controlled, and sustained release of drugs and bioactive molecules. Drug-releasing mechanisms in hydrogels are mainly classified into diffusion-controlled, swelling-controlled, or degradation-controlled [23]. Studies have shown the feasibility to develop stimuli-responsive hydrogels, wherein the gel properties can be modulated by changing the external environmental conditions such as pH, temperature, and ionic strength. These smart hydrogel systems are useful as cell or drug delivery systems [24]. Hydrogels used in biomedical applications can also be classified in terms of whether they are derived from natural or synthetic polymers. Table 2 provides a list of some of the natural and synthetic polymers used for hydrogel preparation. Hydrogels made from natural polymers have the advantages of bioactivity, biocompatibility, and biodegradability but suffer in terms of low mechanical strength, poor reproducibility, and the potential presence of pathogens that may lead to immune responses. On the other hand, hydrogels derived from synthetic polymers can be effectively and consistently tailored to provide desired mechanical properties and degradability but lack inherent bioactivity [23].

Table 2 Natural and synthetic polymers used in hydrogel fabrication [30, 31]

Natural polymers	Synthetic polymers
Agarose	Poly(ethylene glycol) (PEG) and its derivatives
Chitosan	Pluronics
Cellulose	Poly(acrylamide) (PAAm)
Hyaluronic acid	Poly(vinylpyrrolidone) (PVP)
Elastin	Poly(acrylic acid) (PAA)
Collagen	Poly(*N*-isopropylacrylamide) (PNIPAm)
Gelatin	Poly(vinyl alcohol) (PVA)
Chondroitin sulfate	Poly(hydroxyethyl methacrylate) (PHEMA)

2 Chitosan Hydrogels

Chitosan has been widely investigated for a variety of applications in the biomedical industry. The beneficial properties of chitosan are its biocompatibility, biodegradability [32], antimicrobial activity [33], mucoadhesivity [34], wound healing and hemostatic properties [35], and low toxicity [32].

Biodegradability of chitosan is dependent on different factors such as the degree of deacetylation, distribution of amine groups, the presence of acetyl groups, and molecular weight of the polymer [7]. Chitosan can be degraded using enzymes such as lysozyme, which is a glycosidic hydrolase present in the human body. Lysozyme is reported to hydrolyze the β(1-4) linkages between *N*-acetylglucosamine and glucosamine [36]. Therefore, the degree of acetylation plays an important role in enzyme-mediated degradation of chitosan. Chitosan with higher degree of deacetylation undergoes limited degradation, whereas increasing acetylation results in higher degradation [37]. Antimicrobial property of chitosan stems from its interaction with the negatively charged cell surfaces by affecting cellular permeability or by chitosan's interaction with DNA, thereby inhibiting microbial RNA synthesis [38]. Mucoadhesive property of chitosan also arises from its positively charged amine groups that can interact with the negatively charged groups in the mucin molecule [34]. Hemostatic property is due to the presence of positively charged groups in chitosan that interact with the negatively charged surfaces of blood cells [7]. Chitosan-based hemostatic bandage called Hemcon® has been FDA approved [39]. Due to these unique biological properties, extensive research has gone to develop chitosan-based hydrogels for biomedical applications. Both physical and chemical crosslinking methods can be used to develop chitosan hydrogels.

2.1 *Physically Crosslinked Chitosan Hydrogels*

Physically crosslinked hydrogels are formed by physical interactions such as electrostatic, hydrophobic, or hydrogen bonding between the polymer chains. Hydrogel formation can be induced by mixing the constituents under suitable conditions to initiate the gelation. Physical crosslinking is usually triggered by stimuli such as pH and temperature. Controlling the concentration of chitosan with respect to that of other components and thereby controlling the polymer interactions has been shown to significantly control the gel properties [7, 40]. Different methods of physical crosslinking of chitosan hydrogels are discussed below.

2.1.1 Thermogelling Hydrogels

Thermogelling chitosan hydrogels undergo gelation in response to changes in temperature [41]. Thermogelling

chitosan hydrogels have been investigated as injectable carriers for biomedical applications [42]. One of the most extensively studied thermogelling chitosan formulations is chitosan/β-glycerophosphate (β-GP) system that can undergo sol–gel transition at or near physiological temperature. Chitosan/β-GP has been investigated mainly as cell, drug, or growth factor delivery system due to the mild gelation process and the feasibility to deliver them in a minimally invasive manner [43]. For bone regeneration, Niranjan et al. synthesized a novel thermosensitive carrier comprising of chitosan/β-GP doped with zinc. Metals such as zinc have shown to provide thermal resistance and antibacterial property. Briefly, chitosan was dissolved in 0.1 M acetic acid. Zinc sulfate solution was added under stirring at a 1:1 (v/v) ratio followed by the addition of β-GP at a 1:9 (v/v) ratio. Addition of β-GP to the chitosan solution changed the solution pH from 3 to 7. Gelation time of zinc-doped chitosan/β-GP solution was ~5 min. The gelation resulted in a porous hydrogel with a pore size of around 200 μm [44]. In addition to β-GP, inorganic phosphate salts have also been shown to impart thermogelling property to chitosan. Nair et al. prepared an injectable thermogelling chitosan–inorganic phosphate hydrogel. Briefly, chitosan was dissolved in 0.5 % acetic acid solution with pH of ~5.6. Addition of ammonium hydrogen phosphate (AHP) to chilled chitosan solution increased the pH to 7–7.2, and the resultant solution showed effective sol–gel transition at or near physiological temperature. Figure 3 shows the thermogelation of chitosan–AHP solution.

Depending on the amount of AHP, the gelling time was variable from 5 min to 30 h. Gelling time decreased with increase in the concentration of AHP and chitosan. The mechanism of thermogelation is presumed to be a combination of electrostatic as well as hydrophobic attractions similar to that of chitosan/β-GP solution [45]. Liu et al. used chitosan-4-thio-butylamidine (CS-TBA), β-GP, and nano-hydroxyapatite (nano-HA) to develop a thermoresponsive composite hydrogel that can gel in situ. Thiolated chitosan can dissolve in neutral pH unlike

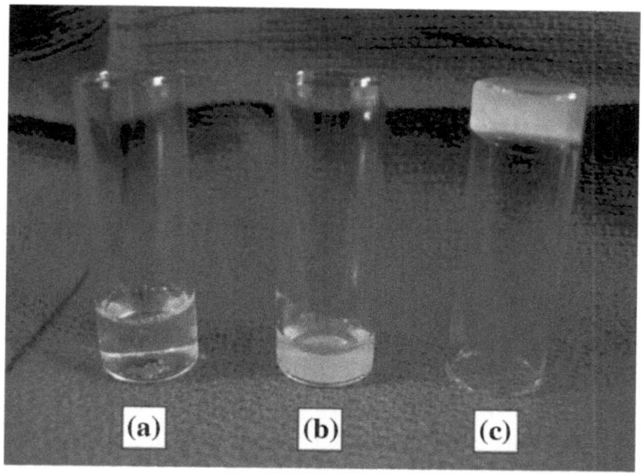

Fig. 3 Thermogelation of chitosan–AHP solution. **a** Chitosan solution in 0.5 % acetic acid. **b** Chitosan–AHP solution. **c** Thermogelled chitosan–AHP at 37 °C [45]

unmodified chitosan. The hydrogel was prepared by adding nano-HA powder to CS-TBA solution under continuous stirring followed by drop-wise addition of β-GP solution. The final solution was kept at 37 °C for 10 min for gel formation [46]. The physical crosslinking has been attributed to the increased hydrogen bonding and hydrophobic attraction between chitosan chains at higher pH [47, 48]. Additionally, thiol groups of CS-TBA also contribute to the gel formation through intra-/inter-molecular disulfide bond formation, and the process is found to be time-dependent and temperature-independent [49]. The gelation process was found to be dependent on factors such as concentration of the gel constituents and degree of deacetylation of the chitosan [46]. For instance, increase in deacetylation of chitosan has shown to significantly decrease the gelation time [50]. Scanning electron microscopy (SEM) analysis showed that the hydrogel had a pore size of 40–80 μm. The presence of thiol group led to higher storage modulus (G′) and loss modulus (G″) for CS-TBA/β-GP/HA hydrogels in comparison to the unmodified CS/β-GP/HA hydrogel due to the additional disulfide linkages [46]. Miguel et al. prepared a hydrogel by blending filtered deacetylated chitosan and agarose. Agarose has inherent thermogelling properties, and therefore combining chitosan and agarose could provide a system that can gel in response to temperature. Briefly, chitosan was dissolved in 1 % acetic acid, and agarose powder was added under stirring at 50 °C. The chitosan–agarose mixture underwent gelation when the temperature was reduced from 50 to 37 °C. SEM analysis showed that the hydrogel had a pore diameter of 90–400 μm. Swelling study indicated that the hydrogel exhibited increase in water uptake after 12 h incubation in tris buffer (pH 5) with subsequent increase in pore diameter. As discussed above, in this case, the gelling is dependent on the thermoreversible property of agarose without any change in mechanical and thermal properties of the gel [29]. The mechanism of gelation is that the agarose exists as random coils at high temperature and, upon decreasing the temperature, forms double helices and aggregates that can behave as physical junctions [51]. *N*-isoproplyacrylamide (NIPAM) is another material with inherent thermoresponsive properties. Chen et al. developed a thermoresponsive chitosan hydrogel in which chitosan acted as the backbone on which poly(*N*-isopropylacrylamide) (PNIPAM) with a carboxylic acid end group was grafted. PNIPAM is reported to remain soluble under its lower critical solution temperature (LCST) but forms a gel when the temperature is increased above LCST. This mechanism can be explained as follows: increasing temperature above LCST causes the release of water molecules attached to the isopropyl moieties of the polymer, leading to a compact form with increase in inter-/intra-molecular hydrophobic attraction between isopropyl groups [12, 52, 53]. Due to this property, PNIPAM grafting to natural polymers is widely used to impart thermogelling properties to the polymers. The grafting of PNIPAM to chitosan was done using 1-ethyl-3-(3-dimethylaminopropyl) carbodiimide hydrochloride (EDC)/*N*-hydroxysulfosuccinimide (NHS) chemistry, wherein the carboxylic acid group of PNIPAM-COOH was linked to the amine group of the chitosan. The reaction was carried out at 25 °C for 12 h followed by purification using thermoprecipitation and dialysis. A porous hydrogel was then prepared by re-dissolving the polymer and incubating at 37 °C. Gel formation was observed to be faster with an increase in concentration of NIPAM-grafted chitosan. SEM analysis of the hydrogels showed a pore size of 10–40 μm. The study also revealed that phase transition of the hydrogel was completely reversible, implying that the conjugated PNIPAM retained its property in the gel [54].

2.1.2 pH-Mediated Gelation

In addition to temperature, the pH of the solution can also trigger polymer gelation. Many of these systems are pH-responsive and can undergo changes in shape, swelling, and release properties depending on the pH of the environment. pH-responsive hydrogels are mostly used in drug delivery and biosensor applications [55]. Chitosan is known to exhibit pH-sensitive behavior because of the amine groups, and when protonated, they become soluble in acidic media. On the other hand, this makes chitosan insoluble in basic and neutral pH [56]. Studies have shown that injectable in situ-forming hydrogels can be developed using the pH-responsive properties of chitosan derivatives. Chiu et al. designed an injectable hydrogel from *N*-palmitoyl chitosan (NPCS) that exhibits pH-responsive changes in properties within a pH range of 6.5–7. The polymer solution at pH 6.5 exhibits sol–gel transition when injected using a needle as droplets into a saline solution at pH 7.4. The immediate gelation has been attributed to the physical crosslinking of the outer layer of the palmitoyl group at pH 7.4, while the inner core still remained at pH 6.5. With time, the physical crosslinking proceeded throughout the gel, thereby making the transparent gel opaque. The entire bead has shown to be crosslinked within 30 min. In vivo study confirmed that when sterile *N*-palmitoyl chitosan solution was injected subcutaneously in wistar rat, hydrogel formation was observed in the injected area, demonstrating its utility as an injectable delivery system. Figure 4 shows the synthesis of NPCS and its gelation in vivo. The

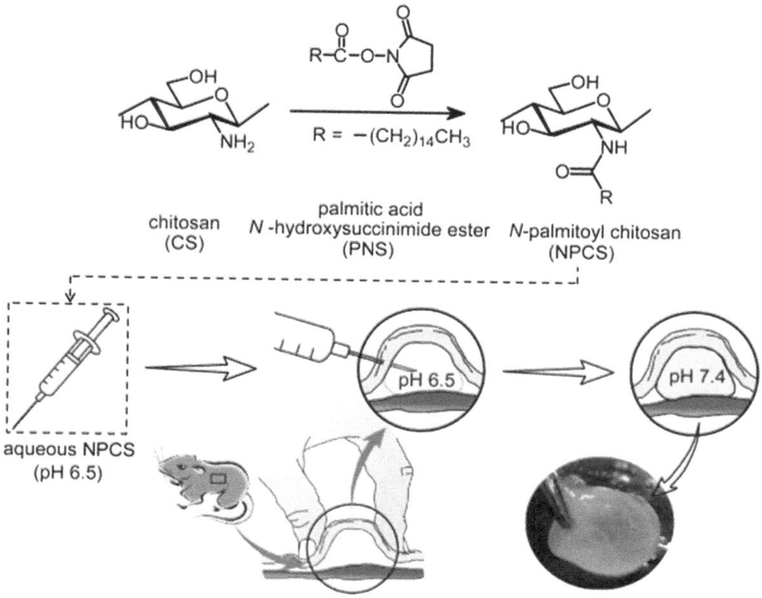

Fig. 4 Schematic showing the synthesis of *N*-palmitoyl chitosan (NPCS) and the pH-triggered hydrogelation of aqueous NPCS injected into the subcutaneous space of a rat model [57]

mechanism behind the gelation is that at lower pH, the charge repulsion between the protonated amine groups is more prominent. As pH increases, the hydrophobic interaction of the palmitoyl groups becomes dominant, which causes the polymer to condense, thereby increasing the intermolecular physical crosslinking [57].

2.1.3 Polyelectrolyte Complexes (PECs)

PECs are formed between oppositely charged molecules due to electrostatic forces. PEC formation is common in the natural environment in enzymatic catalysis and antigen–antibody interactions [58, 59]. Polymers that form polyelectrolytes are soluble in aqueous media and exhibit a net charge (either positive or negative) at neutral pH. Since PECs are formed without the use of any crosslinking agents, toxicity associated with these materials is considered low. PECs have been extensively investigated as drug delivery systems for DNA and protein extraction and other pharmaceutical applications. Due to its aqueous solubility and polycationic nature, chitosan exhibits properties that enable it to form PEC with a wide range of polyanionic polymers [59, 60]. Ji et al. designed a PEC hydrogel prepared from chitosan (CS)/polyglutamate (PG) incorporated with hydroxyapatite particles (HAp) for soft tissue augmentation. HAp was incorporated in the hydrogel to decrease matrix degradation and improve the mechanical properties. CS powder was dissolved in 1 % acetic acid, and PG was added to the solution at a concentration range of 1–6 wt%. Following this, the pH of the solution was adjusted to 6.8 and the solution was incubated at room temperature for 24 h. The hydrogel showed slow degradation and maintained structural integrity when incubated in PBS for 60 days [61]. Li et al. prepared a PEC hydrogel made of chitosan and phosphorylated chitosan as osteoblast carrier. Briefly, 0.173 wt% of phosphorylated chitosan solution in deionized water was mixed with equal volume of 1 wt% chitosan dissolved in acetic acid and incubated overnight. The mixture was semi-translucent at first which subsequently became opaque with time upon forming a compact PEC. The mechanism of PEC formation has been reported to be via the electrostatic interaction between the amine group of chitosan and the phosphate group (PO_4^{3-}) of phosphorylated chitosan. PEC formation has shown to be dependent on factors such as pH of the solutions, method of mixing the solutions, and strength of the ions as well as temperature. SEM analysis showed that the hydrogel had anastomosing, porous, sponge-like structure with micropores and macropores. Micropores were ~100–120 nm in diameter and macropores were ~5–100 μm in diameter. The 3D network of the hydrogel showed similarity to the physiochemical environment of extracellular matrix (ECM) and exhibited good osteocompatibility [58]. Chang et al. synthesized a PEC-based hydrogel prepared from chitosan–poly (γ-glutamic acid) (γPGA). γPGA is a polyanionic polypeptide which is biodegradable and water soluble. Equimolar mixture of polyanion γPGA solution and polycation chitosan solution resulted in PEC hydrogel. The preparation involved adding chitosan powder to the γPGA solution at 4 wt% followed by the addition of 1 % acetic acid to the solution. Chitosan powder dissolved completely due to the presence of its protonated amine group, which subsequently led to the formation of a homogenous PEC hydrogel due to the interaction of the amine and carboxylic acid groups. The formed hydrogel was later dipped in 1 N NaOH solution and pH was adjusted to 7. Solid, porous matrices can be obtained from the gel upon freeze drying [62].

As discussed before, the absence of exogenous crosslinkers has the potential to significantly reduce the toxicity of physically crosslinked hydrogels making, them preferred candidates for biological applications [40]. Despite the advantages, the scope for the utility of physically crosslinked chitosan hydrogels is limited owing to drawbacks related to low mechanical strength and poor reproducibility of properties such as pore size, dissolution rate, and functionalization with chemical groups. These limitations can be addressed to a great extent using chemical crosslinking methods, wherein the crosslinking molecules can be used to enable covalent bonding between the chitosan chains to form the hydrogel [33].

2.2 Chemically Crosslinked Hydrogels

Chemically crosslinked hydrogels are formed as a result of covalent bonding [7]. Chemical crosslinking of chitosan can be achieved using different crosslinkers or by modifying the $-NH_2$ or $-OH$ groups present in the polymer [40]. Different methods of chemical crosslinking of chitosan to form hydrogels are discussed below.

2.2.1 Chemically Crosslinked Hydrogels Using Exogenous Crosslinkers

Glutaraldehyde (GA) is one of the most commonly used crosslinking agent to form chemically crosslinked chitosan hydrogel. For instance, Azab et al. developed a biodegradable chemically crosslinked chitosan hydrogel for brachytherapy by adding GA solution to chitosan in 1 M acetic acid. Due to the high reactivity of GA, highly crosslinked gel can be obtained by this process [63]. Vaghani et al. developed a pH-responsive hydrogel from carboxymethyl chitosan, which is known to have antibacterial properties and show solubility over a wide range of pH [64, 65]. Carboxymethylchitosan can be prepared by a high-temperature reaction of chitosan with monochloroacetic acid in sodium hydroxide/isopropyl alcohol. The swelling properties of carboxymethyl chitosan hydrogel can be tuned by adjusting the pH due to the presence of amine and carboxylic groups in the polymer. The GA-crosslinked carboxymethyl chitosan hydrogel has been reported to show differential swelling properties at different pH solutions. For instance, the hydrogel showed 12 % swelling at pH 1.2, 97 % swelling at pH 6.8, and 118 % swelling at pH 7.4. Also, the study demonstrated that the degree of deacetylation and the extent of carboxyl substitution can also affect the extent of swelling [65]. This property makes it an attractive candidate as stimuli-responsive drug delivery system. Lin et al. developed a pH-responsive *N*-(2-carboxybenzyl) chitosan (CSBC) hydrogel for drug delivery applications to colon. CSBC was dissolved in distilled water with 1 % GA as the crosslinking agent. Hydrogel formation occurred under stirring for 2 min. The hydrogel was frozen at −18 °C followed by freeze drying. The degree of substitution of 2-carboxybenzyl in the position of $-NH_2$ can result in a relational twisting of polymer chain, leading to stereospecific blockage preventing complete reaction with GA. The hydrogel showed reversible swelling properties between pH 1.0 and pH 7.4. Drug release was faster at pH 7.4 than at pH 1. Swelling property of the hydrogel can also be explained on the basis of dissociated carboxylate group ($-CO_2^-$), which is a dominant charge in this hydrogel. At higher pH (7.4–9.0), the alkalinity causes an increase in the concentration of $-CO_2^-$ inside the gel, causing an increase in osmotic pressure and leading to higher gel swelling. However, when the pH is reduced, the $-CO_2^-$ concentration inside the gel is reduced, thereby reducing the osmotic pressure and resulting in hydrogel shrinkage [66].

Although GA is commonly used for crosslinking, the high toxicity of GA has a cytotoxic effect, and hence, alternatives such as polyaldehydes from starch and dextran are also explored as crosslinking agents for polysaccharides [67].

N, *N*-methylenebisacrylamide (MBA) is a commonly used crosslinking agent in free radical polymerization to form hydrogels and has been reported to be biocompatible with low toxicity toward cells. Ranjha et al. developed a chitosan-based interpenetrating polymer network (IPN) in the presence of acrylic acid (AA) and MBA for drug delivery applications. Briefly, MBA and benzyl peroxide dissolved in AA was added to acetic acid solution of chitosan and allowed to polymerize at room temperature. SEM analysis showed that the hydrogels had spongy porous structure. Chitosan and AA concentration had significant effect on the porosity of hydrogel, and hydrogels with higher porosity showed higher swelling. Porosity decreased upon increasing the content of the MBA crosslinker because the crosslinker increased molecular entanglement, thereby decreasing the pore size. Moreover, the dynamic and equilibrium swelling ratios for these hydrogels were found to be higher between pH 6.5 and 7.5 [68].

Glyoxal has been investigated as a crosslinking agent to prepare chitosan hydrogels. It is considered to have lower toxicity when compared to GA. Wang et al. used glyoxal as the crosslinking agent for preparing a composite hydrogel of chitosan and collagen along with β-GP for bone regenerative engineering. In addition to the physical crosslinking of chitosan by β-GP, the glyoxal-mediated chemical crosslinking method resulted in porous scaffolds with higher stability. Chitosan and collagen solutions containing β-GP were mixed in various ratios followed by the addition of glyoxal solution. The mixture was allowed to gel for 30 min at 37 °C. SEM analysis showed that the crosslinked hydrogel had larger pores with plate-like structure. Glyoxal crosslinking showed sixfold increase in the stiffness of the hydrogel compared to the β-GP crosslinked gel. Good cytocompatibility of the hydrogel was also demonstrated [69].

Genipin is a crosslinking agent that is extracted from geniposide, which is used in Chinese medicine. It has been extensively studied as a crosslinking agent for polymers containing amine groups and is known to be biocompatible and have lower toxicity than GA [70, 71]. The genipin-based crosslinking reaction is mediated by the reaction between the amine group of chitosan and the carboxymethyl group of genipin to form a secondary amide [72]. Silva et al. designed a composite scaffold for cartilage tissue regeneration by crosslinking chitosan and silk fibroin using genipin. The hydrogel was prepared by mixing desired ratio of chitosan dissolved in acetic acid and silk fibroin in water followed by addition of genipin powder and incubating the mixture for about 24 h at 37 °C. Genipin crosslinking gave rise to a covalently crosslinked blend hydrogel matrix. Environmental SEM analysis indicated that the hydrogel had pore sizes ranging between 29 and 167 μm. Pore size was variable depending on the content of chitosan and silk fibroin, and decreased with increase in silk fibroin content. The hydrogels also showed a pH-dependent swelling behavior [70]. Fiejdasz et al. developed an in situ-gelling hydrogel based on chitosan and collagen using genipin as the crosslinking agent. Briefly, the hydrogel was prepared by mixing chitosan solution in 1 % acetic acid with collagen at desired ratio followed by the addition of genipin at 37 °C. The extent of crosslinking in the hydrogel was determined by ninhydrin assay. SEM analysis showed that the hydrogel had an open network structure. Swelling studies indicated that, as chitosan content increased, the gel showed decrease in swelling. Microviscosity measurement of chitosan–collagen hydrogel showed that they were viscous in nature, and viscosity increased with increase in chitosan content. The study implied that the physiochemical properties of the hydrogel can be tailored by varying chitosan and genipin content during hydrogel preparation [73].

2.2.2 Photocrosslinked Hydrogels

Photocrosslinkable hydrogels have been extensively investigated as scaffolds for tissue engineering [74], drug delivery [75], as bioadhesives [76], and for endoscopic treatments [77]. Photocrosslinkable hydrogels can be prepared in situ and therefore are potential candidates to develop minimally invasive delivery systems [78]. Tsuda et al. prepared a photocrosslinked chitosan hydrogel (Az-CS-LA) using chitosan (CS) incorporated with azide (Az) and lactose (LA) functional groups. The lactose group made the Az-CS-LA polymer water soluble. Briefly, the hydrogels were prepared by exposing a solution of Az-CH-LA 20 mg/mL to UV light for 30 s [79]. Gelation time was dependent on the intensity of UV radiation; the higher the intensity, the lower the gelation time. The mechanism behind the gelation is that during UV irradiation, the azide group ($-N_3$) releases N_2 and gets converted to nitrene, which is a very reactive group. These nitrene groups either interact with other nitrene groups or with the amino groups of chitosan, resulting in the formation of azo groups ($-N=N-$) causing gelation [80]. The hydrogels showed higher sealing strength when compared to fibrin glue. Apart from this, light-mediated free radical polymerization has also been extensively investigated to develop chitosan hydrogels. Zhou et al. used a water-soluble (methacryloyloxy) ethyl carboxymethyl chitosan (MAOECECS) for the preparation of photocrosslinkable hydrogels for tissue engineering applications. MAOECECS was synthesized using Michael addition reaction between $-NH_2$ group of chitosan and C=C of acrylate group. Briefly, solution of MAOECECS in water was prepared and mixed with the photoinitiator D-2959. The solution was exposed to UV light of wavelength 320–480 nm for 15 min to obtain the hydrogel. D-2959 was used as a photoinitiator due to its low cytotoxicity. SEM analysis showed spongy macroporous structure. Crosslinking density played a key role in determining the pore structure of the hydrogel. Similarly, the swelling behavior of these hydrogel depends on the concentration of MAOECECS. Degradation of hydrogel in the presence of lysozyme can also be modulated by varying the crosslinking density. Higher crosslinking density caused lower degradation whereas, lower crosslinking density showed higher degradation because lower crosslinking makes the hydrogel more accessible for the enzymatic attack as well as for enzyme penetration [78]. Arakawa et al. prepared a photocrosslinkable hydrogel using methacrylated glycol chitosan (MeGC) and collagen (Col). For the fabrication of the hydrogel, MeGC solution was prepared in phosphate-buffered saline (PBS). Col was added to the solution to form a semi-interpenetrating network. Riboflavin was used as the photoinitiator for crosslinking. The hydrogel was formed when the composite solution was exposed to visible blue light of wavelength 400–500 nm in the presence of the photoinitiator. Use of visible light is more beneficial as it is less harmful, less mutagenic, does not generate heat, and penetrates deeper into the tissue. The gelation time decreased with increase in riboflavin content. MeGC-Col hydrogel had a compressive strength of ~1.8 kPa, which is higher than that of the MeGC gel. Higher compressive strength for the hydrogel has been attributed to the formation of semi-interpenetrating network [81].

2.2.3 Enzymatically Crosslinked Hydrogels

Crosslinking of polymers to form hydrogels can be performed in the presence of enzymes. This type of crosslinking is beneficial for hydrogel formation because of the mild activity of the enzymes [82]. In situ hydrogel systems can be easily developed using enzyme-mediated reaction as physiological pH and temperature are optimum for many enzymatic reactions. Enzymes commonly used for crosslinking chitosan to form hydrogels are peroxidase [83], transglutaminase [84], and tyrosinase [85].

Horse-radish peroxidase (HRP) enzyme is the most commonly used peroxidase enzyme for enzymatic crosslinking. Peroxidase enzyme oxidizes phenols and generates free radicals that mediate the crosslinking reaction. Sakai et al. designed chitosan hydrogels using chitosan conjugated with 3-(4-hydroxyphenyl) propionic acid (HPP) prepared via standard EDC coupling reaction. The hydrogels were prepared by mixing the conjugated chitosan with HRP followed by chilled H_2O_2 solution. Gelation time of the hydrogel was found to be dependent on many factors such as concentration of HRP, H_2O_2, phenol groups present in the polymer, and temperature. Cytocompatibility of the gel was confirmed by growing L929 fibroblast cells on the hydrogel surface [83]. Jin et al. developed a biodegradable injectable hydrogel prepared from chitosan-*graft*-glycolic acid (GlA) and phloretic acid (PA) using enzymatic crosslinking in the presence of HRP and H_2O_2 as scaffold for cartilage regeneration. Briefly, the hydrogel was prepared by adding HRP and H_2O_2 to the CS-GlA/PA solution. CS-GlA/PA concentration has shown to significantly affect the gelation time; increase in the polymer concentration from 1 to 3 wt% decreased the gelation time from 4 min to 10 s. Similarly, the water uptake of the hydrogel decreased with increase in polymer concentration and has been attributed to the increase in crosslinking density at higher polymer concentration. The hydrogel also exhibited pH-dependent swelling behavior. Storage modulus of the hydrogel was in the range between 1.3 and 5.5 kPa for 1 and 2 wt% polymer solutions. Thus, the study indicated that the physiochemical and mechanical properties of the hydrogel can be tailored by varying the initial polymer concentration [86]. Brittain recently reported the feasibility of developing enzymatically crosslinked injectable hydrogel using water-soluble glycol chitosan. Glycol chitosan (GC) modified with HPP using EDC/NHS chemistry was treated with HRP and H_2O_2 to initiate the oxidative coupling of phenol moieties. Similar to the previous study, the GC gel also showed increase in storage modulus with increase in polymer concentration. SEM analysis showed the porous microstructure of the hydrogels [87].

Transglutaminase enzyme catalyzes the formation of isopeptide bond between glutamine's γ-amine group and the amine group of lysine. Transglutaminase is found in microbes and other living organisms. Da Silva et al. used microbial transglutaminase (mTGase) to crosslink chitosan and gelatin for preparing hydrogel. The enzyme was used to form the bond between chitosan's glucosamine and gelatin's glutamine. Chitosan was dissolved in acetic acid and gelatin was added to prepare the chitosan–gelatin solution. Solution pH was kept at 5, where mTGase shows optimal activity. Chemical crosslinking of chitosan and gelatin was performed at 37 °C by adding mTGase in the concentration range of 10–40 enzymatic units depending on the amount of gelatin used. Thermal deactivation of mTGase was then performed at 70 °C. Gelation time was found to be dependent on mTGase concentration; the higher the enzyme concentration, the lower the gelation time. Mechanical property of the hydrogel was found to be dependent on the concentration of chitosan and mTGase. Increase in chitosan concentration led to gels with higher moduli. Instead of chemical crosslinking alone, physical-*co*-chemical gelation of the solution led to the formation of homogenous gels with higher modulus and gel transparency [84].

Tyrosinase is an enzyme derived from both plants and animals [82]. It is known to catalyze the oxidation of phenol compounds containing tyrosine and other residues into *o*-quinones [85]. Kang et al. used tyrosinase enzyme for crosslinking chitosan and silk fibroin. Chitosan solution was prepared by dissolving chitosan granules in water and adjusting the pH to 2 using hydrochloric acid. Prior to crosslinking, pH of the chitosan solution was adjusted from 5 to 5.5, which is required for tyrosinase activity. Silk fibroin solution was dissolved in a solution comprising calcium chloride, ethanol, and water followed by dialysis and filtration. Chitosan and silk fibroin solutions were mixed together in different ratios, and tyrosinase was added to the solution to initiate gelation. Chemical crosslinking was confirmed using UV spectroscopy. The mechanism behind the crosslinking is that tyrosinase enzyme converts tyrosyl residues in silk fibroin to reactive *o*-quinones. The amine group in the chitosan then reacts with the *o*-quinones in silk fibroin through Michael addition reaction [85].

3 Chitosan Hydrogel for Regenerative Engineering Applications

Chitosan has been widely used for regenerative applications due to its ability to induce scarless wound healing, tissue regeneration, and its potential as a drug/protein/gene delivery vehicle [88, 89]. Chitosan accelerates wound healing by modulating macrophage response [90], by attracting appropriate cells to the wound site, and by supporting granulation tissue formation [91]. Besides this, other properties of chitosan such as antimicrobial activity, biodegradability, and biocompatibility also contribute to its use in tissue engineering. This section briefly discusses the different regenerative applications of chitosan hydrogels.

3.1 Orthopedic Regenerative Engineering

Orthopedic regenerative engineering is an area that is concerned with treating injuries, trauma, and diseases affecting the musculoskeletal system of the body [92]. Current treatments mainly involve autografts and allografts. Autograft is widely used due to histocompatibility and osteoinductivity but it requires a second surgery for obtaining bone grafts leading to donor site morbidity. Allografts can elicit immune reaction and are less osteoinductive. Since they are obtained from a donor, they may also serve as a source of infection. Orthopedic regenerative engineering aims at addressing the limitations of autografts and allografts by creating a matrix that can present cells and growth factors and also support vascularization and cell homing [93]. Due to osteocompatibility, chitosan has been extensively investigated for orthopedic regenerative engineering [6]. Arakawa et al. developed an injectable photocrosslinked hydrogel from methacrylated GC and collagen as bone marrow stromal cells (BMSCs) delivery system to support bone regeneration [81]. Similarly, a composite of chitosan and collagen matrix prepared by glyoxal crosslinking also showed efficacy to support osteogenic differentiation of human BMSCs in vitro [69]. Chitosan hydrogel has also been investigated to prevent excessive bone formation. Using a photocrosslinkable chitosan hydrogel (Az-CS-LA), Tsuda et al. demonstrated the ability of the hydrogel to effectively prevent ectopic bone formation for up to 8 weeks in a rat model with calvarium and fibula defects [79]. In native ECM, the presence of sulfated polysaccharides such as chondroitin sulfate plays an essential role to bind and retain growth factors to support cell functions. To mimic this ECM composition, Li et al. prepared a hydrogel comprising chitosan, gelatin, and carrageenan via PEC formation and covalent crosslinking. The system could mimic the native ECM due to the presence of various functional groups such as $-COOH$, $-NH_2$, $-OH$, and $-SO_3H$ that can help in cell attachment and differentiation. The hydrogel supported osteogenic differentiation of adipose-derived mesenchymal stem cells (ADMSCs) along with neovascularization [94]. Similarly, glyoxal-crosslinked superporous chitosan and its composite hydrogels have shown to enhance bone regeneration [95]. In addition, physically crosslinked thermosensitive hydrogels have also been investigated as matrices to support bone regeneration. Dessi et al. developed a thermosensitive β-tricalcium phosphate–chitosan hydrogel in the presence of β-GP and glyoxal to impart physical and chemical gelations. The presence of β-tricalcium phosphate in the in situ-gelling hydrogel showed to significantly increase cell attachment and proliferation [96].

For enhancing bone repair and regeneration, drug and growth factors can also be incorporated in hydrogels for localized delivery. Growth factors such as bone morphogenic protein (BMP) have been widely studied due to their potent

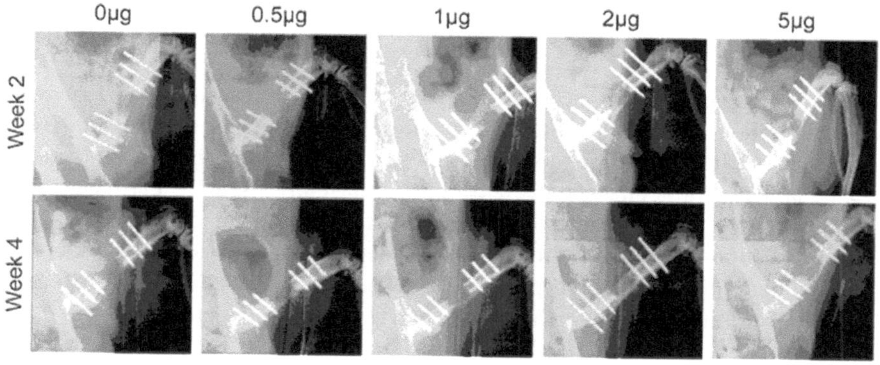

Fig. 5 Representative radiographs of the bone defects implanted with CLF hydrogels containing different concentrations of BMP-2 (0, 0.5, 1, 2 or 5 μg) at 2 and 4 weeks [100]

osteoinductive properties [97]. Most commonly used BMPs are BMP-2, BMP-4, and BMP-7. They are reported to heal ectopic, orthotopic, and critical bone defects. Besides these, angiogenic factors such as vascular endothelial growth factor (VEGF), fibroblast growth factor (FGF), and platelet-derived growth factor (PDGF) have shown to improve bone healing. Inflammatory factors such as interleukins, interferon-γ, and prostaglandins, that are known to induce migration and differentiation of osteogenic cells, are also widely studied [98]. Dong et al. used chitosan hydrogels as nucleic acid delivery systems for the treatment of endotoxin-induced osteolysis. The composite hydrogel prepared from chitosan and gelatin was used to deliver antisense oligonucleotide that can target tumor necrosis factor-α (TNF-α) and thereby silence its expression. The suppression of TNF-α plays an effective role in supporting bone formation as it is associated with osteoclastogenesis. In vitro and in vivo studies demonstrated the ability of the hydrogel to deliver nucleic acid to the surrounding tissues and cells leading to effective blocking of TNF-α [99]. Recently, Kim et al. developed a photocrosslinked hydrogel made of chitosan-lactide-fibrinogen (CLF) as a delivery system for BMP-2 [100]. Chitosan and lactide in the hydrogel controlled the physiochemical properties of the hydrogel [101]. Fibrinogen, which contains heparin-binding domains, was added to this system to improve BMP-2 binding. The incorporation of growth factor-binding ligands to the hydrogel has shown to maintain protein bioactivity and provide sustained release of BMP-2 over a period of time. Efficiency of the hydrogel delivery system was evaluated in vitro as well as in vivo in a rat defect model. Figure 5 shows the radiographs of the bone defects implanted with CLF hydrogels containing different concentrations of BMP-2. The results indicated increased cell viability, neo-osteogenesis, and bone healing. The responses were dependent on the dose of the growth factor [100].

3.2 Cartilage Regenerative Engineering

Cartilage regenerative engineering is another emerging research area that deals with the treatment of injuries, diseases, and defects occurring in cartilage tissues [102]. Osteochondral allograft/autograft, autologous chondrocyte implantation (ACI), and microfracture surgery are options currently used for cartilage repair and restoration. These methods are effective but have associated limitations [103]. Novel regenerative strategies are currently under investigation to develop the optimal treatment [104]. Chitosan plays a very significant role in cartilage regeneration as it can mimic the glycosaminoglycan (GAG) component of cartilage tissue [6]. Tan et al. developed an injectable in situ composite hydrogel from *N*-succinyl-chitosan and aldehyde hyaluronic acid as a matrix to support cartilage tissue regeneration. Encapsulation of bovine articular chondrocytes in the hydrogel demonstrated its ability to retain cell morphology and cellular viability [105, 106]. The potential of injectable thermosensitive chitosan–pluronic hydrogel as a chondrocyte delivery system was demonstrated by encapsulating bovine chondrocytes. The encapsulated cells in chitosan gel showed enhanced cell proliferation and glycosaminoglycan production for 28 days when compared to cells encapsulated in alginate hydrogel [105]. A chitosan/starch/β-GP composite injectable matrix was also studied for chondrocyte encapsulation. These gels also maintained chondrocyte cell phenotype and supported the expression of collagen-II molecule [107]. Other chitosan-based hydrogels investigated for chondrocyte encapsulation include alginate/lactose-modified chitosan hydrogels [108], photocrosslinked chitosan–gelatin scaffold incorporated in sodium alginate hydrogel [109], chitosan hydrogel grafted with methacrylic acid and lactic acid [110], and chitosan/silk fibroin sponges [70].

Growth factors that have shown to promote chondrogenesis include transforming growth factors TGF-β1, 2, and 3. Different BMP molecules such as BMP-2, 4, 6, 7, and 9 are also reported to be capable of inducing chondrogenic differentiation of stem cells. Besides these, other factors such as IGF-1 and FGF-2 have shown to play key roles in cartilage homeostasis [111]. For the delivery of TGF-β1, Faikrua et al. investigated the efficacy of chitosan/starch/β-GP hydrogel as a cell and growth factor delivery system. The hydrogel showed controlled release of the growth factor for up to 14 days and maintained proper functioning of the chondrocytes. Subcutaneous implantation of the hydrogel with chondrocytes and TGF-β1 in rat model showed the expression of collagen and aggrecan transcripts that are important ECM molecules [112]. To increase the bioactivity of chitosan-based hydrogels, Choi et al. developed a photocrosslinked chitosan hydrogel containing ECM constituents such as collagen II and chondroitin sulfate for cartilage repair. The hydrogel demonstrated the ability to promote chondrogenesis and increase cell–matrix interactions [113]. For repairing focal chondral injuries, a uniquely designed RGD-grafted *N*-methacrylated GC hydrogel was used to encapsulate ADSCs. Co-delivery of growth factors was achieved by incorporating TGF-β3 and BMP-6 in microspheres and encapsulating in the hydrogel along with the cells. The study showed that the sustained release of the growth factors was very effective in inducing higher expression of chondrogenic markers and chondrogenic differentiation [114]. These studies demonstrate the potential of chitosan-based hydrogels as a matrix to support cartilage tissue regeneration.

3.3 Neural Regenerative Engineering

Neural regenerative engineering pertains to the repair and rejuvenation of damaged nerves. The field assumes greater significance because damages to nerves are normally very difficult to restore [6]. Currently used strategies to treat nerve defects include autografts and allogenic grafts from cadavers. However, autografts are in short supply and have issues with functional recovery of nerves, while allogenic grafts have issues with immune rejection. To address the limitations of biological grafts, new methods such as acellular nerve grafting are currently under investigation. Another approach to support regeneration of neural systems is designing 3D scaffolds using biomaterials that can mimic the structure and function of the ECM and support cell growth and regeneration [115]. Due to its biocompatibility and biodegradability and its ability to mimic GAG, chitosan has been investigated as matrix to support neural tissue regeneration [6]. Pfister et al. demonstrated that chitosan hydrogels can be used to develop nerve conduits to treat peripheral nerve problems. Nerve conduits were prepared from chitosan and alginate PEC using a spinning mandrel technology. The study demonstrated that the hydrogel exhibited biodegradability, mechanical strength, and permeability appropriate for designing nerve conduits [116]. Another group investigated the potential of hydrogel blends prepared from agarose and methylcellulose to support neuronal cell attachment and extension. The study showed that the elastic moduli and surface charge of the blend hydrogel can be further controlled by adding chitosan and dextran [117]. Kwon et al. fabricated a 3D chitosan hydrogel and demonstrated that stem cells derived from rat muscles (rMDSCs) can be differentiated into neuronal cells in the presence of valproic acid. Valproic acid is a drug that is used to treat disorders like epilepsy. It also acts as histone deacetylase inhibitor and is reported to cause differentiation of stem cells to neural lineage [118]. Figure 6 shows the morphology of chitosan hydrogel alone and hydrogel seeded with rMDSCs in the presence of valproic acid. The cells exhibited bipolar and multipolar morphologies in the presence of valproic acid. This study may have implications in developing new treatment strategies for neurodegenerative disorders like Alzheimer's and Parkinson's disease. Another study investigated the potential of thermosensitive chitosan hydrogel for the 3D culture of neuronal cells. The neuronal cells cultured in a poly-D-lysine (PDL)-immobilized chitosan/glycerophosphate hydrogel showed good viability with larger cell bodies [119]. Photocrosslinked chitosan hydrogels are also employed in neural tissue engineering. Reports suggest that these hydrogels can enhance the differentiation and extension of neuritis and neural stem cells [120]. Chitosan hydrogels are also implemented for neurosurgical applications. For example, Rickett et al. used

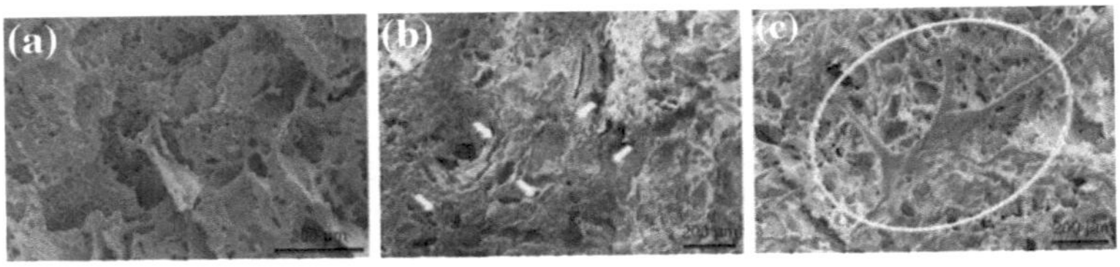

Fig. 6 SEM micrographs taken on day 4 showing the morphology of **a** the chitosan hydrogel alone **b** chitosan hydrogel seeded with rMDSCs in the presence of valproic acid. The *circle* indicates rMDSCs on the chitosan hydrogel. The magnifications are (**a**) 500× and (**b**) 300×, and the *scale bars* (**a** and **b**) represent 200 μm [118]

a photocrosslinked chitosan hydrogel conjugated with 4-azidobenzoic acid as an adhesive for treating peripheral nerve anastomosis. Cell culture study revealed the hydrogel to be non-toxic with excellent mechanical properties and qualities that are essential for an efficient bioadhesive [121].

Apart from cells, growth factors play an important role in nerve tissue repair and regeneration. Growth factors such as nerve growth factor (NGF) and neurotropic factors are investigated for neural regeneration [115]. In situ-gelling thermosensitive hydrogels made from chitosan/β-GP are used as carriers for lentiviral vector that expresses neurotropin-3. The use of the lentiviral system may have potential for repairing injuries to the central nervous system by providing sustained release of the coded protein for longer durations [122]. 3D chitosan hydrogel incorporated with immobilized biotin-rIFN-γ has shown to promote neuronal cell growth and differentiation of adult neural stem/progenitor cells [123]. Li et al. demonstrated the use of growth factor-immobilized photopolymerized methacrylamide chitosan (MAC) hydrogel for controlling cell signaling and differentiation of neural stem/progenitor cells (NPSCs). Growth factor used for the study was a fusion protein with biotin tag containing IFN-γ, PDGF-AA, and BMP-2. IFN-γ was for providing neuronal-specific differentiation of stem cells, PDGF-AA to support oligodendrocyte specification, and BMP-2 for astrocyte specification of cells. Subcutaneous implantation of the chitosan/growth factor/cell construct in the back of rat demonstrated that the hydrogel with growth factor exhibited spatial control for maintaining neural lineage in the cells for four weeks [124].

3.4 Corneal Regenerative Engineering

Cornea is a transparent layer of the eyeball that covers the iris and the pupil. It provides vision by refracting light to the retina and lens, functions as a physical barrier against microbes and dirt, and prevents damage to the eye by absorbing harmful UV light [125]. Availability of corneas from donors is limited, so research is currently underway to develop artificial substrates that have the properties required for transplanting corneal cells. Ozcelik et al. used ultrathin chitosan–PEG hydrogel films prepared by epoxy-amine chemistry for culturing ovine corneal endothelial cells (CECs). The chitosan films demonstrated the ability to support the attachment and proliferation of ovine CECs and also had excellent mechanical, optical, biocompatible properties appropriate for corneal regenerative engineering [126]. Similarly, Liang et al. showed that a composite biodegradable hydrogel made of chitosan and sodium alginate can be used as a scaffold for CEC proliferation and are useful for the reconstruction of the corneal endothelium [127]. Grolik et al. developed a novel hydrogel scaffold based on a chitosan–collagen blend crosslinked with genipin for culturing corneal epithelial cells [125]. Chitosan was used for preparing the membrane due to its ability to promote proliferation and differentiation of cells. The blend hydrogel made from collagen and chitosan also exhibited better mechanical and elastic properties, optical properties, permeability, and suturability compared to non-blended hydrogel. In vivo study demonstrated the biocompatibility of the blend membranes implanted in pig cornea for 12 months and membranes supported the regeneration of corneal epithelium, nerve, and stroma [128]. Chitosan hydrogels have also been investigated to repair the cornea using stem cells. Chein et al. used an injectable, thermogelling, amphipathic, carboxymethyl-hexanoyl chitosan hydrogel as a delivery vehicle for human keratinocyte reprogramed to induced pluripotent stem cells (iPSCs). The study demonstrated the efficacy of the hydrogels for cell-based healing of corneal wounds [129].

Besides these, drug delivery systems based on chitosan hydrogels are also used for treating ocular diseases. Chitosan-based ocular drug delivery devices have shown to have properties such as ocular tolerance, non-toxicity, better physiochemical properties, permeability, and bioadhesion [130]. In situ-forming thermosensitive chitosan/β-GP hydrogel was investigated as a carrier for liposomes containing ofloxacin drug. The carrier provided extended transcorneal penetration of the drug, resulting in decreased side effects caused by the drug and reduced the frequency at which the drug was administered to the patient [131].

3.5 Intestinal Regenerative Engineering

The gastrointestinal (GI) tract is a complex system of the human body that performs various functions such as digestion of food, absorption of nutrients, and elimination of waste [132]. Surgery is the commonly used method for the treatment of intestinal diseases but it has the limitation of not being a long-term solution. Several regenerative strategies are currently under investigation [133]. Rabbani et al. used an innovative approach to repair intestinal fistulas using chitosan hydrogels. The study revealed that the wound healing as well as antibacterial properties of chitosan hydrogel was very effective in healing cecal fistulas in rats [134]. Falabella et al. demonstrated the efficacy of in situ-crosslinked oxidized dextran/*N*-carboxyethyl chitosan hydrogel in reducing intra-abdominal adhesion in rats by forming a physical barrier between the injured areas [135]. Another study showed the efficacy of chitosan–dextran hydrogel to reduce the peritoneal adhesion without any sensitization in porcine hemicolectomy model [136]. Xu et al. used a bio-inspired method to show that the mucoadhesive properties of chitosan hydrogels in wet conditions can be improved by adding catechol compounds (e.g., hydrocaffeic acid (HCA), dopamine, 3,4-dihydroxy-L-phenylalanine (DOPA)). Chitosan hydrogel containing HCA showed twofold increase in mucoadhesion in rabbit's intestine. Pre-oxidation of catechol has shown to reduce the mucoadhesive property of hydrogel. However, oxidation of HCA chitosan hydrogel during contact with intestinal mucosa made it a better mucoadhesive. The oxidation of catechol on contact enhanced mucoadhesion due to the formation of covalent bonding with the cysteine group of mucin [137].

The bioadhesive property of chitosan hydrogel has also been utilized for the localized delivery of drugs to the intestinal tissue. Chitosan-based superporous hydrogel composite has been investigated for the delivery of metoprolol succinate. The study showed that the composite remained adhered to the intestine for up to 8 h and showed an anomalous non-fickian release mechanism [138]. Stimuli-sensitive chitosan hydrogels are also useful in intestinal drug delivery. Yadav et al. demonstrated the efficacy of carboxymethyl chitosan hydrogel for the intestinal delivery of the drug theophylline in basic pH environment. In vitro and in vivo studies showed that the drug was released slowly and in a controlled manner for a prolonged period [139]. Several growth factors such as FGF-2, VEGF, EGF, PDGF-BB, and TGF-β have also been investigated to support intestinal tissue regeneration [132]. Maeng et al. used a chitosan hydrogel containing EGF for endoscopic applications for treating GI peptic ulcers and mucosectomy-induced ulcers. In vivo gastric ulcers were induced in pig and rabbit models and used endoscopic catheters to apply EGF–chitosan gel to the ulcer area. The chitosan hydrogel protected the region from the harsh gastric environment, and EGF was released slowly to the mucosal defects. Histological staining and endoscopic imaging revealed that the hydrogel-mediated growth factor delivery accelerated the healing of the ulcer [140].

3.6 Adipose Regenerative Engineering

Adipose regenerative engineering is a developing area that deals with soft tissue replacements in case of injuries, defects, and other age-related adipose tissue losses [141, 142]. Currently used treatment involves autologous soft tissue transplants but the surgery causes a decrease in the volume of tissue and leads to other problems at the donor site [141]. Chitosan hydrogels have been investigated as matrices for filling the defects and for the delivery of adipocytes and proteins. GA-crosslinked collagen–chitosan hydrogels were studied to deliver pre-adipocytes. In vitro and in vivo studies demonstrated the biocompatibility of the hydrogel and its ability to support vascularization and regeneration of adipose tissue in rats [143]. Additionally, collagen–chitosan hydrogels have shown to stimulate signaling molecules such as nitric oxide for tissue healing when used as a delivery system for ADSCs [144]. In addition to preformed scaffolds, injectable biodegradable hydrogels are also been investigated for adipose tissue regeneration. Jaikumar et al. developed an injectable alginate-*o*-carboxymethyl chitosan/nanofibrin blended composite hydrogel as ADSC carrier. In vitro studies showed adherence, proliferation, and differentiation of the cells in the scaffold [145].

Insulin is one of the most extensively studied growth factor to support adipose tissue regeneration. Tan et al. used a glucose-responsive chitosan–hyaluronic acid hydrogel as a controlled insulin delivery system. The hydrogel was prepared from *N*-succinyl-chitosan and aldehyde hyaluronic acid conjugated with the enzymes glucose oxidase and catalase. The conversion of glucose reduces the pH of the hydrogel environment, leading to higher swelling and subsequent release of insulin [146].

3.7 Liver Regenerative Engineering

The liver, which is the largest organ in the human body, aids in diverse functions such as detoxification, digestion, and metabolism. Liver failure can lead to multiple organ failure and eventually death. Liver transplantation is the only treatment option available currently and has issues related to limited donor availability. Liver regenerative engineering aims to overcome these issues using scaffolds for hepatocyte delivery as well as developing biomaterial-based vehicles for drug/protein delivery [147]. Seo et al. fabricated a synthetic ECM mimic porous hydrogel scaffold made up of alginate (AL)/galactosylated chitosan (GCS)/heparin as cell carrier. The hydrogel had pore sizes in the range of 150–200 μm and showed hepatocyte spheroid formation. Cell viability was higher in AL/GCS and AL/GCS/heparin sponges. E-cadherin and connexin 32 gene expression confirmed cell-to-cell contact in the scaffolds. The data suggests the suitability of the composite chitosan matrix for designing bioartificial liver devices [148].

Fibrin-coated collagen fleece is commonly used as a hemostatic agent for various surgical applications but fibrin obtained from humans has limitations related to availability as well as contamination. Horio et al. prepared a blend of photocrosslinkable chitosan hydrogel mixed with photocrosslinked chitosan sponges (PCSM-S) and used it for treating liver injury in rat model. In vivo analysis in heparinized rats with penetrating wound confirmed that PCSM-S showed increased hemostatic effect and had no adverse effects [149]. Microwave ablation is a treatment method for liver cancers and has side effects that can cause post-operative complications. Zhang et al. used the insulating effect of chitosan-based thermosensitive hydrogels for microwave-assisted ablation of liver tissue. In vivo experiments performed in rabbits showed that the in situ-formed hydrogel has the ability to protect the nearby stomach wall during microwave ablation of liver tissues [150].

Cationic polymers are widely investigated as non-viral gene delivery systems for increasing gene transfection efficiency with low toxicity. Chitosan hydrogels have been investigated as gene delivery system. Injectable, thermosensitive poly(organophosphazene) hydrogel loaded with galactosylated chitosan-*g*-PEI/DNA complexes showed excellent transfection in vitro in HepG2 and in vivo in BALB/c mice. Results indicated that the galactosylated hydrogel complex was less toxic and was able to accumulate in the liver over time, indicating better hepatocyte targeting [151].

3.8 Intervertebral Disk Regenerative Engineering

Intervertebral disks (IVDs) consist of two regions—an inner soft region called nucleus pulposus (NP) and an outer region called annulus fibrosus that surrounds the nucleus region [152]. Regenerative strategies that are under investigation for disk replacement comprise of selection of suitable biomaterial scaffolds, stem cell-based therapies, and development of noninvasive treatment such as injectable hydrogels [153]. For IVD replacement, the hydrogels must have properties such as the ability to support NP cell function, absorb shock, and tunable mechanical properties [26]. Chitosan hydrogel, due to its ability to maintain NP cell phenotype, is an attractive candidate for IVD regenerative engineering [154]. Degeneration of IVDs is a complex issue that can affect the disk as well as the area surrounding the region [153]. An attractive treatment strategy will be to design a bioactive scaffold that can temporarily replace the disk. Mwale et al. investigated the efficacy of a genipin-crosslinked hydrogel developed from chitosan chlorides and chitosan glutamates in supporting disk regeneration. The study showed that 2.5 % chitosan glutamate hydrogel (Protasan UP G213) with 5 % genipin crosslinking was optimal in showing 95 % viability of disk cells. The hydrogel showed excellent biocompatibility in mice model. Using human cadaveric IVDs, the study also showed the feasibility of the gel to be injected into human NP without leakage [155]. Research is also ongoing to find suitable candidates for restoring the structure and mechanical functions of the disk. Smith et al. developed chitosan, dextran, and teleostean triple-interpenetrating network as a potential matrix with NP mimic properties. NP cells showed viability and proliferation when cultured for 14 days in the hydrogel. Mechanical properties of the hydrogel were similar to that of human NP [156]. Chitosan–GP hydrogel was also investigated as a potential candidate for repairing degenerated intervertebral disk. Human mesenchymal stem cells (MSCs) were cultured in the hydrogel for 4 weeks, and gene expression profile was evaluated to compare the behavior of the MSCs to NP cells and articular chondrocytes. The study showed that the amounts of proteoglycan and collagen produced by the MSCs encapsulated in the chitosan hydrogel were similar to that of NP cells [157]. For NP replacement, Sasson et al. fabricated chitosan hydrogels and studied their hyperelastic mechanical behavior using mechanical testing and computational analysis. Results revealed that Ogden hyperelastic model can be used for representing the nonlinear behavior of chitosan hydrogel, and the model can be used for biomechanical simulation for the replacement of NP [154].

Chitosan hydrogels have also been investigated for the controlled release of molecules to NP. Cheng et al. used an injectable thermosensitive chitosan/gelatin/glycerol phosphate hydrogel for the controlled delivery of ferulic acid (FA). FA is an antioxidant and can relieve H_2O_2-induced oxidative stress on the NP cells. Controlled delivery of FA from the hydrogel decreased the oxidative stress on NP cells and showed a decrease in cellular apoptosis [158].

3.9 Cardiac Regenerative Engineering

Cardiac tissue regeneration pertains to repairing heart tissue and restoring the functioning ability of the heart [26]. Current treatments include transplants and artificial devices, as the natural capability of the human heart to repair itself is generally low. However, due to the limited availability of donor hearts, there is significant focus on developing novel regenerative strategies to support cardiac tissue growth. Due to its biocompatibility and biodegradability, chitosan-based hydrogels have been investigated for cardiac tissue repair and regeneration [159]. Reis et al. used a thermoresponsive hydrogel system composed of chitosan and collagen I incorporated with QHREDGS peptide (derived from angiopoietin-1) for cardiomyocytes delivery to treat myocardial infarction (MI) in mouse model. Increase in bioactive peptide concentration showed increase in cardiomyocyte cell viability, morphology, and metabolism [160]. Lu et al. demonstrated the use of temperature-responsive chitosan hydrogels prepared in the presence of β-GP for delivering embryonic stem cells (ESCs) for the treatment of MI in rat model. The study was performed by injecting hydrogel solution with ESCs into the heart wall of the mice. The results showed improved heart functioning and increased wall thickness and density of microvessels [43]. Wang et al. investigated cardiac cell differentiation of brown adipose-derived stem cells (BADSCs) in chitosan hydrogels for the treatment of MI. The study demonstrated that chitosan hydrogels with BADSCs have the ability to preserve heart functioning as well as increase angiogenesis, making it a potential candidate for treating MI [159]. Recently, photo-crosslinked collagen–chitosan hydrogels were used to design a contractile cardiac tissue in vitro. Microgrooves 10, 20, and 100 μm were formed on the hydrogel using polydimethylsiloxane (PDMS) molds. The cardiomyocytes showed oriented growth with respect to the microgrooves in vitro, and electrical stimulations caused further alignment of cardiomyocytes. The study showed that cardiomyocytes seeded in 3D hydrogels with smaller grooves (~10 μm) produced beating heart tissue in vitro, indicating the efficacy of the smaller features in promoting cell-to-cell communication. Additionally, electrical stimulation improved cell density and helped in attaining tissue morphology that can mimic the native tissue upon implantation. Finally, the biodegradability of the hydrogel can contribute toward translating this engineered technique to clinical applications [161]. Pok et al. developed a biodegradable cardiac patch for the treatment of congenital cardiac repair using a scaffold consisting of multiple layers of chitosan and gelatin hydrogels with a polycaprolactone core [162].

Growth factors commonly investigated to promote cardiac tissue regeneration include FGF, VEGF, granulocyte colony-stimulating factor (GCSF), IGF-1, and stem cell factor (SCF) [163]. Fujitha et al. synthesized a photocrosslinked chitosan hydrogel for delivering FGF-2 for treating chronic MI in rabbit model [164]. Significant improvement in systolic pressure at the left ventricle with increased angiogenesis and formation of viable myocardium was observed. Another group used a collagen–chitosan hydrogel for the delivery of thymosin β4 (Tβ4). The study in infarcted rat model showed that the hydrogel with Tβ4 promoted vascularization, reduced tissue loss, and kept the ventricular wall thickness intact [165]. These studies demonstrate the potential of chitosan hydrogel as a suitable biomaterial to support cardiac tissue regeneration.

3.10 Skin Regenerative Engineering

Skin is a major part of the human body and acts as a protective outer layer. Skin regenerative engineering deals with healing skin defects and wounds caused due to burn, injury, pathology, and trauma. Skin availability for autologous transplantation is limited, so tissue engineering and regenerative therapies are needed to address skin-related issues [166]. Studies have shown that chitosan can play a major role in skin regeneration and wound healing due to its antimicrobial property and ability to induce wound healing and promote vascularization. Additionally, its ability to reduce scarring has also been reported [6] Miguel et al. evaluated a thermoresponsive chitosan–agarose hydrogel for wound dressing application. In vitro and in vivo studies indicated that the hydrogel had the ability to provide wound healing environment by preventing water loss and also had bactericidal properties that are essential for 'ideal' wound dressing material [29]. Figure 7 shows histological section of tissue samples isolated from the saline control group and chitosan–agarose hydrogel. The presence of hydrogel led to complete re-epithelialization of the wound without signs of acute inflammation.

Fluorinated methacrylamide chitosan hydrogels are developed as oxygen carriers that can uptake oxygen and supply them at an optimum level to promote wound healing. The hydrogels, based on perfluorocarbon substitution, were capable of taking up oxygen for 2–6 h and releasing them for 12–120 h, which can help in enhancing the wound-healing process [167]. For regenerating the skin after third degree burns, Boucard et al. investigated bilayered chitosan hydrogels in which the first rigid layer was for providing protection as well as exchange of gas, and the second soft layer was for wound healing. Experiments were performed on pig dorsal deep burn model, and results showed that the chitosan hydrogel was capable of regenerating new skin with vascularized skin tissue [168]. Kiyozumi

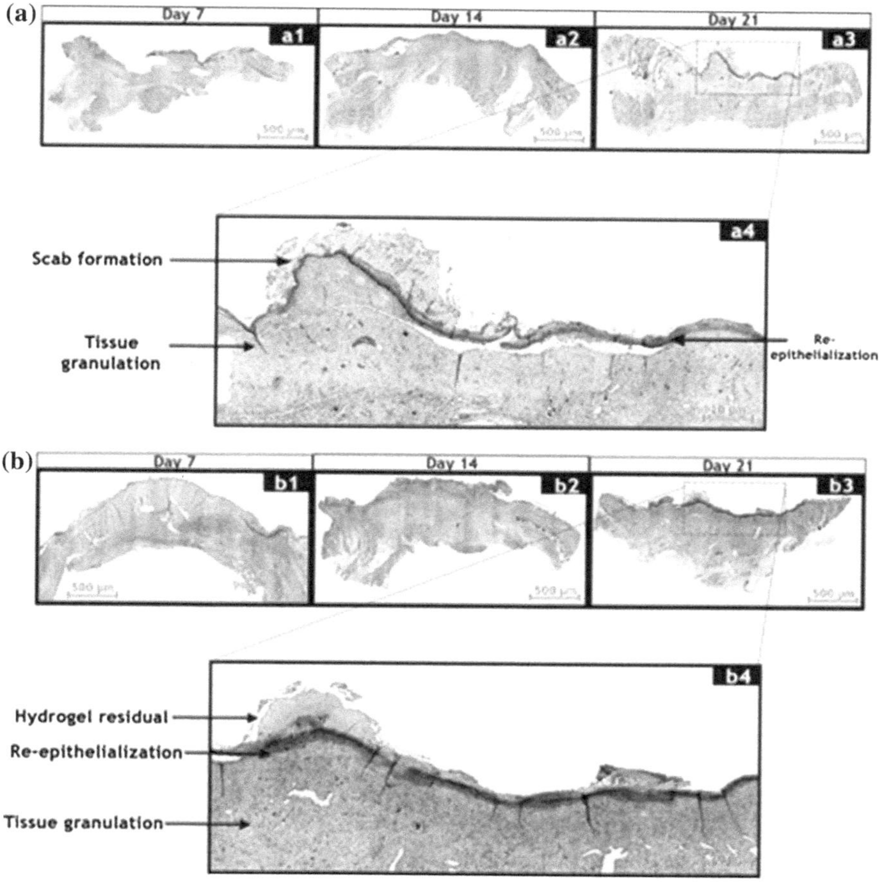

Fig. 7 Hematoxylin and eosin-stained sections of tissue samples after 7, 14, and 21 days **a** treated with saline solution **b** treated with CAH (3 % ag–1.5 % Ch) [29]

et al. compared the efficiency of a photocrosslinked chitosan hydrogel and collagen sponge for treating deep dermal burns. The experiment was conducted in wistar rats, and chitosan hydrogel showed neovascularization and degradability and helped in forming thicker granulation tissue compared to the collagen sponge [169].

With regards to drug delivery applications, it has been reported that the hydrogel made of chitosan and PVA can enhance wound healing by delivering silver sulfadiazine in vivo. The results showed excellent wound healing property in hydrogel containing 7.5 % PVA, 0.75 % chitosan, and 1 % silver sulfadiazine [90]. Similarly, Cui et al. demonstrated the delivery of lysostaphin using chitosan–collagen hydrogel to prevent methicillin-resistant s*taphylococcus aureus* infection. The study showed that the hydrogel was effective in wound healing by delivering the drug to the third degree burn area and that there was no trace of bacteria in the wound site by the second week [170].

Growth factors play important roles in different stages of wound healing. To promote skin regeneration, growth factors such as IGF, basicFGF (bFGF), EGF, VEGF, and PDGF have been investigated [171]. Fujitha et al. reported the potential of chitosan hydrogel for the delivery of FGF-2 to induce angiogenesis [172]. Similarly, bFGF/heparin incorporated in chitosan and pluronic hydrogel has been reported for fibroblast encapsulation and protein delivery [173]. Apart from this, hydrogels based on chitosan and polyacrylamide were investigated for the controlled delivery of the antibiotic piperacillin–tazobactam and EGF to promote wound healing [174].

3.11 Ligament Regenerative Engineering

Ligaments are fibrous connective tissue and connect bones with one another. Ligaments are known to have poor regeneration capacity. Strategies currently under investigation include artificial ligaments, use of biodegradable materials that can mimic the natural environment of ligaments, stem cell-based treatments, and growth factor-mediated therapy [175]. However, artificial ligament replacements present challenges due to less than optimal long-term behavior. Hayami et al. investigated the potential of a cell-based therapy using a biodegradable composite scaffold composed of electrospun poly(ε-caprolactone-*co*-D,L-lactide) (PCLDLLA) fibers in photocrosslinked *N*-methacrylated glycol chitosan (MeGC) hydrogel containing cells. MeGC is non-cell adherent, so for improving cell adhesion, base etching was performed on PCLDLLA fibers. In vitro study showed that ligamentum fibroblast cells attached on to the aligned PCLDLLA fibers and also maintained ligament-cell phenotype inside the hydrogel. The study indicates that composite scaffold design may help in generating in vivo natural environment for the cells in vitro [176]. Deepthi et al. used another approach for designing scaffolds for ligament regeneration. In this study, a bilayered poly(caprolactone) scaffold was coated with chitosan–hyaluronic hydrogel. Coating the poly(caprolactone) fibers with the hydrogel provided an environment that can mimic natural ECM and can also hold the cells on the fibers. The study reported high cell survival, cell elongation, and protein adsorption, demonstrating the potential of the hydrogel carrier for ligament regeneration [177].

4 Conclusions

Chitosan hydrogels have been widely researched in the field of regenerative engineering due to its favorable properties such as biocompatibility, biodegradability, antimicrobial activity, mucoadhesivity, and low toxicity. The development of water-soluble chitosan derivatives led to the development of wide range of chitosan hydrogels as cell and/protein/drug delivery vehicles. As discussed in the chapter, chitosan-based hydrogels have been investigated as potential biomaterials for a wide range of tissues, indicating the versatility of the polymer system. There is still ample scope for research to be conducted on chitosan hydrogels, especially with regards to the development of injectable hydrogels to develop biologically active substrates to support efficient tissue regeneration.

References

1. Ravi Kumar MNV (2000) A review of chitin and chitosan applications. React Funct Polym 46:1–27
2. Rinaudo M (2006) Chitin and chitosan: properties and applications. Prog Polym Sci 31:603–632
3. Narayanan D, Jayakumar R, Chennazhi KP (2014) Versatile carboxymethyl chitin and chitosan nanomaterials: a review. WIREs Nanomed Nanobiotechnol 6:574–598
4. Tokura S, Itoyama K, Nishi N, Nishimura S, Saiki I, Azuma I (1994) Selective sulfation of chitin derivatives for biomedical functions. J Macromol Sci A 31:1701–1718
5. Pillai CKS, Paul W, Sharma CP (2009) Chitin and chitosan polymers: chemistry, solubility and fiber formation. Prog Polym Sci 34:641–678
6. Dash M, Chiellini F, Ottenbrite RM, Chiellini E (2011) Chitosan—a versatile semi-synthetic polymer in biomedical applications. Prog Polym Sci 36:981–1014
7. Croisier F, Jerome C (2013) Chitosan-based biomaterials for tissue engineering. Eur Polym J 49:780–792
8. Yuan Y, Chesnutt BM, Haggard WO, Bumgardner JD (2011) Deacetylation of chitosan: material characterization and in vitro evaluation via albumin adsorption and pre-osteoblastic cell cultures. Materials 4:1399–1416
9. Cartier N, Domard A, Chanzy H (1990) Single crystals of chitosan. Int J Biol Macromol 12:289–294
10. Rinaudo M, Milas M, Dung PL (1993) Characterization of chitosan. Influence of ionic strength and degree of acetylation on chain expansion. Int J Biol Macromol 15(5):281–285
11. Roberts GA, Domszy JG (1982) Determination of the viscometric constants for chitosan. Int J Biol Macromol 4:374–377
12. Mourya VK, Inamdar NN (2008) Chitosan—modifications and applications: opportunities galore. React Funct Polym 68:1013–1051
13. Mourya VK, Inamdar NN, Tiwari Ashutosh (2010) Carboxymethyl chitosan and its applications. Adv Mat Lett 1(1):11–33
14. Upadhyay L, Singh J, Agarwal V, Tewari RP (2013) Biomedical applications of carboxymethyl chitosans. Carbohyd Polym 91:452–466
15. Pereira P, Pedrosa SS, Correia A, Lima CF, Olmedo MP, Gonzalez-Fernandez A, Vilanova M, Gama FM (2015) Biocompatibility of a self-assembled glycol chitosan nanogel. Toxicol In Vitro 29:638–646
16. Trapani A, Gioia SD, Ditaranto N, Cioffi N, Goycoolea FM, Carbone A, Garcia-Fuentes M, Conese M, Alonso MJ (2013) Systemic heparin delivery by the pulmonary route using chitosan and glycol chitosan nanoparticles. Int J Pharm 447:115–123
17. Jiang G, Sun J, Ding F (2014) PEG-g-chitosan thermosensitive hydrogel for implant drug delivery: cytotoxicity, in vivo degradation and drug release. J Biomat Sci-Polym E 25(3):241–256

18. Ding K, Wang Y, Wang H, Yuan L, Tan M, Shi X, Lyu Z, Liu Y, Chen H (2014) 6-*O*-sulfated chitosan promoting the neural differentiation of mouse embryonic stem cells. ACS Appl. Mater. Interfaces 6(22):20043–20050
19. Vikhoreva G, Bannikova G, Stolbushkina P, Panov A, Drozd N, Makarov V, Varlamov V, Gal'braikh L (2005) Preparation and anticoagulant activity of a low-molecular-weight sulfated chitosan. Carbohyd Polym 62:327–332
20. Heras A, Rodriguez NM, Ramos VM, Agullo E (2001) *N*-methylene phosphonic chitosan: a novel soluble derivative. Carbohyd Polym 44:1–8
21. Zhu D, Yao K, Bo J, Zhang H, Liu L, Dong X, Song L, Leng X (2010) Hydrophilic/lipophilic *N*-methylene phosphonic chitosan as a promising non-viral vector for gene delivery. J Mater Sci-Mater M 21(1):223–229
22. Sajomsang W, Nuchuchua O, Saesoo S, Gonil P, Chaleawlert-umpon S, Pimpha N, Sramala I, Soottitantawat A, Puttipipatkhachorn S, Ruktanonchai UR (2013) A comparison of spacer on water-soluble cyclodextrin grafted chitosan inclusion complex as carrier of eugenol to mucosae. Carbohyd Polym 92:321–327
23. Lin CC, Metters AT (2006) Hydrogels in controlled release formulations: network design and mathematical modeling. Adv Drug Deliver Rev 58:1379–1408
24. Patel A, Mequanint K (2011) Hydrogel biomaterials. In: Fazel-Rezai R (ed) Chapter 14 in Biomedical Engineering—Frontiers and Challenges, pp 275–296
25. Lee KY, Mooney DJ (2001) Hydrogels for tissue engineering. Chem Rev 101(7):1869–1879
26. Hunt JA, Chen R, van Veen T, Bryan N (2014) Hydrogels for tissue engineering and regenerative medicine. J Mater Chem B 2:5319–5338
27. Ottenbrite RM, Park K, Okano T (eds) (2010) Biomedical applications of hydrogels handbook, p 204
28. Tan H, Marra KG (2010) Injectable, biodegradable hydrogels for tissue engineering applications. Materials 3:1746–1767
29. Miguel SP, Ribeiro MP, Branca H, Coutinho P, Correia IJ (2014) Thermoresponsive chitosan-agarose hydrogel for skin regeneration. Carbohyd Polym 111:366–373
30. Zhu J, Marchant RE (2011) Design properties of hydrogel tissue-engineering scaffolds. Expert Rev Med Devices 8(5):607–626
31. Gibas I, Janik H (2010) Review: synthetic polymer hydrogels for biomedical applications. Chem Chem Technol 4:297–304
32. Berger J, Reist M, Mayer JM, Felt O, Gurny R (2004) Structure and interactions in chitosan hydrogels formed by complexation or aggregation for biomedical applications. Eur J Pharm Biopharm 57:35–52
33. Aziz MA, Cabral JD, Brooks HJL, Moratti SC, Hanton LR (2012) Antimicrobial properties of a chitosan dextran-based hydrogel for surgical use. Antimicrob Agents Ch 56(1):280–287
34. He P, Davis SS, Illum L (1998) In vitro evaluation of the mucoadhesive properties of chitosan microspheres. Int J Pharm 166:75–88
35. Lee YH, Chang JJ, Yang MC, Chien CT, Lai WF (2012) Acceleration of wound healing in diabetic rats by layered hydrogel dressing. Carbohyd Polym 88:809–819
36. Han T, Nwe N, Furuike T, Tokura S, Tamura H (2012) Methods of N-acetylated chitosan scaffolds and its in vitro biodegradation by lysozyme. J. Biomed Sci Eng 5:15–23
37. Verheul RJ, Amidi M, van Steenbergen MJ, van Riet E, Jiskoot W, Hennink WE (2009) Influence of the degree of acetylation on the enzymatic degradation and in vitro biological properties of trimethylated chitosans. Biomaterials 30(18):3129–3135
38. Chung YC, Chen CY (2008) Antibacterial characteristics and activity of acid-soluble chitosan. Bioresource Technol 99:2806–2814
39. Kim SK (ed) (2013) Chitin and chitosan derivatives: advances in drug discovery and developments, p 246
40. Bhattarai N, Gunn J, Zhang M (2010) Chitosan-based hydrogels for controlled, localized drug delivery. Adv Drug Deliver Rev 62:83–99

41. Schuetz YB, Gurny R, Jordan O (2008) A novel thermoresponsive hydrogel based on chitosan. Eur J Pharm Biopharm 68:19–25
42. Sheridan WS, Grant OB, Duffy GP, Murphy BP (2014) The application of a thermoresponsive chitosan/β-GP gel to enhance cell repopulation of decellularized vascular scaffolds. J Biomed Mater Res B 102B:1700–1710
43. Lu WN, Lu SH, Wang HB, Li DX, Duan CM, Liu ZQ, Hao T, He WJ, Xu B, Fu Q, Song YC, Xie XH, Wang CY (2009) Functional improvement of infarcted heart by co-injection of embryonic stem cells with temperature-responsive chitosan hydrogel. Tissue Eng Pt A 15(6):1437–1447
44. Niranjan R, Koushik C, Saravanan S, Moorthi A, Vairamani M, Selvamurugan N (2013) A novel injectable temperature-sensitive zinc doped chitosan/β-glycerophosphate hydrogel for bone tissue engineering. Int J Biol Macromol 54(2013):24–29
45. Nair LS, Starnes T, Ko JWK, Laurencin CT (2007) Development of injectable thermogelling chitosan-inorganic phosphate solutions for biomedical applications. Biomacromolecules 8:3779–3785
46. Liu X, Chen Y, Huang Q, He W, Feng Q, Yu B (2014) A novel thermo-sensitive hydrogel based on thiolated chitosan/hydroxyapatite/beta-glycerophosphate. Carbohyd Polym 110:62–69
47. Chenite A, Buschmann M, Wang D, Chaut C, Kandani N (2001) Rheological characterization of thermogelling chitosan/glycerol-phosphate solutions. Carbohyd Polym 46(1):39–47
48. Han H, Nam D, Seo D, Kim T, Shin B, Choi H (2004) Preparation and biodegradation of thermosensitive chitosan hydrogel as a function of pH and temperature. Macromol Res 12(5):507–511
49. Kafedjiiski K, Krauland AH, Hoffer MH, Bernkop-Schnuech A (2005) Synthesis and in vitro evaluation of a novel thiolated chitosan. Biomaterials 26(7):819–826
50. Khodaverdi E, Tafaghodi M, Ganji F, Abnoos K, Naghizadeh H (2012) In vitro insulin release from thermosensitive chitosan hydrogel. AAPS PharmSciTech 13(2):460–466
51. Jeong B, Kim SW, Bae YH (2012) Thermosensitive sol-gel reversible hydrogels. Adv Drug Deliver Rev 64:154–162
52. Kim JH, Lee SB, Kim SJ, Lee YM (2002) Rapid temperature/pH response of porous alginate-g-poly(Nisopropylacrylamide) hydrogels. Polymer 43:7549–7558
53. Qiu Y, Park K (2001) Environment-sensitive hydrogels for drug delivery. Adv Drug Deliver Rev 53:321–339
54. Chen JP, Cheng TH (2006) Thermo-responsive chitosan-*graft*-poly(N-isopropylacrylamide) injectable hydrogel for cultivation of chondrocytes and meniscus cells. Macromol Biosci 6:1026–1039
55. Mao J, Kondu S, Ji HF, McShane MJ (2006) Study of the near-neutral pH-sensitivity of chitosan/gelatin hydrogels by turbidimetry and microcantilever deflection. Biotechnol Bioeng 95(3):333–341
56. Bostan MS, Senol M, Cig T, Peker I, Goren AC, Ozturk T, Eroglu MS (2003) Controlled release of 5-aminosalicylicacid from chitosan based pH and temperature sensitive hydrogels. Int J Biol Macromol 52:177–183
57. Chiu YL, Chen SC, Su CJ, Hsiao CW, Chen YM, Chen HL, Sung HW (2009) pH-triggered injectable hydrogels prepared from aqueous N-palmitoyl chitosan: in vitro characteristics and in vivo biocompatibility. Biomaterials 30:4877–4888
58. Li QL, Chen ZQ, Darvell BW, Liu LK, Jiang HB, Zen Q, Peng Q, Ou GM (2007) Chitosan-phosphorylated chitosan polyelectrolyte complex hydrogel as an osteoblast carrier. J Biomed Mater Res 82(2):481–486
59. Lankalapalli S, Kolapalli VRM (2009) Polyelectrolyte complexes: a review of their applicability in drug delivery technology. Indian J Pharm Sci 71(5):481–487
60. Coutinho DF, Sant S, Shakiba M, Wang B, Gomes ME, Neves NM, Reis RL, Khademhosseini A (2012) Microfabricated photocrosslinkable polyelectrolyte-complex of chitosan and methacrylated gellan gum. J Mater Chem 22(33):17262–17271

61. Ji DY, Kuo TF, Wu HD, Yang JC, Lee SY (2012) A novel injectable chitosan/polyglutamate polyelectrolyte complex hydrogel with hydroxyapatite for soft-tissue augmentation. Carbohyd Polym 89:1123–1130
62. Chang HH, Wang YL, Chiang YC, Chen YL, Chuang YH, Tsai SJ, Heish KH, Lin FH, Lin CP (2014) A novel chitosan-γPGA polyelectrolyte complex hydrogel promotes early new bone formation in the alveolar socket following tooth extraction. PLoS ONE 9(3):e92362
63. Azab AK, Orkin B, Doviner V, Nissan A, Klein M, Srebnik M, Rubinstein A (2006) Crosslinked chitosan implants as potential degradable devices for brachytherapy: in vitro and in vivo analysis. J Control Release 111:281–289
64. de Abreu FR, Campana-Filho SP (2009) Characteristics and properties of carboxymethylchitosan. Carbohyd Polym 75:214–221
65. Vaghani SS, Patel MM, Satish CS (2012) Synthesis and characterization of pH-sensitive hydrogel composed of carboxymethyl chitosan for colon targeted delivery of ornidazole. Carbohyd Res 347:76–82
66. Lin Y, Chen Q, Luo H (2007) Preparation and characterization of N-(2-carboxybenzyl) chitosan as a potential pH-sensitive hydrogel for drug delivery. Carbohyd Res 342:87–95
67. Gomez-Mascaraque LG, Mendez JA, Fernandez-Gutierrez M, Vazquez B, Roman JS (2014) Oxidized dextrins as alternative crosslinking agents for polysaccharides: application to hydrogels of agarose-chitosan. Acta Biomater 10:798–811
68. Ranjha NM, Ayub G, Naseem S (2010) Ansari MT (2010) Preparation and characterization of hybrid pH-sensitive hydrogels of chitosan-co-acrylic acid for controlled release of verapamil. J Mater Sci-Mater M 21:2805–2816
69. Wang L, Stegemann JP (2011) Glyoxal crosslinking of cell-seeded chitosan/collagen hydrogels for bone regeneration. Acta Biomater 7:2410–2417
70. Silva SS, Motta A, Rodrigues MT, Pinheiro AFM, Gomes ME, Mano JF, Reis RL, Migliaresi C (2008) Novel genipin-cross-linked chitosan/silk fibroin sponges for cartilage engineering strategies. Biomacromolecules 9:2764–2774
71. Bhattarai N, Ramay HR, Gunn J, Matsen FA, Zhang M (2005) PEG-grafted chitosan as an injectable thermosensitive hydrogel for sustained protein release. J Control Release 103:609–624
72. Giri TK, Thakur A, Alexander A, Ajazuddin Badwaik H, Tripathi DK (2012) Modified chitosan hydrogels as drug delivery and tissue engineering systems: present status and applications. Acta Pharmaceutica Sinica B 2(5):439–449
73. Fiejdasz S, Szczubialka K, Lewandowska-Lancucka J, Osyczka AM, Nowakowska M (2013) Biopolymer-based hydrogels as injectable materials for tissue repair scaffolds. Biomed Mater 8:035013
74. Obara K, Ishihara M, Ishizuka T, Fujita M, Ozeki Y, Maehara T, Saito Y, Yura H, Matsui T, Hattori H, Kikuchi M, Kurita A (2003) Photocrosslinkable chitosan hydrogel containing fibroblast growth factor-2 stimulates wound healing in healing-impaired db/db mice. Biomaterials 24:3437–3444
75. Ishihara M, Obara K, Ishizuka T, Fujita M, Sato M, Masuoka K, Saito Y, Yura H, Matsui T, Hattori H, Kikuchi M, Kurita A (2003) Controlled release of fibroblast growth factors and heparin from photocrosslinked chitosan hydrogels and subsequent effect on in vivo vascularization. J Biomed Mater Res Pt A 64A(3):551–559
76. Ono K, Ishihara M, Ozeki Y, Deguchi H, Sato M, Saito Y, Yura H, Sato M, Kikuchi M, Kurita A, Maehara T (2001) Experimental evaluation of photocrosslinkable chitosan as a biologic adhesive with surgical applications. Surgery 130(5):844–850
77. Hayashi T, Matsuyama T, Hanada K, Nakanishi K, Uenoyama M, Fujita M, Ishihara M, Kikuchi M, Ikeda T, Tajiri H (2004) Usefulness of photocrosslinkable chitosan for endoscopic cancer treatment in alimentary tract. J Biomed Mater Res B 71(2):367–372
78. Zhou Y, Ma G, Shi S, Yang D, Nie J (2011) Photopolymerized water-soluble chitosan-based hydrogel as potential use in tissue engineering. Int J Biol Macromol 48:408–413

79. Tsuda Y, Hattori H, Tanaka Y, Ishihara M, Kishimoto S, Amako M, Arino H, Nemoto K (2013) Ultraviolet light-irradiated photocrosslinkable chitosan hydrogel to prevent bone formation in both rat skull and fibula bone defects. J Tissue Eng Regen M 7:720–728
80. Ono K, Saito Y, Yura H, Ishikawa K, Kurita A, Akaike T, Ishihara M (2000) Photocrosslinkable chitosan as a biological adhesive. J Biomed Mater Res 49(2):289–295
81. Arakawa C, Ng R, Tan S, Kim S, Wu B, Lee M (2014) Photopolymerizable chitosan-collagen hydrogels for bone tissue engineering. J Tissue Eng Regen Med (in press)
82. Teixeira LSM, Feijen J, van Blitterswijk CA, Dijkstra PJ, Karperien M (2012) Enzyme-catalyzed crosslinkable hydrogels: emerging strategies for tissue engineering. Biomaterials 33:1281–1290
83. Sakai S, Yamada Y, Zenke T, Kawakami K (2009) Novel chitosan derivative soluble at neutral pH and in-situ gellable via peroxidase-catalyzed enzymatic reaction. J Mater Chem 19:230–235
84. da Silva MA, Bode F, Drake AF, Goldoni S, Stevens MM, Dreiss CA (2014) Enzymatically cross-linked gelatin/chitosan hydrogels: tuning gel properties and cellular response. Macromol Biosci 14:817–830
85. Kang GD, Lee KH, Ki CS, Nahm JH, Park YH (2004) Silk fibroin/chitosan conjugate crosslinked by tyrosinase. Macromol Res 12(5):534–539
86. Jin R, Teixeira LSM, Dijkstra PJ, Karperien M, van Blitterswijk CA, Zhong ZY, Feijen J (2009) Injectable chitosan-based hydrogels for cartilage tissue engineering. Biomaterials 30:2544–2551
87. Brittain SB (2013) Development and characterization of a bioactive injectable chitosan hydrogel for bone repair. University of Connecticut, Master's Thesis
88. Shi C, Zhu Y, Ran X, Wang M, Su Y, Cheng T (2006) Therapeutic potential of chitosan and its derivatives in regenerative medicine. J Surg Res 133(2):185–192
89. Mansouri S, Lavigne P, Corsi K, Benderdour M, Beaumont E, Fernandes JC (2004) Chitosan-DNA nanoparticles as non-viral vectors in gene therapy: strategies to improve transfection efficacy. Eur J Pharm Biopharm 57(1):1–8
90. Peluso G, Petillo O, Ranieri M, Santin M, Ambrosio L, Calabro D, Avallone B, Balsamo G (1994) Chitosan-mediated stimulation of macrophage function. Biomaterials 15(15):1215–1220
91. Chakavala SR, Patel NG, Pate NV, Thakkar VT, Patel KV, Gandhi TR (2012) Development and in vivo evaluation of silver sulfadiazine loaded hydrogel consisting polyvinyl alcohol and chitosan for severe burns. J Pharm Bioallied Sci 4(Supplement 1):S54–S56
92. Ivkovic A, Marijanovic I, Hudetz D, Porter RM, Pecina M, Evans CH (2011) Regenerative medicine and tissue engineering in orthopaedic surgery. Front Biosci (Elite Edition) 3:923–944
93. Amini AT, Laurencin CT, Nukavarapu SP (2012) Bone tissue engineering: recent advances and challenges. Crit Rev Biomed Eng 40(5):363–408
94. Li J, Yang B, Qian Y, Wang Q, Han R, Hao T, Shu Y, Zhang Y, Yao F, Wang C (2014) Iota-carrageenan/chitosan/gelatin scaffold for the osteogenic differentiation of adipose-derived MSCs in vitro. J Biomed Mater Res B (in press)
95. Beskardes IG, Demirtas TT, Durukan MD, Gumusderelioglu M (2012) Microwave-assisted fabrication of chitosan–hydroxyapatite superporous hydrogel composites as bone scaffolds. J Tissue Eng Regen Med (in press)
96. Dessi M, Borzacchiello A, Mohamed THA, Abdel-Fattah WI, Ambrosio L (2013) Novel biomimetic thermosensitive β-tricalcium phosphate/chitosan-based hydrogels for bone tissue engineering. J Biomed Mater Res A 101(10):2984–2993
97. Dimitriou R, Jones E, McGonagle D, Peter V, Giannoudis PV (2011) Bone regeneration: current concepts and future directions. BMC Med 9:66
98. Vo TN, Kasper FK, Mikos AG (2012) Strategies for controlled delivery of growth factors and cells for bone regeneration. Adv Drug Deliver Rev 64(12):1292–1309

99. Dong L, Huang Z, Cai X, Xiang J, Zhu YA, Wang R, Chen J, Zhang J (2011) Localized delivery of antisense oligonucleotides by cationic hydrogel suppresses TNF-α expression and endotoxin-induced osteolysis. Pharm Res 28:1349–1356
100. Kim S, Bedigrew K, Guda T, Maloney WJ, Park S, Wenke JC, Yang YP (2014) Novel osteoinductive photo-cross-linkable chitosan-lactide-fibrinogen hydrogels enhance bone regeneration in critical size segmental bone defects. Acta Biomater 10(12):5021–5033
101. Kim S, Kang Y, Mercado-Pagan AE, Maloney WJ, Yang Y (2014) In vitro evaluation of photo-crosslinkable chitosan-lactide hydrogels for bone tissue engineering. J Biomed Mater Res, Part B 102B:1393–1406
102. Bhattacharjee M, Coburn J, Centola M, Murab S, Barbero A, Kaplan DL, Martin I, Ghosh S (2014) Tissue engineering strategies to study cartilage development, degeneration and regeneration. Adv Drug Deliver Rev pp. 107−122
103. Ryan JM, Flanigan DC (2013) Emerging technologies: what is the future of cartilage restoration? Hard Tissue 2(2):12
104. Musumeci G, Castrogiovanni P, Leonardi R, Trovato FM, Szychlinska MA, Di Giunta A, Loreto C, Castorina S (2014) New perspectives for articular cartilage repair treatment through tissue engineering: a contemporary review. World J Orthop 5(2):80–88
105. Park KM, Lee SY, Joung YK, Na JS, Lee MC, Park KD (2009) Thermosensitive chitosan-pluronic hydrogel as an injectable cell delivery carrier for cartilage regeneration. Acta Biomater 5:1956–1965
106. Tan H, Chu CR, Payne KA, Marra KG (2009) Injectable in situ forming biodegradable chitosan-hyaluronic acid based hydrogels for cartilage tissue engineering. Biomaterials 30:2499–2506
107. Ngoenkam J, Faikrua A, Yasothornsrikul S, Viyoch J (2010) Potential of an injectable chitosan/starch/β-glycerol phosphate hydrogel for sustaining normal chondrocyte function. Int J Pharm 391:115–124
108. Marsich E, Borgogna M, Donati I, Mozetic P, Strand BL, Salvador SG, Vittur F, Paoletti S, Alginate/lactose-modified chitosan hydrogels: a bioactive biomaterial for chondrocyte encapsulation. J Biomed Mater Res A 84(2):364–376
109. Zhao P, Deng C, Xu H, Tang X, He H, Lin C, Su J (2014) Fabrication of photo-crosslinked chitosan-gelatin scaffold in sodium alginate hydrogel for chondrocyte culture. Bio-Med Mater Eng 24:633–641
110. Hong Y, Song H, Gong Y, Mao Z, Gao C, Shen J (2007) Covalently crosslinked chitosan hydrogel: properties of in vitro degradation and chondrocyte encapsulation. Acta Biomater 3:23–31
111. Lam J, Lu S, Kasper FK, Mikos AG (2014) Strategies for controlled delivery of biologics for cartilage repair, Adv Drug Deliv Rev pp. 123−134
112. Faikrua A, Wittaya-areekul S, Oonkhanond B, Viyoch J (2013) In vivo chondrocyte and transforming growth factor-β1 delivery using the thermosensitive chitosan/starch/β-glycerol phosphate hydrogel. J Biomater Appl 28(2):175–186
113. Choi B, Kim S, Lin B, Wu BM, Lee M (2014) Cartilaginous extracellular matrix-modified chitosan hydrogels for cartilage tissue engineering. ACS Appl Mater Interfaces 6:20110–20121
114. Sukarto A, Yu C, Flynn LE, Amsden BG (2012) Co-delivery of adipose-derived stem cells and growth factor-loaded microspheres in RGD-grafted N-methacrylate glycol chitosan gels for focal chondral repair. Biomacromolecules 13:2490–2502
115. Subramanian A, Krishnan UM, Sethuraman S (2009) Development of biomaterial scaffold for nerve tissue engineering: biomaterial mediated neural regeneration. J Biomed Sci 16:108–118
116. Pfister LA, Papaloizos M, Merkle HP, Gander B (2007) Hydrogel nerve conduits produced from alginate/chitosan complexes. J Biomed Mater Res A 80(4):932–937

117. Zuidema JM, Pap MM, Jaroch DB, Morrison FA, Gilbert RJ (2011) Fabrication and characterization of tunable polysaccharide hydrogel blends for neural repair. Acta Biomater 7:1634–1643
118. Kwon JS, Kim GH, Kim DY, Yoon SM, Seo HW, Kim JH, Min BH, Kim MS (2012) Chitosan-based hydrogels to induce neuronal differentiation of rat muscle-derived stem cells. Int J Biol Macromol 51:974–979
119. Crompton KE, Goud JD, Bellamkonda RV, Gengenbach TR, Finkelstein DI, Horne MK, Forsythe JS (2007) Polylysine-functionalised thermoresponsive chitosan hydrogel for neural tissue engineering. Biomaterials 28:441–449
120. Valmikinathan CM, Mukhatyar VJ, Jain A, Karumbaiah L, Dasari M, Bellamkonda RV (2012) Photocrosslinkable chitosan based hydrogels for neural tissue engineering. Soft Matter 8:1964–1976
121. Rickett TA, Amoozgar Z, Tuchek CA, Park J, Yeo Y, Shi R (2011) Rapidly photo-cross-linkable chitosan hydrogel for peripheral neurosurgeries. Biomacromolecules 12:57–65
122. McMahon SS, Nikolskaya N, Choileain SN, Hennessy N, O'Brien T, Strappe PM, Gorelov A, Rochev Y (2011) 0 Thermosensitive hydrogel for prolonged delivery of lentiviral vector expressing neurotrophin-3 in vitro. J Gene Med 13:591–601
123. Leipzig ND, Wylie RG, Kim H, Shoichet MS (2011) Differentiation of neural stem cells in three-dimensional growth factor-immobilized chitosan hydrogel scaffolds. Biomaterials 32:57–64
124. Li H, Koenig AM, Sloan P, Leipzig ND (2014) In vivo assessment of guided neural stem cell differentiation in growth factor immobilized chitosan-based hydrogel scaffolds. Biomaterials 35:9049–9057
125. Grolik M, Szczubialka K, Wowra B, Dobrowolski D, Orzechowska-Wylegala B, Wylegala E, Nowakowska M (2012) Hydrogel membranes based on genipin-cross-linked chitosan blends for corneal epithelium tissue engineering. J Mater Sci-Mater Med 23:1991–2000
126. Ozcelik B, Brown KD, Blencowe A, Daniell M, Stevens GW, Qiao GG (2013) Ultrathin chitosan-poly(ethylene glycol) hydrogel films for corneal tissue engineering. Acta Biomater 9:6594–6605
127. Liang Y, Liu W, Han B, Yang C, Ma Q, Song F, Bi Q (2011) An in situ formed biodegradable hydrogel for reconstruction of the corneal endothelium. Colloid Surface B 82:1–7
128. Rafat M, Li F, Fagerholm P, Lagali NS, Watsky MA, Munger R, Matsuura T, Griffith M (2008) PEG-stabilized carbodiimide crosslinked collagen-chitosan hydrogels for corneal tissue engineering. Biomaterials 29:3960–3972
129. Chien Y, Liao YW, Liu DM, Lin HL, Chen SJ, Chen HL, Peng CH, Liang CM, Mou CY, Chiou SH (2012) Corneal repair by human corneal keratocyte-reprogrammed iPSCs and amphiphatic carboxymethyl-hexanoyl chitosan hydrogel. Biomaterials 33:8003–8016
130. Alonso MJ, Sanchez A (2003) The potential of chitosan in ocular drug delivery. J Pharm Pharmacol 55(11):1451–1463
131. Hosny KM (2009) Preparation and evaluation of thermosensitive liposomal hydrogel for enhanced transcorneal permeation of ofloxacin. AAPS PharmSciTech 10(4):1336–1342
132. Bitar KN, Raghavan S (2012) Intestinal tissue engineering: current concepts and future vision of regenerative medicine in the gut. Neurogastroent Motil 24(1):7–19
133. Bitar KN, Zakhem E (2013) Tissue engineering and regenerative medicine as applied to the gastrointestinal tract. Curr Opin Biotechnol 24(5):909–915
134. Rabbani S, Rabbani A, Mohagheghi MA, Mirzadeh H, Amanpour S, Alibakhshi A, Anvari MS, Ghazizadeh Y (2010) Novel approach for repairing of intestinal fistula using chitosan hydrogel. J Biomater Appl 24(6):545–553
135. Falabella CA, Melendez MM, Weng L, Chen W (2010) Novel macromolecular crosslinking hydrogel to reduce intra-abdominal adhesions. J Surg Res 159:772–778
136. Lauder CI, Strickland A, Maddern GJ (2012) Use of a modified chitosan-dextran gel to prevent peritoneal adhesions in a porcine hemicolectomy model. J Surg Res 176:448–454

137. Xu J, Soliman GM, Barralet J, Cerruti M (2012) Mollusk glue inspired mucoadhesives for biomedical applications. Langmuir 28:14010–14017
138. Chavda H, Modhia I, Mehta A, Patel R, Patel C (2013) Development of bioadhesive chitosan superporous hydrogel composite particles based intestinal drug delivery system. BioMed Res Int 2013:Article ID 563651
139. Kumar Singh Yadav H, Shivakumar HG (2012) In vitro and in vivo evaluation of pH-sensitive hydrogels of carboxymethyl chitosan for intestinal delivery of theophylline. ISRN Pharm. 2012:Article ID 763127
140. Maeng JH, Bang BW, Lee E, Kim J, Kim HG, Lee DH, Yang SG (2014) Endoscopic application of EGF-chitosan hydrogel for precipitated healing of GI peptic ulcers and mucosectomy-induced ulcers. J Mater Sci-Mater M 25:573–582
141. Choi JH, Gimble JM, Lee K, Marra KG, Rubin JP, Yoo JJ, Vunjak-Novakovic G, Kaplan DL (2010) Adipose tissue engineering for soft tissue regeneration. Tissue Eng Pt B-Rev 16(4):413–426
142. Cheung HK, Han TT, Marecak DM, Watkins JF, Amsden BG, Flynn LE (2014) Composite hydrogel scaffolds incorporating decellularized adipose tissue for soft tissue engineering with adipose-derived stem cells. Biomaterials 35:1914–1923
143. Wu X, Black L, Santacana-Laffitte G, Patrick CW (2007) Preparation and assessment of glutaraldehyde crosslinked collagen-chitosan hydrogels for adipose tissue engineering. J Biomed Mater Res A 81(1):59–65
144. Zotarelli Filho IJ, Frascino LF, Greco OT, de Araujo JD, Bilaqui A, Kassis EN, Ardito RV, Bonilla-Rodriguez GO (2013) Chitosan-collagen scaffolds can regulate the biological activities of adipose mesenchymal stem cells for tissue engineering. J Regen Med Tissue Eng 2:12
145. Jaikumar D, Sajesh KM, Soumya S, Nimal TR, Chennazhi KP, Nair SV, Jayakumar R (2014) Injectable alginate-O-carboxymethyl chitosan/nano fibrin compositehydrogels for adipose tissue engineering. Int J Biol Macromol 74:318–326
146. Tan H, Rubin JP, Marra KG (2010) Injectable in situ forming biodegradable chitosan-hyaluronic acid based hydrogels for adipose tissue regeneration. Organogenesis 6(3):173–180
147. Palakkan AA, Hay DC, Anil Kumar PR, Kumary TV, Ross JA (2013) Liver tissue engineering and cell sources: issues and challenges. Liver Int 33:666–676
148. Sj Seo, Choi YJ, Akaike T, Higuchi A, Cho CS (2006) Alginate/galactosylated chitosan/heparin scaffold as a new synthetic extracellular matrix for hepatocytes. Tissue Eng 12(1):33–44
149. Horio T, Ishihara M, Fujita M, Kishimoto S, Kanatani Y, Ishizuka T, Nogami Y, Nakamura S, Tanaka Y, Morimoto Y, Maehara T (2010) Effect of photocrosslinkable chitosan hydrogel and its sponges to stop bleeding in a rat liver injury model. Artif Organs 34:342–347
150. Zhang D, Xie D, Wei X, Zhang D, Chen M, Yu X, Liang P (2014) Microwave ablation of the liver abutting the stomach: insulating effect of a chitosan-based thermosensitive hydrogel. Int J Hyperther 3(2):126–133
151. Jiang HL, Kim YK, Lee SM, Park MR, Kim EM, Jin YM, Arote R, Jeong HJ, Song SC, Cho MH, Cho CS (2010) Galactosylated chitosan-g-PEI/DNA complexes-loaded poly(organophosphazene) hydrogel as a hepatocyte targeting gene delivery system. Arch Pharm Res 33(4):551–556
152. Urban JPG, Roberts S, Ralphs JR (2000) The nucleus of the intervertebral disc from development to degeneration. Am Zool 40(1):53–61
153. Richardson SM, Mobasheri A, Freemont AJ, Hoyland JA (2007) Intervertebral disc biology, degeneration and novel tissue engineering and regenerative medicine therapies. Histol Histopathol 22:1033–1041
154. Sasson A, Patchornik S, Eliasy R, Robinson D, Haj-Ali R (2012) Hyperelastic mechanical behavior of chitosan hydrogels for nucleus pulposus replacement-experimental testing and constitutive modeling. J Mech Behav Biomed 8:143–153

155. Mwale F, Iordanova M, Demers CN, Steffen T, Roughley P, Antoniou J (2005) Biological evaluation of chitosan salts cross-linked to genipin as a cell scaffold for disk tissue engineering. Tissue Eng 11(1–2):130–140
156. Smith LJ, Gorth DJ, Showalter BL, Chiaro JA, Beattie EE, Elliott DM, Mauck RL, Chen W, Malhotra NR (2014) In vitro characterization of a stem-cell-seeded triple-interpenetrating-network hydrogel for functional regeneration of the nucleus pulposus. Tissue Eng Pt A 20(13–14):1841–1849
157. Richardson SM, Hughes N, Hunt JA, Freemont AJ, Hoyland JA (2008) Human mesenchymal stem cell differentiation to NP-like cells in chitosan-glycerophosphate hydrogels. Biomaterials 29(1):85–93
158. Cheng YH, Yang SH, Lin FH (2011) Thermosensitive chitosan-gelatin-glycerol phosphate hydrogel as a controlled release system of ferulic acid for nucleus pulposus regeneration. Biomaterials 32:6953–6961
159. Wang H, Shi J, Wang Y, Yin Y, Wang L, Liu J, Liu Z, Duan C, Zhu P, Wang C (2014) Promotion of cardiac differentiation of brown adipose derived stem cells by chitosan hydrogel for repair after myocardial infarction. Biomaterials 35:3986–3998
160. Reis LA, Chiu LL, Liang Y, Hyunh K, Momen A, Radisic M (2012) A peptide-modified chitosan–collagen hydrogel for cardiac cell culture and delivery. Acta Biomater 8:1022–1036
161. Chiu LL, Janic K, Radisic M (2012) Engineering of oriented myocardium on threedimensional micropatterned collagen-chitosan hydrogel. Int J Artif Organs 35(4):237–250
162. Pok S, Myers JD, Madihally SV, Jacot JG (2013) A multi-layered scaffold of a chitosan and gelatin hydrogel supported by a PCL core for cardiac tissue engineering. Acta Biomater 9(3):5630–5642
163. Beohar N, Rapp J, Pandya S, Losordo DW (2010) Rebuilding the damaged heart: the potential of cytokines and growth factors in the treatment of ischemic heart disease. J Am Coll Cardiol 56(16):1287–1297
164. Fujita M, Ishihara M, Morimoto Y, Simizu M, Saito Y, Yura H, Matsui T, Takase B, Hattori H, Kanatani Y, Kikuchi M, Maehara T (2005) Efficacy of photocrosslinkable chitosan hydrogel containing fibroblast growth factor-2 in a rabbit model of chronic myocardial infarction. J Surg Res 126:27–33
165. Chiu LL, Reis LA, Radisic M (2012) Controlled delivery of thymosin β4 for tissue engineering and cardiac regenerative medicine. Ann NY Acad Sci 1269:16–25
166. Priya SG, Jungvid H, Kumar A (2008) Skin tissue engineering for tissue repair and regeneration. Tissue Eng Pt B-Rev 14(1):105–118
167. Wijekoon A, Fountas-Davis N, Leipzig ND (2013) Fluorinated methacrylamide chitosan hydrogel systems as adaptable oxygen carriers for wound healing. Acta Biomater 9:5653–5664
168. Boucard N, Viton C, Agay D, Mari E, Roger T, Chancerelle Y, Domard A (2007) The use of physical hydrogels of chitosan for skin regeneration following third-degree burns. Biomaterials 28:3478–3488
169. Kiyozumi T, Kanatani Y, Ishihara M, Saitoh D, Shimizu J, Yura H, Suzuki S, Okada Y, Kikuchi M (2007) The effect of chitosan hydrogel containing DMEM/F12 medium on full-thickness skin defects after deep dermal burn. Burns 33(5):642–648
170. Cui F, Li G, Huang J, Zhang J, Lu M, Lu W, Huan J, Huang Q (2011) Development of chitosan-collagen hydrogel incorporated with lysostaphin (CCHL) burn dressing with anti-methicillin-resistant staphylococcus aureus and promotion wound healing properties. Drug Deliv 18(3):173–180
171. Gainza G, Villullas S, Pedraz JL, Hernandez RM, Igartua M (2015) Advances in drug delivery systems (DDSs) to release growth factors for wound healing and skin regeneration. Nanomedicine (in press)
172. Fujita M, Ishihara M, Shimizu M, Obara K, Nakamura S, Kanatani Y, Morimoto Y, Takase B, Matsui T, Kikuchi M, Maehara T (2007) Therapeutic angiogenesis induced by controlled release of fibroblast growth factor-2 from injectable chitosan/ non-anticoagulant heparin hydrogel in a rat hindlimb ischemia model. Wound Repair Regen 15(1):58–65

173. Choi JS, Yoo HS (2013) Chitosan/pluronic hydrogel containing bFGF/heparin for encapsulation of human dermal fibroblasts. J Biomat Sci-Polym E 24(2):210–223
174. Pulat M, Kahraman AS, Tan N, Gumusderelioglu M (2013) Sequential antibiotic and growth factor releasing chitosan-PAAm semi-IPN hydrogel as a novel wound dressing. J Biomat Sci-Polym E 24(7):807–819
175. Yilgor C, Huri PY, Huri G (2012) Tissue engineering strategies in ligament regeneration. Stem Cells Int 2012:Article ID 374676
176. Hayami JW, Surrao DC, Waldman SD, Amsden BG (2010) Design and characterization of a biodegradable composite scaffold for ligament tissue engineering. J Biomed Mater Res A 92(4):1407–1420
177. Deepthi S, Jeevitha K, Sundaram MN, Chennazhi KP, Jayakumar R (2015) Chitosan-hyaluronic acid hydrogel coated poly(caprolactone) multiscale bilayer scaffold for ligament regeneration. Chem Eng J 260:478–485

Prospects of Bioactive Chitosan-Based Scaffolds in Tissue Engineering and Regenerative Medicine

M. Prabaharan and P.R. Sivashankari

Abstract Chitosan, a natural-based polymer obtained by alkaline deacetylation of chitin, is non-toxic, biocompatible, and biodegradable. Due to its desired properties, chitosan-based materials are widely considered to fabricate scaffolds for tissue engineering and regenerative medicine. These scaffolds provide characteristic advantages, such as preservation of cellular phenotype, binding and enhancement of bioactive factors, control of gene expression, and synthesis and deposition of tissue-specific extracellular matrix (ECM), to tissue regeneration. Therefore, the scaffolds based on chitosan and its composites have potential to be used in bone, cartilage, liver, nerve, and musculoskeletal tissue engineering.

Keywords Chitosan · Scaffolds · Tissue engineering · Bioactivity · Regenerative medicine

1 Introduction

The objective of tissue engineering is to regenerate and repair damaged tissues and organs of the human and animal body. In tissue engineering, three-dimensional (3-D) porous scaffolds are used for cell adhesion, proliferation, and differentiation and development of an extracellular matrix (ECM) [1]. The tissue scaffolds may have an ability to load bioactive/therapeutic substances and to release them at a controlled manner in the defected sites. This phenomenon can be used to improve the bioactivity of the scaffolds and therapeutic effects in the injured tissues [2]. For the betterment of tissue engineering approaches, more appropriate materials with the suitable cells and bioactive molecules need to be considered for

M. Prabaharan (✉) · P.R. Sivashankari
Department of Chemistry, Hindustan Institute of Technology and Science, Padur, Chennai 603 103, India
e-mail: mprabaharan@yahoo.com

P.K. Dutta (ed.), *Chitin and Chitosan for Regenerative Medicine*, Springer Series on Polymer and Composite Materials, DOI 10.1007/978-81-322-2511-9_2

the fabrication of scaffolds. In this regard, chitosan, a biopolymer consisting of glucosamine and *N*-acetylglucosamine repeating units, has received much attention to develop the scaffolds due to its desired properties such as low cost, large-scale availability, antimicrobial activity, low toxicity, and biodegradability. Using chitosan, a range of scaffolds with different microstructures (hydrogels, open-pore structures, fibrous matrices, etc.) have been prepared for the specific tissue engineering application. These scaffolds permit the development of normal tissue regeneration due to their minimal foreign body reaction and ability to attain hemostasis [3, 4].

In recent years, a variety of hybrid scaffolds based on chitosan and other biodegradable and/or biocompatible materials have been developed. These hybrid scaffolds have an improved microstructure, swelling ability, mechanical strength, compression modulus, and biocompatibility due to their counter part of chitosan. The materials used for the preparation of hybrid scaffolds play an important role on the properties and applications of hybrid scaffolds in tissue engineering. The objective of this chapter is to consolidate and discuss the recent advancements of chitosan-based materials as scaffolds for bone, cartilage, liver, nerve, and musculoskeletal tissue regeneration.

2 In Bone Tissue Engineering

Bone tissue engineering is an important field of research for developing new three-dimensional (3D) scaffolds with highly interconnected porous structure. These scaffolds should match the characteristics of the tissue that is to be replaced. The tissue engineering scaffolds need to be biocompatible, osteoinductive, osteoconductive, and mechanically well-suited in order to restore bones which have been lost or damaged. In this respect, artificial scaffolds based on chitosan, inorganic materials, and/or synthetics polymers have received much attention in recent years due to their desirable properties for bone tissue engineering application.

2.1 Chitosan–Synthetic Polymer Hybrid Scaffolds

Poly(L-lactic acid) (PLLA)/chitosan hybrid scaffolds were developed using chitosan solution and previously prepared PLLA scaffolds for bone tissue engineering as shown in Fig. 1 [5]. The shape and size of microstructure of the hybrid scaffolds were found to be depended on the concentration of chitosan solution used to soak the PLLA scaffolds. It was found that PLLA/chitosan hybrid scaffolds can produce calcium phosphate precursors on its structure on dipping them into alternate phosphorous and calcium solutions. The formation of apatite layers within the hybrid scaffolds was also observed by bioactivity tests. Santo et al. [6] prepared a hybrid scaffolds based on poly (D, L-lactic acid) (PDLLA) impregnated with

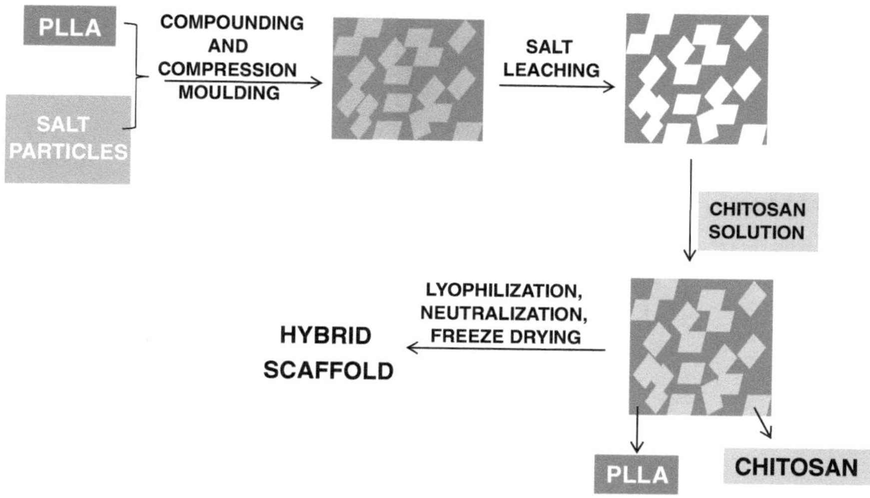

Fig. 1 Schematic representation for the preparation of PLLA/chitosan hybrid scaffolds

chitosan/chondroitin sulfate nanoparticles (NPs) with drug release capabilities. Due to the presence of chitosan/chondroitin sulfate NPs, these hybrid scaffolds showed higher swelling characters and adequate mechanical properties for cell adhesion and support for bone tissue engineering. Using thermally induced phase separation and lyophilization techniques, hybrid scaffolds based on chitosan and PDLLA-*co*-glycolide were developed for bone tissue engineering applications [7]. FT-IR and field-emission SEM studies confirmed the formation of apatite layers on the hybrid scaffolds after impregnated with stimulated body fluid (SBF). These studies revealed the potential of chitosan-based hybrid scaffolds in bone tissue engineering.

Niu et al. [8] reported the properties of chitosan microsphere-loaded porous PLLA scaffolds as a carrier for BMP-2-derived synthetic peptide. There were strong hydrogen bonds between the PLLA and chitosan component observed by FT-IR. When the chitosan microspheres' contents increased from 0 to 50 %, the compressive strength of the PLLA scaffolds was increased from 0.48 to 0.66 MPa, while the compressive modulus increased from 7.29 to 8.23 MPa. The insertion of chitosan microspheres into the PLLA scaffolds was found to neutralize the acidity of PLLA degradation products. Release studies showed that PLLA /chitosan hybrid scaffolds presented a controlled release of loaded peptide when compared with control chitosan microspheres. The release pattern was found to be depending on the degradation of PLLA matrix. This result indicates that PLLA/chitosan scaffolds can be used to deliver bioactive factors for a range of non-loaded bone regeneration.

Santo et al. [9] developed a hybrid scaffolds based on PDLLA containing chitosan–chondroitin sulfate NPs loaded with Platelet Lysate (PL) by supercritical fluid foaming technique as shown in Fig. 2. Release studies demonstrated that PL was released in a sustained manner from the hybrid scaffolds. Due to the presence of PL and hASCs, this hybrid scaffolds can be used as multi-functional materials for bone tissue

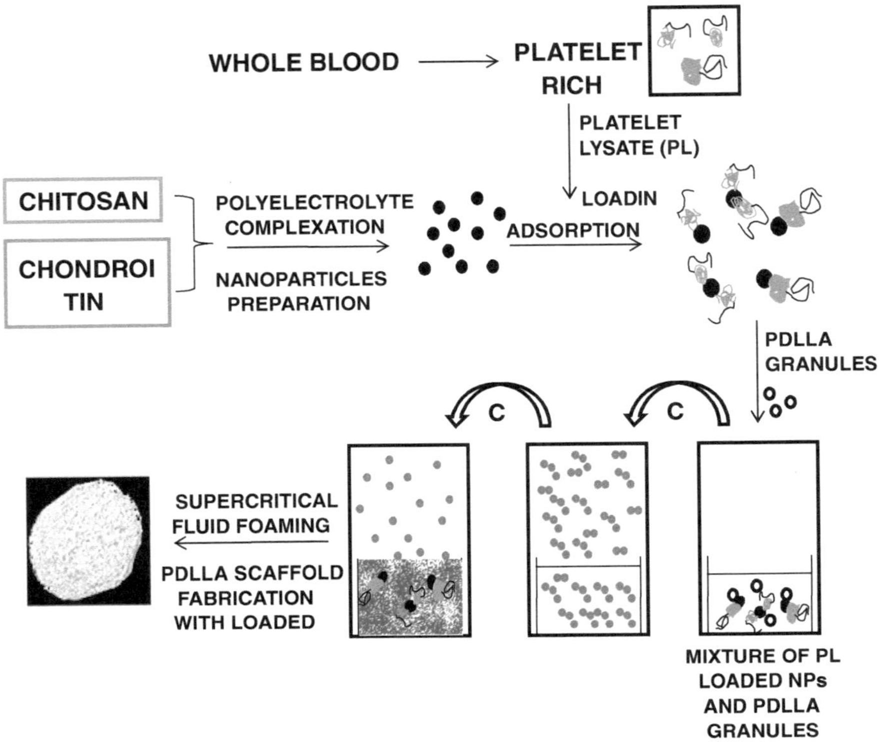

Fig. 2 Schematic representation of experimental procedure for the preparation of PDLLA scaffolds incorporating the PL-loaded chitosan–chondroitin sulfate NPs

engineering applications. Recently, L-lactide-methoxy PEG–tetrandrine nanosphere-loaded chitosan–gelatin hybrid scaffolds were prepared using freeze-drying method for bone tissue engineering [10]. Tetrandrine is a plant derivative, which can be used as a modifier to hybrid scaffolds to promote chondrocyte differentiation and secrete type-II collagen. Since tetrandrine-loaded nanospheres implanted within chitosan–gelatin scaffolds, sustained release of tetrandrine was observed from the hybrid scaffolds.

2.2 *Chitosan–Calcium Phosphate Hybrid Scaffolds*

Hybrid scaffolds based on calcium phosphate cements (CPCs) and chitosan have been widely used as bone graft substitutes due to their in situ-setting ability and bioactivity. Zhao et al. [11] studied human umbilical cord mesenchymal stem cells (hUCMSCs) delivery of CPC–chitosan–polyglactin fiber scaffolds for bone tissue engineering. The fatigue resistance of CPC–chitosan–polyglactin fiber scaffolds was found to be increased due to the presence of chitosan and polyglactin fibers. In addition, it was found that the CPC–chitosan–polyglactin fiber scaffolds supported hUCMSCs attachment and proliferation. hUCMSCs showed well distribution and anchored on the polyglactin fibers in scaffolds via cytoplasmic extensions. These results propose that CPC–chitosan–polyglactin fiber scaffold may be appropriate for stem cell delivery and bone tissue engineering. Wen et al. [12] fabricated an iron foam coated with calcium phosphate/chitosan using electrophoretic deposition method for bone tissue engineering. The deposition of calcium phosphate/chitosan on iron foam improved the interfacial bonding strength and the in vitro bone-forming bioactivity. Moreover, it was observed that the bioactivity of the implant was not affected by the iron foam coated with calcium phosphate/chitosan. Recently, Meng et al. [13] prepared bioactive cement based on CPC containing chitosan microspheres as an injectable material for the bone regeneration. The bioactive cement containing 10 % (w/w) chitosan microspheres had a compressive strength of 14.78 ± 0.67 MPa. In this study, CPC/chitosan microsphere and α-TCP/CPC (control group) were implanted into the bone defects in both femoral condylar regions of New Zealand white rabbits. SEM and histological examination after implantation showed the formation of more new bones and degradation of the bioactive cement in the bone defects. These studies show the potential application of CPC/chitosan hybrid scaffolds in bone regeneration.

2.3 Chitosan–Bioactive Glass Hybrid Scaffolds

Bioactive glasses are used for bone regeneration due to their osteoconductive and biodegradable properties. Degradation products of bioactive glasses could stimulate the production of growth factors and cell proliferation, and activate the gene expression of osteoblast [14]. In recent years, bioactive glasses combined with chitosan have been extensively considered to fabricate hybrid macroporous scaffolds for the improved bone repair. Mansur and Costa studied the physical, mechanical, and biological properties of hybrid scaffolds consisting of poly(vinyl alcohol), chitosan, and bioactive glass [15]. The results revealed that these hybrid scaffolds can be used for bone tissue engineering applications due to their appropriate mechanical, morphological, and cell viability properties. Couto et al. [16] developed injectable hybrid scaffolds based on chitosan–β-glycerophosphate and bioactive glass NPs for orthopedic reconstructive and regenerative medicine applications. The formation of apatite layers on the hybrid scaffold was observed after soaking them with SBF solution due to the presence of bioactive glass NPs in the scaffolds. The thickness of the apatite layer formed on the scaffold was found to be increased with increasing bioactive glass content and soaking time in SBF.

A hybrid scaffold composed of chitosan and bioactive glass ceramic NPs (nBGC) was fabricated by blending nBGC with chitosan solution using freeze-drying technique [17]. This hybrid scaffold showed adequate swelling and degradation properties due to the presence of hydrophilic chitosan and nBGC. The

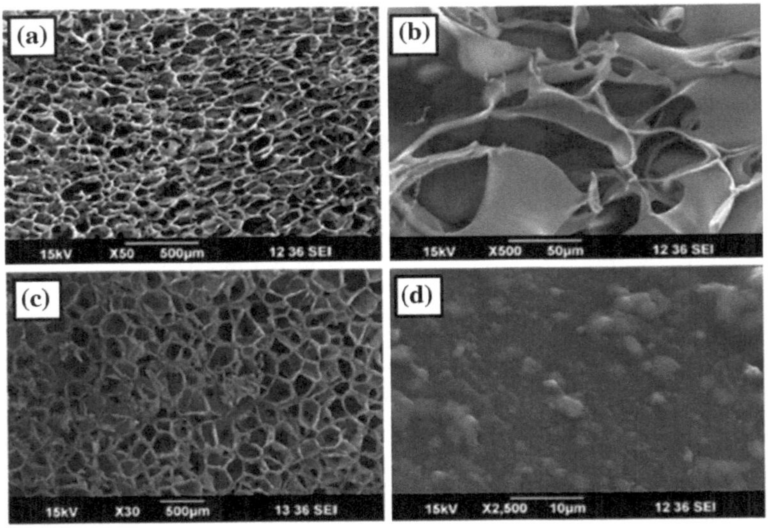

Fig. 3 SEM images of chitosan–gelatin (**a**, **b**) and chitosan–gelatin/nBGC composite scaffolds (**c**, **d**)

bioactivity of chitosan/nBGC hybrid scaffold was confirmed by the biomineralization studies. Peter et al. [18] developed composite scaffolds with pore size ranging from 150 to 300 μm based on chitosan–gelatin/nBGC for alveolar bone tissue engineering (Fig. 3). In this study, the scaffold properties and biocompatibility were analyzed in order to understand the role nBGC in the scaffold matrix. The results showed that the degradation and swelling behavior of the composite scaffolds were decreased with the addition of nBGC. Biomineralization studies showed that mineral deposits on the nano-composite scaffold were increased with the increase in time of incubation. In vitro studies demonstrated that the composite scaffolds are capable to provide a healthier environment for cell attachment and spreading. Yang et al. [19] fabricated alginate and chitosan-reinforced bioactive glass scaffolds with superior mechanical properties and structural stability. In this study, the bioactive glass scaffolds were developed using microsphere replication method. Alginate- and chitosan-reinforced bioactive glass scaffolds showed an enormous improvement in compressive strength and nearly 30 % of shrinkage in wet state when compared to blank bioactive glass scaffold. This biopolymer-reinforced bioactive glass scaffolds presented an excellent strain tolerance during the prolonged immersion in simulated body fluid. These results indicated that biopolymer-reinforced bioactive glass scaffolds can be used for bone repair.

Multi-functional hybrid scaffolds with drug loading and releasing abilities have received considerable interest for bone tissue engineering due to their efficacy to improve drug delivery, healing, and regeneration. A scaffold with controlled drug release ability can be developed by integrating biomolecules within biodegradable carriers and further inclusion of such carriers into tissue-engineered scaffolds. Nazemi et al. [20] developed chitosan–bioactive glass scaffolds loaded with poly(lactic-*co*-glycolic acid) NPs by freeze-drying method. In this study, the mechanical properties of the hybrid scaffolds were found to be improved due to the presence of poly(lactic-*co*-glycolic acid) NPs in the scaffolds. It was observed that the presence of poly(lactic-*co*-glycolic acid) NPs did not affect the morphology of the hybrid scaffolds. Recently, porous poly(caprolactone) and vancomycin-loaded chitosan-coated hybrid scaffolds have been developed by the replication technique using 45S5 Bioglass® (BG) powder [21]. The mechanical properties and compressive strength of the poly(caprolactone) and vancomycin-loaded chitosan-coated scaffolds were found to be improved when compared with uncoated scaffolds. The coated scaffolds presented a sustained release of encapsulated drug for a period of 11 days. This result suggests the potential of the caprolactone and vancomycin-loaded chitosan-coated scaffolds as bone tissue scaffolds. Soundrapandian et al. [22] formulated porous scaffolds based on BGZ and MBG bioactive glasses. In this study, these scaffolds were loaded with the model drug, gatifloxacin, by vacuum infiltration technique. Thereafter, the drug-loaded scaffolds were coated with 0.5–1 % chitosan solution. The results demonstrated that 63–66 % porous and 5–50 μm porous MBG and BGZ bioactive glass scaffolds were capable of releasing drugs effectively for prolonged periods. In addition, the coating of chitosan on the scaffolds decreased the release of drug. The scaffolds based on MBG bioactive glass were found to be bioactive, biocompatible, non-cytotoxic, and exhibited excellent wound healing potential.

Pon-On et al. [23] fabricated poly(vinyl alcohol)–chitosan–collagen hybrid scaffolds loaded with bioactive glass by three mechanical freeze–thaw followed by freeze-drying methods. The porosity and compressive strength of the hybrid scaffolds were found to be controlled by the weight ratio of poly(vinyl alcohol) and mixtures of chitosan–collagen. Formation of apatite layers on the scaffold surface was observed after seven days of incubation in SBF. MTT assay revealed that there is no cytotoxicity of hybrid scaffolds on UMR-106 cells. The drug release studies showed that poly(vinyl alcohol)–chitosan–collagen hybrid scaffolds loaded with bioactive glass hybrid scaffolds presented a controlled release for about a month. The effect of poly(lactic-*co*-glycolic) acid NPs on a chitosan–bioactive glass scaffold was reported recently [24]. In this work, two types of chitosan–bioactive glass scaffolds, with and without poly(lactic-*co*-glycolic) acid NPs, were prepared. The mechanical strength of the hybrid scaffolds was found to be improved while adding the poly(lactic-*co*-glycolic) acid NPs. However, the swelling behavior of the hybrid scaffolds was marginally decreased as a result of adding NPs. The results demonstrated that these hybrid scaffolds can be used as a controlled-release platform of model drugs to the bone regeneration.

2.4 Chitosan–Hydroxyapatite Hybrid Scaffolds

The ECM present in bone tissue contains a porous composite of interpenetrating phases of type-I and type-III collagen and hydroxyapatite. In recent years,

Fig. 4 SEM images for showing the mineralization of **a** chitosan–gelatin, **b** chitosan–gelatin/nanohydroxyapatite and **c** chitosan–gelatin/nanohydroxyapatite–montmorillonite composite scaffolds immersed in SBF after 14 days of incubation

chitosan, gelatin, and hydroxyapatite in different combinations have been developed as hybrid scaffolds for bone regeneration because of their chemical similarity to natural bone ECM. Oliveira et al. [25] prepared hybrid scaffolds based on dexamethasone-loaded carboxymethyl chitosan/poly(amidoamine) dendrimer NPs and the mixture of hydroxyapatite and starch–poly(caprolactone). In this work, the effect of these hybrid scaffolds on the proliferation and osteogenic differentiation of rat bone marrow stromal cells (RBMSCs) was studied. It was observed that RBMSCs seeded onto the surface of both hydroxyapatite and starch–poly(caprolactone) scaffolds differentiate into osteoblasts when cultured in the presence of 0.01 mg ml^{-1} dexamethasone-loaded carboxymethyl chitosan/poly(amidoamine) dendrimer NPs. The dexamethasone-loaded carboxymethyl chitosan/poly(amidoamine) dendrimer NPs combined with the hydroxyapatite was also found to enhance osteogenesis by increasing alkaline phosphatase activity and mineralization of the ECM.

Olad and Azhar developed highly porous chitosan–gelatin/nanohydroxyapatite–montmorillonite hybrid scaffolds with the pore size of 100–350 μm using freeze–drying method for use in bone tissue engineering [3]. Bioactivity study conducted with the SBF showed a decreased degradation rate and increased biomineralization of the hybrid scaffolds due to the presence of nanohydroxyapatite and montmorillonite as shown in Fig. 4. Recently, Zhu et al. [26] developed a biomimetic hybrid scaffolds which consists of hydroxyapatite and human bone marrow mesenchymal stem cells (BMSCs)-loaded chitosan hydrogel. The breast cancer adhesion and proliferation of the hybrid scaffolds were found to be influenced by the amount of nanohydroxyapatite present in the scaffolds. The maximum breast cancer adhesion and proliferation was found on 10 % nanohydroxyapatite–chitosan scaffold.

3 In Cartilage Tissue Engineering

In recent years, a wide variety of chitosan-based biomaterials have been developed as scaffolds for cartilage tissue engineering due to their desirable properties such as biocompatible, biodegradable, highly porous, suitable for cell attachment, proliferation and differentiation, osteoconductive, non-cytotoxic, flexible and elastic, and nonantigenic.

3.1 Chitosan-Based Fibrous Scaffolds

Iwasaki et al. [27] fabricated alginate–chitosan hybrid fibrous scaffolds for cartilage tissue engineering. These scaffolds presented superior adhesion ability with chondrocytes when compared with alginate fiber. SEM studies showed that the presence of the distinctive round morphology of the chondrocyte and the formation of type-II collagen fibers by the chondrocytes in the hybrid scaffolds. A stratified composite scaffold based on chitosan nanofibrous layer on a porous 45S5 bioactive glass was developed for osteochondral segment regeneration by Liverani et al. [28]. In this study, chitosan and alginate were used for constructing the interface between the scaffold and the soft cartilage. A chitosan-based electrospun nanofibrous membrane was used for constructing the upper layer of the scaffold. This composite scaffold had good resistance to layer delamination, preservation of the bioactivity, and improvement of the mechanical properties.

3.2 Chitosan-Based Scaffolds

Chitosan-based scaffolds can deliver ECM components such as type-II collagen and chondroitin sulfate in a controlled fashion, which endorses the in-growth and biosynthetic capability of chondrocytes. Choi et al. [29] developed chitosan hydrogel containing type-II collagen and chondroitin sulfate for the management of cartilage defects. The addition of type-II collagen and chondroitin sulfate into chitosan hydrogels was found to increase chondrogenesis. In particular, type-II collagen was found to be responsible for the improved chondrogenesis. Kim et al. [30] stabilized transforming growth factor (TGF-β1) signaling in the chitosan hydrogel for use in cartilage regeneration. In this study, TGF-β1 was linked to chitosan with preserving type-II collagen in order to reduce the burst release of protein in a complex biological environment of serum and cells. The ability of TGF-β1 linked chitosan to promote cartilage regeneration in a rat partial-thickness chondral defect model was confirmed.

3.3 Chitosan-Based Composite Scaffolds

Bi et al. [31] prepared a biphasic scaffold based on collagen–chitosan and bioactive glass collagen by combination of sol–gel, freeze-drying, and cross-linking techniques. In this scaffold, cross-linking agents were used to connect the collagen–chitosan and bioactive glass collagen phases. This scaffold presented interconnected porous structures and precipitation of hydroxyapatite grains after being immersed into SBF. As shown in Fig. 5, BMSCs were found to be anchored on this scaffold with healthy spreading. This result shows that this

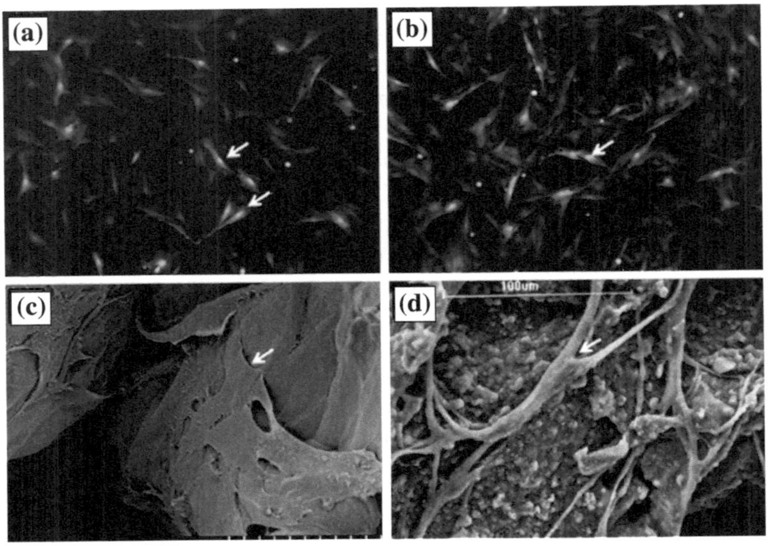

Fig. 5 Fluorescence and SEM images of BMSCs on the biphasic scaffold **a** chandral phase (fluorescence of 7th day), **b** osseous phase (fluorescence of 7th day), **c** chandral phase (SEM of 7th day), and **d** osseous phase (SEM of 7th day)

biphasic scaffold can be used for osteochondral tissue engineering. Silva et al. [32] developed chitosan and chondroitin sulfate 3-D nanostructures to support the attachment and proliferation of bovine chondrocytes. The obtained 3-D nanostructure had a high porosity and water uptake capacity of about 300 %. The results of this study showed that cells were attached, proliferated, and metabolically active over the entire 3-D chitosan and chondroitin sulfate nanostructure scaffold.

Porous scaffolds composed of poly(L-glutamic acid) and chitosan was prepared for the repair of articular cartilage defects using a freeze-drying method [33]. The scaffolds fabricated from 2 % poly(L-glutamic acid)/chitosan content and at a freezing temperature of −20 °C exhibited an interconnected porous structure with average pore size between 150 and 200 μm, the contact angle of less than 75°, and high swelling ratio about 700 %. In vitro culture of rabbit adipose-derived stem cells indicated that poly(l-glutamic acid)/chitosan porous scaffolds supported cell attachment and growth. Lee et al. [34] developed macroporous poly(vinyl alcohol)-carboxylmethyl chitosan–poly(ethylene glycol) hybrid scaffolds for cartilage tissue engineering. The MTT, immunohistochemistry, SEM, and TEM analyses confirmed that these scaffolds promoted cell attachment and proliferation in vitro. It was observed that the chondrocyte-poly(vinyl alcohol)-carboxylmethyl chitosan–poly(ethylene glycol) scaffolds secreted glycosaminoglycan (GAG) and collagen type-II. Moreover, these scaffolds were not shown any adverse effects on the host tissue.

Chen et al. [35] prepared histidine-*graft*-chitosan/PLLA scaffolds using a dual-phase separation technique by changing the weight ratio of histidine-*graft*-chitosan and PLLA. In this study, the chemical structure, morphology, and mechanical properties of NHCS/PLLA hybrid scaffolds were characterized through FT-IR, WXRD, thermal gravimetric analyzer, and field-emission SEM. The results showed that the pore size of histidine-*graft*-chitosan/PLLA hybrid scaffolds decreased with the decrease of the weight ratio of histidine-*graft*-chitosan and PLLA. The pore size and porosity of the scaffolds were found to be about 12–25 μm and >92 %, respectively. The comprehensive strength and the comprehensive modulus were found to be 0.33–0.78 MPa and 1.75–5.28 MPa, respectively, which indicates the potential application of histidine-*graft*-chitosan/PLLA hybrid scaffolds in cartilage tissue engineering. Kamoun et al. [36] developed an injectable hydrogel based on *N*-succinyl chitosan cross-linked with water-soluble dialdehyde starch (DAS) for tissue engineering and cartilage repair. In this study, *N*-succinyl chitosan content was found to have an important role for the formation of highly cross-linked hybrid hydrogels. However, the strength of the cross-linked hybrid hydrogels was found to be decreased when increasing the concentration of DAS. It was found that the content of *N*-succinyl chitosan in the hybrid hydrogels influenced the model drug, curcumin, release profile, and adherence of HGF cells on the hydrogels.

4 In Liver Tissue Engineering

The objective of liver tissue engineering is to construct an artificial liver tissue for the replacement of the failure liver function in patients. Liver tissue engineering strategies can be used to overcome the drawbacks of liver transplantation such as requirement of immunosuppressive drugs, donor organ and its storage and high cost, etc. Using chitosan-based materials, a variety of scaffold systems were developed for liver tissue engineering applications.

4.1 Chitosan–Collagen Matrices

Chitosan–collagen composite system was synthesized using coupling agents 1-ethyl-3-(3-dimethylaminopropyl)-carbodiimide and *N*-hydroxysuccinimide as implantable artificial livers by Wang et al. [37]. The platelet deposition and hepatocyte culture studies revealed that chitosan–collagen composite had adequate mechanical properties, outstanding blood, and cell compatibility. The highly porous hybrid scaffolds with adequate blood compatibility based on collagen/chitosan/heparin were also fabricated by Wang et al. [38]. Hepatocytes cultured on these hybrid scaffolds presented maximum urea and triglyceride discharge after 25 days from their seeding, which indicates the potential of collagen/chitosan/heparin hybrid scaffolds in liver tissue engineering.

4.2 Alginate–Chitosan Composite Matrices

Yang et al. [39] fabricated porous alginate/galactosylated chitosan hybrid scaffolds for liver tissue engineering. The hybrid scaffolds seeded with primary hepatocytes presented greater cell attachment and viability due to the definite interactions between the asialoglycoprotein receptors on hepatocyte and galactose ligands on hybrid scaffolds. Recently, Chen et al. [40] reported a scaffold with average pore size of 50–150 μm and interconnected pore structure based on galactosylated chitosan cross-linked with oxidized alginate for liver tissue engineering. The porosity and compressive modulus of the scaffolds was determined as about 70 % and 4.2–6.3 kPa, respectively. The equilibrium swelling and in vitro degradation rate of the scaffolds were found to be decreased with the increase of the oxidized alginate content. The hepatocytes seeded on the scaffolds showed multi-cellular aggregates with a characteristic spheroidal morphology and great interactions with the scaffolds.

4.3 Chitosan-Based Microfibers

Lee et al. [41] reported the chemical, mechanical, and diffusion properties of microfluidic chitosan microfibers for liver tissue engineering applications. In order to assess the potential of the chitosan microfibers as scaffolds for liver tissue regeneration, hepatoma HepG2 cells were seeded onto microfibers. These microfibers presented albumin secretion and urea synthesis. Fan et al. [42] fabricated highly porous hybrid scaffolds which consist of galactosylated hyaluronic acid and chitosan using freeze-drying method for the enhanced function of hepatocytes in vitro. Due to the presence of hyaluronic acid, the hybrid scaffolds had an improved hydrophilicity and mechanical strength. In this study, rat primary hepatocytes seeded in the hybrid scaffolds demonstrated the multi-cellular spheroid morphologies.

5 In Nerve Tissue Engineering

In recent years, a wide variety of biocompatible materials have been considered for the construction of artificial tubes for nerve repair. Among these, chitosan-based materials are more promising for nerve regeneration due to their desired properties such as biocompatibility and biodegradability, and ability to provide a cellular and molecular framework for Schwann cells and neurite migration across the nerve gap.

5.1 Chitosan-Based Membranes

Yang et al. [43] studied the effects of chitosan–collagen hybrid membrane on the behavior of rat neural stem cells. The hybrid membranes were found to be more appropriate for the co-culture with rat neural stem cells because of their low cytotoxicity and supporting ability for the cell survival. Recently, tissue engineering scaffolds based on conductive polymers combined with electrical stimulation are considered as prospective materials for the treatment of neural injuries. Huang et al. [44] reported a conductive hybrid membrane based on polypyrrole (2.5 %) and chitosan (97.5 %) in order to electrically stimulate Schwann cells. These membranes showed superior cell adhesion, spreading, and proliferation with or without electrical stimulation. Due to the electrical stimulation, these hybrid membranes expressed the emission of NGF and BDNF when compared with control cells. Wrobel et al. [45] developed chitosan film for nerve tissue engineering application. To study the biocompatibility of film, various types of Schwann cells were seeded onto chitosan film and found that all cell types were viable on the chitosan film. Moreover, different types of metabolic activities and proliferation behavior were observed on the of Schwann cell-seeded chitosan films. Recently, Morelli et al. [46] prepared hybrid membranes which consist of chitosan, poly(caprolactone), and poly(urethane) using phase-inversion techniques. The efficacy of these membranes to enhance the adhesion and differentiation of neuronal cells was determined. The results showed that neural cell responses of the hybrid membrane were found to be depending on the type and properties of the polymers used for the fabrication of membranes.

5.2 Chitosan-Based Hydrogels

Hydrogels based on chitosan can be suitable for neural tissue engineering due to their physicochemical and mechanical properties to hold neurite extension and assist transplantation of cells. Freier et al. [47] developed chitin and chitosan hydrogel tubes using mold casting method. Both chitin and chitosan hydrogel tubes facilitated adhesion and differentiation of primary chick dorsal root ganglion neurons in vitro. However, chitosan hydrogel tube showed improved nerve cell adhesion and neurite outgrowth, which represents the potential of this material in nerve tissue engineering. Valmikinathan et al. [48] prepared a photo cross-linked chitosan hydrogel as shown in Fig. 6. This hydrogel showed less cytotoxicity against hMSCs. An improved neurite differentiation from primary cortical neurons and neurite extension from dorsal root ganglia was observed on the chitosan hydrogel when compared to the control agarose hydrogen under similar conditions. Moreover, neural stem cells seeded on the chitosan hydrogels assisted differentiation into tubulin-positive neurons and astrocytes.

6 In Musculoskeletal Tissue Engineering

In the recent years, chitosan-based functional materials are widely considered to fabricate the scaffolds for regeneration of severely damaged tissues. Masuko et al. [49] prepared chitosan-RGDSGGC peptide complex by reacting thiolated chitosan with RGDSGGC peptide containing RGDS groups as shown in Fig. 7. In this

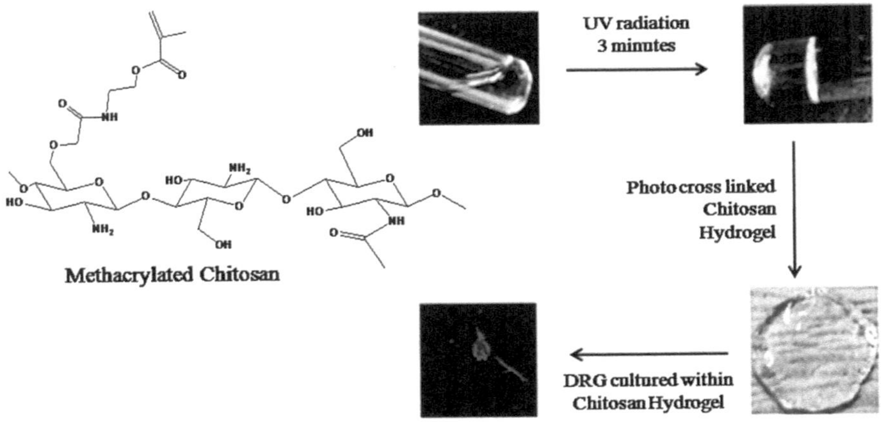

Fig. 6 Synthesis of photo cross-linkable hydrogel

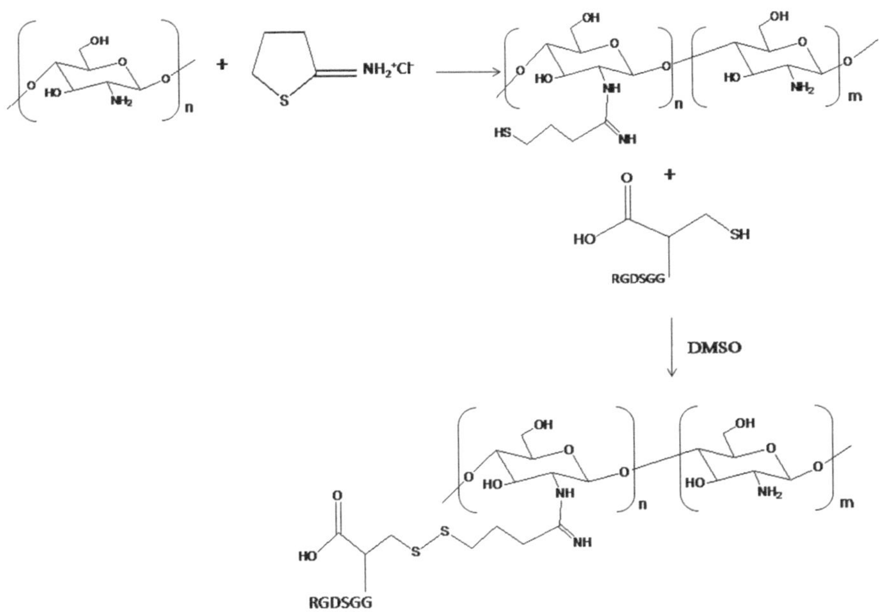

Fig. 7 Synthesis of chitosan–RGDSGGC conjugates

study, the effects of RGDSGGC peptide to thiolated chitosan on cell adhesion and proliferation activity of chondrocytes and fibroblasts were determined. The results showed that chitosan-RGDSGGC peptide complex can improve both cell adhesion and cell proliferation of chondrocytes and fibroblasts.

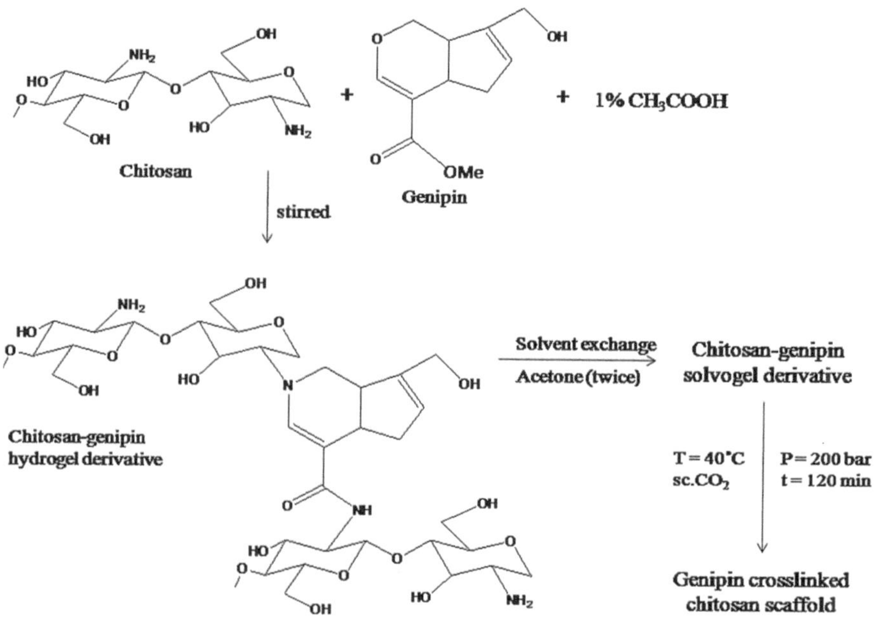

Fig. 8 Preparation of chitosan–genipin scaffold

Rinki and Dutta fabricated supercritical carbon dioxide-assisted porous chitosan scaffolds using chitosan in aqueous acetic acid and genipin by as a cross-linking agent (Fig. 8) [50]. These scaffolds showed an improved bioactivity in SBF and cellular attachment with MG63 osteoblastic cells. These results indicate that supercritical carbon dioxide-assisted porous chitosan scaffolds can be used for musculoskeletal tissue engineering application. Zhang et al. [51] recently prepared conducting glutaraldehyde cross-linked hydrogels based on carboxymethyl chitosan and aniline. These hydrogels presented a controlled release of encapsulated model drug diclofenac sodium. Live/Dead assay and Alamar blue assay proved the biocompatibility of the conducting hydrogels by C2C12 myoblast cells.

Adipose tissue engineering is considered as a promising technique for reconstructive and cosmetic applications in plastic surgery. In this context, Cheung et al. [52] developed photo cross-linked methacrylated glycol chitosan and methacrylated chondroitin sulfate scaffolds integrated with bioactive decellularized adipose tissue. The results showed that these scaffolds improved hASCs viability due to the presence of adipose tissue as a cell-supportive matrix. In addition, methacrylated chondroitin sulfate-based scaffolds presented a better implant integration and adipogenesis, with allogenic hASCs promoting cell infiltration, angiogenesis and eventually, fat formation. Recently, Martel-Estrada et al. [53] prepared porous chitosan/mimosa tenuiflora hybrid scaffolds by thermally induced phase separation and lyophilization methods. Due to the existence of more amine groups, apatite layer was formed efficiently on the hybrid scaffolds, indicating the potential application of these hybrid scaffolds in tissue regeneration.

7 Concluding Remarks

Using chitosan and its composites with inorganic and/or polymeric materials, a variety of porous scaffolds have been fabricated for tissue engineering and regenerative medicine. These chitosan-based scaffolds are non-toxic, biocompatible, and biodegradable. Due to the desirable physicochemical and biological properties, chitosan-based scaffolds are widely considered as potential biomaterials for bone, cartilage, liver, nerve, and musculoskeletal tissue regeneration. However, for the efficient bone and cartilage tissue engineering applications, chitosan-based scaffolds still require adequate stability and mechanical strength. To fulfill these requirements, different types of hybrid scaffolds based on chitosan and other biocompatible materials have been developed. These hybrid scaffolds have highly porous structure with inter-connectivity and enough mechanical properties for cell adhesion and support for bone and cartilage tissue engineering. Due to the presence of bioactive and functional materials, chitosan-based hybrid scaffolds are bioactive and can be used to deliver bioactive materials for exciting cell differentiation and proliferation or drug molecules to provoke therapeutic effects in tissue engineering approaches.

Acknowledgments The authors thank DST-Nano Mission, Department of Science and Technology, Government of India for their financial support.

References

1. Prabaharan M, Rodriguez-Perez MA, de Saja JA, Mano JF (2007) Preparation and characterization of poly(L-lactic acid)-chitosan hybrid scaffolds with drug release capability. J Biomed Mater Res B Appl Biomater 81:427–434
2. Han DK, Park KD, Hubbell JA, Kim YH (1998) Surface characteristics and biocompatibility of lactide-based poly (ethylene glycol) scaffolds for tissue engineering. J Biomater Sci Polym Ed 9:667–680
3. Olad A, Azhar FF (2014) The synergetic effect of bioactive ceramic and nanoclay on the properties of chitosan–gelatin/nanohydroxyapatite–montmorillonite scaffold for bone tissue engineering. Ceram Int 40:10061–10072
4. Suh FJK, Matthew HWT (2000) Application of chitosan-based polysaccharide biomaterials in cartilage tissue engineering: a review. Biomaterials 21:2589–2598
5. Mano JF, Hungerford G, Ribelles JLG (2008) Bioactive poly (L-lactic acid)-chitosan hybrid scaffolds. Mater Sci Eng, C 28:1356–1365
6. Santo VE, Duarte ARC, Gomes ME, Mano JF, Reis RL (2010) Hybrid 3D structure of poly(D, L-lactic acid) loaded with chitosan/chondroitin sulfate nanoparticles to be used as carriers for biomacromolecules in tissue engineering. J Supercrit Fluids 54:320–327

7. Martel-Estrada SA, Martínez-Pérez CA, Chacón-Nava JG, García-Casillas PE, Olivas-Armendáriz I (2011) In vitro bioactivity of chitosan/poly (D, L-lactide-co-glycolide) composites. Mater Lett 65:137–141
8. Niu X, Feng Q, Wang M, Guo X, Zheng C (2009) In vitro degradation and release behavior of porous poly(lactic acid) scaffolds containing chitosan microspheres as a carrier for BMP-2-derived synthetic peptide. Polym Degrad Stab 94:176–182
9. Santo VE, Duarte ARC, Popa EG, Gomes ME, Mano JF, Reis RL (2012) Enhancement of osteogenic differentiation of human adipose derived stem cells by the controlled release of platelet lysates from hybrid scaffolds produced by supercritical fluid foaming. J Control Release 162:19–27
10. Xiaoyan A, Jun Y, Min W, Haiyue Z, Li C, Kangdec Y, Fanglian Y (2008) Preparation of chitosan–gelatin scaffold containing tetrandrine-loaded nano-aggregates and its controlled release behavior. Int J Pharm 350:257–264
11. Zhao L, Burguera EF, Xu HHK, Amin N, Ryou H, Arola DD (2010) Fatigue and human umbilical cord stem cell seeding characteristics of calcium phosphate–chitosan–biodegradable fiber scaffolds. Biomaterials 31:840–847
12. Wen Z, Zhang L, Chen C, Liu Y, Wu C, Dai C (2013) A construction of novel iron-foam-based calcium phosphate/chitosan coating biodegradable scaffold material. Mater Sci Eng, C 33:1022–1031
13. Meng D, Dong L, Wen Y, Xie Q (2015) Effects of adding resorbable chitosan microspheres to calcium phosphate cements for bone regeneration. Mater Sci Eng, C 47:266–272
14. Rezwan K, Chen QZ, Blaker JJ, Boccaccini AB (2006) Biodegradable and bioactive porous polymer, inorganic composite scaffold for bone tissue engineering. Biomaterials 27:3413–3431
15. Mansur HS, Costa HS (2008) Nanostructured poly (vinyl alcohol)/bioactive glass and poly (vinyl alcohol)/chitosan/bioactive glass hybrid scaffolds for biomedical applications. Chem Eng J 137:72–83
16. Couto DS, Hong Z, Mano JF (2009) Development of bioactive and biodegradable chitosan-based injectable systems containing bioactive glass nanoparticles. Acta Biomater 5:115–123
17. Peter M, Binulal NS, Soumya S, Nair SV, Furuike T, Tamura H, Jayakumar R (2010) Nanocomposite scaffolds of bioactive glass ceramic nanoparticles disseminated chitosan matrix for tissue engineering applications. Carbohydr Polym 79:284–289
18. Peter M, Binulal NS, Nair SV, Selvamurugan N, Tamurac H, Jayakumar R (2010) Novel biodegradable chitosan–gelatin/nano-bioactive glass ceramic composite scaffolds for alveolar bone tissue engineering. Chem Eng J 158:353–361
19. Yang G, Yang X, Zhang L, Lin M, Sun X, Chen X, Gou Z (2012) Counterionic biopolymers-reinforced bioactive glass scaffolds with improved mechanical properties in wet state. Mater Lett 75:80–83
20. Nazemi K, Azadpour P, Moztarzadeh F, Urbanska AM, Mozafari M, Tissue-engineered chitosan/bioactive glass bone scaffolds integrated with PLGA nanoparticles: a therapeutic design for on-demand drug delivery. Mater Lett 138:16–20
21. Yao Q, Nooeaid P, Roether JA, Dong Y, Zhang Q, Boccaccini AR (2013) Bioglass®-based scaffolds incorporating polycaprolactone and chitosan coatings for controlled vancomycin delivery. Ceram Int 39:7517–7522
22. Soundrapandian C, Mahato A, Kundu B, Datta S, Sa B, Basu D (2014) Development and effect of different bioactive silicate glass scaffolds: In vitro evaluation for use as a bone drug delivery system. J Mech Behav Biomed Mater 40:1–12
23. Pon-On W, Charoenphandhu N, Teerapornpuntakit J, Thongbunchoo J, Krishnamra N, Tang IM (2014) Mechanical properties, biological activity and protein controlled release by poly(vinyl alcohol)–bioglass/chitosan–collagen composite scaffolds: a bone tissue engineering applications. Mater Sci Eng, C 38:63–72
24. Nazemi K, Azadpour P, Moztarzadeh f, Urbanska AM, Mozafari M (2015) Tissue-engineered chitosan/bioactive glass bone scaffolds integrated with PLGA nanoparticles: A therapeutic design for on-demand drug delivery. Mater Lett 138:16–20

25. Oliveira JM, Sousa RA, Kotobuki N, Tadokoro M, Hirose M, Mano JF, Reis RL, Ohgushi H (2009) The osteogenic differentiation of rat bone marrow stromal cells cultured with dexamethasone-loaded carboxymethylchitosan/poly(amidoamine) dendrimer nanoparticles. Biomaterials 30:804–813
26. Zhu W, Wang M, Fu Y, Castro NJ, Fu SW, Zhang LG (2015) Engineering a biomimetic three-dimensional nanostructured bone model for breast cancer bone metastasis study. Acta Biomater 14:164–174
27. Iwasaki N, Yamane ST, Majima T, Kasahara Y, Minami A, Harada K, Nonaka S, Maekawa N, Tamura H, Tokura S, Shiono M, Monde K, Nishimura S (2004) Feasibility of polysaccharide hybrid materials for scaffolds in cartilage tissue engineering: evaluation of chondrocyte adhesion to polyion complex fibers prepared from alginate and chitosan. Biomacromolecules 5:828–833
28. Liverani L, Roether JA, Nooeaid P, Trombetta M, Schubert DW, Boccaccini AR (2012) Simple fabrication technique for multilayered stratified composite scaffolds suitable for interface tissue engineering. Mater Sci Eng, A 557:54–58
29. Choi B, Kim S, Lin B, Wu BM, Lee M (2014) Cartilaginous extracellular matrix-modified chitosan hydrogels for cartilage tissue engineering. ACS Appl Mate Interfaces 6(22):20110–20121
30. Kim SE, Park JH, Cho YW, Chung H, Jeong SY, Lee EB, Kwon IC (2003) Porous chitosan scaffold containing microspheres loaded with transforming growth factor-β_1: implications for cartilage tissue engineering. J Control Release 91:365–374
31. Bi L, Li D, Liu J, Hu Y, Yang P, Yang B, Yuan Z (2011) Fabrication and characterization of a biphasic scaffold for osteochondral tissue engineering. Mater Lett 65:2079–2082
32. Silva JM, Georgi N, Costa R, Sher P, Reis RL, van Blitterswijk CA, Karperien M, Mano JF (2013) Nanostructured 3D constructs based on chitosan and chondroitin sulphate multilayers for cartilage tissue engineering. PLoS ONE 8(2):e55451
33. Yan S, Zhang K, Liu Z, Zhang X, Gan L, Cao B, Chen X, Cui L, Yin J (2013) Fabrication of poly(L-glutamic acid)/chitosan polyelectrolyte complex porous scaffolds fortissue engineering. J Mater Chem B 1(11):1541–1551
34. Lee SY, Wee AS, Lim CK, Abbas AA, Selvaratnam L, Merican AM, Ahmad TS, Kamarul T (2013) Supermacroporous poly(vinyl alcohol)-carboxylmethyl chitosan-poly(ethylene glycol) scaffold: An in vitro and in vivo pre-assessments for cartilage tissue engineering. J Mater Sci Mater Med 24(6):1561–1570
35. Chen ZX, Li MC, Xin MH, Chen XD, Mao YF (2015) Preparation and characterization of histidine-grafted-chitosan/ poly(L-lactide) scaffolds. J Funct Mater 46(5):05118–05122
36. Kamoun EA (2015) N-succinyl chitosan-dialdehyde starch hybrid hydrogels for biomedical applications. J Adv Res doi:10.1016/j.jare.2015.02.002
37. Wang XH, Li DP, Wang WJ, Feng QL, Cui FZ, Xu YX, Song XH, van der Werf M (2003) Crosslinked collagen/chitosan matrix for artificial livers. Biomaterials 24:3213–3220
38. Wang X, Yan Y, Lin F, Xiong Z, Wu R, Zhang R, Lu Q (2005) Preparation and characterization of a collagen/chitosan/heparin matrix for an implantable bioartificial liver. J Biomater Sci Polym Ed 16:1063–1080
39. Yang J, Cung TW, Nagaoka M, Goto M, Cho CS, Akaike T (2001) Hepatocyte-specific porous polymer-scaffolds of alginate/galactosylated chitosan sponge for liver-tissue engineering. Biotechnol Lett 23:1385–1389
40. Chen F, Tian M, Zhang D, Wang J, Wang Q, Yu X, Zhang X, Wan C (2012) Preparation and characterization of oxidized alginate covalently cross-linked galactosylated chitosan scaffold for liver tissue engineering. Mater Sci Eng, C 32:310–320
41. Lee KH, Shin SJ, Kim CB, Kim JK, Cho YW, Chung BG, Lee SH (2010) Microfluidic synthesis of pure chitosan microfibers for bio-artificial liver chip. Lab Chip 10:1328–1334
42. Fan J, Shang Y, Yang J, Yuan Y (2009) Preparation of galactosylated hyaluronic acid/chitosan scaffold for liver tissue engineering. Sheng Wu Yi Xue Gong Cheng Xue Za Zhi 26:1271–1275

43. Yang ZY, Mo LH, Duan HM, Li XG (2010) Effects of chitosan/collagen substrates on the behavior of rat neural stem cells. Sci China Life Sciences 53(2):215–222
44. Huang J, Hu X, Lu L, Ye Z, Zhang Q, Luo Z (2010) Electrical regulation of Schwann cells using conductive polypyrrole/chitosan polymers. J Biomed Mater Res A 93(1):164–174
45. Wrobel S, Serra SC, Ribeiro-Samy S, Sousa N, Heimann C, Barwig C, Grothe C, Salgado AJ, Haastert-Talini K (2014) In vitro evaluation of cell-seeded chitosan films for peripheral nerve tissue engineering. Tissue Eng Part A 20:2339–2349
46. Morelli S, Piscioneri A, Messina A, Salerno S, Al-Fageeh MB, Drioli E, Bartolo LD (2015) Neuronal growth and differentiation on biodegradable membranes. J Tissue Eng Regen Med 9(2):106–117
47. Freier T, Montenegro R, Shan Koh H, Shoichet MS (2005) Chitin-based tubes for tissue engineering in the nervous system. Biomaterials 26:4624–4632
48. Valmikinathan CM, Mukhatyar VJ, Jain A, Karumbaiah L, Dasari M, Bellamkonda RV (2012) Photocrosslinkable chitosan based hydrogels for neural tissue engineering. Soft Matter 8:1964–1976
49. Masuko T, Iwasaki N, Yamane S, Funakoshi T, Majima T, Minami A, Ohsuga N, Ohta T, Nishimura SI (2005) Chitosan–RGDSGGC conjugate as a scaffold material for musculoskeletal tissue engineering. Biomaterials 26:5339–5347
50. Rinki K, Dutta PK (2010) Physicochemical and biological activity study of genipin-crosslinked chitosan scaffolds prepared by using supercritical carbon dioxide for tissue engineering applications. Int J Biol Macromol 46:261–266
51. Zhang L, Li Y, Li L, Guo B, Ma PX (2014) Non-cytotoxic conductive carboxymethyl-chitosan/aniline pentamer Hydrogels. React Funct Polym 82:81–88
52. Cheung HK, Han TTY, Marecak DM, Watkins JF, Amsden BG, Flynn LE (2014) Composite hydrogel scaffolds incorporating decellularized adipose tissue for soft tissue engineering with adipose-derived stem cells. Biomaterials 35:1914–1923
53. Martel-Estrada SA, Olivas-Armendáriz I, Santos-Rodríguez E, Martínez-Pérez CA, García-Casillas PE, Hernández-Paz J, Rodríguez-González CA, Chapa-González C (2014) Evaluation of in vitro bioactivity of chitosan/mimosa tenuiflora composites. Mater Lett 119:146–149

Chitosan-Based Scaffolds for Cartilage Regeneration

Xuezhou Li, Jianxun Ding, Xiuli Zhuang, Fei Chang, Jincheng Wang and Xuesi Chen

Abstract Intra-joint trauma often accompanies cartilage damage, as one of the main reasons of osteoarthritis, which often induce severe pain and limited joint function in the final stage. Because of the poor regenerative capacity, cartilage repair has been on the top list of regenerative medicine from decades ago. Recently, the researches of cartilage regeneration are mainly focused on the development of novel scaffolds, which can provide spatial frame and logistic template for stem cells, other progenitor cells, or chondrocytes to proliferate or differentiate into cartilage-like tissues. Among the dazzling scaffolds, chitosan-based systems, including physical hydrogels, chemically cross-linked hydrogels, or porous scaffolds, show great potential in cartilage tissue regeneration. Chitosan possesses superior characteristics, such as biocompatibility, biodegradability, bioabsorbability, low immunogenicity, and intrinsic antibacterial nature, for potential applications in tissue engineering. Specially, the chemical structure of chitosan is similar with various glycosaminoglycans (GAGs), which play important roles in chondrocyte morphology modulation, differentiation, and function. In addition, appropriate mechanical properties and porosity, excellent cell adhesion, and even control release of functional growth factors are achieved in chitosan-based scaffolds. In this chapter, the advancements of different types of chitosan-based scaffolds for cartilage regeneration are systemically summarized, and the future directions are predicted.

Keywords Cartilage regeneration · Chitosan · Chondrocytes · Hydrogel · Porous scaffolds · Stem cells · Three-dimensional

---

X. Li · F. Chang · J. Wang (✉)
Department of Orthopedics, The Second Hospital of Jilin University, Changchun 130041, People's Republic of China
e-mail: jinchengwang@hotmail.com

X. Li · J. Ding (✉) · X. Zhuang · X. Chen
Key Laboratory of Polymer Ecomaterials, Changchun Institute of Applied Chemistry, Chinese Academy of Sciences, Changchun 130022, People's Republic of China
e-mail: jxding@ciac.ac.cn

P.K. Dutta (ed.), *Chitin and Chitosan for Regenerative Medicine*, Springer Series on Polymer and Composite Materials, DOI 10.1007/978-81-322-2511-9_3

1 Introduction

The features of cartilage, such as avascularity and low cellularity, make the damaged cartilage hard to spontaneously repair [1]. Therefore, even the tiny lesion can further aggravate the degeneration of articular cartilage, which results in progressive osteoarthritis, and cause severe pain and disability. Owing to the serious consequences caused by cartilage damage, interventions must be taken to control or restore the damage. The mostly used clinical intervening strategies include microfracture (MF), autologous chondrocyte implantation (ACI), and biomaterials implantation.

As a highly cost-effective, minimally invasive, and relatively simple procedure, the MF technic under arthroscopy has become to be the most widely used strategy for cartilage defect repair [2]. First, focal cleaning is applied, that is, the damaged cartilage tissues are cleared until to the subchondral bone. Then, several holes are drilled to bone marrow, which allow the progenitor cells in bone marrow to go through and arrive at the defect area to repair the defective cartilage. However, most of the new tissues are fibrous ones under the administration of MF, which cannot compensate for the damaged compact cartilage. The early prognosis of cartilage after MF treatment looks fine, while the outcome cannot satisfy the patients as the follow-up is prolonged [3]. Besides MF, ACI is also introduced to treat cartilage defect. ACI emphasizes the effect of implanted normal autologous cartilage, which is harvested from the non-weight bearing area. Although better outcomes are observed, ACI cannot be used in the large defect of articular cartilage due to the limitation of available normal cartilage for implantation [2].

The biomimetic scaffolds of cartilage tissue engineering can be fabricated by physical or chemical methods from natural or synthetic polymers. The biomimetic implants can provide better cell intergrowth spatial structure and proper mechanical support. Compared with those originated from synthetic polymers, the scaffolds made of natural macromolecules are more attractive due to their good biocompatibility, unique chemical structures, and some inherent biological activities. These scaffolds have been widely used in cartilage restoration in the forms of hydrogels [4] and three-dimensional (3D) porous scaffolds [5].

Among so many natural polymers, chitosan is particularly impressive. As depicted in Fig. 1, chitosan is a deacetylated derivative of chitin, and the second most abundant natural biopolymer commonly found in the shells of marine crustaceans and cell walls of fungi. Chitosan exhibits similar structure to glycosaminoglycans (GAGs) and possesses many superior characteristics, including biocompatibility, biodegradability, bioabsorbability, intrinsic antibacterial nature, and chondro-conducive and chondro-integrative properties [6]. Comparative studies demonstrate that the chitosan hydrogels can promote and maintain a better chondrogenic phenotype in mesenchymal stem cells (MSCs) compared to those originated from alginate and fibrin [7]. Also as a composite platform, the chitosan-based scaffold seeding chondrocytes have the greatest cell surface enhancement compared to that from hyaluronic acid (HA), fibrin, or collagen type I (Col I) [8].

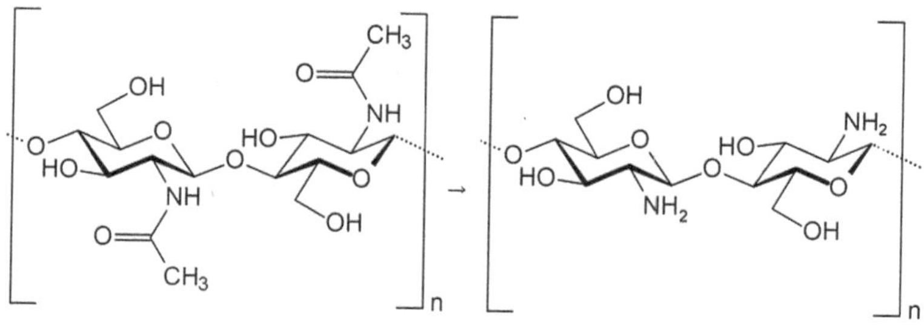

Fig. 1 Synthesis of chitosan from deacetylation of chitin

Chitosan also bears several drawbacks, including low water solubility due to its rigid structure and crystallinity, poor mechanical strength, instability, and so on [6, 9–13]. The scaffolds based on chitosan can inherit its advantages and overcome the drawbacks through the complexation with other materials, such as gelatin [14], alginate [15], collagen [16], and blood clot [17], or special processing procedures. As for processing procedures, different scaffolds, such as hydrogels [18–20], 3D porous scaffolds [21–23], 3D printed scaffolds [24, 25], or coating technology [26, 27], are designed and fabricated.

According to the merits of chitosan, the chitosan-based scaffolds have been widely applied in the regeneration of various tissues, for example, nervous system [28–30], bone tissue [31–35], and wounds healing [36]. Especially for cartilage repairing, chitosan has the biggest space to play. It is not just because of its structural similarity to GAGs, which are one of the major extracellular matrix (ECM) components, but also due to their "shapeshifter magic", which means they can be easily manufactured into many phases under mild condition.

In this chapter, the characteristics of cartilage damage and reconstruction, and the superiorities and development of chitosan-based scaffolds for cartilage tissue engineering are summarized, and also the challenges and opportunities are discussed.

2 Characteristics of Cartilage Damage and Reconstruction

Cartilage injury means the incomplete of cartilage structure. For fabrication of biomimetic scaffolds, attentions must be paid to the components and ingredients of cartilage and their mechanical properties. Hyaline cartilage is the most abundant type of cartilage in human body, especially in joints. It is composed of chondrocytes dispersed in an abundant ECM, which mainly consists of collagen type II (Col II) and proteoglycan (aggrecan).

As described above, the defect of cartilage is an acute loss of partial or whole layer of cartilage. All of the ingredients, including chondrocytes and ECM, need to be gradually repaired. As aforementioned the traditional therapeutic methods, like bone marrow stimulation and osteochondral grafting, all cannot provide satisfied prognosis [2]. With the development of tissue engineering, it provides an alternative way to realize the reconstruction. Three essential elements are needed for all the tissue engineering reconstruction, that is, cells, scaffolds, and growth factors. Scaffold as spatial structure provides support for cell proliferation, migration, and differentiation. All the biological factors to realize their functions must be on the basis of scaffold support. So the exploration of biomimetic, cartilage strength similar, biocompatible, and biodegradable scaffolds grab enormous attention of researchers.

3 Superiorities of Chitosan-Based Scaffolds for Cartilage Regeneration

Chitosan, as a suitable candidate for cartilage regeneration, is known for its good biocompatibility, biodegradability, bioabsorbability, intrinsic antibacterial nature, and structural similarity to GAGs, which are the basic ingredients of ECM.

The most important property of tissue engineering scaffolds is biocompatibility. Only in this way, it can be used as a spatial support for cell migration in vitro and implanted into body. The biocompatibility of chitosan has been well studied in numerous studies. The results show that the chitosan-based biomaterials did not cause serious inflammatory or allergic reaction following implantation, injection, topical application, or ingestion in preclinical studies or human body [12, 37, 38].

As for the degradation dynamics, chitosan can be degraded in vivo by enzymatic hydrolysis [39, 40]. The degradability of chitosan-based scaffolds can promise a suitable process of cell growth and replacement to achieve tissue regeneration. Different tissues need various repair times, so the scaffolds should be capable of an adjustable degradability, that is, indeed one of the advantages of chitosan-based scaffolds. First, the deacetylation degree of chitosan directly decides the degree of crystallinity and inversely affects the degradation rate. A low degradation rate, even last for several months, can be obtained in the scaffold with highly deacetylated chitosan. In contrast, lowly deacetylated chitosan-based scaffold exhibits a shorter degradation period [41]. Second, when chitosan is combined with different other materials, the degradation rate can also be adjusted by the supplemented materials and their interactions. It is worth noting that the degradation rate inherently affects both solubility and mechanical properties.

The antibacterial property of chitosan or chitosan derivatives makes them suitable for tissue healing. Machul et al. reported that chitosan occurred to be effective in the reduction of bacterial populations for both planktonic- and biofilm-related Pseudomonas cells [42]. In addition, the chitosan-based scaffolds are also found to inhibit *Escherichia coli*, *Proteus mirabilis*, and *Pseudomonas aeruginosa* [43]. The antibacterial mechanism of chitosan is still not clearly confirmed. One possible way of antibacterial capability is the electrostatic interaction between polycationic chitosan and negatively charged bacteria membrane components, which can lead to loosening of bacteria membrane and consequent leakage of internal components, and finally cause the death of bacteria [44]. Moreover, chitosan may directly penetrate into bacteria and interact with DNA to inhibit the transcription of DNA and the subsequential protein synthesis [45, 46].

Previous studies demonstrated that the carboxymethylated chitosan can suppress the mRNA expression of matrix metalloproteinase-1/-3 (MMP-1/-3) during the early stage of osteoarthritis by intra-articular injection [47]. It also can inhibit the activation and expression of MMP-2 in primary human dermal fibroblasts, which all can lead to the inhibition of local inflammation [48]. Besides, the thermo-sensitive hydrogels made by chitosan are injectable, which can be delivered into particular tissues or organs by a minimal invasive manner [49, 50]. Furthermore, the covalent and ionic modifications allow extensive adjustment of mechanical and biological properties of chitosan-based scaffolds.

Although chitosan exhibits several drawbacks, the choice of chitosan as a tissue support matrix is governed among others by multiple ways.

4 Chitosan-Based Hydrogels for Cartilage Regeneration

Hydrogels are networks of hydrophilic polymers cross-linked by physical interactions or chemical bonds, which can swell but do not dissolve in water. Chitosan, as one of the mostly used matrices, can form hydrogels under mild condition.

4.1 Physically Cross-Linked Hydrogels with Chitosan as Matrix

The physical cross-linking for preparation of hydrogels can avoid introducing of exogenous cross-linking agents, which may bear potential toxicity. Although physical cross-linking may sacrifice a part of mechanical strength, it is easy to be processed without complex synthesis steps compared with the chemical one.

Thermo-sensitive chitosan-based hydrogel is one of the most commonly used scaffolds for tissue engineering. Researchers usually set the trigger temperatures between 30 and 37 °C. When below trigger temperature, the liquid sol allows for the encapsulation of cells and/or growth factors. Once injected into animal or human body, the liquid solution can transform into gel status with the ascent of temperature to provide spatial structure for cell proliferation and differentiation. Previous studies showed that several synergistic forces were favorable to thermogel formation, such as hydrogen bonding, electrostatic interaction, and hydrophobic interaction [51].

Chitosan-based thermogel have been widely applied in cartilage repair. For fabrication process, chitosan is dissolved in acetic acid to obtain chitosan solution and blended with β-glycerophosphate solution to obtain a final chitosan–β-glycerophosphate solution. The trigger temperature is adjusted to 37 °C by regulating the ratio of the two solutions for a proper gel formation in vivo [52]. In addition, other materials, like gelatin, can be dissolved with chitosan in acetic acid, and then β-glycerophosphate solution is added to form chitosan/gelatin/β-glycerophosphate hydrogel. The gelation temperature of this gel at neutral pH can be controlled at around 33 °C. Cells, like nucleus pulposus cells, can be encapsulated into above hydrogel. After periods of culture in vitro, the scaffold showed good production of sulfated GAGs, up-regulation of relative gene expression, and high synthesis of ECM in nucleus pulposus [53].

Pure chitosan hydrogel also can be formulated to be the support of other delivery carriers, such as nanoparticles [54] or nanofibers [55]. The combination can improve the mechanical property of composite scaffolds and provide a better control release behavior.

Another great advantage of thermo-sensitive chitosan-based hydrogel is injectable. The injectable hydrogels allow for minimally invasive implantation, filling a desired shape, and easy incorporation of various therapeutic agents [49]. Chenite et al. found that polyol salt can transform purely pH-dependent chitosan solution into temperature-controlled pH-dependent chitosan hydrogel, which could be injected into tissues for biological compound delivery [51]. Furthermore, the thermogel can deliver active growth factor leading to neo-cartilage formation in vivo.

4.2 Chitosan-Based Chemically Cross-Linked Hydrogels

Chemical cross-linking will formulate the scaffolds with the bridges of covalent bonds, so more tough scaffolds with low degradation rate can be obtained to meet the demand of cartilage repair. Cross-linker, as the spirit agent of whole chemical cross-linking procedure, plays an important role in the binding process. The widely used chemical cross-linking agents in chitosan-based hydrogels include genipin [56–58], glutaraldehyde [59, 60], and acrylate group [61], which react with amine group in chitosan. Moreover, many abnormal cross-linkers are explored in cross-linking of chitosan-based hydrogels, like sodium tripolyphosphate [62], citrate [63], horseradish peroxidase (HRP) and hydrogen peroxide (H_2O_2) [64], and water-soluble carbodiimide (WSC) [65].

Among them, citrate-cross-linked chitosan hydrogels possess strain reversibility and viscoelastic behavior. The results demonstrate early cartilage formation and woven bone deposition after being implanted into a rabbit osteochondral defect model [63]. Tripolyphosphate, as a polyanionic cross-linker with quick gelling capability and non-toxic property, can cross-link the amino groups at the C-2 site of chitosan unit [62]. The HRP- and H_2O_2-based enzymatic cross-linking of glycolic acid (GA)- and phloretic acid (PA)-grafted chitosan can provide a fast and efficient way to obtain an in situ formed hydrogel under physiological condition [64]. WSC is also found to be the most suitable cross-linker, and chondrocytes are able to proliferate well and secrete abundant ECM in the chitosan–gelatin complex scaffold cross-linked by WSC compared to the non-cross-linked one [65].

The photo-cross-linked chitosan-based hydrogels are also widely applied in cartilage tissue engineering. Similar as thermogel, this kind of hydrogel is micro-environment friendly without any small molecule cross-linking agents.

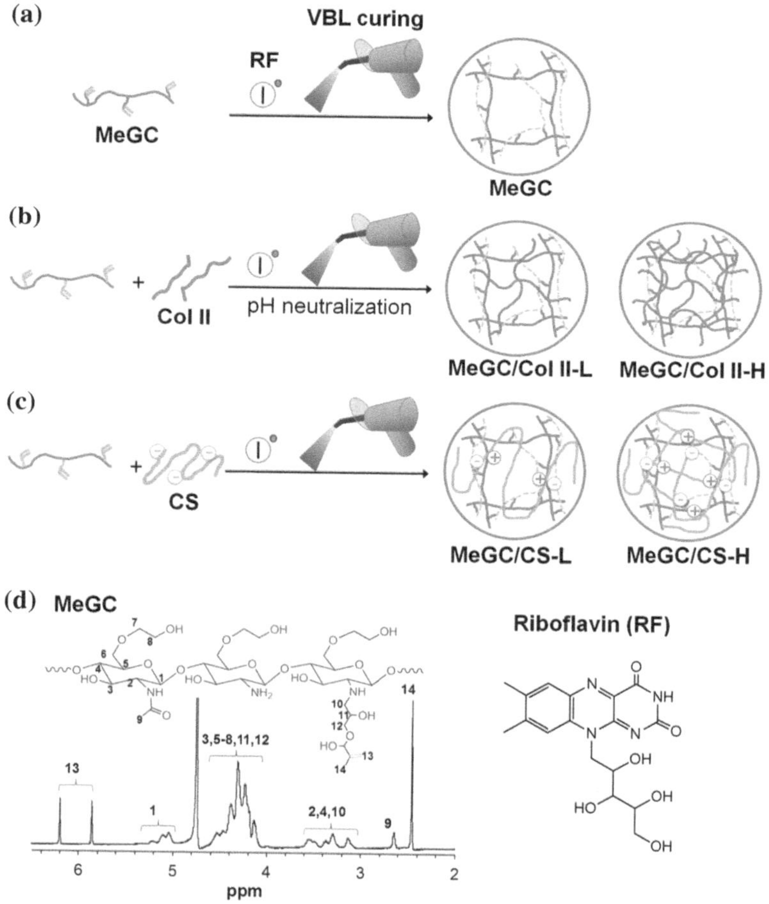

Fig. 2 Design of VBL cross-linked chitosan/ECM composite hydrogels [19]. **a** Chitosan hydrogel is prepared by cross-linking of MeGC via VBL irradiation with RF. **b** The mixture of MeGC with Col II and RF is pH neutralized for initiating Col II gelation and followed by VBL cross-linking to form a hydrogel (MeGC/Col II). The fibrous structure of hydrogel can be varied in a Col II concentration-dependent manner. **c** The ionically interacted MeGC and CS are cross-linked using VBL irradiation with RF (MeGC/CS). **d** ^{1}H NMR spectrum and structure of MeGC in D_2O. Structure of RF

Hu et al. combined photo-polymerizable chitosan (i.e., methacrylated glycol chitosan (MeGC)) and riboflavin (RF) to achieve gelation in situ by visible blue light (VBL) irradiation, and this hydrogel was injectable [66]. Recently, the formulation of photo-cross-linked hydrogel is improved by adding Col II (Fig. 2), which is the natural compound of cartilage. The modified hydrogels are proved to further increase chondrogenesis by encapsulating both chondrocytes and MSCs [19]. When injected into superficial tissues, the mixed solution of MeGC and HA can be triggered into gels by photo-cross-linking with a riboflavin photo-initiator under visible light [67]. Furthermore, to provide high modulus scaffolds, an two-phase and bicontinuous scaffold was synthesized from MeGC and a hydrolyzable, hydrophobic, and acrylated poly(ε-caprolactone-*co*-D,L-lactide) or poly(ε-caprolactone-*co*-trimethylene carbonate). By chondrocyte seeding test, the author found that the chitosan-based poly(ε-caprolactone-*co*-trimethylene carbonate) bicontinuous scaffold accumulated more ECM after 56 days of culture in vitro [68].

Although various cross-linkers are applied in hydrogel formation, the most important property of them is non-toxic. Only based on this principle, the chemically cross-linked hydrogels can be used as implants in vivo for tissue regeneration or scaffolds in vitro for cells culture.

5 3D Porous Chitosan-Based Scaffolds

As aforementioned, the 3D porous chitosan-based scaffold is one of the most important implant for cartilage repair. It can provide porous structure to load cells and support their proliferation and migration. When compared to soft hydrogel, the solid porous scaffolds possess more mechanical strength for repairing cartilage. The 3D porous chitosan-based scaffolds are fabricated in various phases (Fig. 3).

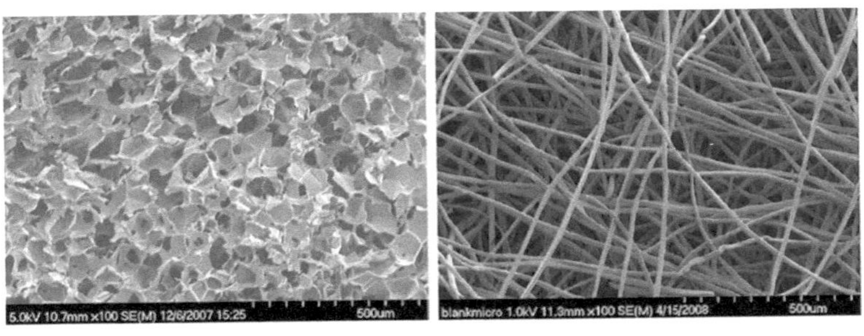

Fig. 3 Typical SEM micrographs of 3D porous chitosan-based scaffolds: (*Left*) chitosan sponge scaffold and (*Right*) chitosan fibrous scaffold [13]. Magnification: 100×

5.1 Normal 3D Porous Chitosan-Based Scaffolds

The normal 3D porous chitosan-based scaffolds can be obtained by freeze-drying method without or with chemical cross-linking. The most used chemical cross-linking agents include 1-ethyl-3-(3-dimethylaminopropyl) carbodiimide hydrochloride (EDC·HCl) and *N*-hydroxysulfosuccinimide (NHS) [69, 70], genipin [71, 72], and calcium chloride [73], and the choice of cross-linkers is based on the type of chemical reaction and the demand of mechanical property.

To overcome the drawbacks and enhance the physical properties, chitosan often be blended with other nature materials, like alginate [74, 75], HA [76], collagen [77, 78], and gelation [79]. Collagen is a natural ECM, which possesses good cell-binding property. HA is also one important component of ECM and can enhance cell adhesion, proliferation, and differentiation. Gelatin is a partial derivative of collagen, which exhibits better nonimmunogenic property compared to its precursor and also can promote cell behaviors. Through the combination, the composite scaffolds can achieve better cross-linking strength and biocompatibility.

Tri-copolymer scaffolds, like the ones based on collagen–chitosan–hyaluronan [77, 80–82], gelatin–chitosan–GAG [69], and chitosan–alginate-hyaluronate [80], also are created to combine more advantages from other natural materials.

5.2 Fibrous Scaffolds with Chitosan as Matrix

The nanofiber-based scaffolds possess adjustable and highly interconnected porous structures, which is similar to the ECM of articular cartilage. The interconnected open porous structure allows the migration of cells into the core of the scaffolds. Microfibers have high surface to volume ratio, which can enhance the interaction between cells and material. Further, fibrous scaffolds elicit compressive mechanical strength, and it is not easy to be damaged due to their flexible nature and structural integrity. Chitosan as the versatile material for different phasic scaffold syntheses can also be used in fibrous scaffolds fabrication.

To fabricate fibrous scaffolds, the chitosan singly or combined with other materials, like alginate [83], HA [84], and synthetic polymers [85], was dissolved in solvent. Then, electrospinning technique was applied to spin the nanofiber into scaffold layer by layer. The combination was not just to improve the mechanical strength, but also could impair the inherent drawbacks that chitosan possessed. The fibrous scaffolds could be fabricated in oriented and chaotic phases, and both of the phases would affect the behaviors of cells seeded on them. To change the condition of electrospinning technique, different diameters of fibers and various porous scaffolds could be obtained according to the demands.

It is reported that the GAG quantification and Col II mRNA expression of chitosan fibrous scaffolds were superior to chitosan sponge scaffolds [13]. Noriega et al. also found that cells cultured on chitosan fibrous scaffold with a mean fiber diameter of 300 nm was 2 folds higher than the cells cultured on sponge-like scaffold fabricated by lyophilization [86]. Usually, chitosan often combines polymer materials to fabricate fibrous scaffolds, because the polymers, like poly(ε-caprolactone) (PCL) and poly(lactide-co-glycolide) (PLGA), can provide better tension strength, and the ratios between them will affect the tension strength of scaffolds and the efficiency of neo-cartilage formation [87].

5.3 Chitosan–Silk Scaffolds

Silk fibroin (SF) is an attractive natural fibrous protein, which possesses excellent biological compatibility, mechanical strength, and biodegradation. It has been used as a promising biomaterial for a number of biomedical and biotechnological applications.

To overcome the insufficient mechanical properties of chitosan scaffolds, SF as a material with robust mechanical strength can be added into chitosan-based scaffolds. After being blended, the scaffolds showed higher compressive strength and modulus than the ones with individual components. SF facilitates cell adhesion and growth to overcome the chitosan's drawbacks of lower cell adhesion. Besides, highly crystallized SF possesses low protease susceptibility, which can prolong the degradation times of blended scaffolds according to the ratio of chitosan to SF [88].

A lot of studies have been performed on the chitosan–SF porous scaffolds. The blended scaffolds can be chemically cross-linked, and the widely used cross-linkers include of poly(ethylene glycol) dimethacrylate [89], EDC·HCl and NHS [90], genipin [91], etc. The solid-phase porous blended scaffolds of chitosan and SF also can be achieved by freeze-drying method [22, 23, 88, 92, 93]. Before lyophilization, SF must be degummed firstly. Besides the solid-phase porous scaffolds, the thermo-sensitive hydrogels made of chitosan and silk also have been studied. Mirahmadi et al. dispersed chopped degummed silk fiber into the pre-prepared chitosan-based thermo-sensitive hydrogel solution to enhance the mechanical strength of blended scaffold and prolong its degradation time [94].

The cross-linked scaffold of chitosan and SF was applied in cartilage defect models of rabbit. After the estimations of gross observation, hematoxylin–eosin and toluidine blue staining, Col II immunohistochemistry, and Wakitani histology, the results showed that the cartilage defect was nearly repaired, the scaffold was absorbed, and immunohistochemistry was positive [90]. The DNA content and GAG quantification of ATDC5 cells seeded on genipin-cross-linked SF-chitosan scaffold were higher than those of control group [91]. The blended lyophilized scaffold of chitosan and SF can support cell attachment and proliferation. There were more accumulation of GAGs, and collagens were observed in the blended scaffolds [92] (Fig. 4). Real-time polymerase chain reaction (RT-PCR) showed

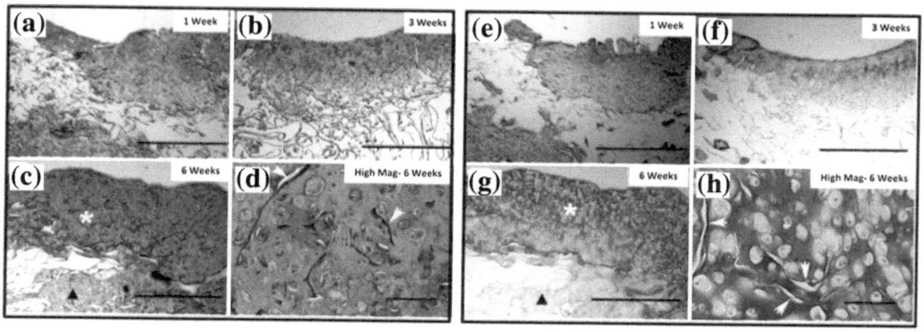

Fig. 4 ECM deposition by chondrocytes in SF-CS scaffold in vitro. (*Left*) Representative images of Movat's staining of cell-scaffold constructs at different time points: **a** 1 week; **b** 3 weeks; **c** 6 weeks (the scaffold was stained red, and the GAG was dyed blue). (*Right*) Immunostaining with anti-Col II of cell–scaffold constructs: **e** 1 week; **f** 3 weeks; **g** 6 weeks (Col II was detected as brown) [92]

well-chondrogenic differentiation of MSCs seeded on the scaffolds in vitro [22, 23]. In Mirahmadi's study, the scaffold was evaluated by assessing the cell viability and proliferation, and the production of GAGs and Col II. The results showed that the scaffold can serve as a promising agent for cartilage repair [94].

5.4 3D Printed Chitosan Scaffolds

3D printing also known as a rapid prototyping (RP) technology has been used in scaffold fabrication. The 3D printing technique was applied in scaffold fabrication mainly in two forms: first, printing dissolvable molds [25], and second, fabricating spatial scaffolds directly [24] (Fig. 5).

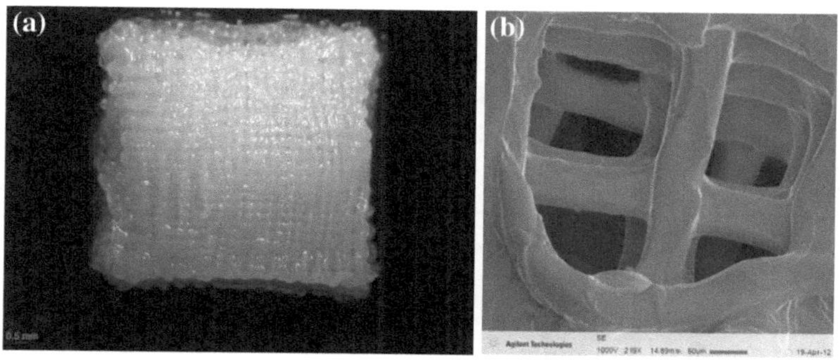

Fig. 5 3D printed chitosan scaffolds: **a** macroscopic image of a 3D printed chitosan scaffold and **b** SEM microimage showing the lattice network of chitosan fiber [24]

Modeling dissolvable molds by 3D printing is easy to be achieved, and a lot of materials are available for the printing of molds. Build wax (protobuild) and support wax (protosupport) both can serve as dissolvable molds for fabrication materials [25]. As for directly fabricating scaffolds, different materials have been tested in various printers to create ideal compatible and mimetic spatial scaffolds used in the reconstruction of various tissues. Most of them are made up of polymers, such as PCL, PLGA, etc. Chitosan as material for 3D printing possesses some attractive advantages, and the most important one is plasticity. Ye et al. created a 3D printing method to directly fabricate scaffolds by extruding chitosan solution onto a glass slide immersed in a precipitating bath of isopropyl alcohol, and infrapatellar fat pad adipose stem cells can successfully undergo chondrogenesis on the surface of the scaffolds [24].

Bio 3D printing as a novel technology has not been well developed, and the source of materials are the main limitation. Chitosan-based materials may provide a possible option by processing further.

5.5 *Dynamic Compression Chitosan Scaffolds*

A novel way of tissue engineering for cartilage regeneration is dynamic compression culture, which plays an important role in stem cell proliferation and chondrogenic differentiation (Fig. 6). It also can stimulate the synthesis and deposition of ECM and lead to better reconstruction of articular cartilage [95].

Chitosan is already applied in the fabrication of dynamic compression supporting scaffolds [57, 78, 96, 97]. Almost all of the scaffolds are chemically cross-linked for more compression strength. In 2008, Jung et al. reported that dynamic compression of elastic scaffolds could encourage chondrocytes to maintain their phenotypes and enhance the production of GAGs, which would improve the quality of cartilaginous tissue formation both in vitro and in vivo [98]. Chitosan/gelatin scaffolds cross-linked by EDC·HCl/NHS combined insulin-like growth factor-I (IGF-I) elicited good elastic property. It can induce the chondrogenesis of rabbit adipose-derived stem cells (rADSCs) by activating calcium signaling pathway and

Fig. 6 Resilience test of culture medium-presoaked chitosan/gelatin scaffolds [97]

up-regulating the expressions of Sox-9, Col II, and aggrecan. Besides, it also can promote cell proliferation and depositions of proteoglycan and collagen, which benefit articular cartilage regeneration [96]. Chitosan–collagen scaffold cross-linked by genipin was also tested by loading rabbit chondrocytes and cultured under cyclic compression of 40 % strain, 0.1 Hz, and 30 min/day for 2 weeks. Cell proliferation and total GAG deposition were enhanced [57].

5.6 Aggregated Chitosan–Particle Scaffolds

Particle agglomeration method is an innovative methodology based on the random package of microsphere with further aggregation by physical or thermal mean to create a 3D porous structure. Chitosan particle-agglomerated scaffolds are not the mainstream in the whole scaffold research for cartilage regeneration, but they also showed possibility in cartilage regeneration.

Malafaya et al. reported a method to create a chitosan-based particle-agglomerated scaffold [99]. First, non-sintered hydroxylapatite was homogeneously dispersed in chitosan solution. The prepared solution was extruded to form chitosan droplet into a sodium hydroxide (NaOH) precipitation bath where particle with regular diameter formed. After cross-linked by glutaraldehyde, the particle was subsequently pressed into mold and left to dry. Finally, the cylinder-shaped

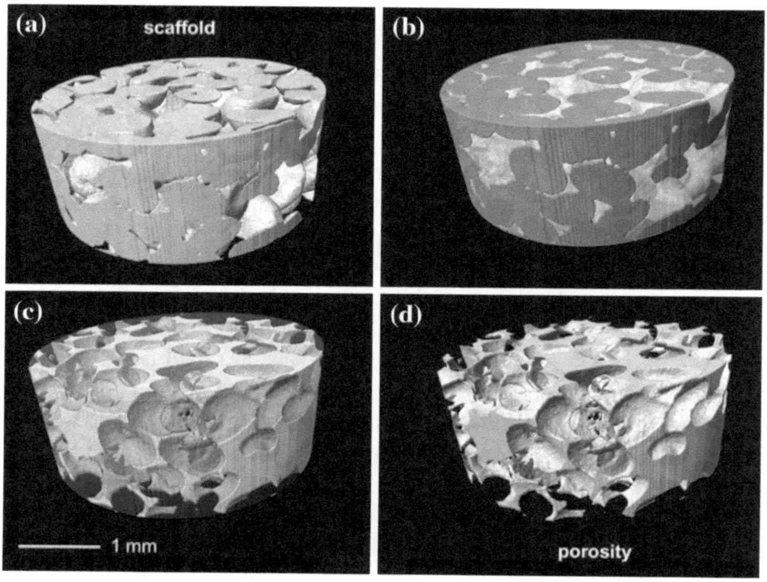

Fig. 7 3D virtual models of chitosan particle-aggregated scaffolds: **a** gradual transition, **b**, **c** inverse 3D model, and **d** porous morphology [101]

scaffold was obtained. The chitosan solution can also be mixed with other solutions, like insulin protein solution, to obtain a composite particle-aggregated scaffold [100]. The interconnected porous structure was obtained using particle agglomeration, because the particle was sphere and they cannot paste together tightly. The interconnected interspace allows cell intergrowth and can provide better cartilage integration [101] (Fig. 7). For cartilage regeneration repair test, higher DNA and GAG contents, and up-regulated Sox-9 and aggrecan were observed in the insulin-loaded chitosan-agglomerated scaffold after 4 weeks of culture [100].

6 Other Scaffolds with Chitosan as Matrix

Furthermore, the following novel chitosan-based scaffolds are also applied in cartilage repair.

6.1 Blood Implant in Marrow Stimulation

Based on the blood coagulation property, the autologous blood clot serves as a scaffold, which can be safely implanted into host body. But the low intensity and easy degradation limit its applications. The chitosan-based hybrid clot is demonstrated to possess practical advantages over normal blood clot. Chitosan can physically stabilize the blood clot from lysis and rapid degradation in serum after clot [102].

There are two main kinds of chitosan-based implants for now, chitosan–sodium chloride (NaCl)/blood implants and chitosan–glycerol phosphate (GP)/blood implants, both of which are applied in cartilage regeneration. The thrombogenic mixture of chitosan–GP solution and peripheral whole blood of sheep was first reported in 2005 by Hoemann [103], and the solidification mechanism of chitosan–GP/blood implant was studied deeply by Marchand [102]. It demonstrated that thrombin generation, platelet activation, and fibrin polymerization led to dual fibrin-polysaccharide clot scaffold. Chitosan–NaCl/blood implant was found to be a more rapid coagulation than chitosan–GP/blood mixture in vitro, and the detailed mechanism was still not clear [104]. The pre-solidified chitosan–NaCl/blood implant elicited reproducible early osteochondral wound repair, promoted bone plate resorption, enhanced repair tissue integration, and delayed matrix deposition [17, 104]. Besides, chitosan implants solidified with thrombin elicit shorter coagulation time and long-term durability [105].

Since 2005, the chitosan–GP/blood implant has been demonstrated to improve cartilage regeneration compared with MF alone, and the following studies also proved the results that the chitosan-based blood implant could produce hyaline and structurally integrated osteochondral unit [17, 104–108].

6.2 Chitosan-Based Cryogels

Cryogels are gel matrices that are synthesized at subzero temperatures using monomeric or polymeric precursors, and they are well summarized by Lozinsky [109] (Fig. 8). The properties of cryogels, such as osmotic, chemical, and mechanical stabilities, make them to be ideal agents in bioseparation. Besides, polymeric cryogels are also efficient carriers of biomolecules and cells in tissue regeneration.

Chitosan-based cryogels can provide efficient cell adhesion and proliferation, and ECM secretion. The scaffolds possess excellent biocompatibility, biodegradability, interconnected macroporous structure, mechanical stability, and elasticity, which are all urgently needed in cartilage regeneration [110]. In animal studies, chitosan combined with gelatin [110], HA [111], or agarose [14, 112] to fabricate scaffolds and was tested for cartilage repair. After freeze-drying, the cryogels could be implanted into the focal cartilage defect without or with cells and/or growth factors. After proper healing time, the cartilage-defected femoral condyles were harvested and evaluated by histology and morphology methods. The results showed that cryogel scaffolds can promote the repair of osteochondral defect at a load-bearing site [14], the syntheses of GAGs, and COL II, and providing elastic modulus neo-cartilage similar to the native one [111].

6.3 Chitosan-Coated Scaffolds

Another alternative application way of chitosan in preparing scaffolds is coating. If the researchers do not want to change the mechanical properties of initial scaffolds from polymers, but want to obtain a surface of good biocompatibility, hydrophilicity, and biological interactions, the technique of chitosan coating provides a possible option.

Magalhaes et al. reported that chitosan-coated poly(L-lactide) scaffold exhibited a better water uptake capacity and elastic modulus [113]. Two chitosan forms were used to coat the surface, that is, hydrogel-like coating and nanofiber coating. The results showed that the hydrogel-like coating supported the formation

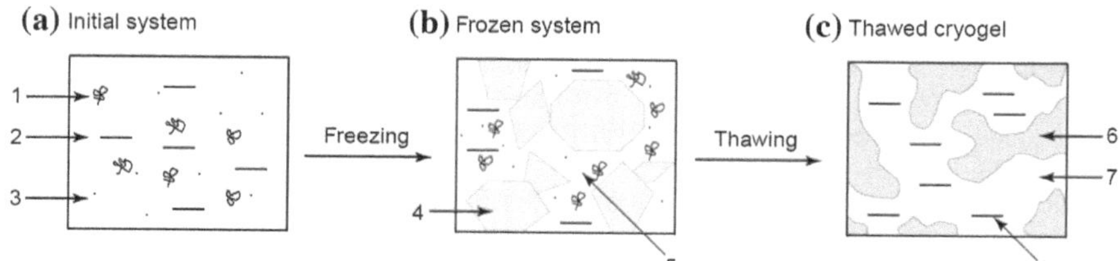

Fig. 8 Conceptual diagram of polymeric cryogel formation. *1* macromolecule in a solution; *2* solvent; *3*, low-molecular solute; *4*, polycrystal of frozen solvent; *5* unfrozen liquid microphase; *6* polymeric framework of cryogel; *7* macropore; *8* solvent [109]

of heterogeneous tissue, whereas the fiber-like coating produced a more stable hyaline-like cartilage tissue. Chen et al. also reported that the chitosan-modified PLGA scaffold had the significantly enhanced properties [26]. When cultured in vitro, the scaffold surface was covered by galectin-1 efficiently and spontaneously, which can promote the adhesion and intergrowth of chondrocytes compared to the control group. All the results indicated that the chitosan coating technique could be a promising method for surface modification of hydrophobic materials.

7 Summary and Future Directions

Chitosan, as one of the promising materials for cartilage tissue engineering, possesses outstanding properties compared to the synthetic polymers. Although chitosan also bears several drawbacks, the chitosan-based scaffolds combined with other natural or synthetic materials really have achieved great advances as we summarized. In this chapter, we reveal various chitosan-based scaffolds in the forms of hydrogels, porous solid scaffolds, and nanofibrous scaffolds and films. Chitosan can also serve as an ingredient of 3D printing materials. Although poor mechanical property of chitosan limit its application to some extent, the drawbacks can be minimized or diminished by chemical modification or blending other robust mechanical materials. In future, it is sure that chitosan will still be one of the main candidates in cartilage tissue engineering.

Acknowledgments This work was financially supported by the National Natural Science Foundation of China (Nos. 51303174, 51273196, 51203153, 51233004, 51390484, and 51321062), the Scientific Development Program of Jilin Province (Nos. 20140520050JH and 20140309005GX), and the Science and Technology Planning Project of Changchun City (No. 14KG045).

References

1. Huey DJ, Hu JC, Athanasiou KA (2012) Unlike bone, cartilage regeneration remains elusive. Science 338:917–921
2. Oussedik S, Tsitskaris K, Parker D (2015) treatment of articular cartilage lesions of the knee by microfracture or autologous chondrocyte implantation: A systematic review. Arthrosc J Arthrosc Relat Surg Off Publ Arthrosc Assoc North Am Int Arthrosc Assoc 31:732–744
3. Goyal D, Keyhani S, Lee EH, Hui JHP (2013) Evidence-based status of microfracture technique: A systematic review of level I and II studies. Arthrosc J Arthrosc Relat Surg 29:1579–1588
4. Qi BW, Yu AX, Zhu SB, Zhou M, Wu G (2013) Chitosan/poly(vinyl alcohol) hydrogel combined with Ad-hTGF-beta 1 transfected mesenchymal stem cells to repair rabbit articular cartilage defects. Exp Biol Med 238:23–30
5. Wang W, Li B, Li Y, Jiang Y, Ouyang H, Gao C (2010) In vivo restoration of full-thickness cartilage defects by poly(lactide-co-glycolide) sponges filled with fibrin gel, bone marrow mesenchymal stem cells and DNA complexes. Biomaterials 31:5953–5965

6. Dutta PK, Rinki K, Dutta J (2011) Chitosan: A promising biomaterial for tissue engineering scaffolds. Chitosan Biomater II 244:45–79
7. Sheehy EJ, Mesallati T, Vinardell T, Kelly DJ (2014) Engineering cartilage or endochondral bone: A comparison of different naturally derived hydrogels. Acta Biomater 13:245–253
8. Schagemann JC, Kurz H, Casper ME, Stone JS, Dadsetan M, Yu-Long S, Mrosek EH, Fitzsimmons JS, O'Driscoll SW, Reinholz GG (2010) The effect of scaffold composition on the early structural characteristics of chondrocytes and expression of adhesion molecules. Biomaterials 31:2798–2805
9. El Kadib A, Bousmina M, Brunel D (2014) Recent progress in chitosan bio-based soft nanomaterials. J Nanosci Nanotechnol 14:308–331
10. Sarvaiya J, Agrawal YK (2014) Chitosan as a suitable nanocarrier material for anti-Alzheimer drug delivery. Int J Biol Macromol 72C:454–465
11. Kim IY, Seo SJ, Moon HS, Yoo MK, Park IY, Kim BC, Cho CS (2008) Chitosan and its derivatives for tissue engineering applications. Biotechnol Adv 26:1–21
12. Wan AC, Tai BC (2013) CHITIN–a promising biomaterial for tissue engineering and stem cell technologies. Biotechnol Adv 31:1776–1785
13. Ragetly GR, Griffon DJ, Lee HB, Fredericks LP, Gordon-Evans W, Chung YS (2010) Effect of chitosan scaffold microstructure on mesenchymal stem cell chondrogenesis. Acta Biomater 6:1430–1436
14. Gupta A, Bhat S, Jagdale PR, Chaudhari BP, Lidgren L, Gupta KC, Kumar A (2014) Evaluation of three-dimensional chitosan-agarose-gelatin cryogel scaffold for the repair of subchondral cartilage defects: an in vivo study in a rabbit model. Tissue Eng Part A 20:3101–3111
15. Tigli RS, Guemuesderelioglu M (2009) Evaluation of alginate-chitosan semi IPNs as cartilage scaffolds. J Mater Sci-Mater Med 20:699–709
16. Bi L, Cao Z, Hu Y, Song Y, Yu L, Yang B, Mu J, Huang Z, Han Y (2011) Effects of different cross-linking conditions on the properties of genipin-cross-linked chitosan/collagen scaffolds for cartilage tissue engineering. Journal of materials science. Mater Med 22:51–62
17. Guzman-Morales J, Lafantaisie-Favreau CH, Chen G, Hoemann CD (2014) Subchondral chitosan/blood implant-guided bone plate resorption and woven bone repair is coupled to hyaline cartilage regeneration from microdrill holes in aged rabbit knees. Osteoarthr Cartil/OARS, Osteoarthr Res Soc 22:323–333
18. Walker KJ, Madihally SV (2015) Anisotropic temperature sensitive chitosan-based injectable hydrogels mimicking cartilage matrix. J Biomed Mater Res Part B, Appl Biomater 103B:1149–1160
19. Choi B, Kim S, Lin B, Wu BM, Lee M (2014) Cartilaginous extracellular matrix-modified chitosan hydrogels for cartilage tissue engineering. ACS Appl Mater Interfaces 6:20110–20121
20. Zhang J, Yang Z, Li C, Dou Y, Li Y, Thote T, Wang DA, Ge Z (2013) Cells behave distinctly within sponges and hydrogels due to differences of internal structure. Tissue Eng Part A 19:2166–2175
21. Fang J, Zhang Y, Yan S, Liu Z, He S, Cui L, Yin J (2014) Poly(L-glutamic acid)/chitosan polyelectrolyte complex porous microspheres as cell microcarriers for cartilage regeneration. Acta Biomater 10:276–288
22. Bhardwaj N, Kundu SC (2012) Chondrogenic differentiation of rat MSCs on porous scaffolds of silk fibroin/chitosan blends. Biomaterials 33:2848–2857
23. Bhardwaj N, Nguyen QT, Chen AC, Kaplan DL, Sah RL, Kundu SC (2011) Potential of 3-D tissue constructs engineered from bovine chondrocytes/silk fibroin-chitosan for in vitro cartilage tissue engineering. Biomaterials 32:5773–5781
24. Ye K, Felimban R, Traianedes K, Moulton SE, Wallace GG, Chung J, Quigley A, Choong PFM, Myers DE (2014) Chondrogenesis of infrapatellar fat pad derived adipose stem cells in 3D Printed Chitosan Scaffold. PloS ONE 9:e99410

25. Manjubala I, Woesz A, Pilz C, Rumpler M, Fratzl-Zelman N, Roschger P, Stampfl J, Fratzl P (2005) Biomimetic mineral-organic composite scaffolds with controlled internal architecture. J Mater Sci-Mater Med 16:1111–1119
26. Chen SJ, Lin CC, Tuan WC, Tseng CS, Huang RN (2010) Effect of recombinant galectin-1 on the growth of immortal rat chondrocyte on chitosan-coated PLGA scaffold. J Biomed Mater Res Part A 93A:1482–1492
27. Wu YC, Shaw SY, Lin HR, Lee TM, Yang CY (2006) Bone tissue engineering evaluation based on rat calvaria stromal cells cultured on modified PLGA scaffolds. Biomaterials 27:896–904
28. Yu LMY, Kazazian K, Shoichet MS (2007) Peptide surface modification of methacrylamide chitosan for neural tissue engineering applications. J Biomed Mater Res Part A 82A:243–255
29. Kuo YC, Chiu KH (2011) Inverted colloidal crystal scaffolds with laminin-derived peptides for neuronal differentiation of bone marrow stromal cells. Biomaterials 32:819–831
30. Kuo YC, Yeh CF (2011) Effect of surface-modified collagen on the adhesion, biocompatibility and differentiation of bone marrow stromal cells in poly(lactide-co-glycolide)/chitosan scaffolds. Colloids Surf B-Biointerfaces 82:624–631
31. Jiang L, Li Y, Wang X, Zhang L, Wen J, Gong M (2008) Preparation and properties of nano-hydroxyapatite/chitosan/carboxymethyl cellulose composite scaffold. Carbohydr Polym 74:680–684
32. Jiang L, Li Y, Xiong C (2009) Preparation and biological properties of a novel composite scaffold of nano-hydroxyapatite/chitosan/carboxymethyl cellulose for bone tissue engineering. J Biomed Sci 16:65
33. Oliveira JM, Costa SA, Leonor IB, Malafaya PB, Mano JF, Reis RL (2009) Novel hydrox yapatite/carboxymethylchitosan composite scaffolds prepared through an innovative "autocatalytic" electroless coprecipitation route. J Biomed Mater Res Part A 88A:470–480
34. Thein-Han WW, Misra RDK (2009) Biomimetic chitosan-nanohydroxyapatite composite scaffolds for bone tissue engineering. Acta Biomater 5:1182–1197
35. Tigli RS, Akman AC, Guemuesderelioglu M, Nohutcu RM (2009) In vitro release of dexamethasone or bfgf from chitosan/hydroxyapatite scaffolds. J Biomater Sci-Polym Ed 20:1899–1914
36. Miguel SP, Ribeiro MP, Brancal H, Coutinho P, Correia IJ (2014) Thermoresponsive chitosan-agarose hydrogel for skin regeneration. Carbohydr Polym 111:366–373
37. Venkatesan J, Kim SK (2010) Chitosan composites for bone tissue engineering–an overview. Mar Drugs 8:2252–2266
38. Shukla SK, Mishra AK, Arotiba OA, Mamba BB (2013) Chitosan-based nanomaterials: a state-of-the-art review. Int J Biol Macromol 59:46–58
39. Zainol I, Ghani SM, Mastor A, Derman MA, Yahya MF (2009) Enzymatic degradation study of porous chitosan membrane. Mater Res Innovations 13:316–319
40. Pechsrichuang P, Yoohat K, Yamabhai M (2013) Production of recombinant Bacillus subtilis chitosanase, suitable for biosynthesis of chitosan-oligosaccharides. Bioresour Technol 127:407–414
41. Lee KY, Ha WS, Park WH (1995) Blood compatibility and biodegradability of partially N-acylated chitosan derivatives. Biomaterials 16:1211–1216
42. Machul A, Mikolajczyk D, Regiel-Futyra A, Heczko PB, Strus M, Arruebo M, Stochel G, Kyziol A (2015) Study on inhibitory activity of chitosan-based materials against biofilm producing Pseudomonas aeruginosa strains. J Biomater Appl doi: 10.1177/0885328215578781
43. Wang R, Neoh KG, Kang ET (2015) Integration of antifouling and bactericidal moieties for optimizing the efficacy of antibacterial coatings. J Colloid Interface Sci 438:138–148
44. Martinez LR, Mihu MR, Han G, Frases S, Cordero RJ, Casadevall A, Friedman AJ, Friedman JM, Nosanchuk JD (2010) The use of chitosan to damage Cryptococcus neoformans biofilms. Biomaterials 31:669–679

45. Goy RC, Britto Dd, Assis OBG (2009) A review of the antimicrobial activity of chitosan. Polímeros 19:241–247
46. Martins AF, Facchi SP, Follmann HD, Pereira AG, Rubira AF, Muniz EC (2014) Antimicrobial activity of chitosan derivatives containing N-quaternized moieties in its backbone: A review. Int J Mol Sci 15:20800–20832
47. Liu SQ, Qiu B, Chen LY, Peng H, Du YM (2005) The effects of carboxymethylated chitosan on metalloproteinase-1, -3 and tissue inhibitor of metalloproteinase-1 gene expression in cartilage of experimental osteoarthritis. Rheumatol Int 26:52–57
48. Muzzarelli RAA (2009) Chitins and chitosans for the repair of wounded skin, nerve, cartilage and bone. Carbohydr Polym 76:167–182
49. Gutowska A, Jeong B, Jasionowski M (2001) Injectable gels for tissue engineering. Anat Rec 263:342–349
50. Lu JX, Prudhommeaux F, Meunier A, Sedel L, Guillemin G (1999) Effects of chitosan on rat knee cartilages. Biomaterials 20:1937–1944
51. Chenite A, Chaput C, Wang D, Combes C, Buschmann MD, Hoemann CD, Leroux JC, Atkinson BL, Binette F, Selmani A (2000) Novel injectable neutral solutions of chitosan form biodegradable gels in situ. Biomaterials 21:2155–2161
52. Huang H, Zhang X, Hu X, Dai L, Zhu J, Man Z, Chen H, Zhou C, Ao Y (2014) Directing chondrogenic differentiation of mesenchymal stem cells with a solid-supported chitosan thermogel for cartilage tissue engineering. Biomed Mater 9:035008
53. Cheng YH, Yang SH, Su WY, Chen YC, Yang KC, Cheng WT, Wu SC, Lin FH (2010) Thermosensitive chitosan-gelatin-glycerol phosphate hydrogels as a cell carrier for nucleus pulposus regeneration: An in vitro study. Tissue Eng Part A 16:695–703
54. Zhu M, Zhu Y, Zhang L, Shi J (2013) Preparation of chitosan/mesoporous silica nanoparticle composite hydrogels for sustained co-delivery of biomacromolecules and small chemical drugs. Sci Technol Adv Mater 14:045005
55. Cui Z, Wright LD, Guzzo R, Freeman JW, Drissi H, Nair LS (2013) Poly(d-lactide)/poly(caprolactone) nanofiber-thermogelling chitosan gel composite scaffolds for osteochondral tissue regeneration in a rat model. J Bioact Compat Polym 28:115–125
56. Zheng L, Lu HQ, Fan HS, Zhang XD (2013) Reinforcement and chemical cross-linking in collagen-based scaffolds in cartilage tissue engineering: A comparative study. Iran Polym J 22:833–842
57. Wang PY, Tsai WB (2013) Modulation of the proliferation and matrix synthesis of chondrocytes by dynamic compression on genipin-crosslinked chitosan/collagen scaffolds. J Biomater Sci-Polym Ed 24:507–519
58. Sarem M, Moztarzadeh F, Mozafari M (2013) How can genipin assist gelatin/carbohydrate chitosan scaffolds to act as replacements of load-bearing soft tissues? Carbohydr Polym 93:635–643
59. Fabela-Sanchez O, Zarate-Trivino DG, Elizalde-Pena EA, Garcia-Carvajal Z, Sanchez IC, Parra-Cid C, Gomez-Garcia R, Ibarra C, Garcia-Gaitan B, Zavala R, Guevara-Olvera L, Villasenor-Ortega F, Munoz-Sanchez CI, Perez-Perez C, Herrera-Perez S, Velasquillo C, Luna-Barcenas G, Solis L (2009) Mammalian Cell Culture on a Novel Chitosan-Based Biomaterial Crosslinked with Gluteraldehyde. Macromol Symp 283–284:181–190
60. Chen H, Liu Y, Jiang Z, Chen W, Yu Y, Hu Q (2014) Cell-scaffold interaction within engineered tissue. Exp Cell Res 323:346–351
61. Hong Y, Song H, Gong Y, Mao Z, Gao C, Shen J (2007) Covalently crosslinked chitosan hydrogel: properties of in vitro degradation and chondrocyte encapsulation. Acta Biomater 3:23–31
62. Zhu Y, Wan Y, Zhang J, Yin D, Cheng W (2014) Manufacture of layered collagen/chitosan-polycaprolactone scaffolds with biomimetic microarchitecture. Coll Surf B-Biointerfaces 113:352–360
63. Ghosh P, Rameshbabu AP, Dhara S (2014) Citrate cross-linked gels with strain reversibility and viscoelastic behavior accelerate healing of osteochondral defects in a rabbit model. Langmuir 30:8442–8451

64. Jin R, Moreira Teixeira LS, Dijkstra PJ, Karperien M, van Blitterswijk CA, Zhong ZY, Feijen J (2009) Injectable chitosan-based hydrogels for cartilage tissue engineering. Biomaterials 30:2544–2551
65. Whu SW, Hung KC, Hsieh KH, Chen CH, Tsai CL, Hsu SH (2013) In vitro and in vivo evaluation of chitosan-gelatin scaffolds for cartilage tissue engineering. Mater Sci Eng C-Mater Biol Appl 33:2855–2863
66. Hu J, Hou Y, Park H, Choi B, Hou S, Chung A, Lee M (2012) Visible light crosslinkable chitosan hydrogels for tissue engineering. Acta Biomater 8:1730–1738
67. Park H, Choi B, Hu J, Lee M (2013) Injectable chitosan hyaluronic acid hydrogels for cartilage tissue engineering. Acta Biomater 9:4779–4786
68. Hayami JWS, Waldman SD, Amsden BG (2013) Injectable, high modulus, and fatigue resistant composite scaffold for load-bearing soft tissue regeneration. Biomacromolecules 14:4236–4247
69. Yan J, Qi N, Zhang Q (2007) Rabbit articular chondrocytes seeded on collagen-chitosan-GAG scaffold for cartilage tissue engineering in vivo. Artif Cells Blood Substit Biotechnol 35:333–344
70. Wu YN, Yang Z, Hui JH, Ouyang HW, Lee EH (2007) Cartilaginous ECM component-modification of the micro-bead culture system for chondrogenic differentiation of mesenchymal stem cells. Biomaterials 28:4056–4067
71. Kuo YC, Hsu YR (2009) Tissue-engineered polyethylene oxide/chitosan scaffolds as potential substitutes for articular cartilage. J Biomed Mater Res Part A 91A:277–287
72. Kuo YC, Ku IN (2008) Cartilage regeneration by novel polyethylene oxide/chitin/chitosan scaffolds. Biomacromolecules 9:2662–2669
73. Li Z, Gunn J, Chen MH, Cooper A, Zhang M (2008) On-site alginate gelation for enhanced cell proliferation and uniform distribution in porous scaffolds. J Biomed Mater Res Part A 86A:552–559
74. Li ZS, Zhang MQ (2005) Chitosan-alginate as scaffolding material for cartilage tissue engineering. J Biomed Mater Res Part A 75A:485–493
75. Pound JC, Green DW, Roach HI, Mann S, Oreffo ROC (2007) An ex vivo model for chondrogenesis and osteogenesis. Biomaterials 28:2839–2849
76. Frenkel SR, Bradica G, Brekke JH, Goldman SM, Ieska K, Issack P, Bong MR, Tian H, Gokhale J, Coutts RD, Kronengold RT (2005) Regeneration of articular cartilage—Evaluation of osteochondral defect repair in the rabbit using multiphasic implants. Osteoarthr Cartil 13:798–807
77. Yan J, Li X, Liu L, Wang F, Zhu TW, Zhang Q (2006) Potential use of collagen-chitosan-hyaluronan tri-copolymer scaffold for cartilage tissue engineering. Artif Cells Blood Substit Immobil Biotechnol 34:27–39
78. Wang PY, Tsai WB (2013) Modulation of the proliferation and matrix synthesis of chondrocytes by dynamic compression on genipin-crosslinked chitosan/collagen scaffolds. J Biomater Sci Polym Ed 24:507–519
79. Guo T, Zhao JN, Chang JB, Ding Z, Hong H, Chen JN, Zhang JF (2006) Porous chitosan-gelatin scaffold containing plasmid DNA encoding transforming growth factor-beta 1 for chondrocytes proliferation. Biomaterials 27:1095–1103
80. Hsu SH, Whu SW, Hsieh SC, Tsai CL, Chen DC, Tan TS (2004) Evaluation of chitosan-alginate-hyaluronate complexes modified by an RGD-containing protein as tissue-engineering scaffolds for cartilage regeneration. Artif Organs 28:693–703
81. Yan JH, Li XM, Liu LR, Wang FJ, Zhu TW, Zhang QQ (2006) Potential use of collagen-chitosan-hyaluronan tri-copolymer scaffold for cartilage tissue engineering. Artif Cells Blood Substit Biotechnol 34:27–39
82. Tan H, Gong Y, Lao L, Mao Z, Gao C (2007) Gelatin/chitosan/hyaluronan ternary complex scaffold containing basic fibroblast growth factor for cartilage tissue engineering. J Mater Sci-Mater Med 18:1961–1968

83. Iwasaki N, Yamane ST, Majima T, Kasahara Y, Minami A, Harada K, Nonaka S, Maekawa N, Tamura H, Tokura S, Shiono M, Monde K, Nishimura SI (2004) Feasibility of polysaccharide hybrid materials for scaffolds in cartilage tissue engineering: Evaluation of chondrocyte adhesion to polyion complex fibers prepared from alginate and chitosan. Biomacromolecules 5:828–833
84. Yamane S, Iwasaki N, Majima T, Funakoshi T, Masuko T, Harada K, Minami A, Monde K, Nishimura S (2005) Feasibility of chitosan-based hyaluronic acid hybrid biomaterial for a novel scaffold in cartilage tissue engineering. Biomaterials 26:611–619
85. Shim IK, Lee SY, Park YJ, Lee MC, Lee SH, Lee JY, Lee SJ (2008) Homogeneous chitosan-PLGA composite fibrous scaffolds for tissue regeneration. J Biomed Mater Res Part A 84A:247–255
86. Noriega SE, Hasanova GI, Schneider MJ, Larsen GF, Subramanian A (2012) Effect of fiber diameter on the spreading, proliferation and differentiation of chondrocytes on electrospun chitosan matrices. Cells Tissues Organs 195:207–221
87. Neves SC, Teixeira LSM, Moroni L, Reis RL, Van Blitterswijk CA, Alves NM, Karperien M, Mano JF (2011) Chitosan/Poly(epsilon-caprolactone) blend scaffolds for cartilage repair. Biomaterials 32:1068–1079
88. Bhardwaj N, Kundu SC (2011) Silk fibroin protein and chitosan polyelectrolyte complex porous scaffolds for tissue engineering applications. Carbohydr Polym 85:325–333
89. Saengthong S, Piroonpan T, Tangthong T, Pasanphan W (2014) Fabrication of microporous chitosan/silk fibroin as a scaffold material using electron beam. Macromol Res 22:717–724
90. Deng J, She R, Huang W, Dong Z, Mo G, Liu B (2013) A silk fibroin/chitosan scaffold in combination with bone marrow-derived mesenchymal stem cells to repair cartilage defects in the rabbit knee. J Mater Sci-Mater Med 24:2037–2046
91. Silva SS, Motta A, Rodrigues MT, Pinheiro AFM, Gomes ME, Mano JF, Reis RL, Migliaresi C (2008) Novel genipin-cross-linked chitosan/silk fibroin sponges for cartilage engineering strategies. Biomacromolecules 9:2764–2774
92. Zang M, Zhang Q, Davis G, Huang G, Jaffari M, Rios CN, Gupta V, Yu P, Mathur AB (2011) Perichondrium directed cartilage formation in silk fibroin and chitosan blend scaffolds for tracheal transplantation. Acta Biomater 7:3422–3431
93. She Z, Zhang B, Jin C, Feng Q, Xu Y (2008) Preparation and in vitro degradation of porous three-dimensional silk fibroin/chitosan scaffold. Polym Degrad Stab 93:1316–1322
94. Mirahmadi F, Tafazzoli-Shadpour M, Shokrgozar MA, Bonakdar S (2013) Enhanced mechanical properties of thermosensitive chitosan hydrogel by silk fibers for cartilage tissue engineering. Mater Sci Eng C, Mater Biol Appl 33:4786–4794
95. Waldman SD, Couto DC, Grynpas MD, Pilliar RM, Kandel RA (2007) Multi-axial mechanical stimulation of tissue engineered cartilage: Review. Eur Cells Mater 13:66–73; discussion 73–74
96. Li J, Zhao Q, Wang E, Zhang C, Wang G, Yuan Q (2012) Dynamic compression of rabbit adipose-derived stem cells transfected with insulin-like growth factor 1 in chitosan/gelatin scaffolds induces chondrogenesis and matrix biosynthesis. J Cell Physiol 227:2003–2012
97. Wang PY, Chow HH, Lai JY, Liu HL, Tsai WB (2009) Dynamic compression modulates chondrocyte proliferation and matrix biosynthesis in chitosan/gelatin scaffolds. J Biomed Mater Res Part B-Appl Biomater 91B:143–152
98. Jung Y, Kim SH, Kim SH, Kim YH, Xie J, Matsuda T, Min BG (2008) Cartilaginous tissue formation using a mechano-active scaffold and dynamic compressive stimulation. J Biomater Sci Polym Ed 19:61–74
99. Pp BM, Pedro AJ, Peterbauer A, Gabriel C, Redl H, Reis RL (2005) Chitosan particles agglomerated scaffolds for cartilage and osteochondral tissue engineering approaches with adipose tissue derived stem cells. J Mater Sci-Mater Med 16:1077–1085
100. Malafaya PB, Oliveira JT, Reis RL (2010) The effect of insulin-loaded chitosan particle-aggregated scaffolds in chondrogenic differentiation. Tissue Eng Part A 16:735–747

101. Malafaya PB, Santos TC, van Griensven M, Reis RL (2008) Morphology, mechanical characterization and in vivo neo-vascularization of chitosan particle aggregated scaffolds architectures. Biomaterials 29:3914–3926
102. Marchand C, Rivard GE, Sun J, Hoemann CD (2009) Solidification mechanisms of chitosan-glycerol phosphate/blood implant for articular cartilage repair. Osteoarthr Cartil/OARS, Osteoarthr Res Soc 17:953–960
103. Hoemann CD, Hurtig M, Rossomacha E, Sun J, Chevrier A, Shive MS, Buschmann MD (2005) Chitosan-glycerol phosphate/blood implants improve hyaline cartilage repair in ovine microfracture defects. J Bone Joint Surg Am 87:2671–2686
104. Lafantaisie-Favreau CH, Guzman-Morales J, Sun J, Chen G, Harris A, Smith TD, Carli A, Henderson J, Stanish WD, Hoemann CD (2013) Subchondral pre-solidified chitosan/blood implants elicit reproducible early osteochondral wound-repair responses including neutrophil and stromal cell chemotaxis, bone resorption and repair, enhanced repair tissue integration and delayed matrix deposition. BMC Musculoskelet Disord 14:27
105. Marchand C, Chen G, Tran-Khanh N, Sun J, Chen H, Buschmann MD, Hoemann CD (2012) Microdrilled cartilage defects treated with thrombin-solidified chitosan/blood implant regenerate a more hyaline, stable, and structurally integrated osteochondral unit compared to drilled controls. Tissue Eng Part A 18:508–519
106. Chevrier A, Hoemann CD, Sun J, Buschmann MD (2011) Temporal and spatial modulation of chondrogenic foci in subchondral microdrill holes by chitosan-glycerol phosphate/blood implants. Osteoarthr Cartil/OARS, Osteoarthr Res Soc 19:136–144
107. Hoemann CD, Sun J, McKee MD, Chevrier A, Rossomacha E, Rivard GE, Hurtig M, Buschmann MD (2007) Chitosan-glycerol phosphate/blood implants elicit hyaline cartilage repair integrated with porous subchondral bone in microdrilled rabbit defects. Osteoarthr Cartil/OARS, Osteoarthr Res Soc 15:78–89
108. Chevrier A, Hoemann CD, Sun J, Buschmann MD (2007) Chitosan-glycerol phosphate/blood implants increase cell recruitment, transient vascularization and subchondral bone remodeling in drilled cartilage defects. Osteoarthr Cartil/OARS, Osteoarthr Res Soc 15:316–327
109. Lozinsky VI, Galaev IY, Plieva FM, Savina IN, Jungvid H, Mattiasson B (2003) Polymeric cryogels as promising materials of biotechnological interest. Trends Biotechnol 21:445–451
110. Kathuria N, Tripathi A, Kar KK, Kumar A (2009) Synthesis and characterization of elastic and macroporous chitosan-gelatin cryogels for tissue engineering. Acta Biomater 5:406–418
111. Kuo CY, Chen CH, Hsiao CY, Chen JP (2015) Incorporation of chitosan in biomimetic gelatin/chondroitin-6-sulfate/hyaluronan cryogel for cartilage tissue engineering. Carbohydr Polym 117:722–730
112. Bhat S, Tripathi A, Kumar A (2011) Supermacroprous chitosan-agarose-gelatin cryogels: In vitro characterization and in vivo assessment for cartilage tissue engineering. J Royal Soc, Interf/Royal Soc 8:540–554
113. Magalhaes J, Lebourg M, Deplaine H, Gomez Ribelles JL, Blanco FJ (2015) Effect of the physicochemical properties of pure or chitosan-coated poly(L-lactic acid)scaffolds on the chondrogenic differentiation of mesenchymal stem cells from osteoarthritic patients. Tissue Eng Part A 21:716–728

Composite Chitosan-Calcium Phosphate Scaffolds for Cartilage Tissue Engineering

Anuhya Gottipati and Steven H. Elder

Abstract Cartilage covering the articulating surfaces of bones in diarthrodial joints provides for almost frictionless motion. If this tissue is damaged either due to traumatic injury or disease, it lacks the ability of self-repair. The goal of cartilage tissue engineering is to regenerate healthy hyaline cartilage by combining chondrocytes or stem cells with a variety of natural and synthetic scaffold materials. Chitin is one of the most abundant naturally occurring polysaccharides. Its deacetylated derivative chitosan is biocompatible, biodegradable, and may retain functional characteristics that promote site-appropriate tissue reconstruction. This chapter includes methods for fabrication of chitosan-calcium phosphate (CHI–CaP) composite scaffolds, scaffold physical characteristics, as well as techniques for creation of cartilage/CHI–CaP biphasic constructs. Coating CHI–CaP scaffolds with type I collagen facilitates formation of a continuous layer of neocartilage with approximately uniform thickness over the cell-seeded area. Aspects of the scaffold's degradation are also discussed.

Keywords Chitosan · Calcium phosphate · Biphasic constructs · Cartilage tissue engineering

1 Articular Cartilage

Articular cartilage, also known as hyaline cartilage, is one of the three types of cartilages found in the human body. It is the smooth, white tissue that covers the articular surface of bones in synovial joints. It absorbs shock, distributes load, and provides a smooth lubricating surface for nearly frictionless joint movements. The extracellular matrix of articular cartilage is mainly composed of water, collagen

A. Gottipati · S.H. Elder (✉)
Department of Agricultural and Biological Engineering, Mississippi State University, Starkville, Mississippi 39762, USA
e-mail: selder@abe.msstate.edu

P.K. Dutta (ed.), *Chitin and Chitosan for Regenerative Medicine*, Springer Series on Polymer and Composite Materials, DOI 10.1007/978-81-322-2511-9_4

type II, and the large, aggregating proteoglycan called aggrecan. The relative concentrations of these components, as well as their organization, vary with tissue depth from the articular surface. Compressive loading causes some matrix consolidation and fluid flow out of the tissue in the regions closest to the articular surface, which contributes to joint lubrication. Under loading of short duration, the tissue in the deeper region adjacent to bone experiences almost no deformation as load is supported by interstitial fluid pressure. The structure of the extracellular matrix must be preserved in order to maintain this functionality. Unfortunately, cartilage is susceptible to injury through a number of mechanisms, including trauma (e.g., patellar dislocation or hyperflexion) and congenital anatomical abnormality. Articular cartilage is devoid of blood vessels, lymphatics, and nerves. The normal wound-healing process is not provoked and the tissue has a limited capacity of intrinsic healing due to the absence of a vascular system. Not only can injuries to articular cartilage be symptomatic (e.g., pain and swelling), but also they can advance to osteoarthritis if left untreated. Osteoarthritis is one of the leading disabilities in United States, and there is no reliable treatment. Current therapies for articular cartilage lesions, such as microfracture, often result in the formation of mechanically inferior fibrocartilage. The emerging field of cartilage tissue engineering shows promising results in regenerating healthy hyaline cartilage, which can alleviate pain and restore the functions of normal tissue.

2 Tissue Engineering

The loss or failure of an organ or tissue is one of the most frequent, disastrous, and costly problems in human healthcare. A new field, tissue engineering, applies the principles of biology and engineering to the development of functional substitutes for damaged repair [1]. Most tissue engineering approaches involve the combination of two main components—cells to regenerate the lost or damaged tissue and scaffolds (or matrices) that organize the regenerated tissue in three dimensions. Tissue engineering is quite challenging where the engineered tissue should contain the extracellular matrix components of the target native tissue in similar proportions and structural arrangements and also provide similar biomechanical functions. Major research is being conducted to engineer cartilage tissue using various scaffold materials [2–4]. The typical cartilage tissue engineering strategy is to form new tissue and allow for some maturation in vitro so that it will be able to withstand joint loading upon implantation. In this chapter, we will discuss the formation of cartilage by chondrocytes and marrow-derived mesenchymal stem cells on a chitosan-calcium phosphate (CHI–CaP) scaffold. The scaffold is intended to provide a mechanism for anchorage of the cartilage within the joint and also to serve as a platform for regeneration of subchondral bone. The cartilage itself is formed through self-assembly on top of, rather than within, the CHI–CaP phase. Thus, we refer to a biphasic construct, where the CHI–CaP phase is designed to support the formation of bone in vivo and the cells constitute the second phase and form new cartilage in vitro.

3 Chitosan-Calcium Phosphate Scaffold

Chitosan is a deacetylated derivative of chitin, which is found in the exoskeleton of marine crustaceans. It is a biocompatible and osteoconductive polymer with enhanced wound-healing capability. Chitosan is biodegradable and even the oligosaccharide degradation products, liberated primarily by enzymatic hydrolysis of the acetylated residues, do not cause any damage. Chitosan exhibits antimicrobial properties which could reduce the potential for bacterial infection upon implantation [5, 6]. Chitosan has been widely used in various biomedical applications—tissue engineering, wound dressing, drug delivery, and cancer diagnosis [7]. Previous research also revealed that chitosan has hemostatic and cholesterol lowering properties [8]. Furthermore the material properties of chitosan can be regulated by varying the molecular weight and degree of deacetylation (DDA) [9–11]. Combining chitosan with calcium phosphate, a calcium salt found in the inorganic phase of bone mainly as hydroxyapatite ($Ca_{10}(PO_4)_6(OH)_2$), improves osteoconductivity, strength, and hardness of the scaffold [12]. With all these advantages added to chitosan, it was used as a scaffold material for cartilage tissue engineering [13, 14]. Chitosan and its mixtures have been shown to support cell attachment for various cell types including human fetal osteoblast cells [15], human embryonic palatal mesenchymal cells [16], chondrocytes [17], and human umbilical cord stem cells [18]. We have previously demonstrated that composite CHI–CaP scaffolds can support attachment and chondrogenesis of chondrocytes and mesenchymal stem cells [19].

Composite CHI–CaP scaffolds are fabricated using a co-precipitation method [16]. Chitosan solution is prepared by dissolving 3.57 gm of 78.7 % DDA chitosan powder (Vanson Halosource, Remond, WA) in 84 ml of 2 wt% acetic acid. This chitosan solution is left for overnight on an electromagnetic stirrer to obtain a homogeneous solution. 10 ml of 1 M $CaCl_2$ in 2 % acetic acid and 6 ml of 1 M NaH_2PO_4 in 2 % acetic acid is added to the chitosan solution to achieve a final Ca:P ratio of 1.67, which is approximately the Ca:P ratio in healthy bone [20]. The precipitation solution is made of 20 % NaOH, 30 % methanol, and 50 % DI water at a pH 13. The chitosan solution is added drop-wise to the stirred precipitation solution by means of a syringe pump (30 cc syringe with 18 G needle) at rate of 15 ml per hour as shown in Fig. 1. Each drop forms an individual CHI–CaP bead. The beads are left in stirring in the precipitating solution for 24 h to allow for the formation of crystalline hydroxyapatite as proposed by Rusu et al. [21].

$$10CaHPO_4 + 12OH^- \rightarrow Ca_{10}(PO_4)_6(OH)_2 + 10H_2O + 4PO_4^{3-}$$
$$PO_4^{3-} + ACP + OH^- \rightarrow Ca_{10}(PO_4)_6(OH)_2$$

After 24 h the precipitation solution is replaced by DI water. The beads are washed regularly in DI water until the pH reaches 7. Transmission electron microscopy has been used to demonstrate that CaP crystal formation is evident within 10 min of stirring in the precipitating solution (Fig. 1). This CaP deposition was similar to those shown by Chesnutt et al. [16].

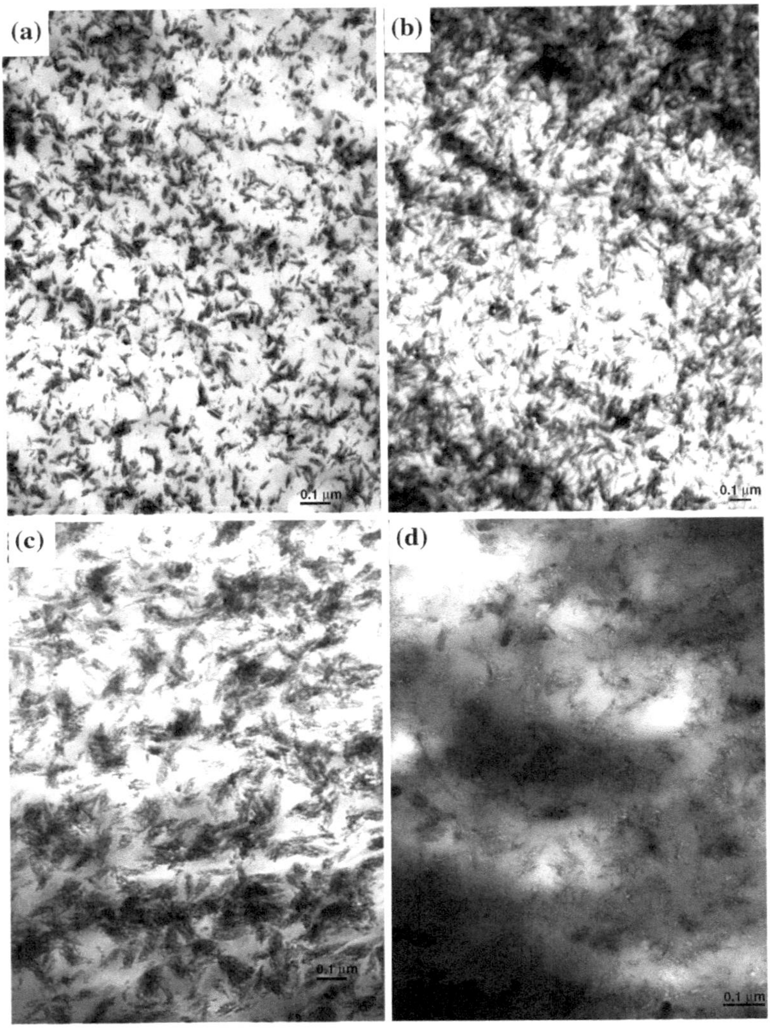

Fig. 1 Transmission electron micrograph of CHI–CaP scaffolds showing the presence of calcium phosphate crystals at **a** 10,000× after 10 min in precipitate solution; **b** 10,000× after 24 h in precipitate solution; **c** 50,000× after 10 min in precipitate solution; **d** 50,000× after 24 h in precipitate solution

Neutralized beads are spaced apart on a Pyrex dish and air dried overnight at room temperature. As they dry, they shrink to about 60 % of their original size. Once dry, the beads formed in this are typically 800–900 μm in diameter. The size of the beads can be varied by varying the size of CHI–CaP drop (Fig. 2). This can be achieved by changing needle sizes and/or by focusing a jet of air directly at the tip of the needle in order to dislodge each CHI–CaP droplet from the needle before

Fig. 2 CHI–CaP beads of various sizes

it grows large enough to fall under its own weight. Dried beads are packed into custom cylindrical molds (Fig. 3). The fixture consists of two separate plates, a solid flat plate clamped to another 7 mm-thick plate containing an array of 6.5 mm diameter holes. Beads are packed into the holes and fused into scaffolds by brief exposure to 1 % acetic acid for 30 s under manually applied pressure. Scaffolds are immediately rinsed with DI water to remove residual acetic acid and then allowed to air dry for 24 h at room temperature. They are then frozen at −20°C and lyophilized overnight. Previous study has shown that lyophilization enhances porosity, surface texture, and protein adsorption [12]. Lyophilized scaffolds can be sterilized using ethylene oxide. Incubation of the scaffolds in culture medium for up to 28 days had no noticeable effect on their compressive modulus of approximately 5 MPa, which approaches the lower range of modulus for human trabecular bone, as shown in Fig. 4.

Solid composite disks can be created if a smooth surface is desired, for example to measure the water contact angle. They have the same composition as porous scaffolds and are fabricated by adding the chitosan-phosphate solution to a transwell culture insert and exposing the top and bottom surfaces to the precipitation solution in a 6-well plate. The solidified disk is held in a custom sandwich press during drying to prevent wrapping.

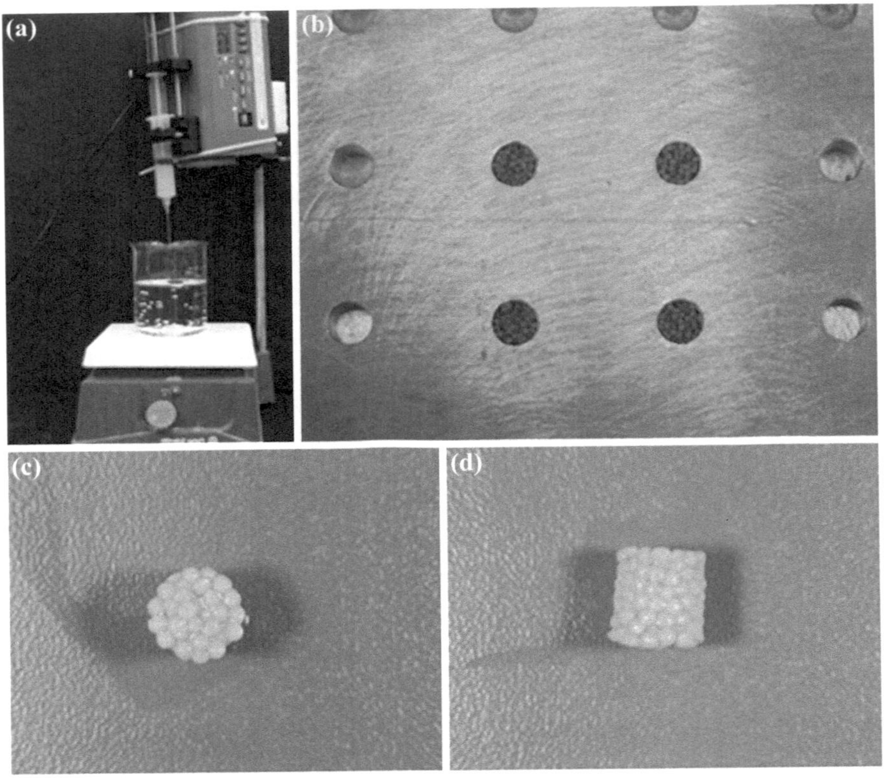

Fig. 3 CHI–CaP scaffold fabrication. **a** Formation of CHI–CaP beads in precipitate solution; **b** fusing CHI–CaP beads in a cylindrical mold to form scaffolds; **c** top view rehydrated scaffold; **d** side view of rehydrated scaffold

4 Formation of Biphasic Constructs

Two techniques for creating biphasic constructs have been investigated. The first technique involved the use of a transwell tissue culture insert and placement of the scaffold on top of a layer of cells. A suspension containing 10^7 first passage primary porcine chondrocytes was pipetted into a Ø6.5 mm transwell culture insert with polyethylene terephthalate filter and 3 μm pores (Millicell Hanging Cell Culture Insert, Millipore, Billerica, MA). They were cultured in defined chondrogenic medium: high glucose DMEM containing 1 % ITS + Premix (BD Biosciences, San Jose, CA), 0.1 mM dexamethasone, 50 μg/mL ascorbate-2 phosphate, 1 mM sodium pyruvate, 40 μg/mL L-proline, 1 % antibiotic-antimycotic solution (Sigma-Aldrich, St. Louis, MO), and 10 ng/ml human recombinant transforming growth factor-β3 (PeproTech, Rock Hill, NJ) [22]. After 14 days of culture, additional 10^7 chondrocytes were added to each insert. As soon as the cells

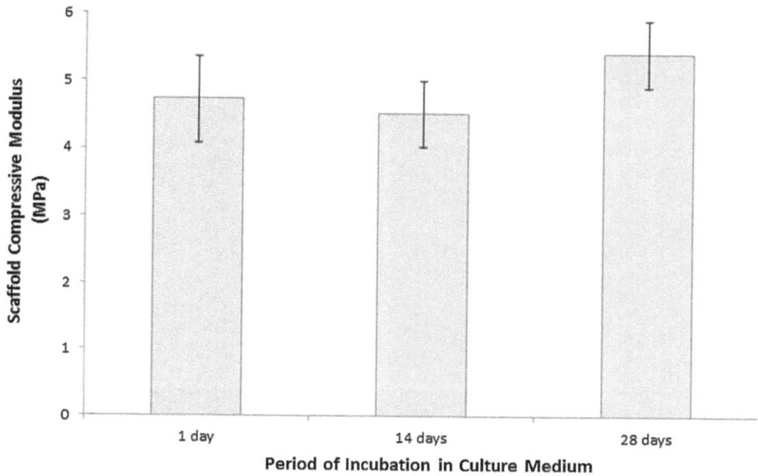

Fig. 4 Compressive modulus of rehydrated scaffolds in unconfined compression ($n = 5$). Differences among groups were not statistically significant by one-way ANOVA ($p > 0.05$)

had settled, a CHI–CaP scaffold which had been rehydrated for 24 h in DMEM and 10 % fetal bovine serum was gently pressed into the bottom of the insert (Fig. 5a). The constructs were cultured for an additional 42 days to allow for tissue maturation and integration with the scaffold. The cells formed a dense, hyaline cartilage-like tissue which did not tightly adhere to CHI–CaP scaffold (Fig. 6). It required minimal force to completely detach the cartilage layer from the scaffold. We obtained similar results using porcine bone marrow mesenchymal stem cells and a similar technique [19]. Thus, an alternative technique designed to promote better scaffold tissue-scaffold integration was devised.

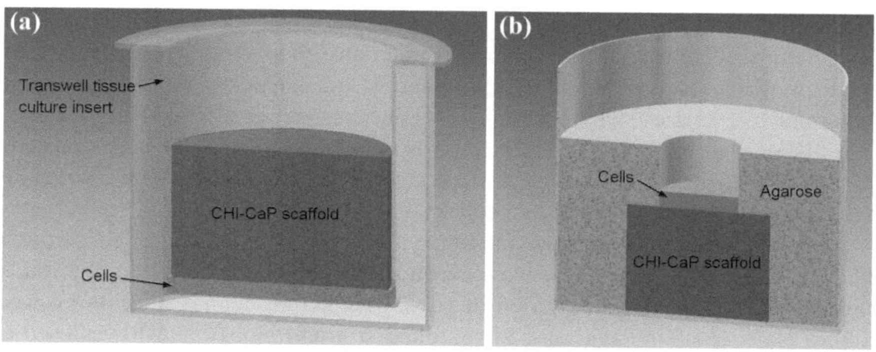

Fig. 5 Schematic diagram showing the formation of biphasic constructs using **a** articular chondrocytes in a Millicell hanging transwell and **b** human mesenchymal stem cells seeded on top of a scaffold resting in agarose well

Fig. 6 Biphasic constructs formed using porcine chondrocytes and CHI–CaP scaffolds in a Millicell transwell. **a** A scaffold free layer of cartilage tissue on a CHI–CaP scaffold; **b** top and **c** side view of cartilage tissue and scaffold

The second technique involves seeding the cells directly on top of the CHI–CaP scaffold. Agarose is used to confine the cells to an area on the scaffold's flat surface. In addition, collagen type I coating of the scaffold was explored as a means to further improve cell adhesion. Collagen is roughly a third of the total body protein, and abundant in cartilage, bone, blood vessels, skin, and many other tissues. It has both mechanical and physiological functions. Stem cells have receptors for collagen attachment. $\alpha 2\beta 1$ and $\alpha 11\beta 1$ integrins, transmembrane proteins, play an important role in mesenchymal stem cell adhesion and proliferation on collagen type I [23]. The scaffolds were fabricated as mentioned earlier and were soaked in 0.05 % of collagen type I in PBS for 3 h and air dried overnight. These scaffolds were then EtO gas sterilized before cell seeding.

Frozen human bone marrow mesenchymal stem cells of passage 1 from a 22 year-old healthy male donor were expanded in monolayer culture to the fourth passage. Dry scaffolds were buried in 1.5 % low gelling temperature in a 6-well culture plate. Once the agarose had solidified, a 4 mm biopsy punch was used to cut a cylindrical-shaped hole right above the scaffold (Fig. 5b). The overlying agarose was removed by Pasteur pipette under vacuum. Approximately, 4×10^6 cells were pipetted onto each scaffold. They were allowed to settle for 60 min before

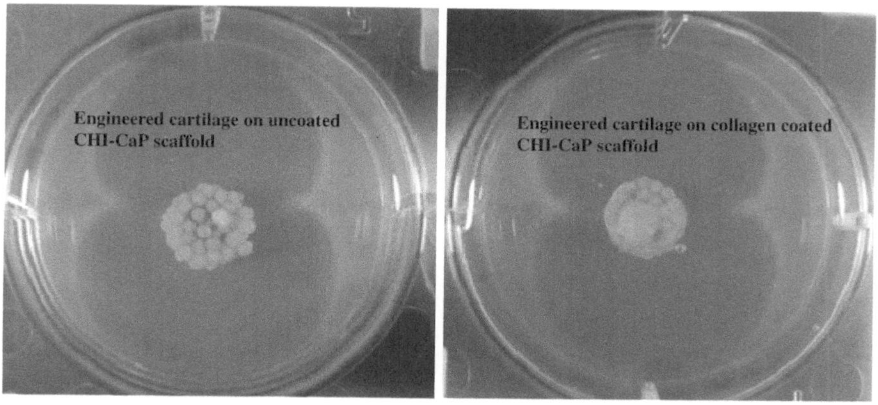

Fig. 7 Engineered cartilage tissue formed using human mesenchymal stem cells after culturing for 28 days in DCM on uncoated CHI–CaP scaffold got contracted into a circular shape whereas the tissue on collagen-coated CHI–CaP scaffold, extended into the pores of the scaffold

the wells were flooded with medium. The constructs were cultured for 28 days in defined chondrogenic medium. A tissue similar to hyaline cartilage was formed on all scaffolds. The tissue on collagen-coated scaffolds extended between the pores, covered the initial cell-seeded area, and was also well integrated with the CHI–CaP scaffold (Fig. 7). However, the tissue formed on uncoated scaffolds was contracted into a bolus at one place. The area of the tissue covered on collagen-coated scaffolds (11.09 ± 2.25 mm^2) was five times more than the area covered on uncoated scaffolds (2.76 ± 0.38 mm^2).

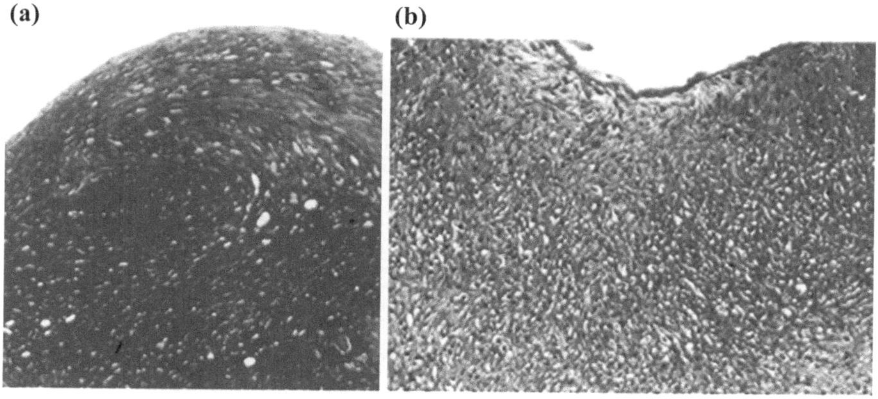

Fig. 8 Toluidine blue staining of histological sections of engineered cartilage formed from human mesenchymal stem cells on top of **a** uncoated and **b** collagen-coated CHI–CaP scaffolds at 10× magnification

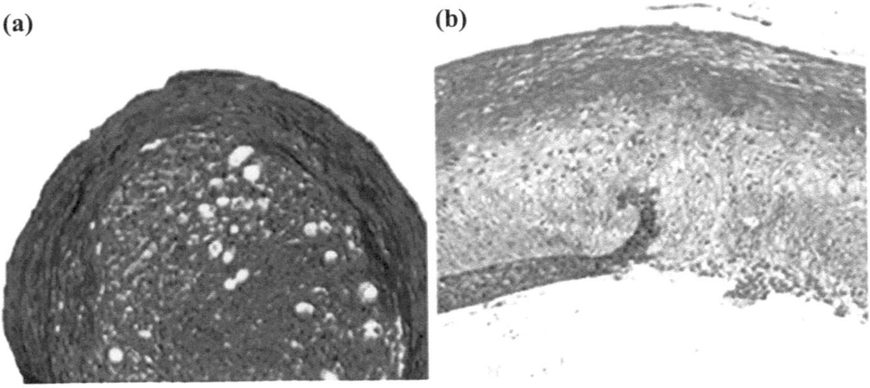

Fig. 9 Picrosirius red staining of histological sections of engineered cartilage formed from human mesenchymal stem cells on top of **a** uncoated and **b** collagen-coated CHI–CaP scaffolds at 5× magnification. Collagen fiber orientation varies with depth of the tissue formed on collagen-coated scaffold

Formalin fixed biphasic constructs were processed for histology. Paraffin- and plastic-embedded sections were stained with toluidine blue for proteoglycan detection and with picro-sirius red to stain collagen. The tissue on both collagen-coated and uncoated scaffolds demonstrated strong metachromatic toluidine blue staining, especially in the deep zone (Fig. 8). The tissue contained abundant collagen (Fig. 9), but immunostaining revealed that only the collagen in the coated constructs contained a much higher proportion of collagen type II. Furthermore, the collagen in the tissue on coated constructs displayed a preferential fiber alignment in the superficial and deep zones (0.30° vs. 95.5° with respect to the top surface, $p < 0.05$), a pattern which is similar to that observed in native articular cartilage.

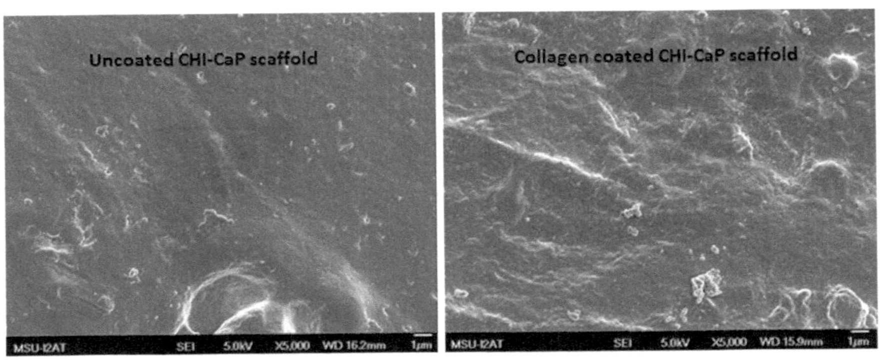

Fig. 10 Scanning electron micrograph of uncoated and collagen-coated CHI–CaP scaffold at 5,000× magnification

5 Collagen-Coated Scaffold Characteristics

Because coating the CHI–CaP scaffold with type I collagen had such a profound influence on cell adhesion and chondrogenesis, the effects of the coating were more thoroughly investigated. Scanning electron microscopy of CHI–CaP scaffolds showed a slightly bumpy surface, which was not significantly affected by collagen coating (Fig. 10). Porosity, as determined by the displacement of methanol, was approximately 37 % for both coated and uncoated scaffolds. Micro-computed tomography confirmed the interconnected porous structure (Fig. 11).

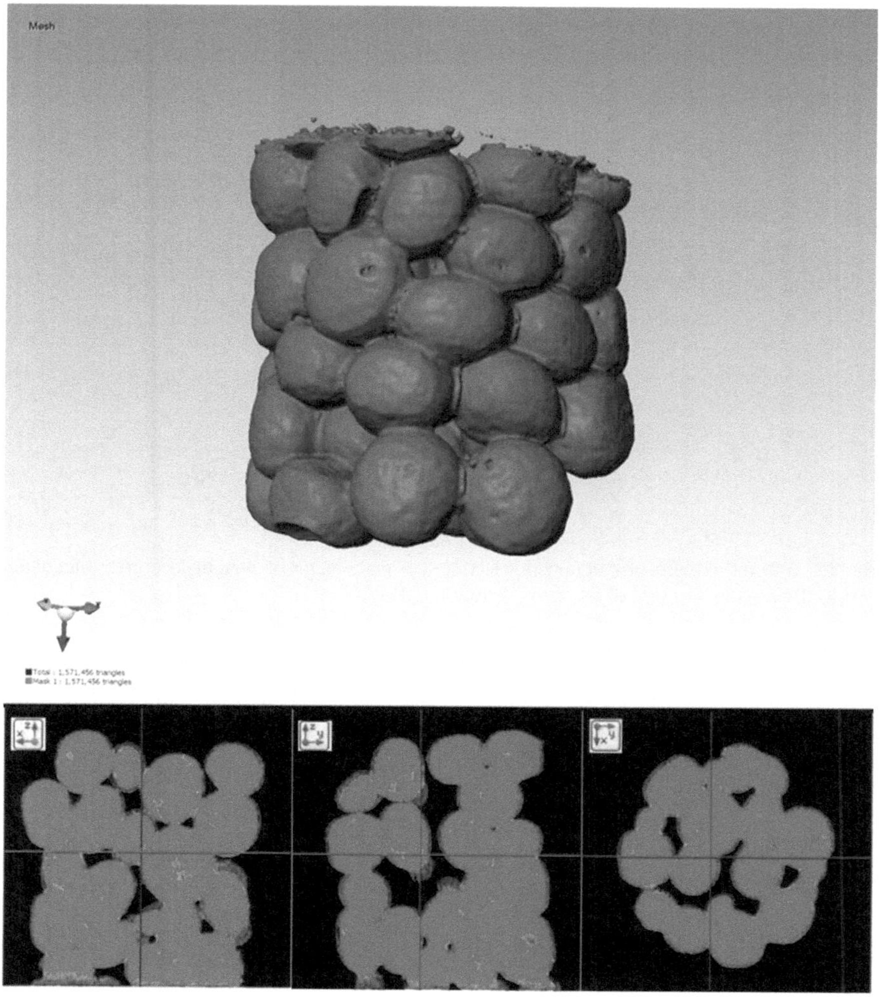

Fig. 11 Micro-CT image of CHI–CaP beads fused into a cylindrical-shaped scaffold

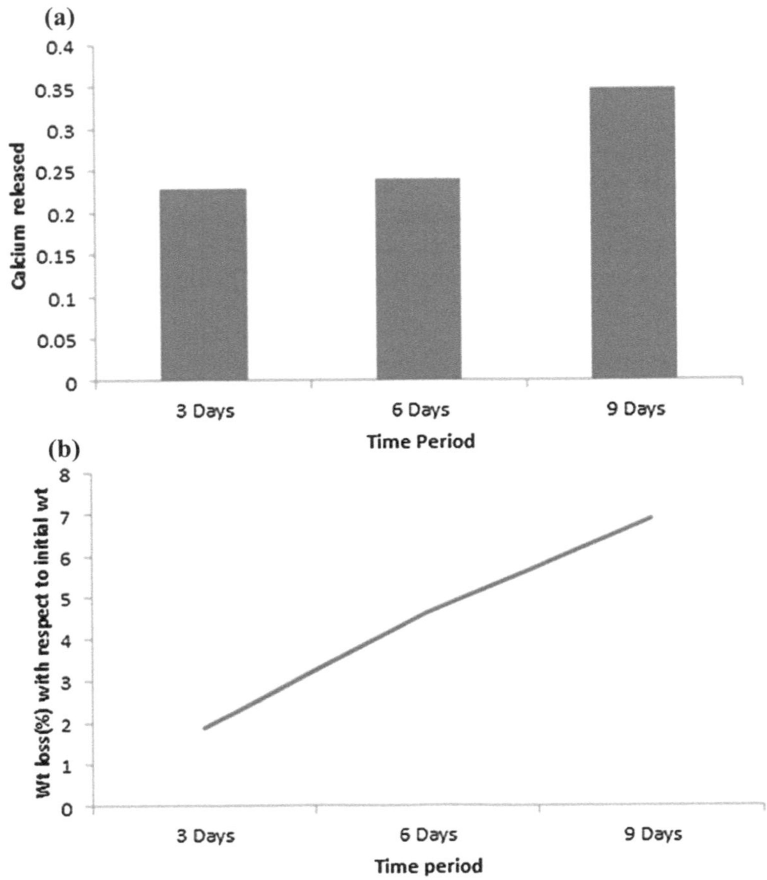

Fig. 12 **a** Total amount of calcium released and **b** percentage of weight loss after incubating CHI–CaP beads in 1 mg/ml of lysozyme solution in PBS ($n = 1$)

Swelling ratio was around 75 % and was likewise unaffected by coating. The water contact angle for collagen-coated scaffolds was more hydrophilic (78.20 ± 8.71) compared to uncoated scaffolds (107.81 ± 13.09).

6 Degradation

In vivo, the CHI–CaP scaffold would be embedded in bone and would ideally degrade at the same rate as new bone formation takes place. Lysozyme plays an important role in chitosan degradation. The rate of degradation will depend on

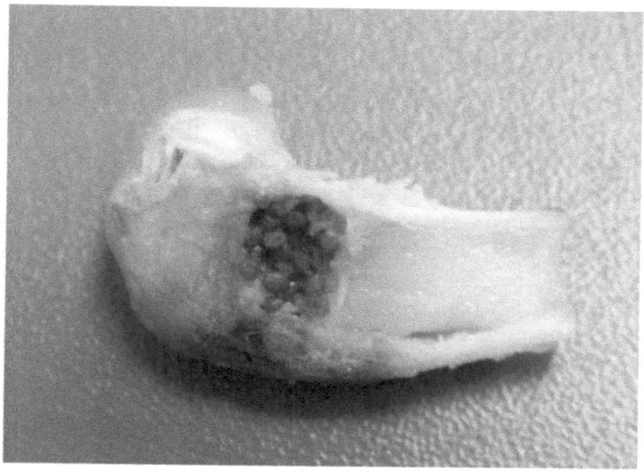

Fig. 13 CHI–CaP beads start falling apart after 12 weeks of implantation of biphasic constructs containing engineered cartilage on collagen-coated CHI–CaP scaffold in a rabbit knee defect

several factors like DDA, crystalline nature, and pore size [16]. CHI–CaP beads made from chitosan with a 77 % DDA were incubated in a 1 mg/ml lysozyme solution in PBS for up to 9 days. The rate of calcium release and weight loss are shown in Fig. 12. This translates to a very slow rate of degradation in vivo.

CHI–CaP scaffolds were tightly fit into surgically created defects in the stifle joints of skeletally mature rabbits. There were no obvious signs of degradation 12 weeks postoperatively. However, it did appear that many beads were no longer fused to the scaffolds, which might be an initial stage of degradation (Fig. 13).

7 Conclusion

Chitosan is a widely used scaffold material for tissue engineering. Our goal was to combine chitosan with calcium phosphate to enhance its strength and osteoconductivity. Microbeads were made by co-precipitating chitosan with calcium and phosphate. These beads were fused together to form cylindrically shaped scaffolds with sufficient mechanical strength and porosity for skeletal tissue engineering applications. Coating the scaffold with type I collagen made the surface hydrophilic and improved cell adhesion. Biphasic cartilage/CHI–CaP constructs were successfully created by high-density seeding of human bone marrow mesenchymal stem cells onto coated scaffolds. Observation of degradation in vivo, as well as in vitro experimentation using lysozyme, suggests that a slow rate of scaffold degradation is a current limitation of the composite scaffold.

References

1. Ahmed TA, Hincke MT (2014) Mesenchymal stem cell-based tissue engineering strategies for repair of articular cartilage. Histol Histopathol 29(6):669–689
2. Toh WS, Lee EH, Guo X-M et al (2010) Cartilagr repair using hyaluronan hydrogel-encapsulated human embryonic stem cell-derived chondrogenic cells,". Biomaterials 31(27):6968–6980
3. Popa E, Reis R, Gomes M (2012) Chondrogenic phenotype of different cells encapsulated in κ-carrageenan hydrogels for cartilage regeneration strategies. Biotechnol Appl Biochem 59(2):132–141
4. Kandel RA, Grynpas M, Pilliar R et al (2006) Repair of osteochondral defects with biphasic cartilage-calcium polyphosphate constructs in a sheep model. Biomaterials 27:4120–4131
5. Felt O, Carrel A, Baehni P et al (2000) Chitosan as a tear substitute: a wetting agent endowed with antimicrobial efficacy. J Ocul Pharmacol Ther 16(3):261–270
6. Tsai G, Su W (1999) Antibacterial activity of shrimp chitosan againt Escherichia coli. J Food Prot 62(3):239–243
7. Jayakumar R, Menon D, Manzoor K et al (2010) Biomedical applications of chitin and chitosan based nanomaterials—a short review. Carbohydr Polym 82(2):227–232
8. Zhang J, Xia W, Liu Ping et al (2010) Chitosan modification and pharmaceutical/biomedical applications. Open Access Marine Drugs 8:1962–1987
9. Majd S, Yuan Y, Mishra S et al (2009) Effects of material property and heat treatment on nanomechanical properties of chitosan films. J Biomed Mater Res B Appl Biomater 90(1):283–289
10. Tangsadthakun C, Kanokpanont S, Sanchavanakit N et al (2007) The influence of molecular weight of chitosan on the physical and biological properties of collagen/chitosan scaffolds. J Biomater Sci Polymer Ed 18(2):147–163
11. Nettles DL, Elder SH, Gilbert JA (2002) Potential use of chitosan as a cell scaffold material for cartilage tissue engineering. Tissue Eng 8(6):1009–1016
12. Reves B, Bumgardner J, Cole J, Yang Y et al (2009) Lyophilization to improve drug delivery for chitosan-calcium phosphate bone scaffold construct: a preliminary investigation. J Biomed Mater Res B Appl Biomater 90(1):1–10
13. Croisier F, Jerome C (2013) Chitosan-based biomaterials for tissue engineering. Eur Polymer J 49(4):780–792
14. Di Martino A, Sittinger M, Risbud MV (2005) Chitosan: a versatile biopolymer for orthopaedic tissue-engineering. Biomaterials 26(30):5983–5990
15. Chesnutt B, Yuan Y, Buddington K et al (2009) Composite chitosa/nano-hydroxyapatite scaffolds induce osteocalcin production by osteoblasts in vitro and support bone formation in vivo. Tissue Eng Part A 15(9):2571–2579
16. Chesnutt B, Viano A, Yuan Y et al (2009) Design and characterization of a novel chitosan/nanocrystalline calcium phosphate composite scaffold for bone regeneration. J Biomed Mater res A 88(2):491–502
17. Suh JK, Matthew HW (2000) Application of chitosan-based polysaccharide biomaterials in cartilage tissue engineering: a review. Biomaterials 21(24):2589–2598
18. Oliveira JM, Rodrigues MT, Silva SS et al (2006) Novel hydroxyapatita/chitosan bilayered scaffold for osteochondral tissue-engineering applications: Scaffold design and its performance when seed with goat bone marrow stromal cells. Biomaterials 27:6123–6137
19. Elder SH, Gottipati A, Zelenka H et al (2013) Attachment, proliferation, and chondroinduction of mesenchymal stem cells on porous chitosan calcium phosphate scaffolds. The Open orthopaedics Journal 7:275–281
20. Tzaphlidou Margaret (2008) Bone Architecture: Collagen Structure and Calcium/Phosphorus Maps. J Biol Phys 34:39–49

21. Rusu V et al (2005) Size-controlled hydroxyapatite nanoparticles as self-organized organic-inorganic composite materials. Biomaterials 26(6):5414–5426
22. Elder S, Cooley A, Borazjani A et al (2009) Production of hyaline-like cartilage by bone marrow mesenchymal stem cells in a self-assembly model. Tissue Eng Part A 15(10):3025–3036
23. Radic T, Haasters F (2011) Integrins α2β1 and α11β1 regulate the survival of mesenchymal stem cells on collagen I. Cell Death Dis 2(7) July 2011

Chitosan-Gelatin Composite Scaffolds in Bone Tissue Engineering

M. Nivedhitha Sundaram, S. Deepthi and R. Jayakumar

Abstract Regenerative medicine focuses on repair/replacement of the damaged tissue or organ in our body. This is done by growing cells on scaffold materials which help in its attachment, migration and proliferation. Chitosan being natural polymer has many unique properties such as being biocompatible, biodegradable and also has antibacterial and wound-healing abilities. Gelatin a derivative of collagen, which is widely present in our body, has been used as a composite with chitosan for promoting cell attachment, proliferation, and differentiation. Composite scaffolds have also shown better mechanical and functional properties because these composites are made of polymer and inorganic/organic blenders. Overall this review focuses on the role of chitosan-gelatin-based composite scaffolds in bone tissue engineering.

Keywords Chitosan · Gelatin · Nanocomposites · Bone regeneration · Scaffolds

1 Introduction

Natural bone has a hierarchical structure consisting of organic and inorganic materials that exist in the form of nanocomposites. Collagenous protein like Type I collagen forms the organic part of the bone and the inorganic part is composed of hydroxyapatite in its carbonated form. These collagen fibrils and carbonated hydroxyapatite are in the range of few nanometers (Fig. 1). The arrangement of these carbonated hydroxyapatites parallel to the arrangement of the collagen fibrils provides the required structural strength to the bone [1].

M. Nivedhitha Sundaram · S. Deepthi · R. Jayakumar (✉)
Amrita Center for Nanosciences and Molecular Medicine,
Amrita Institute of Medical Sciences and Research Centre, Kochi 682041, India
e-mail: rjayakumar@aims.amrita.edu; jayakumar77@yahoo.com

P.K. Dutta (ed.), *Chitin and Chitosan for Regenerative Medicine*, Springer Series on Polymer and Composite Materials, DOI 10.1007/978-81-322-2511-9_5

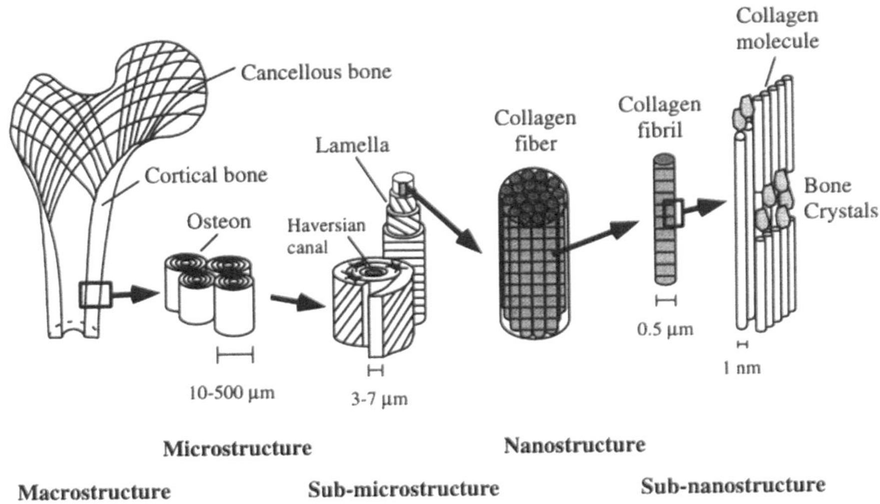

Fig. 1 Hierarchical structure of bone [2]

There are two types of bone structures whose properties mainly depend on the porosity of the bone. They are the cortical bone (compact) and cancellous bone (spongy) [3]. Unlike other tissues, bone has the unique property to heal when there is a tissue damage. An immobilized bone after fracture can heal spontaneously over time. But if there is a critical size defect where there is minimal tissue integrity the self-healing process does not take place properly. Bone grafts such as auto grafts and allografts have been in use for bone regeneration. But these grafts face few limitations such as its limited supply, immune rejection, and cost associated with the use of these grafts and also the possibility to transmit diseases [4].

Bone tissue engineering has come up as an alternative to allografts and autografts-based treatments. Tissue engineering helps regeneration of the damaged tissue at the site of damage by the use of constructs like scaffold along with cells and growth factors that will aid in tissue regeneration [5]. There are few criteria that a scaffold should satisfy in order to aid regeneration of the damaged tissue. This is mainly because the developed construct should be comparable to the native tissue that it replaces. The scaffold should be biocompatible and have good porosity for cells to attach, infiltrate, and proliferate. It should also have mechanical strength comparable to native tissue, good 3D shape, and biodegradability to help regenerate the damaged tissue [6].

Natural or synthetic materials are being subjected to different processing techniques for scaffold preparation. Synthetic polymers like polycaprolactone (PCL) [7], poly(lactic acid) (PLA) [8], and poly(lactic-co-glycolic acid) (PLGA) [9] and natural polymers like collagen [10], gelatin [11], and chitosan [12] are being used for bone regeneration applications. Natural polymers are mostly biodegradable, biocompatible, and they can easily mimic the extra cellular matrix of the damaged tissue. The main criterion that a material should satisfy to be used for bone regeneration is that the material should be an osteoconductive and osteoinductive material [13]. Bioceramic materials have shown good osteoconductive and osteoinductive properties that are mainly required for bone regeneration [14]. The ceramics that are commonly used for bone regeneration are bioactive ceramics such as hydroxyapatite (HAp) [15], tricalcium phosphate [16], and bioglass [17].

The main drawback in using bioceramics for bone regeneration is that they are very brittle and do not process the desired strength. Composite scaffolds which are developed by blending polymers and ceramics would help improve the mechanical stability and cell interaction property of the as-developed scaffold [18]. This composite scaffold overcomes the drawbacks faced by natural polymers and bioceramics. So this review mainly focuses on the contribution of natural polymer-based composite scaffolds for bone regeneration.

2 Chitosan

Chitosan, deacetylated form of chitin is extracted from crustaceans' exoskeletons and cell wall of fungi. Chitosan is a linear polysaccharide composed of β(1–4) linked glucosamine and *N*-acetyl glucosamine, whose ratio determines its degree of deacetylation (Fig. 2). Molecular weight of chitosan ranges from 30 to 1000 kDa and degree of deacetylation from 50 to 95 %; this mostly depends on its source and the method of preparation. Chitosan is soluble in acidic medium due to the protonation of its free amino group in the glucosamine component. Being

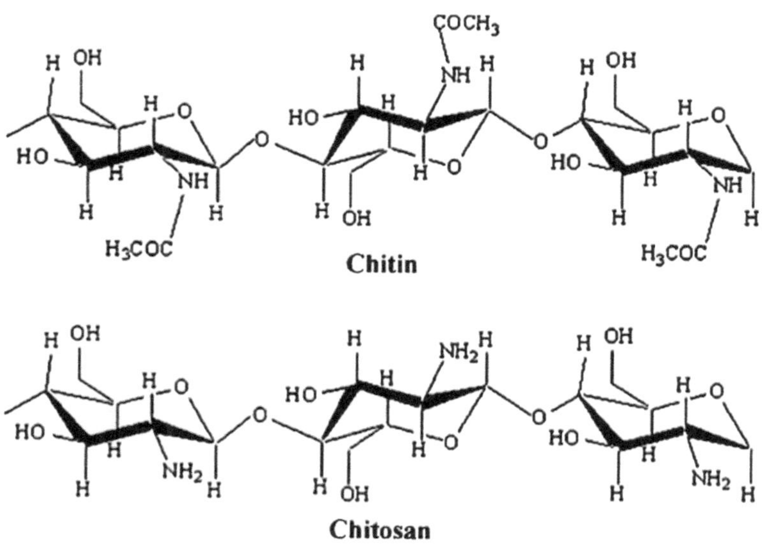

Fig. 2 Structure of chitin and chitosan [20]

cationic in nature chitosan electrostatically binds to glycosaminoglycan (GAG) molecules that are predominantly linked to many cytokines and growth factors present in our body [19].

It possesses various properties such as biocompatibility, biodegradability, due to its cationic nature it possesses hemostatic property that can help in wound healing. It has also got antimicrobial activity [20]. Chitosan together with other natural polymers and bioceramics has been in use for bone tissue regeneration.

Zhensheng et al. has developed chitosan/alginate scaffold for tissue engineering of bone. Highly porous chitosan/alginate scaffold was prepared by coacervation followed by liquid solid separation. It has been reported that addition of alginate to chitosan has increased the mechanical strength of the scaffold which is due to the stable bond formation between alginate and chitosan. In vivo study showed the deposition of calcified matrix after 4 weeks of scaffold implantation [21]. Qiao et al. has developed an injectable cell paste. This group has reported the development and implantation of an injectable calcium phosphate cement (CPC) containing alginate-chitosan-coated (AC) microspheres encapsulating osteoblast (MC3T3-E1) (Fig. 3). The construct (AC microcapsules in CPC) was injected into a nude mice at the dorsal subcutaneous area where it showed new bone formation, collagen deposition, and angiogenesis within 4 weeks. This study reports the use of chitosan-coated cell encapsulating microspheres for successful bone regeneration [22].

Calcium phosphate (CaP) has been in use for reconstruction of bone defects. Fabrication of CaP bioceramics with macro-porous that contain chitosan sponge has been reported by Zhang et al. CaP bioceramics of 300–600 microns was developed using porogen burn out technique and then the pores were filled with chitosan solution. This upon lyophilization yielded the final construct [23].

Osteoblastic cells were seeded on to the construct and cell attachment and differentiation was studied. The results showed that the ALP levels of the cells

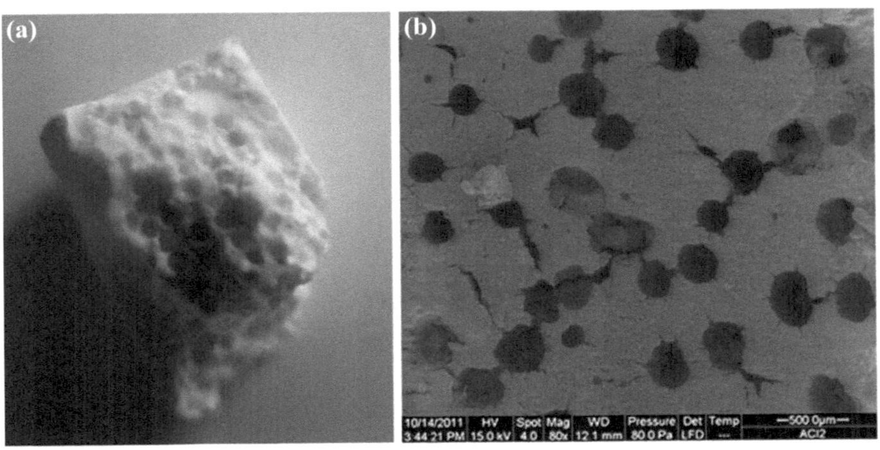

Fig. 3 AC microcapsules in CPC **a** Photographic image. **b** SEM image of the construct [22]

seeded on to the construct were comparable to the control (cells alone) proving its biocompatibility and osteoconductive nature due to the presence of CaP in the developed construct [23]. Composite scaffold of CaP-chitosan for bone regeneration has been developed by Zhang et al. Here CaP glass in soluble form is mixed with chitosan to form the composite scaffold. This construct has shown an enhanced level of ALP and osteocalcin within 7 days of culture compared to cells alone. It was also reported that the addition of CaP glass to chitosan has reduced its fast degradation leading to enhanced osteodifferentiation of cells seeded on the construct [24]. Tanase et al. have developed a CaP-chitosan scaffold by co-precipitation technique and have shown that the scaffold has got a compressive strength comparable to the native human trabecular bone. The scaffold was found to be biocompatible and has also caused an increase in the level of vascular endothelial growth factor (VEGF) during osteoblastic differentiation showing its feature of inducing osteogenesis and angiogenesis for an effective bone regeneration [25].

Growth factors play a major role in cell proliferation and differentiation. Growth factors incorporated in scaffold can aid in their control release leading to better cell proliferation. Transforming growth factor (TGF-β1) plays a major role in osteoblast cell proliferation. Lee et al. have developed a collagen/chitosan-micro-granule containing TGF-β1 for bone regeneration. The sustained release of this growth factor enhanced osteoblast proliferation. The ALP level was enhanced in the growth factor incorporated scaffold. In vitro cell attachment study has shown good attachment of cells on the micro-granules. In vivo study on rabbit calvarial defeat showed better bone regeneration within 4 weeks in the scaffold implanted site compared to control [26].

Park et al. have developed chitosan scaffold with platelet-derived growth factor-BB (PDGF-BB) for periodontal bone regeneration. PDGF-BB has been reported to enhance bone regeneration. This chitosan scaffold with PDGF-BB has shown better osteoblast proliferation due to the sustained release of this growth factor from the scaffold. In rat calvarial defeat, increased ossification was observed in the defeat site at 4 weeks [27]. This group has also developed a chitosan-chondroitin sulfate sponge containing PDGF-BB for guided bone regeneration [28]. Similarly, Lee et al. have developed a PDGF-BB-loaded chitosan-tricalcium phosphate scaffold which showed enhanced bone regeneration when implanted in vivo in a rat calvarial model [29]. Poly(epsilon-caprolactone) (PCL)/chitosan scaffold with PDGF-BB was developed by Im et al. and the results showed better healing of bone defeat. Chitosan increased the hydrophilicity of the scaffold [30].

Application of bioceramics in nano(n)-form such as the use of nanoparticles of CaP [31], hydroxyapatite (HAp) [32], and bioglass (BG) [7] for bone tissue engineering has increased in the past few years. Bioceramics in nanoparticulate form possess high surface area to volume ratio which improves its osteoconductive and osteoinductive nature. Sowjanya et al. have developed a chitosan/alginate/nSiO_2 composite scaffold for bone tissue engineering. Chitosan and alginate was blended with nSiO_2 and was freeze dried to make the composite scaffold. The porosity of the scaffold is about 20–100 microns. Due to the presence of nSiO_2, the composite scaffold showed better apatite formation on its surface [33]. Kavitha et al. developed a titanium-chitosan nano-composite scaffold of 4.5–10.5 nm using sol-gel technique. The developed nano-composite has good cell attachment, better apatite formation, and also shows antibacterial activity [34]. Nano-CaP (nCaP)-chitosan composite scaffold formed by fusion of microspheres was developed by Chesnutt et al. It was reported that better mechanical strength, cell attachment, and proliferation were observed in the composite scaffold due to the presence of nCaP [35].

HAp crystals are naturally present in the native bone. Scaffold containing HAp has improved the scaffold's osteoconductive property and hence aids in bone regeneration. Nanoparticles of hydroxyapatite (nHAp) due to its increased surface area to volume ratio have been widely used in bone tissue engineering. Thein-Han et al. have developed chitosan-nHAp composite scaffold of 50–120 microns porosity. This composite scaffold has increased compressive strength, better pre-osteoblast cell attachment, and proliferation than chitosan scaffold; thus, it acts as a potential candidate for bone regeneration [32]. Chestnutt et al. have reported that the chitosan-nHaP composite scaffold showed increase in osteocalcin level (late marker of osteoblastic differentiation) when human fetal osteoblasts were cultured for 21 days on the scaffold [36]. The composite scaffold has also reported to have better cell attachment.

Jiang et al. have proposed a spiral-cylindrical scaffold made of chitsan-cellulose containing nHAp, and in vivo studies have shown good cell attachment and infiltration leading to good osteointegration at the defeat site [37]. The efficiency of chitosan-nHAp and chitosan-micro-HAp scaffolds (Fig. 4) for bone regeneration was examined in vivo by Lee et al. Of the two scaffolds,

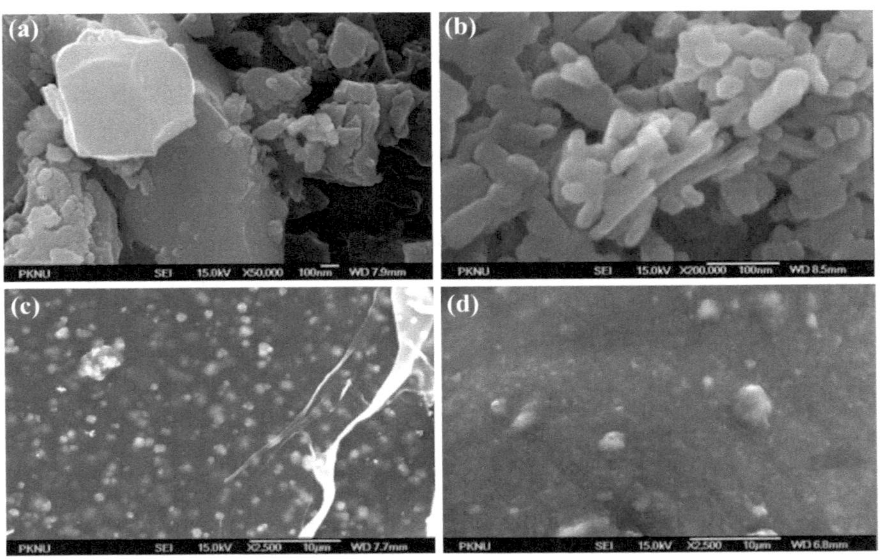

Fig. 4 SEM images of **a** micro-HAp **b** nHAp **c** chitosan-micro-Hap scaffold **d** chitosan-nHAp scaffold [38]

chitosan-nHAp scaffold showed more trabecular number, thickness, and volume proving that the HAp in nano-form has better osteoconductive property [38].

HAp-polymer composite scaffolds were developed by Xiao et al. [39]. In this study, poly(ε-caprolactone) (PCL)-chitosan scaffold containing nHAp—poly(vinyl alcohol) (PVA) was prepared. Similarly, Wang et al. evaluate the osteogenic differentiation of human umbilical cord mesenchymal stem cells in vitro and in vivo on nHaP-chitosan-poly(lactide-co-glycolide) scaffolds [40].

Venkatesan et al. have developed chitosan-HAp composite scaffold grafted with multi walled carbon nanotube (MWCNT-chitosan/HAp) for bone regeneration (Fig. 5). The composite scaffold showed interconnected porosity with uniform

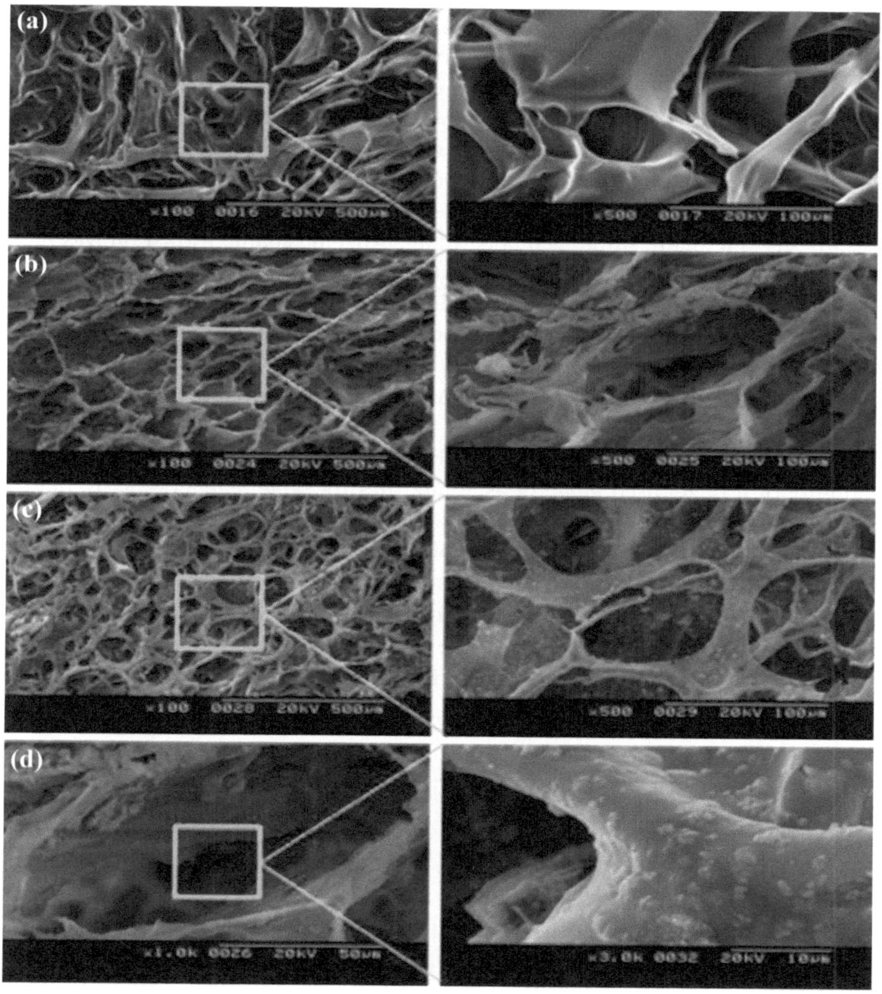

Fig. 5 SEM images of **a** chitosan scaffold **b** chitosan with HAp scaffold **c** MWCNT grafted chitosan-HAp scaffold **d** HAp dispersed in chitosan scaffold [41]

distribution of MWCNT and HAp which was confirmed by SEM. The water retention and degradation rates of the scaffold have decreased due to the addition of MWCNT. MG-63 cells attached and proliferated well on the composite scaffold showing its biocompatibility [41].

Electrospun membranes with bioceramics have shown promising results in bone regeneration. Mohammadi et al. developed nanofibrous membrane of PCL-PVA-chitosan hybrid scaffolds and observed the osteoblastic differentiation of mesenchymal stem cells on this scaffold [42]. Shalumon et al. have compared nano-fibrous scaffold (100–200 nm) of PCL-chitosan-incorporated nHap and nano-bioglass (nBG) for bone regeneration. Osteoblast cells cultured on this membrane showed enhanced ALP level. Of the two, PCL-chitosan nBG-incorporated nano-fibrous scaffold was found to be a better candidate for bone tissue engineering. Surface modification of bone implants with chitosan coating has been reported to increase HaP formation on these surfaces that would aid in better regeneration of bone [7].

Implants used for bone tissue engineering have been modified by surface coating to increase its efficiency in healing bone defeats. Sutha et al. have coated 316 L stainless steel implants with chitosan-silicon-HAp nanoparticulate composite. Results showed that in stimulated body fluid (SBF) the apatite formation was more on the coated samples compared to uncoated implant [43].

Lin et al. studied the osteoblastic differentiation on chitosan, alginate-chitosan, and pectin-chitosan coated on Ti-6Al-4V surface (Ti64). ALP levels were more in the coated samples when compared to bare Ti64. The coating increased the wettability of the surface further enhancing cell attachment and proliferation (Fig. 6) [44].

The ECM mimicking properties of chitosan has remarkably improved the compatibility of it being used alone or as composites for bone regeneration. Being the second most abundant natural polymer, the room for modification by blending or surface modifications can be widely explored. The platform for scaffold fabrication, in vitro and in vivo studies, can be greatly exploited to understand the role of this natural polymer, chitosan, in reducing the gap between the wound and injury to complete tissue regeneration. Moreover, this polymer has also found significant place in scaffold fabrication for not only bone tissue regeneration but also soft tissues like cartilage, ocular, and nerve.

3 Gelatin

Gelatin is a natural polymer obtained by the hydrolysis of collagen. It is biocompatible and biodegradable in physiological conditions. Degradation products of gelatin are found to show low antigenicity than collagen. It is a low cost natural polymer that is soluble in aqueous solvent. Due to the presence of RGD sequence in gelatin, it promotes attachment and proliferation of cells. Gelatin when crosslinked with other polymers the mechanical property of it may get altered

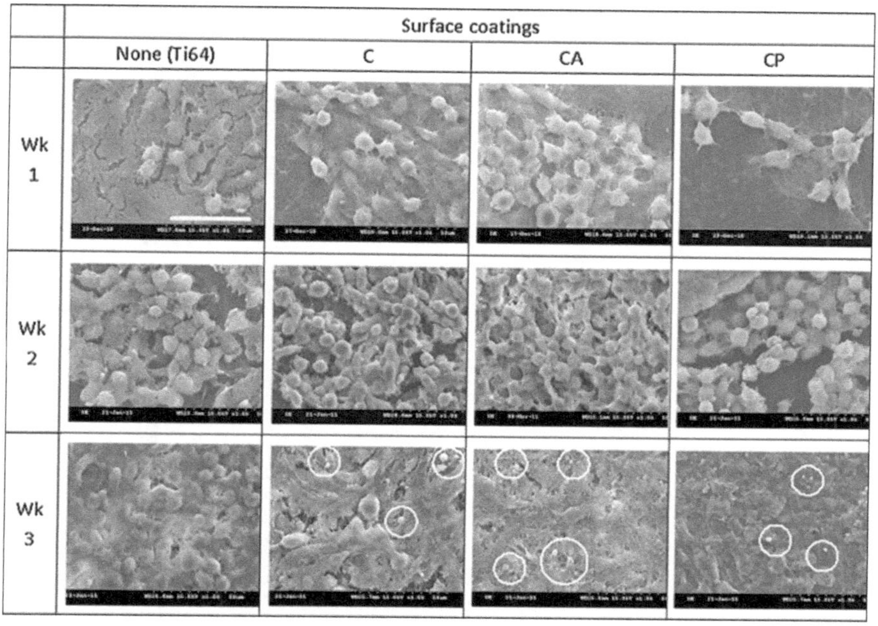

Fig. 6 SEM images of cells attached on Ti-6Al-4 V, chitosan, chitosan-alginate, and pectin-chitosan-coated Ti-6Al-4V surfaces for 3 weeks [44]

[45]. Gelatin has aided bone regeneration and along with bioceramics it has shown better results. Most commonly used bioceramics along with gelatin are calcium sulfate, biphasic calcium phosphate, β tricalcium phosphate, hydroxyapatite, bioglass, etc. These bioceramics increase the mechanical strength of the polymer and also increases the osteoconductive property of the polymer.

Gelatin-calcium sulfate composite scaffold was developed by Gao et al. and the scaffold showed human osteoblast cells attached and spread over the entire scaffold. Preliminary in vivo study showed new bone formation [46]. Gelatin-pectin scaffold (GE-P-BCP) with nano-biphasic calcium phosphate (BCP) was developed and studied by Nguyen et al. Developed scaffold showed enhanced compressive strength, and increased ALP and osteopontin levels in vitro after 21 days culture. Also in vivo studies by micro-computed tomography and histology showed new bone formation in rabbit models [47]. The same group has also developed a gelatin-hyaluronic acid-BCP composite scaffold for bone regeneration. This composite scaffold has also shown better bone regeneration in vivo [48]. In vivo studies using gelatin-coated HAp foams (Fig. 7) have been reported by Gil-Albarova et al. to enhance osteointegration of the scaffold and complete healing of the defeat when the foam is applied to bone defeat in rabbit femur [49].

Rodriguez et al. have reported gelatin sponge with chitin whiskers, HAp, and platelet-rich plasma (PRP) (PHCE) for bone regeneration. 1-Ethyl-3-[3-dimethylaminopropyl] carbodiimide hydrochloride (EDC) cross-linking was

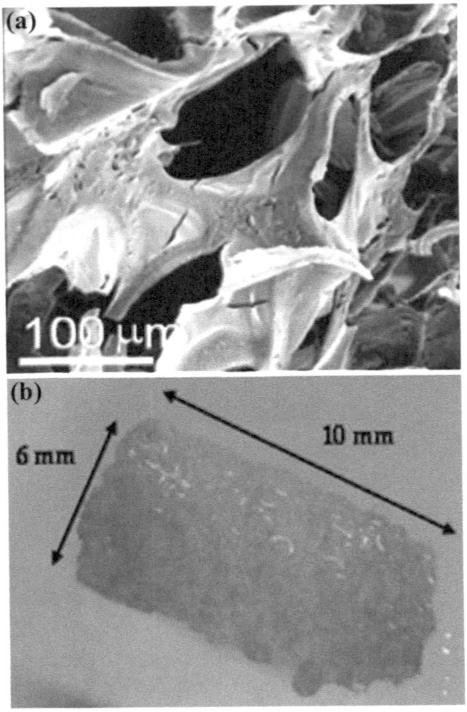

Fig. 7 **a** SEM image of uncoated HAP foam. **b** Photograph of the hydrated Hap foam [49]

done to control gelatin scaffold's degradation rate. MG-63 cells seeded on to PHCE scaffold showed better cell attachment and matrix production. Chitin and HAp were intended to increase the mechanical strength of the scaffold, and the addition of PRP was to enhance angiogenesis and new bone regeneration [50]. Nano-HAp- gelatin composite scaffold with somatic stem cells was developed by Tavakol et al., and the results showed that this composite scaffold enhanced the osteogenic differentiation in vitro and in vivo [51]. Nair et al. have reported the development of graphine oxide (GO)-loaded gelatin-HAp composite scaffold by lyophilization technique. The addition of GO has increased the mechanical strength and it also induced the osteogenic differentiation of human adipose-derived stem cells [52].

For calvaria bone regeneration, Ferreira et al. have developed macroporous gelatin scaffold with titanium dioxide (TiO_2) and HAp along with osteogenically differentiated multipotent progenitor cells. They have reported that the addition of TiO_2 an osteoinductive material along with osteogenically differentiated cells has enhanced the bone formation and osteointegration after 8 and 12 weeks post-implantation in rat calvaria bone defeat [53]. Kovtun et al. performed in vivo studies of critical size defeat in rabbit femur using gelatin-HAp foam enriched with soybean extract. The studies showed that the effect of soybean extract reduced the scaffold's degradation rate and aided bone formation at the defect site [54].

Liu et al. have reported the development of recombinant human bone morphogenetic protein 2 (rhBMP-2)/gelatin/HAp/fibrin glue hybrid scaffold for effective bone regeneration. The hybrid scaffold enhanced the osteogenic differentiation of human bone marrow mesenchymal cells with a sustained release of rhBMP-2 for 40 days. In vivo, this hybrid scaffold aided bone regeneration within 12 weeks [55]. Bioglass, an osteoinductive material, in gelatin composite scaffold has enhanced osteogenic differentiation of human dental pulp stem cells showing increased ALP and matrix deposition. This has been reported by Nadeem et al. [56].

Takaoka et al. have reported about gelatin hydrogel containing lactoferrin and observed that the retention of lactoferrin when this hydrogel was injected subcutaneously was better than it given as injection. In vivo studies showed that the release of lactoferrin from the hydrogel was in a sustained manner and the gelatin hydrogel containing lactoferrin showed better bone regeneration when implanted in defeat created on rat skull bone [57].

Fukui et al. have proposed simvastatin-loaded gelatin hydrogel. After 8 weeks of study, the gelatin hydrogel with simvastatin showed greater angiogenesis and osteogenesis which aided the healing of nonunion femoral fracture in 70 % of the rat model [58]. Similarly, Tanabe et al. developed fluvastatin loaded gelatin hydrogel for bone regeneration [59].

Minocycline (antibacterial agent) releasing HAp-containing gelatin nanocomposite was successfully developed by Doc et al. and was tested in vitro. The in vitro studies showed the possibility of developing an antibacterial agent-loaded gelatin-HAp nanocomposite for bone regeneration [60]. Growth factors-incorporated gelatins scaffold has been used for the treatment of bone damage. Omata et al. have developed gelatin-β tricalcium phosphate composite loaded with bFGF for bone regeneration [61]. Colloidal gelatin gel for the dual delivery of angiogenic (bFGF) and osteoblastic (BMP-2) growth factors has been successfully developed by Wang et al. Gelatin gel was incorporated with cationic nanospheres that were loaded with bFGF and anionic nanospheres that were loaded with BMP-2. The crosslinking density of these nanospheres in the gel matrix controlled the release kinetics of the growth factors. In vivo study in rat femoral condyle defeat was performed and it was reported that the gel with both the growth factors showed an inhibitory effect in the process of osteogenesis [62].

Similarly, Van der stok et al. developed a porous titanium scaffold loaded with nanostructured gelatin gel incorporated with BMP-2, FGF-2, and both in 5:1 ratio. Growth factor release was in a sustained manner from all the three scaffolds. In vivo studies on rat femur were done. BMP-2-loaded scaffold showed early bone formation and FGF-2-loaded scaffold showed late bone regeneration. Scaffold with BMP/FGF showed enhanced bone implant integrity [63].

Nagae et al. showed that gelatin hydrogel with microspheres of platelet-derived growth factor (PRP) promoted the intervertebral disk regeneration in vivo [64].

Hybrid scaffold of PRP in gelatin-hyaluronic acid (HA) gel was loaded on to a biphasic calcium phosphate scaffold which was developed by Son et al. In vitro studies showed better osteoblastic cell attachment and proliferation on the hybrid scaffold [65]. Kim et al. have reported gelatin hydrogel incorporated with macrophage recruitment agent (SEW2871) and PRP for effective bone regeneration. SEW2871 was encapsulated as a micelle with the help of gelatin grafted with l-lactic acid. In vivo study showed increased levels of pro-inflammatory and anti-inflammatory cytokines after 10 days [66]. Gelatin-Poly(L-lactic acid) sponge was developed and its potential for bone regeneration by differentiation of human meshenchymal stromal cells was studied by Mattii et al. [67].

Electrospun fibers containing gelatin and bioceramics materials have aided bone to regenerate. Poly(lactide-co-caprolactone) nanofibers with gelatin-apatite composite electrospun scaffold for bone regeneration were developed by Jegal et al. They have observed that the addition of gelatin-apatite on to the nanofibers improved its tensile strength and cell growth. When implanted into the calvarium of rat model, 6 weeks after implantation, the composite nanofibrous electrospun scaffold when observed under micro-computed tomography showed enhanced bone regeneration [68].

PCL-gelatin electrospun membrane incorporated with stromal cell-derived factor-1α developed by Ji et al. was placed in rat cranial defect. SEM image of the PCL-gelatin membrane showed that the fiber diameter was in the range of 312 ± 146 nm (Fig. 8). After eight weeks, the study showed that membrane with growth factor helped the recruitment of cells to the defect site and helped in the bone regeneration compared to bare membrane [69]. There was a sustained release of the stromal cell-derived factor from the electrospun membrane.

Alvarez Perez et al. developed PCL-gelatin nanofibers for bone regeneration. The nanofibrous scaffold showed better attachment, proliferation, and differentiation of human mesenchymal stem cells to osteoblasts. Increased level of ALP, osteopontin (OPN), osteocalcin, and bone sialoproteins (BSP) in the PCL-gelatin nanofibers was observed after 3 weeks of culture in osteogenic medium [70].

A poly(ε-caprolactone) (PCL)-gelatin electrospun fiber loaded with metronidazole (MNA) was developed by Xue et al. Their study has showed that this electrospun membrane can be used as anti-infective-guided tissue regeneration membrane.

Fig. 8 SEM image of PCL/gelatin membrane [69]

Sustained release of MNA was observed and the PCL-gelatin membrane with MNA showed very less inflammatory response for 8 months from subcutaneously implanted membrane in rabbit model [71]. Non-collagenous protein from bone extra cellular matrix was distributed evenly on gelatin nanofibrous membrane and when osteoblastic cells were cultured on this membrane there was a significant increase in ALP and other osteogenic genes within 14 days of culture in vitro. In vivo study on calvarial defeat has shown new bone formation within 6 weeks post implantation [72].

Linh et al. reported the development of polyvinyl alcohol-gelatin nanofibrous membrane with biphasic calcium phosphate nanoparticles for rat calvaria defeat [73]. Nanofibrous gelatin-apatite composite scaffold was developed by Liu et al. for bone regeneration. Composite scaffold showed enhanced porosity, mechanical strength, and also aided differentiation of osteoblastic cells within 4 weeks [74].

Titanium fabrics was also modified by coating with hydroxyapatite-loaded-gelatin microspheres containing FGF-2 which was reported by Ichinohe et al. The modified titanium implant was then placed on skull defects created in rabbits, and after 6 weeks of implantation the results showed that the Ti-HAp-Gelatin-loaded FGF-2 aided osteointegration of Ti and it also enhanced the bone formation [75].

So far various studies have been reported which demonstrates the use of gelatin for enhancing the biocompatibility of scaffolds and aid in bone regeneration. The use and effect of composites based on chitosan and gelatin for bone regeneration have not been widely explored or compared. Our major goal is to fill these lacunae and throw light upon the use and application of chitosan-gelatin composites.

4 Chitosan-Gelatin Composite Scaffolds

Composite scaffolds are generally made of two or more polymers to produce a new system that will have a property that is unique to both the polymers. Chitosan and gelatin being natural polymers have more advantages such as low cost, abundant in nature, non-immunogenic, and being biocompatible. Composite scaffold has been used for bone tissue regeneration. Both have its own merits for being used in bone regeneration.

Huang et al. have developed a chitosan-gelatin composite scaffold with and without glutaraldehyde cross-linking (Fig. 9) and have also evaluated its potential in vitro for tissue engineering applications. The cell attachment and spread of fibroblast and human umbilical vein endothelial cells on the composite scaffold were studied. The composite scaffold showed better cell attachment and cell spreading than the chitosan scaffold [76].

Chitosan-gelatin scaffold modified with poly(DL-lactic acid) dichloromethane solution in different ratios was developed by Liu et al. The scaffold upon immersion in water after solvent evaporation will swell and finally it is freeze dried to get the final construct. The modified scaffold increased the tensile strength but the attachment and proliferation of human fibroblast cells were comparable to unmodified chitosan-gelatin scaffolds [77].

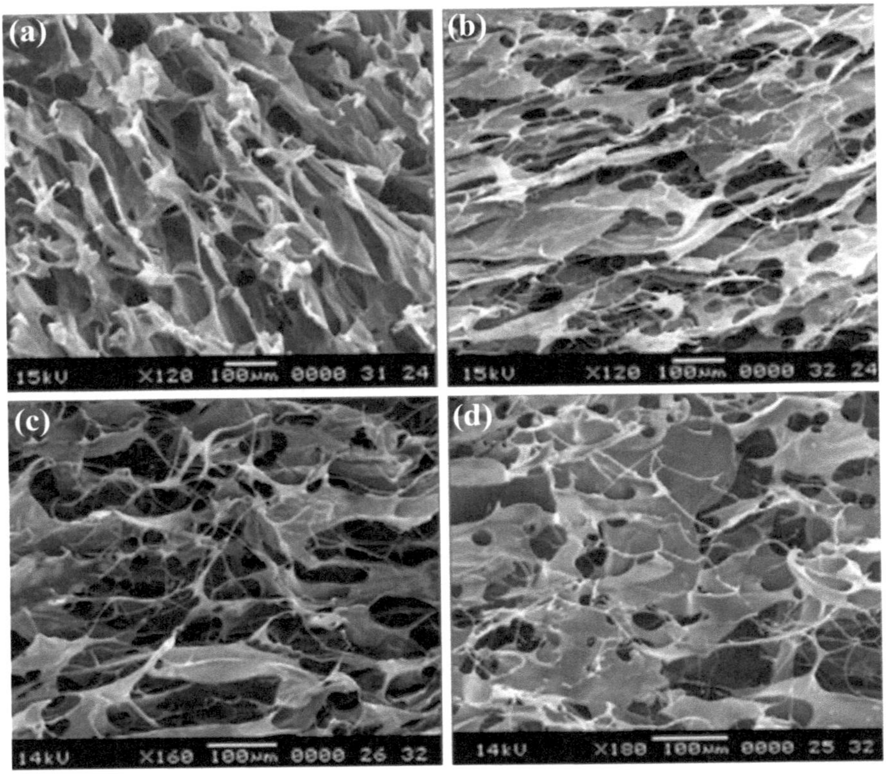

Fig. 9 SEM images of **a** chitosan **b** chitosan-gelatin (1:3) without glutaraldehyde cross-linking **c** chitosan-gelatin (3:1) with glutaraldehyde cross-linking **d** chitosan-gelatin (1:3) after glutaraldehyde cross-linking [76]

Chitosan gel containing BMP-2 with gelatin microspheres containing IGF-1 was developed by Kim et al. It was reported that there was an initial release of BMP-2 from the chitosan gel followed by a slow release of IGF-1 from the gelatin microspheres. In the presence of W-20-17 cells, the chitosan gel with gelatin microspheres has showed enhanced ALP level indicating an early osteoblastic differentiation [78].

Chitosan-gelatin composite scaffold with β-tricalcium phosphate was developed and evaluated by Yin et al. for bone regeneration. The porosity and the compressive strength of the composite scaffold were found to have increased. Subcutaneous implantation of the composite scaffold in rabbits was done and was observed for a period of 12 weeks. Results showed the importance of using this composite scaffold for non-load bearing applications [79].

Hydroxyapatite-chitosan-gelatin (HCG) composite scaffold was developed by Zhao et al. and was compared with chitosan-gelatin scaffold (CG). It was reported that hydroxyapatite in the HCG scaffold was dispersed uniformly in the scaffold.

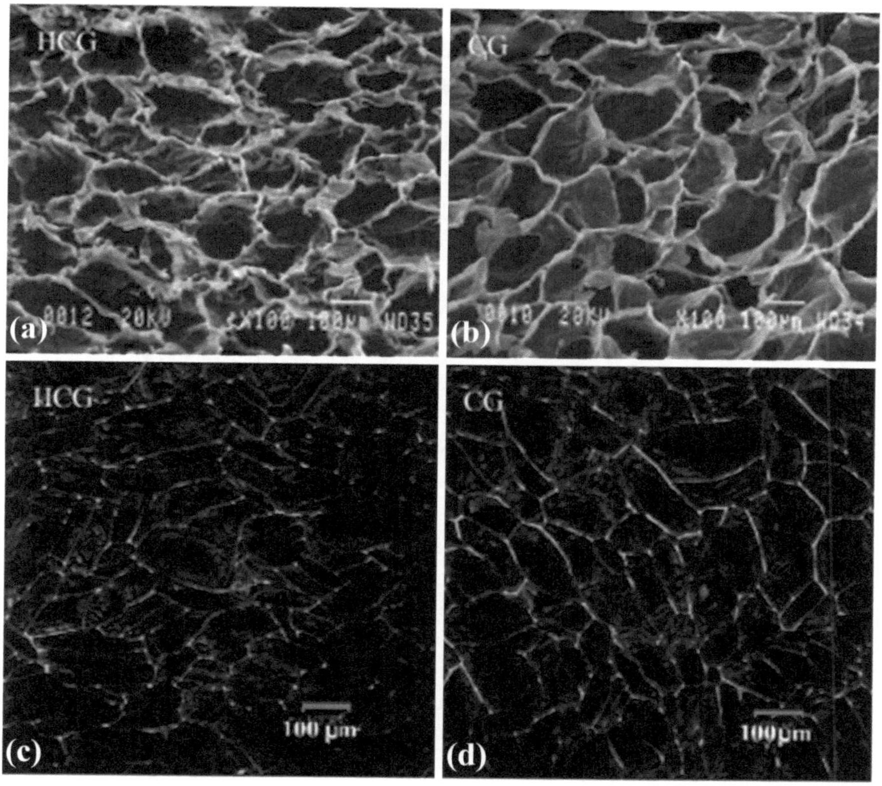

Fig. 10 SEM images of **a** HCG **b** CG scaffold and confocal images of **c** HCG and **d** CG scaffolds [81]

90.6 % porosity was observed in the composite scaffold, and histology, immunohistochemical staining, and scanning electron micrograph showed the attachment and proliferation of rat caldaria osteoblastic cells on the composite scaffold. Collagen I deposition and bone matrix formation were observed during culture condition [80]. Same group has also evaluated the effect of hydroxyapatite in the composite scaffold on the attachment of human mesenchymal stem cells (hMSC) on to the 3D construct. SEM images of the HCG and CG scaffolds showed that the mean pore size was found to be 94 $\pm$ 15 and 98 $\pm$ 24 μm (Fig. 10). Higher proliferation rate of hMSC was observed in the HAp-incorporated composite scaffold and upon osteogenic induction, osteogenic differentiation of hMSC with enhanced ALP level was observed [81].

Sellgren et al. have reported the perfusion condition for hMSC cell proliferation and its osteoblastic differentiation in the 3D construct of hydroxyapatite-chitosan-gelatin composite scaffold [82]. The effect of static and perfusion condition on protein adsorption, cell attachment, and proliferation of hMSC on to the HCG scaffold was analyzed.

Hunter et al. have evaluated the in vitro characteristics of hMSC on hydroxyapatite-chitosan-gelatin membrane for guided tissue regeneration [83]. This membrane showed mechanical and structural property to function as an effective barrier membrane and also the expression of bone marker protein and other specific genes were found to be elevated without any chemical induction.

Miranda et al. have developed chitosan-gelatin scaffold seeded with bone marrow mesenchymal stem cells (BMMSC) from transgenic rat expressing green fluorescent protein (GFP). The in vivo study was done by implantation of cell seeded scaffold on the left socket after upper first molar extraction. BMMSC-loaded chitosan-gelatin scaffold showed enhanced bone mineralization at 21 and 35 days post-implantation [84].

5 Chitosan-Gelatin Nanocomposite Scaffolds

Chitosan-Gelatin along with nano-structured bioceramic can easily mimic the ECM of bone. Nanostructured bioceramics due to their increased surface area to volume ratio might provide better bone regeneration ability.

Bagheri-Khoulenjani et al. have developed micro-spheres of chitosan-gelatin-nanohydroxyapatite nanocomposite for bone regeneration. The effect of different parameters like nanohydroxyapatite to biopolymer ratio, chitosan, gelatin concentration on the surface morphology of the as-developed microspheres was evaluated. Cell studies on the microspheres have shown the biocompatible nature of the as developed microspheres [85].

Mohamed et al. have conducted in vitro studies on chitosan-gelatin-nanohydroxyapatite (nHAp) composite scaffold. Composite scaffolds made with high content of polymer showed higher water absorption property, decreased degradation rate, and also more deposition of hydroxyapatite on the composite scaffold when kept in SBF solution. SEM and Fourier transform infrared spectroscopy (FTIR) confirmed the presence of hydroxyapatite on the surface of the chitosan-gelatin-nHAp scaffold [86].

Peter et al. have developed a composite scaffold of chitosan-gelatin-nHAp for tissue engineering applications (Fig. 11). The composite scaffold showed good swelling property depending on the ratio of chitosan and gelatin, and its degradation rate was reduced due to the addition of nHAp. nHAp also enhanced the attachment and spreading of MG-63 cells on to the composite scaffold. Composite scaffold showed increased biomineralization in stimulated body fluid compared to chitosan-gelatin scaffold [87].

Peter et al. have also developed microporous scaffold of porosity from 150 to 300 μm chitosan-gelatin-nano-bioglass (nBG) scaffold for alveolar bone regeneration. The developed composite scaffold showed reduced swelling rate and degradation but enhanced protein adsorption and cell attachment of MG-63 cells on to the composite scaffold (Fig. 12) [88].

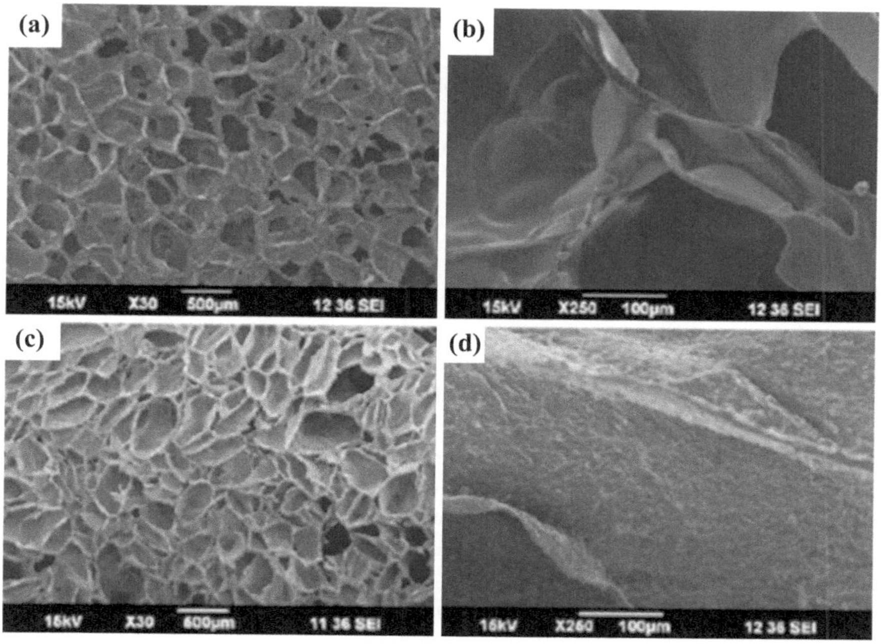

Fig. 11 SEM image of **a**, **b** chitosan-gelatin scaffold, **c**, **d** chitosan-gelatin-nHAp composite scaffold [87]

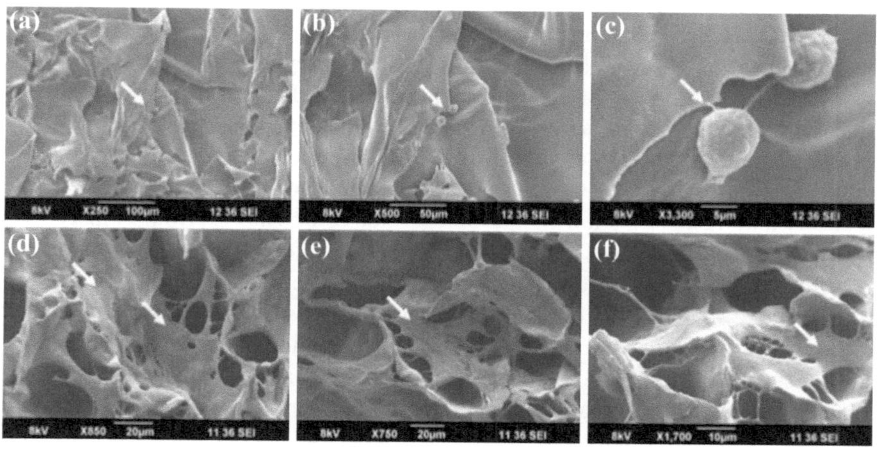

Fig. 12 **a**, **b**, **c** Attachment of MG-63 cells on to the chitosan-gelatin scaffold. **d**, **e**, **f** Attachment of MG-63 cells on to chitosan-gelatin-nBG composite scaffold [88]

Kavya et al. have developed chitosan-gelatin-nano-silica ($nSiO_2$) composite scaffold for bone tissue engineering. The in vitro characterizations of the composite scaffold such as porosity, swelling, degradation, mechanical integrity, and biomineralization on scaffold surface showed the potential of this composite scaffold to be used for the regeneration of bone. The incorporation of $nSiO_2$ has improved the osteoconductive property of the scaffold. Cell attachment and proliferation was more and also there was increased ALP expression on the composite scaffold. These results show the possibility of using this composite scaffold for bone tissue engineering [89].

6 Conclusions

Bone tissue engineering with the help of scaffold, cells, and growth factors for regeneration process can give better results compared to conventionally used bone grafts and filler materials. The literatures cited in this review show that the progress in biomaterials been used for bone regeneration.

Among the different biomaterials used for bone regeneration, composite scaffolds have better effect due to the presence of polymers and ceramic micro- or nano-scale particles. Chitosan and gelatin being natural polymers have its own advantages for bone tissue engineering such as biocompatibility, biodegradability, and easy manufacturing. These natural polymers have been shown to attach osteoblasts and help in its proliferation. Porosity of these scaffolds plays a major role in cell attachment, proliferation, and infiltration. Natural polymers can be manufactured with different porosities. The scaffolds made of these polymers have good mechanical properties but they do not have much of osteoconductive property that is required for bone regeneration. Ceramic particles like calcium phosphate, hydroxyapatite, and bioglass when added to polymeric scaffold improve the osteoconductive property of the composite scaffolds. The ionic dissolutes of these bioceramics induce stem cells to differentiate into osteoblasts under proper culture environment. The major challenge in bone tissue regeneration is the proper cellularization and vascularization of the damaged bone. To accelerate the osteogenesis and vascularization of the defect site, composite scaffolds loaded with growth factors have shown better results. The controlled release of growth factors from these composite scaffolds is a main criteria to be kept in mind while designing a suitable scaffold. On the other hand, bioceramics material like bioglass with controlled degradation rate which helps in the formation of hydroxyapatite in our body. Bioglass has reported to have osteoconductive and osteoinductive property by which it can aid in better bone regeneration. Overall composite scaffolds can aid bone regeneration in a better way because it possesses the beneficial properties of both polymer and bioceramics that constitute the scaffold. Chitosan and gelatin the most available natural polymer have been widely used for bone regeneration. These along with bioceramics and growth factors have shown promising results in bone tissue engineering applications.

Acknowledgments One of the authors S. Deepthi is thankful to the Council of Scientific and Industrial Research for supporting financially under the CSIR-SRF award no: 9/963(0034)2K13-EMR-I

References

1. Landis WJ (1996) Mineral characterization in calcifying tissues: atomic, molecular and macromolecular perspectives. Connect Tissue Res 34:239–246
2. Rho JY (1998) Mechanical properties and the hierarchical structure of bone. Med Eng Phys 20:92–102
3. Van der Meulen MCH, Prendergast PJ (2000) Mechanics in skeletal development, adaptation and disease. Philos Trans R Soc Lond A 358:565–578
4. Giannoudis PV, Dinopoulos H, Tsiridis E (2005) Bone substitutes: an update. Injury 36:20–27
5. Stock UA, Vacanti JP (2001) Tissue engineering: current state and prospects. Annu Rev Med 52:443–451
6. Hutmacher DW (2000) Scaffolds in tissue engineering bone and cartilage. Biomaterials 21:2529–2543
7. Shalumon KT, Sowmya S, Sathish D, Chennazhi KP, Nair SV, Jayakumar R (2013) Effect of incorporation of nanoscale bioactive glass and hydroxyapatite in PCL/Chitosan nanofibers for bone and periodontal tissue engineering. J Biomed Nanotechnol 9:430–440
8. Di Bella C, Farlie P, Penington AJ (2008) Bone regeneration in a rabbit critical-sized skull defect using autologous adipose-derived cells. Tissue Eng Part A 14:483–490
9. Gentile P, Chiono V, Carmagnola l, Hatton PV (2014) An overview of poly(lactic-co-glycolic) acid (PLGA)-based biomaterials for bone tissue engineering. Int J Mol Sci 1:360-59
10. Ferreira AM, Gentile P, Chiono V, Ciardelli G (2012) Collagen for bone tissue regeneration. Acta Biomater 8:3191–3200
11. Binulal NS, Natarajan A, Menon D, Bhaskaran VK, Mony U, Nair SV (2014) PCL–gelatin composite nanofibers electrospun using diluted acetic acid–ethyl acetate solvent system for stem cell-based bone tissue engineering. J Biomater Sci Polym Ed 25:325–340
12. Anitha A, Sowmya S, Sudheesh Kumar PT, Deepthi S, Chennazhi KP, Ehrlich M, Tsurkan Jayakumar R (2014) Chitin and chitosan in selected biomedical applications. Prog Polym Sci 39:1644–1667
13. Khan WS, Rayan F, Dhinsa BS, Marsh D (2012) An osteoconductive, osteoinductive, and osteogenic tissue-engineered product for trauma and orthopaedic surgery: how far are we? Stem Cells Int 2012:236231
14. El-Ghannam A (2005) Bone reconstruction: from bioceramics to tissue engineering. Expert Rev Med Devices 2:87–101
15. Zhou H, Lee J (2011) Nanoscale hydroxyapatite particles for bone tissue engineering. Acta Biomater 7:2769–2781
16. Cao K, Kuboyama N (2010) A biodegradable porous composite scaffold of PGA/beta-TCP for bone tissue engineering. Bone 46:386–395
17. Liu X, Rahaman MN, Fu Q (2013) Bone regeneration in strong porous bioactive glass scaffolds with an oriented microstructure implanted in rat calvarial defects. Acta Biomater 9:4889–4898
18. Wang M (2001) Bioactive ceramic–polymer composites for bone replacement. In: Proceedings of the 13th international conference on composite materials (ICCM-13), Beijing, China. Paper 1541
19. Lan levengood SK, Zhang M (2014) Chitosan scaffolds for bone tissue engineering. J Mater Chem B 2:3161–3184
20. Jayakumar R, Menon D, Manzoor K, Nair SV, Tamura H (2010) Biomedical application of chitin and chitosan nanomaterials. Carbohydr Polym 82:227–232
21. Li Z, Ramay HR, Hauch KD, Xiao D, Zhang M (2005) Chitosan-alginate hybrid scaffolds for bone tissue engineering. Biomaterials 26:3919–3928
22. Qiao P, Wang J, Xie Q, Li F, Dong L, Xu T (2013) Injectable calcium phosphate-alginate-chitosan microencapsulated MC3T3-E1 cell paste for bone tissue engineering in vivo. Mater Sci Eng C Mater Biol Appl 33:4633–4639

23. Zhang Y, Zhang M (2002) Three-dimensional macroporous calcium phosphate bioceramics with nested chitosan sponges for load-bearing bone implants. J Biomed Mater Res 61:1–8
24. Zhang Y, Ni M, Zhang M, Ratner B (2003) Calcium phosphate-chitosan composite scaffolds for bone tissue engineering. Tissue Eng 9:337–345
25. Tanase CE, Sartoris A, Popa MI, Verestiuc L, Unger RE, Kirkpatrick CJ (2013) In vitro evaluation of biomimetic chitosan-calcium phosphate scaffolds with potential application in bone tissue engineering. Biomed Mater 8:025002
26. Lee JY, Kim KH, Shin SY, Rhyu IC, Lee YM, Park YJ, Chung CP, Lee SJ (2006) Enhanced bone formation by transforming growth factor-beta1-releasing collagen/chitosan microgranules. J Biomed Mater Res A 76:530–539
27. Park YJ, Lee YM, Park SN, Sheen SY, Chung CP, Lee SJ (2000) Platelet derived growth factor releasing chitosan sponge for periodontal bone regeneration. Biomaterials 21:153–159
28. Park YJ, Lee YM, Lee JY, Seol YJ, Chung CP, Lee SJ (2000) Controlled release of platelet-derived growth factor-BB from chondroitin sulfate-chitosan sponge for guided bone regeneration. J Control Release 67:385–394
29. Lee YM, Park YJ, Lee SJ, Ku Y, Han SB, Klokkevold PR, Chung CP (2000) The bone regenerative effect of platelet-derived growth factor-BB delivered with a chitosan/tricalcium phosphate sponge carrier. J Periodontol 71:418–424
30. Im SY, Cho SH, Hwang JH, Lee SJ (2003) Growth factor releasing porous poly (epsilon-caprolactone)-chitosan matrices for enhanced bone regenerative therapy. Arch Pharm Res 26:76–82
31. Li B, Liao X, Zheng L, Zhu X, Wang Z, Fan H, Zhang X (2012) Effect of nanostructure on osteoinduction of porous biphasic calcium phosphate ceramics. Acta Biomater 8:3794–3804
32. Thein-Han WW, Misra RD (2009) Biomimetic chitosan-nanohydroxyapatite composite scaffolds for bone tissue engineering. Acta Biomater 5:1182–1197
33. Sowjanya JA, Singh J, Mohita T, Sarvanan S, Moorthi A, Srinivasan N, Selvamurugan N (2013) Biocomposite scaffolds containing chitosan/alginate/nano-silica for bone tissue engineering. Colloids Surf B Biointerfaces 109:294–300
34. Kavitha K, Sutha S, Prabhu M, Rajendran V, Jayakumar T (2013) In situ synthesized novel biocompatible titania-chitosan nanocomposites with high surface area and antibacterial activity. Carbohydr Polym 93:731–739
35. Chesnutt BM, Viano AM, Yuan Y, Yang Y, Guda T, Appleford MR, Ong JL, Haggard WO, Bumgardner JD (2009) Design and characterization of a novel chitosan/nanocrystalline calcium phosphate composite scaffold for bone regeneration. J Biomed Mater Res A 88:491–502
36. Chesnutt BM, Yuan Y, Buddington K, Haggard WO, Bumgardner JD (2009) Composite chitosan/nano-hydroxyapatite scaffolds induce osteocalcin production by osteoblasts in vitro and support bone formation in vivo. Tissue Eng Part A 15:2571–2579
37. Jiang H, Zuo Y, Zou Q, Wang H, Du J, Li Y, Yang X (2013) Biomimetic spiral-cylindrical scaffold based on hybrid chitosan/cellulose/nano-hydroxyapatite membrane for bone regeneration. ACS Appl Mater Interfaces 5:12036–12044
38. Lee JS, Baek SD, Venkatesan J, Bhatnagar I, Chang HK, Kim HT, Kim SK (2014) In vivo study of chitosan-natural nano hydroxyapatite scaffolds for bone tissue regeneration. Int J Biol Macromol 67:360–366
39. Xiao X, Liu R, Huang Q (2008) Preparation and characterization of nano-hydroxyapatite/polymer composite scaffolds. J Mater Sci Mater Med 19:3429–3435
40. Wang F, Zhang YC, Zhou H, Guo YC, Su XX (2014) Evaluation of in vitro and in vivo osteogenic differentiation of nano-hydroxyapatite/chitosan/poly(lactide-co-glycolide) scaffolds with human umbilical cord mesenchymal stem cells. J Biomed Mater Res A 102:760–768
41. Venkatesan J, Qian ZJ, Ryu B, Ashok Kumar N, Kim SK (2011) Preparation and characterization of carbon nanotube-grafted-chitosan: natural hydroxyapatite composite for bone tissue engineering. Carbohydr Polym 83:569–577
42. Mohammadi Y, Soleimani M, Fallahi-Sichani M, Gazme A, Haddadi-Asl V, Arefian E, Kiani J, Moradi R, Atashi A, Ahmadbeigi N (2007) Nanofibrous poly(epsilon-caprolactone)/

poly(vinyl alcohol)/chitosan hybrid scaffolds for bone tissue engineering using mesenchymal stem cells. Int J Artif Organs 30:204–211
43. Sutha S, Kavitha K, Karunakaran G, Rajendran V (2013) In-vitro bioactivity, biocorrosion and antibacterial activity of silicon integrated hydroxyapatite/chitosan composite coating on 316 L stainless steel implants. Mater Sci Eng C Mater Biol Appl 33:4046–4054
44. Lin HY, Chen JH (2013) Osteoblast differentiation and phenotype expressions on chitosan-coated Ti-6Al-4 V. Carbohydr Polym 97:618–626
45. Elzoghby AO (2013) Gelatin-based nanoparticles as drug and gene delivery systems: reviewing three decades of research. J Control Release 172:1075–1091
46. Gao C, Huo S, Li X, You X, Zhang Y, Gao J (2007) Characteristics of calcium sulfate/gelatin composite biomaterials for bone repair. J Biomater Sci Polym Ed 18:799–824
47. Nguyen TB, Min YK, Lee BT (2015) Nanoparticle biphasic calcium phosphate loading on gelatin-pectin scaffold for improved bone regeneration. Tissue Eng Part A 21:1376–1387
48. Nguyen TB, Lee BT (2014) A combination of biphasic calcium phosphate scaffold with hyaluronic acid-gelatin hydrogel as a new tool for bone regeneration. Tissue Eng Part A 20:1993–2004
49. Gil-Albarova J, Vila M, Badiola-Vargas J, Sánchez-Salcedo S, Herrera A, Vallet-Regi M (2012) In vivo osteointegration of three-dimensional crosslinked gelatin-coated hydroxyapatite foams. Acta Biomater 8:3777–3783
50. Rodriguez IA, Sell SA, McCool JM, Saxena G, Spence AJ, Bowlin GL (2013) A preliminary evaluation of lyophilized gelatin sponges, enhanced with platelet-rich plasma, hydroxyapatite and chitin whiskers for bone regeneration. Cells 2:244–265
51. Tavakol S, Azami M, Khoshzaban A, Ragerdi Kashani I, Tavakol B, Hoveizi E, Rezayat Sorkhabadi SM (2013) Effect of laminated hydroxyapatite/gelatin nanocomposite scaffold structure on osteogenesis using unrestricted somatic stem cells in rat. Cell Biol Int 37:1181–1189
52. Nair M, Nancy D, Krishnan AG, Anjusree GS, Vadukumpully S, Nair SV (2015) Graphene oxide nanoflakes incorporated gelatin-hydroxyapatite scaffolds enhance osteogenic differentiation of human mesenchymal stem cells. Nanotechnology 26:161001
53. Ferreira JR, Padilla R, Urkasemsin G, Yoon K, Goeckner K, Hu WS, Ko CC (2013) Titanium-enriched hydroxyapatite-gelatin scaffolds with osteogenically differentiated progenitor cell aggregates for calvaria bone regeneration. Tissue Eng Part A 19:1803–1816
54. Kovtun A, Goeckelmann MJ, Niclas AA, Montufar EB, Ginebra MP, Planell JA, Santin M, Ignatius A (2014) In vivo performance of novel soybean/gelatin-based bioactive and injectable hydroxyapatite foams. Acta Biomater 12:242–249
55. Liu Y, Lu Y, Tian X, Cui G, Zhao Y, Yang Q, Yu S, Xing G, Zhang B (2009) Segmental bone regeneration using an rhBMP-2-loaded gelatin/nanohydroxyapatite/fibrin scaffold in a rabbit model. Biomaterials 30:6276–6285
56. Nadeem D, Kiamehr M, Yang X, Su B (2013) Fabrication and in vitro evaluation of a sponge-like bioactive-glass/gelatin composite scaffold for bone tissue engineering. Mater Sci Eng C Mater Biol Appl 33:2669–2678
57. Takaoka R, Hikasa Y, Hayashi K, Tabata Y (2011) Bone regeneration by lactoferrin released from a gelatin hydrogel. J Biomater Sci Polym Ed 22:1581–1589
58. Fukui T, Ii M, Shoji T, Matsumoto T, Mifune Y, Kawakami Y, Akimaru H, Kawamoto A, Kuroda T, Saito T, Tabata Y, Kuroda R, Kurosaka M, Asahara T (2012) Therapeutic effect of local administration of low-dose simvastatin-conjugated gelatin hydrogel for fracture healing. J Bone Miner Res 27:1118–1131
59. Tanabe K, Nomoto H, Okumori N, Miura T, Yoshinari M (2012) Osteogenic effect of fluvastatin combined with biodegradable gelatin-hydrogel. Dent Mater J 31:489–493
60. Dou XC, Zhu XP, Zhou J, Cai HQ, Tang J, Li QL (2011) Minocycline-released hydroxyapatite-gelatin nanocomposite and its cytocompatibility in vitro. Biomed Mater 6:025002
61. Omata K, Matsuno T, Asano K, Hashimoto Y, Tabata Y, Satoh T (2014) Enhanced bone regeneration by gelatin-β-tricalcium phosphate composites enabling controlled release of bFGF. J Tissue Eng Regen Med 8:604–611

62. Wang H, Zou Q, Boerman OC, Nijhuis AW, Jansen JA, Li Y, Leeuwenburgh SC (2013) Combined delivery of BMP-2 and bFGF from nanostructured colloidal gelatin gels and its effect on bone regeneration in vivo. J Control Release 166:172–181
63. Van der Stok J, Wang H, Amin Yavari S, Siebelt M, Sandker M, Waarsing JH, Verhaar JA, Jahr H, Zadpoor AA, Leeuwenburgh SC, Weinans H (2013) Enhanced bone regeneration of cortical segmental bone defects using porous titanium scaffolds incorporated with colloidal gelatin gels for time- and dose-controlled delivery of dual growth factors. Tissue Eng Part A 19:2605–2614
64. Nagae M, Ikeda T, Mikami Y, Hase H, Ozawa H, Matsuda K, Sakamoto H, Tabata Y, Kawata M, Kubo T (2007) Intervertebral disc regeneration using platelet-rich plasma and biodegradable gelatin hydrogel microspheres. Tissue Eng 13:147–158
65. Son SR, Sarkar SK, Nguyen-Thuy BL, Padalhin AR, Kim BR, Jung HI, Lee BT (2015) Platelet-rich plasma encapsulation in hyaluronic acid/gelatin-BCP hydrogel for growth factor delivery in BCP sponge scaffold for bone regeneration. J Biomater Appl 29:988–1002
66. Kim YH, Furuya H, Tabata Y (2014) Enhancement of bone regeneration by dual release of a macrophage recruitment agent and platelet-rich plasma from gelatin hydrogels. Biomaterials 35:214–224
67. Mattii L, Battolla B, D'Alessandro D, Trombi L, Pacini S, Cascone MG, Lazzeri L, Bernardini N, Dolfi A, Galimberti S, Petrini M (2008) Gelatin/PLLA sponge-like scaffolds allow proliferation and osteogenic differentiation of human mesenchymal stromal cells. Macromol Biosci 8:819–826
68. Jegal SH, Park JH, Kim JH, Kim TH, Shin US, Kim TI, Kim HW (2011) Functional composite nanofibers of poly(lactide-co-caprolactone) containing gelatin-apatite bone mimetic precipitate for bone regeneration. Acta Biomater 7:1609–1617
69. Ji W, Yang F, Ma J, Bouma MJ, Boerman OC, Chen Z, van den Beucken JJ, Jansen JA (2012) Incorporation of stromal cell-derived factor-1α in PCL/gelatin electrospun membranes for guided bone regeneration. Biomaterials 34:735–745
70. Alvarez Perez MA, Guarino V, Cirillo V, Ambrosio L (2012) In vitro mineralization and bone osteogenesis in poly(ε-caprolactone)/gelatin nanofibers. J Biomed Mater Res A 100:3008–3019
71. Xue J, He M1, Liu H, Niu Y, Crawford A, Coates PD, Chen D, Shi R, Zhang L (2014) Drug loaded homogeneous electrospun PCL/gelatin hybrid nanofiber structures for anti-infective tissue regeneration membranes. Biomaterials 35:9395–9405
72. Sun Y, Jiang Y, Liu Q, Gao T, Feng JQ, Dechow P, D'Souza RN, Qin C, Liu X (2013) Biomimetic engineering of nanofibrous gelatin scaffolds with noncollagenous proteins for enhanced bone regeneration. Tissue Eng Part A 19:1754–1763
73. Linh NT, Lee KH, Lee BT (2013) Functional nanofiber mat of polyvinyl alcohol/gelatin containing nanoparticles of biphasic calcium phosphate for bone regeneration in rat calvaria defects. J Biomed Mater Res A. 101:2412–2423
74. Liu X, Smith LA, Hu J, Ma PX (2009) Biomimetic nanofibrous gelatin/apatite composite scaffolds for bone tissue engineering. Biomaterials 30:2252–2258
75. Ichinohe N, Kuboki Y, Tabata Y (2008) Bone regeneration using titanium nonwoven fabrics combined with fgf-2 release from gelatin hydrogel microspheres in rabbit skull defects. Tissue Eng Part A 14:1663–1671
76. Huang Y, Onyeri S, Siewe M, Moshfeghian A, Madihally SV (2005) In vitro characterization of chitosan-gelatin scaffolds for tissue engineering. Biomaterials 26:7616–7627
77. Liu H, Yao F, Zhou Y, Yao K, Mei D, Cui L, Cao (2005) Porous poly (DL-lactic acid) modified chitosan-gelatin scaffolds for tissue engineering. J Biomater Appl 19:303–322
78. Kim S, Kang Y, Krueger CA, Sen M, Holcomb JB, Chen D, Wenke JC, Yang Y (2012) Sequential delivery of BMP-2 and IGF-1 using a chitosan gel with gelatin microspheres enhances early osteoblastic differentiation. Acta Biomater 8:1768–1777
79. Yin Y, Ye F, Cui J, Zhang F, Li X, Yao K (2003) Preparation and characterization of macroporous chitosan-gelatin/beta-tricalcium phosphate composite scaffolds for bone tissue engineering. J Biomed Mater Res A 67:844–855

80. Zhao F, Yin Y, Lu WW, Leong JC, Zhang W, Zhang J, Zhang M, Yao K (2002) Preparation and histological evaluation of biomimetic three-dimensional hydroxyapatite/chitosan-gelatin network composite scaffolds. Biomaterials 23:3227–3234
81. Zhao F, Grayson WL, Ma T, Bunnell B, Lu WW (2006) Effects of hydroxyapatite in 3-D chitosan-gelatin polymer network on human mesenchymal stem cell construct development. Biomaterials 27:1859–1867
82. Sellgren KL, Ma T (2012) Perfusion conditioning of hydroxyapatite-chitosan-gelatin scaffolds for bone tissue regeneration from human mesenchymal stem cells. J Tissue Eng Regen Med 6:49–59
83. Hunter KT, Ma T (2013) In vitro evaluation of hydroxyapatite-chitosan-gelatin composite membrane in guided tissue regeneration. J Biomed Mater Res A 101:1016–1025
84. Miranda SC, Silva GA, Mendes RM, Abreu FA, Caliari MV, Alves JB, Goes AM (2012) Mesenchymal stem cells associated with porous chitosan-gelatin scaffold: a potential strategy for alveolar bone regeneration. J Biomed Mater Res A 100:2775–2786
85. Bagheri-Khoulenjani S, Mirzadeh H, Etrati-Khosroshahi M, Shokrgozar MA (2013) Particle size modeling and morphology study of chitosan/gelatin/nanohydroxyapatite nanocomposite microspheres for bone tissue engineering. J Biomed Mater Res A 101:1758–1767
86. Mohamed KR, Beherei HH, El-Rashidy ZM (2014) In vitro study of nano-hydroxyapatite/chitosan-gelatin composites for bio-applications. J Adv Res 5:201–208
87. Peter M, Ganesh N, Selvamurugan N, Nair SV, Furuike T, Tamura H, Jayakumar R (2010) Preparation and characterization of chitosan–gelatin/nanohydroxyapatite composite scaffolds for tissue engineering applications. Carbohydr Polym 80:687–694
88. Peter M, Binulal NS, Nair SV, Selvamurugan N, Tamura H, Jayakumar R (2010) Novel biodegradable chitosan–gelatin/nano-bioactive glass ceramic composite scaffolds for alveolar bone tissue engineering. Chem Eng J 158:353–361
89. Kavya KC, Jayakumar R, Nair S, Chennazhi KP (2013) Fabrication and characterization of chitosan/gelatin/nSiO2 composite scaffold for bone tissue engineering. Int J Biol Macromol 59:255–263

Chitin and Chitosan Nanocomposites for Tissue Engineering

Arun Kumar Mahanta and Pralay Maiti

Abstract Chitin and chitosan are the most widely used biodegradable and biocompatible materials subsequent to cellulose. Nowadays a wide range of materials, including those classified as organic, inorganic, and biological are used in the synthesis, fabrication, and processing of nanostructures with unique physical properties. The properties of the polymer significantly improve by dispersing a few percentage of nanoparticle in the polymer matrix. In this context, we are focusing on the preparation, characterization, and bioactivity of chitin and chitosan nanocomposite in detail. The morphological changes occur in presence of nanoparticle. The improvement of thermal and mechanical properties including dynamic mechanical behavior of chitin and chitosan in presence of different nanofillers has been discussed in detail with suitable example as potential material for bone and wound tissue engineering applications. We summarize the physicochemical and drug delivery properties of chitin and chitosan composites. The cytocompatibility of the nanocomposites is assessed with improved cell attachment and proliferation using different human cells. This chapter enhances the understanding of biological uses of chitin and chitosan with their improved properties in presence of nanoparticles. A new approach at the intersection of biology and nanotechnology is focused to develop the next promising eco-friendly biopolymer nanocomposites.

Keywords Chitin/chitosan · Nanoparticle · Biomaterial · Drug delivery

1 Introduction

In recent years, tissue engineering has been widely investigated as a promising approach toward regeneration of tissue [1]. Biomaterials are required in tissue engineering strategies for the cell growth, attachment, and proliferation [2, 3].

A.K. Mahanta · P. Maiti (✉)
School of Materials Science and Technology, Indian Institute of Technology
(Banaras Hindu University), Varanasi 221005, India
e-mail: pmaiti.mst@itbhu.ac.in

P.K. Dutta (ed.), *Chitin and Chitosan for Regenerative Medicine*, Springer Series on Polymer and Composite Materials, DOI 10.1007/978-81-322-2511-9_6

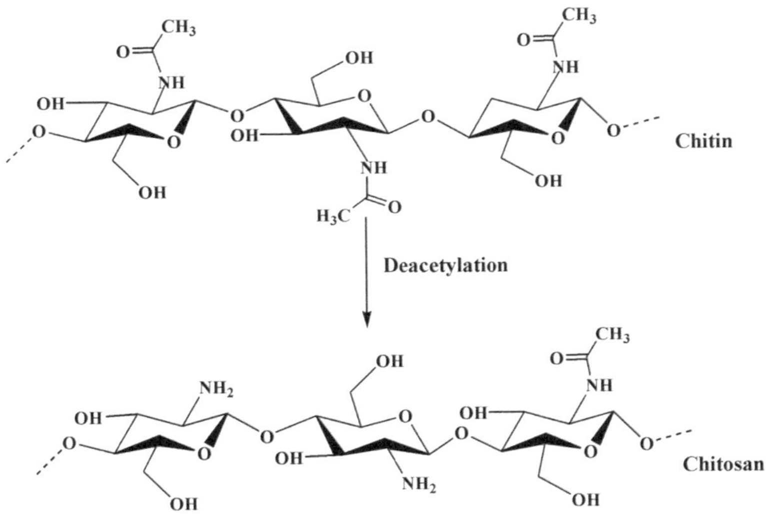

Fig. 1 Chemical structure of chitin and deacetylated chitin (chitosan)

Polysaccharides have the tendency to be extremely bioactive and classified as natural macromolecules. These polysaccharides are generally derived using different biotechnological approaches from agricultural feed stock or crustacean shell waste. In terms of availability, chitin is the second most abundant polymer after cellulose. Chitosan, a deacetylated derivative of chitin has multidirectional application such as in food and nutrition, material science, biotechnology, pharmaceuticals, agricultural and environmental protection [4–9]. Chitin contains two monomeric units D-glucosamine and *N*-acetyl-D-glucosamine which are interconnected by β-glycosidic linkages [10]. It is found in nature as ordered crystalline microfibrils forming structural components in the exoskeleton of arthropods or in the cell walls of fungi and yeast [11]. Chitosan is the most important derivative of chitin, obtained by (partial) deacetylation of chitin in the solid state under alkaline condition (concentrated NaOH) or by enzymatic hydrolysis in the presence of chitin deacetylase [12, 13]. The structure of chitin and chitosan is presented in Fig. 1. Chitin exists in three different polymorphic forms, with varying properties and the different forms are α, β, and γ [14–17]. The sources of α-chitin are crabs and shrimps; β-chitin is squids; and γ-chitin is loligo. The different polymorphic forms differ in their arrangement of polymeric chains like, antiparallel conformation to each other in α-chitin, polymeric chains are arranged in parallel configuration in β-chitin, and polymeric chains are arranged in two parallel chains followed by one antiparallel chain in γ-chitin [17]. Chitin and chitosan are biocompatible, biodegradable, and nontoxic [18–21] and they offer the advantage of being easily processed into gels [22], membranes [23, 24], nanofibers [25], nanofibrils [26], beads [27], microparticles [28], nanoparticles [29, 30], scaffolds [31–33], and sponge-like forms [34]. Owing to their special character and easily processable nature, chitin and chitosan have versatile applications in tissue engineering [34–42], wound healing [43–50], drug delivery [51–56], and gene delivery [57]. Nowadays, nanocomposite of biopolymers and bioactive materials has been extensively used in tissue engineering [58]. The aim of making nanocomposite is achieving a better interaction between the bioactive inorganic phase and the organic phase to get considerably enhanced properties. Nanometer size or nanostructure materials known as nanomaterials, where at least one dimension is in the nanometer scale, are the backbone of nanoscience and nanotechnology. They have unique physicochemical properties compared with the bulk materials of the same composition and capable of changing their properties and applications. Nanomaterials are mostly found as: (1) carbon-based material, consists of carbon atoms and possesses different nanostructures, e.g., single-wall or multiwall carbon nanotubes (CNT), graphene, and fullerene; (2) metallic nanomaterials which is composed of nanomaterials of metals, transition metals or their compounds and composites, e.g., gold, iron oxide, silver, silica, and quantum dots; (3) silicon nanomaterials which consist of silicon or its compound, e.g., silicon or silica nanoparticles and (4) organic nanomaterials, formed via agglomeration or assembly of organic molecules, e.g., dendrimers, polymers biomolecules, or biomacromolecules [59]. Chitin and chitosan both contain hydroxyl groups at C-3 and C-6 positions and in addition to that chitosan contains some extent of amino groups at C-2 position depending on the degree of deacetylation, therefore, they can interact with the nanomaterials through various means. There have been several reports on preparations and applications of chitin and chitosan nanocomposites. In this chapter, an attempt has been made to focus on the application potential on chitin and chitosan-based nanocomposites with different filler nanomaterials for tissue engineering, drug delivery, and antimicrobial activities.

2 Composites Preparation

There are several processes to prepare nanocomposites with the advantages or drawbacks. Some common processes which are frequently used to prepare the chitin and chitosan nanocomposites are illustrated as follows:

(a) *Solution casting method* This is the simplest technique to prepare polymer nanocomposites. In this technique, three steps are required to prepare the nanocomposites. At first, the filler is dispersed/dissolved in the appropriate solvent through mechanical stirring or sonication. The required polymer is dissolved in the same solvent or in two miscible solvents. Finally, the filler solution and polymer solution are mixed together at room temperature or other suitable temperatures. Composites are obtained either by precipitation or casting the solution mixture. Regiel-Futyra et al. have prepared chitosan–gold nanocomposite film through solution casting method [60]. They have poured the chitosan-based gold nanoparticles dispersion into Petri dishes and dried in an air oven at 60 °C until the solvent was evaporated completely.

(b) *In situ technique* This is the most efficient technique to prepare the composites with uniformly distributed filler materials. Filler materials are dispersed in monomer either in presence or absence of solvent followed by the addition of curing agents or the hardener for polymerization at experimental temperature. The major advantage of this technique is that the composites are obtained with improved properties. Sometimes nanomaterials are synthesized in reaction mixture at the same reaction condition. Hebeish et al. prepared chitosan-grafted-poly acrylonitrile silver nanocomposites via in situ chemical reduction of Ag ions in graft copolymerisation of acrylonitrile on to chitosan [61].

(c) *Electro spinning technique* There are three major components in electrospinning instruments: a high-voltage power supply, a spinneret (a metallic needle), and a collector (a grounded conductor). The spinneret is connected to a syringe in which the mixture of polymer and nanomaterial solution is hosted. When high voltage (usually in the range of 1–30 kV), is applied, the pendent drop of solution at the nozzle of spinneret will become highly electrified and the induced charges are evenly distributed over the surface. When a sufficiently high voltage is applied the electrostatic repulsion counteracts the surface tension and the droplet is stretched. Such type of electrified jet then undergoes a stretching and whipping process, leading to the formation of long and thin thread. The threads are collected by the grounded collector placed under the spinneret. With the use of this relatively simple and straightforward technique several polymer composites have been processed as fibers with diameters ranging from tens of nanometers to a few micrometers. Naseri et al. developed electrospun chitosan/polyethylene oxide-based randomly oriented fiber mats reinforced with chitin nanocrystals using electrospinning technique [62].

(d) *Freeze–drying technique* This is another method to fabricate polymer composite scaffolds with variable porosity and pore size and utilizes emulsion/freeze–drying process. The homogenized mixture of polymer and filler solution is poured into copper mold and quenched in liquid nitrogen or in required lower temperature. After quenching, the polymer scaffold is freeze–dried to remove the solvents. The obtained scaffolds have the porosity up to 90 % and median pore sizes in the range of 15–35 μm with an interconnected pore structure. In comparison to solvent casting, the scaffolds offer much higher specific pore surface area as well as the ability to make thick (>1 cm) polymer scaffolds. Such types of scaffolds have drawn very much attention in tissue engineering. Liu et al. have developed novel chitosan–halloysite nanotubes (HNTs) nanocomposite (NC) scaffolds by combining solution-mixing and freeze–drying techniques [63].

3 Dispersion of Nanoparticles

The properties of nanocomposites depend on the extent of dispersion of nanoparticles in polymer matrices. The dispersion can be evaluated through electron microscope; transmission electron microscope (TEM), Scanning Electron

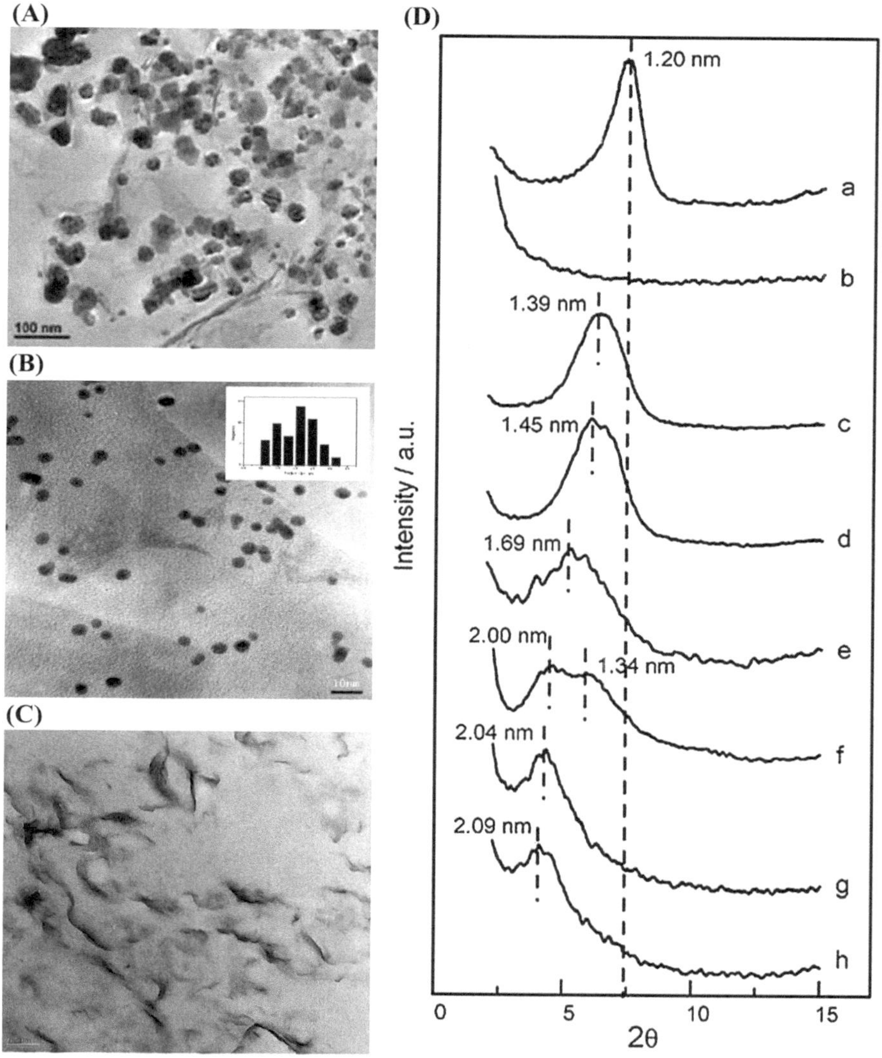

Fig. 2 TEM image of **A** Ag–Fe_3O_4/chitosan nanocomposite, [64] **B** gold/chitosan nanocomposite, [65] and **C** montmorillonite/chitosan nanocomposite, [66] and **D** XRD patterns of **a** Na^+-montmorillonite, **b** chitosan film, and nanocomposites prepared from chitosan–clay ratios of **c** 0.25:1, **d** 0.5:1, **e** 1:1, **f** 2:1, **g** 5:1, and **h** 10:1 [67]

Microscope (SEM), or Atomic Force Microscopy(AFM). The Ag–Fe_3O_4 nanoparticles with spherical shapes having mean diameters of 10–40 nm are dispersed uniformly in the chitin matrix in Fig. 2a [64]. Figure 2b shows the typical TEM image of gold nanoparticle in chitosan–gold nanocomposites showing the histogram of particle size distribution [65]. The image (Fig. 2b) displays monodisperse gold nanoparticles, which are homogeneously distributed in the nanocomposite. The particles are found to be spherical with a narrow size distribution; the average particle diameter is about 4 nm. Gold nanoparticles are found to have a very stable state as no aggregation of gold particles is observed in TEM after 2 months. Wang et al. reported that with the lower content (2.5 wt%) of montmorillonite (MMT) the chitosan–MMT nanocomposite shows the coexistence of both intercalated and exfoliated structures [66]. With the increase in content of MMT (5 wt%) intercalated with occasional flocculation morphology is observed (Fig. 2c). The formation of flocculated structure in chitosan/MMT nanocomposite with higher content of MMT is due to the hydroxylated edge–edge interaction of the silicate layers. The strong interaction between chitosan and MMT through the amino and hydroxyl groups of chitosan with the silicate hydroxylated edge groups, is believed to be the main driving force for the good dispersion of MMT in chitosan matrix to form flocculated structure.

The cationic biopolymer chitosan was intercalated in Na^{+}-montmorillonite to get compact and robust three-dimensional nanocomposites with interesting functional properties [67]. Figure 2d shows the XRD pattern of clay, chitosan, and chitosan–clay composites. The intercalation of chitosan in clay gallery is confirmed by the decrease of 2θ values with the increase of chitosan–clay ratio.

4 Morphology

The morphology of the nanocomposite is investigated using SEM, AFM, and optical microscope. Figure 3a represents the cross-sectional structure of pure chitosan and the chitosan–halloysite nanotubes (HNTs) nanocomposite scaffolds prepared by freeze–drying technique [63]. Both pure chitosan and its nanocomposites display a highly porous, open, and 3D interconnected morphology with the pore size around 200 μm. The chitosan–HNTs nanocomposite scaffolds show more uniform porous structure and less collapse of the pore walls than chitosan which may be attributed to the enhanced pore forming ability of chitosan by HNTs due to the strong interaction. In addition to that, nanocomposite scaffolds with HNTs have slightly larger pore size than that of pure chitosan scaffolds. The interaction between chitosan and HNTs causes absorption of chitosan on the surface of HNTs. It is clear from Fig. 3b (a) that HNTs have cylindrical-shaped tubular morphology and open-ended lumen along the nanotubes. The edge of HNTs is clear and sharp while the edge of chitosan–HNTs is obscure and blurred (Fig. 3b (b)). The SEM image of chitin and chitin–HNTs is shown in Fig. 3c [68]. Both chitin and chitin–HNTs hydrogel display large interconnected pores with pore size of 100–200 μm. The average pore size of chitin and chitin–HNTs hybrid hydrogel with low HNTs concentration are about 200 μm, while the pore walls are found in several micrometers. The hybrid hydrogel exhibits reduced pore size with

(A)

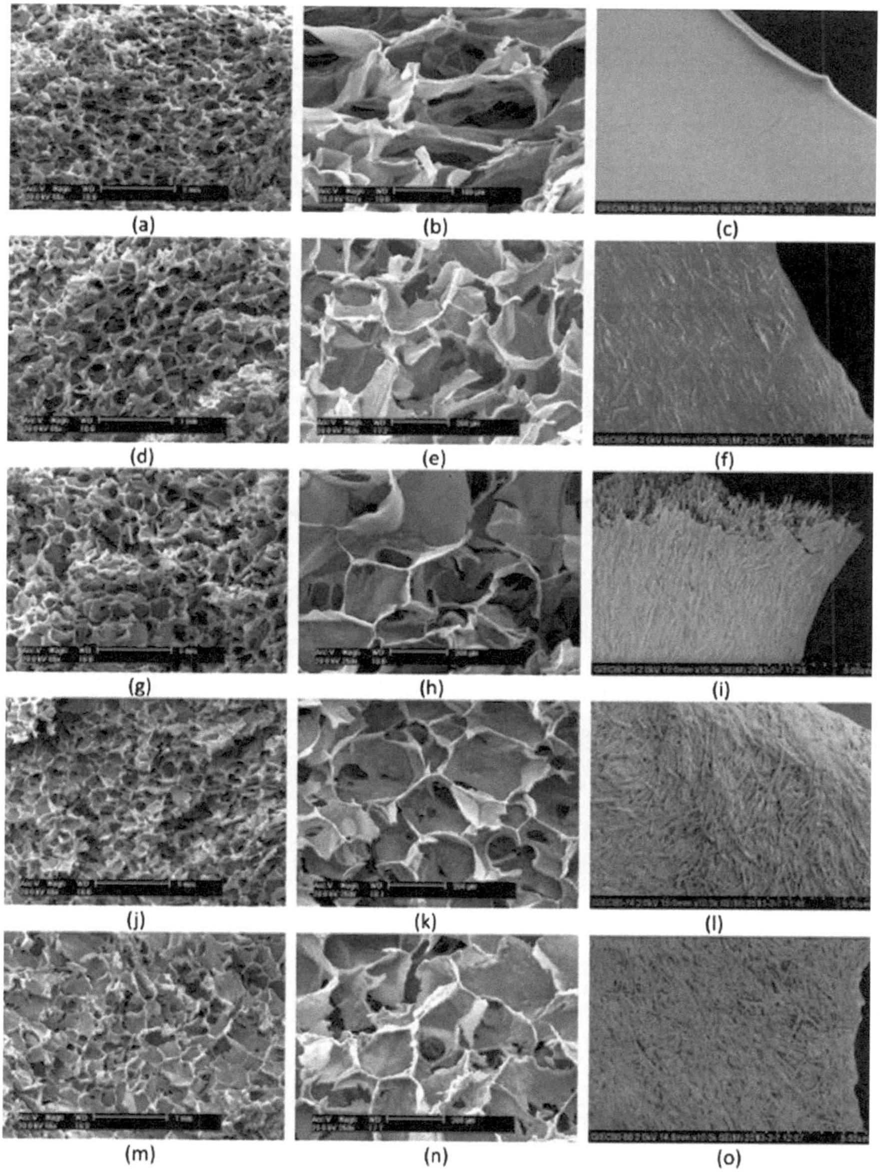

Fig. 3 **A** SEM micrographs of freeze–dried chitosan and chitosan–HNTs NC scaffolds with different magnifications: pure chitosan (**a**–**c**); CS2N1 (**d**–**f**); CS1N1 (**g**–**i**); CS1N2 (**j**–**l**); and CS1N4 (**m**–**o**). [63] **B** AFM images of HNTs and chitosan–HNTs (50 wt% HNTs): (**a** and **c**) HNTs; (**b** and **d**) chitosan–HNTs. The samples were prepared by dipping the dilute HNTs and chitosan–HNTs aqueous dispersions on freshly clean mica, [63] and **C** SEM photos for freeze–dried chitin and chitin–HNTs hybrid hydrogels: **a** chitin; **b** CT2N1; **c** CT1N1; **d** CT1N2; **e** CT1N4 [68]

(B)

Fig. 3 (continued)

high HNTs concentration (CT1N4) as a result of the decrease in water volume in hydrogel. Kumar et al. reported that β-chitin hydrogel/nanohydroxyapatite (nHAp) nanocomposite scaffolds prepared by freeze–drying are porous in nature [69]. They have found that the porosity of the scaffolds is decreased with the increase in the concentration of nHAp. The nHAp particles intact the polymer chain acting as filler and hence cause a reduction in the porosity. In the case of control, the pores were arranged uniformly and interconnected with wall.

(C)

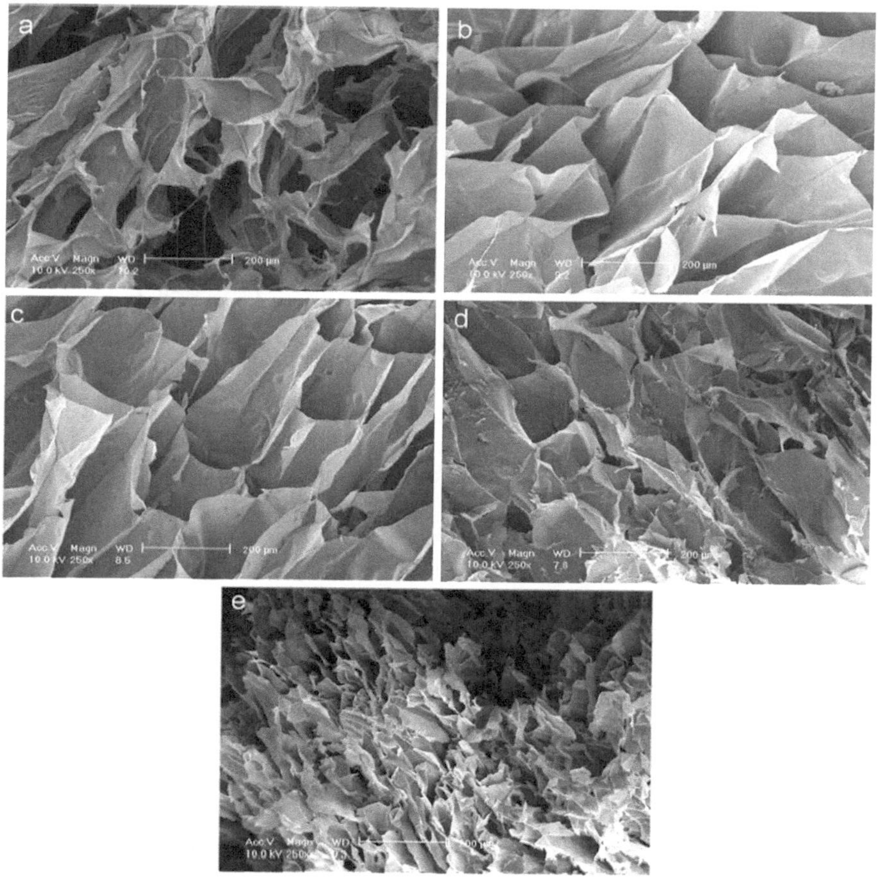

Fig. 3 (continued)

5 Thermal Properties

Thermal stability of nanocomposite is assessed from the weight loss as a function of time. Novel chitosan–halloysite nanotubes (HNTs) nanocomposites (NC) scaffolds were developed by Liu et al. [63]. They have reported that the interaction between HNTs and chitosan can affect the thermal stability of chitosan scaffolds. Figure 4 shows the TGA and DTG curves of chitosan and chitosan–HNTs nanocomposites scaffolds. Three stages of decomposition of chitosan at around 70, 163, and 273 °C are observed, which are related to the loss of free water, loss of bonded water, and the degradation of the chitosan chains, respectively, whereas HNTs show only one peak at 483 °C due to the structural dehydroxylation of HNTs. Chitosan–HNTs nanocomposites show only two peaks which are

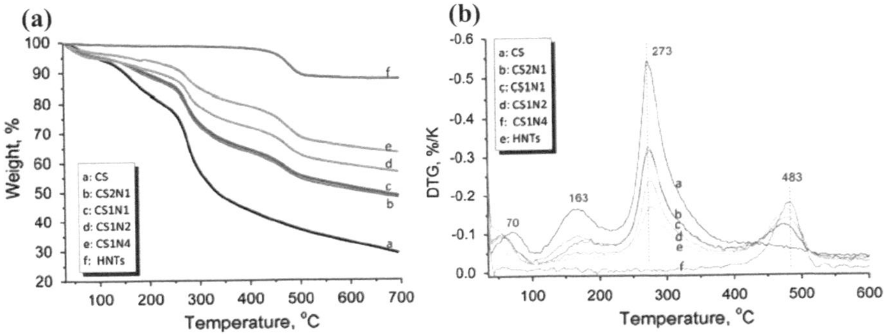

Fig. 4 TGA (**a**) and DTG (**b**) curves for CS, HNTs, and CS–HNTs NC scaffolds [63]

assigned to the degradation of chitosan and HNTs, respectively. It is observed from the DTG curves that the degradation temperature of chitosan component is slightly increased as compared to pure chitosan while the degradation temperature of HNTs component is slightly decreased as compared to pristine HNTs. The degradation temperature of chitosan and HNTs components are 277 and 477 °C in composites where weight ratio of chitosan (CS) and HNTs (N) was 1:4. These degradation temperatures are 4 °C higher and 6 °C lower than pure chitosan and pristine HNTs, respectively. Chitosan–montmorillonite nanocomposites prepared through anion exchange method between water-soluble oligomeric chitosan and a Na^+-montmorillonite shows higher thermal stability than pure chitosan [70]. This thermal resistance arises from the electrostatic interaction between chitosan and silicate layer. Chitosan–hyaluronan (HYA)/nano-chondrotin sulfate (nCS) composites show lower thermal stability than neat chitosan but high temperature resistance than chitosan–HYA [71]. Marroquin et al. reported the thermal degradation of 5 % MWNT/chitosan and 5 % Fe_3O_4/MWNT/chitosan nanocomposite [72]. They observed three major peaks in DTG curves. The first peak around 50–120 °C is associated with the evaporation of physically absorbed and strongly hydrogen-bonded water to chitosan and MWNT. 46 % drop in weight of nanocomposite was observed at around 120–400 °C range. This was associated with depolymerization of chitosan chains through deacetylation and cleavage of glycosidic linkages via dehydration and deamination. The thermal destruction of the pyranose ring produce formic, acetic, and butyric acids as well as a series of lower fatty acids at 400–700 °C. Additional third stage degradation suggested at 750–800 °C corresponds to the reduction of Fe_3O_4 by the reaction with residual carbon. The overall higher thermal stability of Fe_3O_4 containing nanocomposite is due to the incorporation of these nanoparticles into the chitosan matrix which affects the activation energy of the first and major stage of decomposition and also suggests a hindrance of depolymerisation of the main chitosan chains through chemical interactions. Chitosan membranes reinforced by halloysite nanotubes (HNTs) in the concentration range of 2–15 (w/w%) have been prepared by solution casting which shows a drastic enhancement of thermal stability with increasing HNTs concentration in the composite [73]. Enhancement of this thermal stability might be due to two main reasons. The char residue of chitosan/HNTs composite may act as a barrier, therefore, lead to effective delay in heat transport and hinder the volatile escape while help improving the thermal stability. Chitosan/HNTs shows significant thermal stability at higher concentration by adding 15 (w/w%) HNTs into the chitosan matrix. Cellulose nanocrystals (CNCs) enhance the thermal resistance of poly (vinyl alcohol)/chitosan (PVA/CS) bio-nanocomposites films [74]. PVA/CS/CNCs bio-nanocomposite with the 0.5 and 1.0 wt% CNCs show thermal degradation at 288 and 337 °C, respectively. The interactions between the abundant hydroxyl groups of CNCs and the free hydroxyl groups of PVA/CS through hydrogen bonding are key for increasing thermal stability of PVA/CS/CNCs bio-nanocomposite. It is observed that a relatively higher concentration of CNCs like 3 and 5 wt% show degradation at lower temperature at 322 and 311 °C. The decrease in degradation temperature might be due to heterogeneous size distribution and aggregation of nanocrystals within the polymer matrix. Jayakumar et al. reported that with the incorporation of nano-TiO_2 particle into the chitin–chitosan matrix make the chitin–chitosan/nano-TiO_2 composite scaffolds more thermally stable than the control chitin–chitosan scaffolds [75].

6 Mechanical Properties

Wang et al. have prepared nanocomposite of chitosan with the functionalized multiwalled carbon nanotubes (MWCNTs) with different wt% of MWCNTs loading in chitosan matrix [76]. The MWCNTs were functionalized to increase the carboxylic and hydroxyl groups through refluxing in a mixture of concentrated sulphuric acid and nitric acid. They have measured the mechanical properties through uniaxial extension. The typical stress–strain curves (Fig. 5a) showed that tensile modulus and tensile strength of the composite increase about 78 and 94 % with 0.4 wt% loading of MWCTs filler where as it increased to about 93 and 99 %, respectively, with the 0.8 wt% of filler. The increase of mechanical properties of chitosan/MWCNTs nanocomposite as compared to those neat chitosan is due to (a) the reinforcement effect of finely dispersed high-performance MWCNTs nanofillers throughout the chitosan matrix, and, (b) strong interaction of MWCNTs with the chitosan through carboxylic and hydroxyl groups of functionalized MWCNTs with the amino, primary, and secondary hydroxyl groups of chitosan. Yang et al. prepared chitosan/graphene oxide nanocomposite (CS/GO) [77]. The obtained mechanical properties (Fig. 5b) in uniaxial extension indicate that with an increase in GO loading from 0 to 1 wt% not only increase the strength but also elongation at break than those of pure CS. The crystallinity of the chitosan is changed with the addition of GO which is an important factor for the enhancement of tensile properties of the polymers. Casariego et al. showed the higher tensile strength of chitosan–micro/nanoclay (MNC) particle composite films with

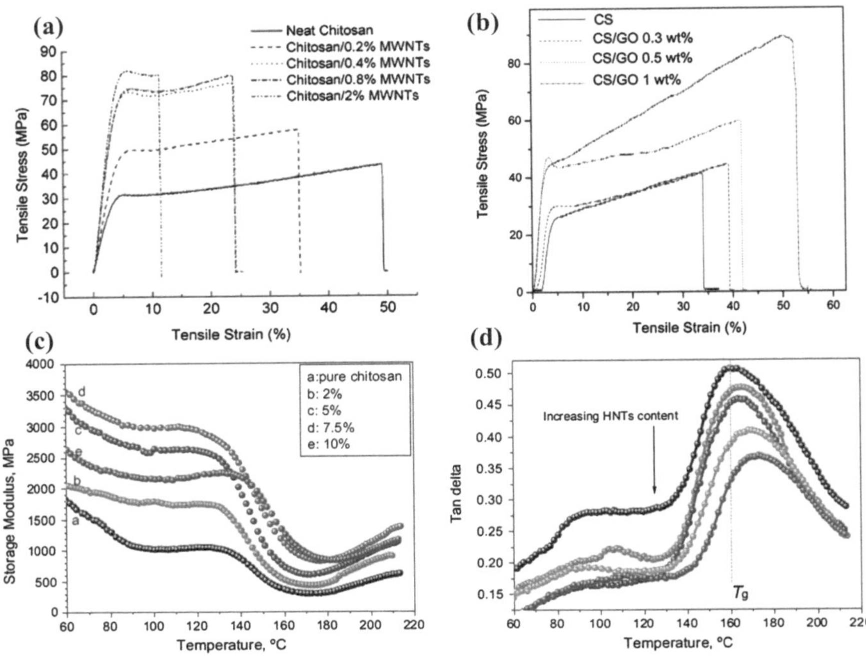

Fig. 5 **a** Typical stress–strain curves of neat chitosan and its MWNTs nanocomposite at a cross-head speed of 5 mm/min, [76] **b** Stress–strain behaviors for the films of CS/GO nanocomposites with different GO loadings [77]. Storage modulus (C) and tan δ (D) versus temperature curves for chitosan/HNTs nanocomposite films [83]

the increase in chitosan concentration whereas the elongation at break decreases [78]. Wang et al. found that hardness and elastic modulus were gradually enhanced with increase in clay concentration in chitosan/montmorillonite nanocomposites [66]. To improve the mechanical properties of chitosan films, chitin whisker is used as an additive [79]. The tensile strength is sharply increased and elongation at break has significantly reduced with the addition of chitin whiskers. It is observed that tensile strength increases exponentially until it reaches a threshold with the increase in whiskers content. Sepiolite (SP) has also been used as reinforce nanofiller to increase the mechanical strength of chitosan (CS)/poly (vinyl alcohol) (PVA) matrix [80]. Tensile strength and tensile modulus were enhanced after introducing the SP into the CS/PVA matrix. It is observed that mechanical properties increase up to certain limit of nanofiller addition. This behavior can be explained by the fact that there exist three-dimensional (3D) network formations through the hydrogen bonds between SP and PVA, PVA and CS. The tensile strength and elongation at break of nanocomposites decrease simultaneously with increasing the SP content. More CS was adhered to the surface of negatively charged SP due to the electrostatic attraction with the increment of SP content. These phenomena will weaken the interaction between CS and PVA, thus, the 3D hydrogen bond network destroys partly. Destruction of the three-dimensional network may lead to the decrease of the mechanical properties of the nanocomposite. The chitin hydrogels are relatively weak which resists their application where high stress is required. Chitin/halloysite nanotubes (HNTs) hybrid hydrogels were prepared with different ration of chitin and HNTs to improve their mechanical property [81]. The Fracture stress of the chitin–HNTs hybrid hydrogels significantly increases in comparison to the pure chitin hydrogels. The mechanical performance is proportional to the content of HNTs in hybrid hydrogels. It is observed that the maximum compressive strength of HNTs hybrid hydrogel is ~300 % higher than pure chitin hydrogels. The chitin–HNTs hybrid hydrogels exhibit a significant increase in mechanical strength as compared to pure chitin and chitosan hydrogels due to the reinforcing effect of HNTs in chitin and the interfacial interactions. The rheological measurement shows that absolute values of the shear modulus (G') for the chitin–HNTs hybrid hydrogels substantially increase with the loading of HNTs especially at high HNTs loadings which is in good agreement with the change of compressive mechanical properties. A significant synergistic effect of Fe_3O_4 and MWNT provides enhanced mechanical properties to chitosan. A 5 wt% loading of Fe_3O_4/MWNT in the nanocomposite enhances the tensile strength and tensile modulus [72]. Tensile strength of 5 % Fe_3O_4/MWNT/chitosan nanocomposite was 159 % higher than that of chitosan. The elastic modulus of the 5 % Fe_3O_4/MWNT/chitosan was 179 % higher than that of chitosan. Variations in storage modulus (E') and tan δ of the 5 % MWNT/chitosan and 5 % Fe_3O_4/MWNT/chitosan nanocomposites are significant. It was found that the 5 % Fe_3O_4/MWNT/chitosan nanocomposite had a higher E' than 5 % MWNT/chitosan nanocomposite which suggests that the interactions between chitosan, MWNT, and Fe_3O_4 are strong and allow an efficient load transfer. The glass transition temperature (T_g) of 5 % Fe_3O_4/MWNT/chitosan is lower than 5 % MWNT/chitosan, which are 161 and 166 °C, respectively. T_g has shifted to lower region and elastic modulus is higher, such type of behavior of Fe_3O_4 is known as antiplasticizing effect. Silva et al. reported that chitosan membrane with 5 (w/w%) halloysite nanotubes (HNTs) has the highest strength as compared to other concentration [73]. The higher strength of chitosan/HNTs is attributed to the interaction of HNTs with the chitosan. The hydroxyl groups on the edge and the surface of HNTs can interact with the hydroxyl and amino groups of the glucosamine rings of the chitosan. This interaction is important for facilitating the stress transfer (generated in the chitosan matrix) to the HNTs which act as a reinforcing agent. The unique synergistic effect of 2D nanotube (CNT) and 1D clay platelet on the reinforcing of chitosan has been demonstrated by Tang et al. [82]. It is observed that the increase of tensile strength is saturated at 3 wt% clay for chitosan/clay and 0.4 wt% CNTs for chitosan/CNTs compopsite which may be influenced by the aggregation of the nanofillers at higher content. The tensile strength and Young's modulus of the nanocomposites are significantly improved by about 171 and 124 %, respectively, as compared with neat chitosan with the incorporation of 3 wt% clay and 0.4 wt% CNTs. The dynamic modulus and tan δ spectra of pure chitosan and chitosan/HNTs nanocomposite films are compared in Fig. 5c, d.

[83]. The modulus of the chitosan/HNTs nanocomposite films increase with the increase in HNTs content until the loading of HNTs reaches at 7.5 %, due to the reinforcing effect of HNTs in chitosan. HNTs are known as inorganic rigid silicate nanotubes with high aspect ratio and hence interfacial reactions taking place between HNTs and chitosan, the chitosan/HNTs nanocomposite films show higher modulus both in glassy as well as rubbery state. Apart from the increased storage modulus, the glass transition temperature of chitosan also increases using HNTs. Composites with 10 wt% HNTs show the maximum T_g at 172 °C, which is 12 °C higher than that of pure chitosan.

7 Tissue Engineering

The cell adhesion, spreading, and cytoskeleton organization are the important parameters in evaluating the cellular compatibility and suitability of biomaterials for a required application. Singh et al. successfully combined the biocompatible nature of chitin with the high electrical conductivity of carbon nanotube to create a biocompatible, electrically conducting scaffold permissive for mesenchymal stem cell (MSCs) function [84]. They have used multiwall carbon nanotubes (MWNTs) coated with carboxymethyl cellulose to prepare composites films with varying weight fraction of MWCNTs. They have performed MTS assay to determine the growth rate and number of MSCs present on the chitin–MWNT films at various MWNT concentrations (Fig. 6a). The MSC viability on chitin–MWNT films was tested using the live/dead viability assay where green fluorescence indicates live cells and red fluorescence shows dead cells as shown in Fig. 6b. The viability of MSCs on chitin–MWNT composite scaffolds was checked by using live and dead assay upto 14 days. The composite film with 0.07 weight fraction of MWNT revealed an increased cell binding capacity. The poor binding of cell of 0.1 weight fraction compared to the 0.07 weight fraction is due to increased hydrophobicity with higher nanotube concentrations. Peter et al. developed chitin/bioactive glass ceramic nanoparticles (nBGC) composite scaffold macro porous in nature of pore size 150–500 μm for tissue engineering applications [85]. Cell culture studies were conducted with osteoblast cells like MG-63. Composites scaffold offers suitable environment to the cells for attachment to the wall of the scaffold composite. The nanocomposite scaffolds are biodegradable and biocompatible and can be used for tissue engineering applications. Gaharwar et al. prepared bio-nanocomposites by the addition of chitosan to silicate (laponite) cross-linked poly(ethelene oxide) (PEO) for creating improved scaffolds for bone repair [86]. The addition of chitosan retards the release of entrapped model macromolecular drug (albumin) from the bio-nanocomposites. The effect is prominent with higher concentration of chitosan and the burst release is suppressed and become sustained after 12 h. An increase in preosteoblast cells adhesion and spreading was observed with the increase in chitosan concentration in chitosan-containing silicate cross-linked PEO bio-nanocomposites. Jayakumar et al. have prepared chitin–chitosan/nano-TiO_2

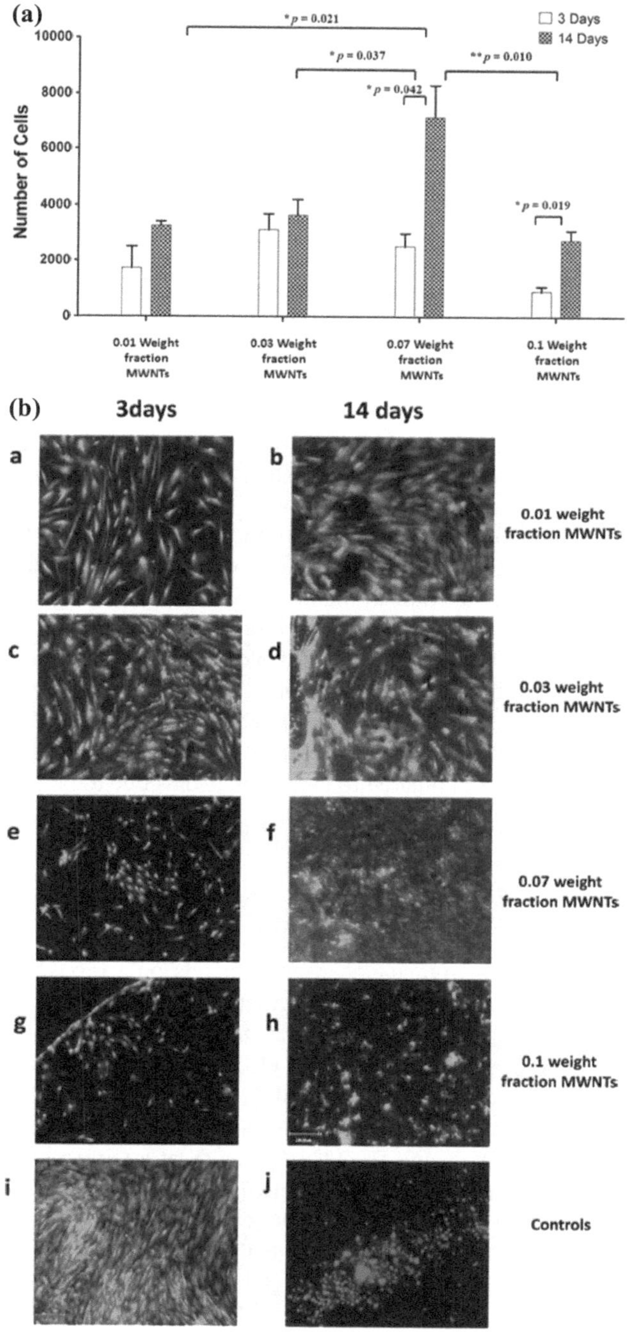
(a)
3 Days
14 Days
Number of Cells
10000
8000
6000
4000
2000
0
*p = 0.021
*p = 0.037
**p = 0.010
*p = 0.042
*p = 0.019
0.01 Weight fraction MWNTs
0.03 Weight fraction MWNTs
0.07 Weight fraction MWNTs
0.1 Weight fraction MWNTs
(b)
3days
14 days
a
b
c
d
e
f
g
h
i
j
0.01 weight fraction MWNTs
0.03 weight fraction MWNTs
0.07 weight fraction MWNTs
0.1 weight fraction MWNTs
Controls

Fig. 6 **a** MTS assay quantifying MSC proliferation on different concentrations of chitin–MWNT films. The positive control in this case is the tissue culture plate (plastic). A cell count below 250 was disregarded from the analysis due to the inherent limitation of the assay. Data are given as mean+standard error. **b** Live and dead viability assay to test the viability of MSCs on the chitin–MWNT film scaffolds. Cells were seeded on the scaffolds and were stained with the Live/Dead viability stain. *Green cells* showed the number of live cells and red cells showed dead cells due to the excitation of fluorescent dye (calcein AM) at 490 nm. The images were obtained after 3 d (*left column*) and 14 d (*right column*) of cells seeding [84]

composites scaffold for bone tissue engineering [75]. They have shown that the cell attachment increases significantly in nanocomposites scaffold as compared to control scaffold. Kumar et al. have prepared β-chitin hydrogel/nano-hydroxyapatite (nHAp) nanocomposites scaffolds with different concentration of nHAp [69]. They have found that more cells were found on the β-chitin composites scaffolds than that of β-chitin control scaffolds. Chen et al. have developed a novel chitosan–gold (CS–Au) hybrid hydrogel as a delivery system for anticancer drug, doxorubicin (DOX) [87]. Encapsulated model drug releases from the hydrogel in a sustained manner over 10 days and the drug releasing property is also affected by the external pH stimuli which showed its potential application for smart drug delivery. Kumar et al. have produced pectin–chitosan nanocomposites scaffold containing $CaCO_3$ nanopowder with an intention to use it in biomedical application such as tissue engineering and drug delivery [88]. The scaffolds have adequate potential for efficient cell adhesion, proliferation and also proved its potency in site-specific drug delivery. Nanocomposites scaffolds of chitin–chitosan with nano-ZrO_2 have been developed by Jayakumar et al. [89]. The nanocomposite scaffold increased the cell attachment significantly as compared to control scaffolds due to large surface area.

8 Drug Release

The drug release kinetics from nanocarriers depends on various factors such as pH, degaradation rate, particle size, and interaction between the drug and the surface of the polymer [90]. Ardeshirzadeh et al. have developed polyethylene oxide (PEO)/chitosan (CS)/graphene oxide (GO) electrospun nanofibrous scaffolds for controlled release of Doxorubicin (DOX) [91]. The different content of GO (0.1, 0.2, 0.5, and 0.7 wt%) was used to prepare the scaffolds. They have studied the DOX release from GO-DOX and PEO/CS/GO/DOX at a temperature of 37 °C in the phosphate buffer solutions (pH = 5.3 and 7.4) and is shown in Fig. 7a. DOX releases from GO-DOX and PEO/CS/GO/DOX nanofibers occurred at slower and controlled manner at pH 7.4 as compared to the release in acidic condition (pH 5.3). Hydrogen-bonded interaction between GO and DOX become stable in neutral condition which results in controlled release of DOX from GO/DOX and PEO/CS/GO/DOX. In acidic condition, such type of hydrogen-bonded interaction becomes weaker resulting faster release of drug. The diffusion of DOX from interconnected

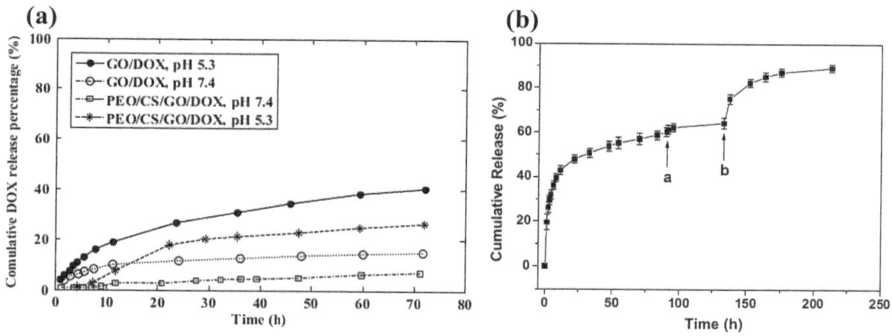

Fig. 7 **a** The DOX release profiles of GO-DOX and PEO/CS/GO/DOX at pH of 5.4 and 7.3. [91]. **b** Cumulative release of DOX·HCl from CS–Au hydrogel at 37 °C and pH 7.4. Point a: increase temperature to 70 °C for 1 h, then reincubate the hydrogel in PBS at 37 °C; Point b: the pH value of PBS was changed to 3.0 [87]

nanofibrous layer exerted an additional barrier for controlled and prolonged release rate. Such type of system could be a promising candidate for controlled and targeted delivery. Chen et al. studied the cumulative doxorubicin (DOX) release from chitosan–gold (CS–Au) hybrid hydrogel at 37 °C in PBS solution (Fig. 7b) [87]. CS–Au hydrogel shows sustained release profile with 60 % of the loaded DOX to the medium within 80 h. No more DOX release was found after that due to the stable gel network which retards the diffusion of DOX from the deep gel pores. Therefore, they have increased the incubation temperature to destroy the hydrogel structure to accelerate the release rate. After 30 h additional incubation at 70 °C cumulative release did not exceed 65 %. Chitosan is in the deprotonated state at pH 7.4 and it is negatively charged while DOX is positively charged. CS and DOX can form complex through the strong electrostatic interaction which is responsible for incomplete release of DOX from CS–Au hydrogel. A burst release was observed when the hydrogel was immersed in PBS with pH 3.0. More than 90 % of loaded DOX was released into the medium after 30 h of incubation at pH 3.0 which confirmed the electrostatic interaction between CS and DOX. The drug release profile is also influenced by external stimuli, such as temperature and pH. So CS–Au hydrogel may be used as a smart drug release system. Kumar et al. have shown the release of Fosamax from the pectin, chitin, and nano-$CaCO_3$ nanocomposite scaffolds [88]. Sustained release of drug was observed from the scaffold. Release rate can be correlated with the hydrophilic nature of Fosamax which will diffuse out in the aqueous medium easily. Nanohydrogel composed of chitosan and montmorillonite (MTT) was prepared for drug release behavior following electrostimulation by Liu et al. [92]. The drug release behavior was strongly influenced by the concentration of MTT under an applied voltage. Vitamin B_{12} displays pseudo-zero-order release kinetics and the release mechanism shifted from being diffusion-controlled to swelling-controlled mode with a low MMT content (1 wt%) but with an MMT content exceeding 1 wt%, both the diffusion exponent '*n*' and the responsiveness to electrical stimulation were decreased. Nivethaa et al. synthesized chitosan/gold nanocomposite through simple chemical reduction method for drug delivery carrier for 5-fluorouracil [93]. The release of 5-fluorouracil shows two phase behavior; one up to 40 h and the next phase is up to 72 h. Slow, sustained, and prolonged release was observed in first phase and a sudden burst release was observed during the second phase. First phase release fitted well to the zero-order release kinetics which refers to the constant release of drug from a drug delivery device and the second phase release fitted well to the Higuchi kinetics.

9 Antimicrobial Activity

Shariatinia et al. have developed nanocomposite film of chitosan (CS)/phosphoramide (Ph)/silver nanoparicles (Ag NPs) containing 1–5 % of Ag NPs and investigated their antibacterial activity [94]. They have performed in vitro antibacterial activities of chitosan (CS), phosphoramide (Ph), Ag nanoparticle (Ag NPs), nanocomposite films containing 1–5 % Ag NPs and nanocomposite films of 1–5 % Ag NPs (as control films without Ph) using four different bacteria: two Gram-positive *Staphylococcus aureus*, *Bacillus cereus* and two Gram-negative *Escherichia coli*, *P. aeruginosa* bacteria. The results are summarized in Table 1. Tested samples are much more effective against gram-positive bacteria than that of gram-negative one because of lack of outer membrane in gram-positive bacteria. Antibacterial effect has increased by increasing the Ag NPs percent from 1 to 5 % in the CS/Ph/Ag Nps nanocomposite films (Fig. 8) comparing zone of inhibition (mm). The control films containing 1–5 % Ag NPs showed less antibacterial

Table 1 The inhibition zone (mm) measured for the antibacterial activities (Mean ($n = 3$) ± standard deviation) [94]

Sample	Bacterium type			
	S. aureus	*B. cereus*	*E. coli*	*P. aeruginos*
Ag NPs	10 ± 0.46	8.8 ± 0.47	7.5 ± 0.50	7.66 ± 0.48
Phosphoramide (Ph)	7.5 ± 0.45	7.2 ± 0.48	6.9 ± 0.48	6.7 ± 0.46
Chitosan (CS)	13.33 ± 0.50	11.66 ± 0.49	9.66 ± 0.49	10.33 ± 0.47
CS+Ph	14.66 ± 0.49	12.66 ± 0.48	10 ± 0.47	11 ± 0.49
CS+AgNPs 1 %	15.50 ± 0.47	13.50 ± 0.50	11.33 ± 0.48	13 ± 0.50
CS+Ag NPs 2 %	15.55 ± 0.48	13.80 ± 0.46	11.50 ± 0.50	13.10 ± 0.48
CS+Ag NPs 3 %	16 ± 0.50	14 ± 0.48	11.66 ± 0.47	13.33 ± 0.49
CS+Ag NPs 4 %	16.33 ± 0.46	14.33 ± 0.49	11.80 ± 0.49	13.50 ± 0.50
CS+AgNPs 5 %	16.66 ± 0.48	14.50 ± 0.50	12 ± 0.48	13.66 ± 0.47
Film 1	16 ± 0.46	14 ± 0.47	11.80 ± 0.50	13.33 ± 0.46
Film 2	16.66 ± 0.47	14.50 ± 0.50	12 ± 0.45	13.50 ± 0.50
Film 3	17.33 ± 0.49	14.80 ± 0.49	12.33 ± 0.48	13.66 ± 0.48
Film 4	17.66 ± 0.48	15 ± 0.48	12.50 ± 0.49	13.70 ± 0.47
Film 5	18.66 ± 0.47	15.66 ± 0.46	12.66 ± 0.48	13.80 ± 0.49

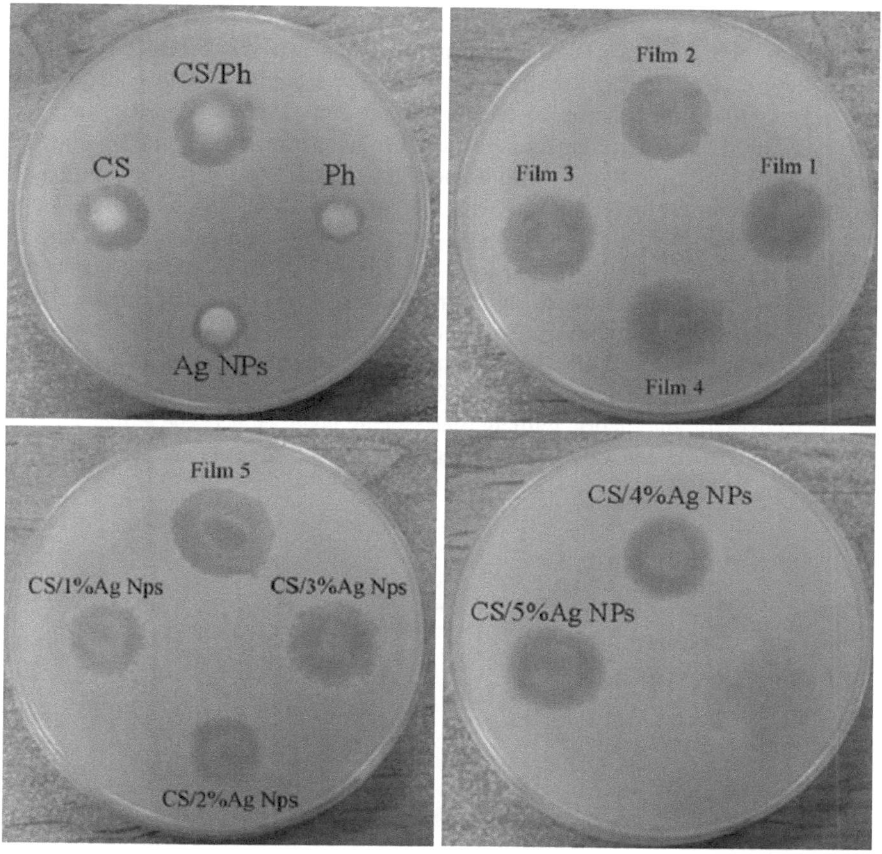

Fig. 8 The digital photographs of antibacterial activities of Ag NPs, chitosan (CS), phosphoramide (Ph), and CS/Ph film, nanocomposite films 1–5 plus CS/1–5 % Ag Nps against *S. aureus* bacterium [94]

activities than those nanocomposoites films. Sharma et al. have prepared silver nanocomposite film using chitosan and alginate (1:1) blend and tested their antibacterial activity against gram-positive and gram-negative bacteria [95]. The blended nanocomposite film showed antibacterial activity against both gram-negative and gram-positive bacteria with more inhibition using gram-positive bacteria. Usually gram-positive bacteria have less negative charge on its cell wall as compared to gram-negative bacteria. The silver nanocomposites have shown overall negative charge which will interact more with the gram-positive bacteria as compared to higher negative charge containing gram-negative bacteria. Hebeish et al. prepared chitosan-grafted-poly acrylonitrile silver nanocomposites (CS-g-PAN/Ag) and performed antibacterial test using *E. coli* and *S. aureus* [61]. The growth of the bacteria was significantly affected in CS-g-PAN/Ag nanocomposite than that of PAN and

CS-g-PAN copolymer. Silver nanocomposites of poly(ethylene glycol-di-aldehyde) cross-linked chitosan have shown good antibacterial activity toward *E. coli* [96]. Regiel-Futyra et al. checked antibacterial activity of chitosan–gold nanocomposite where chitosan of different molecular weight and different degree of deacetylation was used to prepare nanocomposite [60]. Nanocomposite based on chitosan with medium molecular weight with highest gold content had very good antibacterial effect in comparison to chitosan with low and high molecular weight CS-based composites. Chitosan/ZnO nanocomposite films exhibit good antibacterial activity against gram-positive and gram-negative bacteria [97]. Han et al. have performed antibacterial test of pure chitosan and montmorillonite and chitosan–montmorillonite nanocomposites [70]. Interestingly, chitosan–montmorillonite nanocomposites show higher antimicrobial activity than pure chitosan. Chitosan itself shows antibacterial activity due its cationic nature. Cationic charge of chitosan is neutralized via an electrostatic interaction with anionic silicate layer. Hence, the enhanced antimicrobial activity of nanocomposites is due to synergic effect because the chitosan

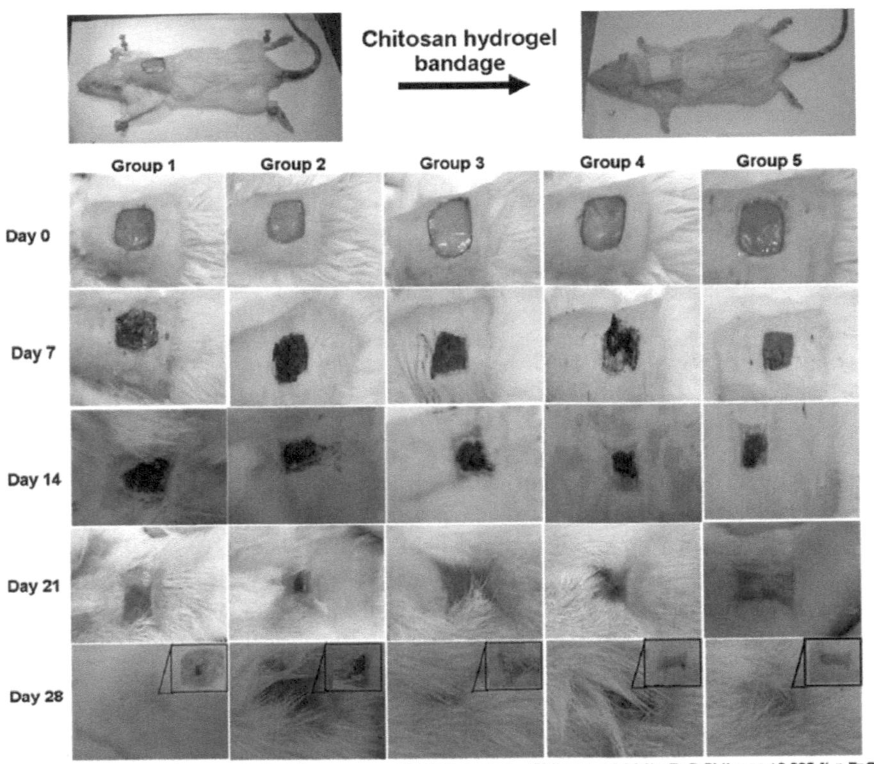

Fig. 9 Photographs of an in vivo wound healing study. Note the extent of wound closure in the wounds treated with chitosan control and CZBs [100]

molecules are evenly distributed through the inorganic matrix. Nanocomposite films of chitosan, starch, glycerin, cyclophosphamide, and Fe_3O_4 nanoparticles have shown good antibacterial activity against gram-positive bacteria as compared to gram-negative one [98]. Wysokowski et al. have shown that β-chitin/ZnO nanocomposite scaffolds exhibit good antibacterial activity against gram-positive Bacillus subtilis but there is no effect on gram-negative *E. coli* [99].

10 Implant Materials

Chitosan nanocomposites are widely used as implant materials. Sudheesh Kumar et al. developed chitosan hydrogel/nano-ZnO nanocomposite bandages (CZBs) via the incorporation of ZnO nanoparticle (nZnO) into chitosan hydrogel [100]. In vivo study was conducted in Sprague-Dawley (SD) rats to prove the enhanced wound healing ability of the prepared CZBs. Figure 9 shows the digital photographs of an in vivo wound healing study. Chitosan control as well as nanocomposites bandages show excellent healing after 1 and 2 weeks, compared to Kaltostat and bare wound. Wounds treated with chitosan control and CZBs achieved significant closure to ~90 % after 2 weeks as compared to Kaltostat-treated wound and bare wounds, which showed 70 % closure (Fig. 10a). There

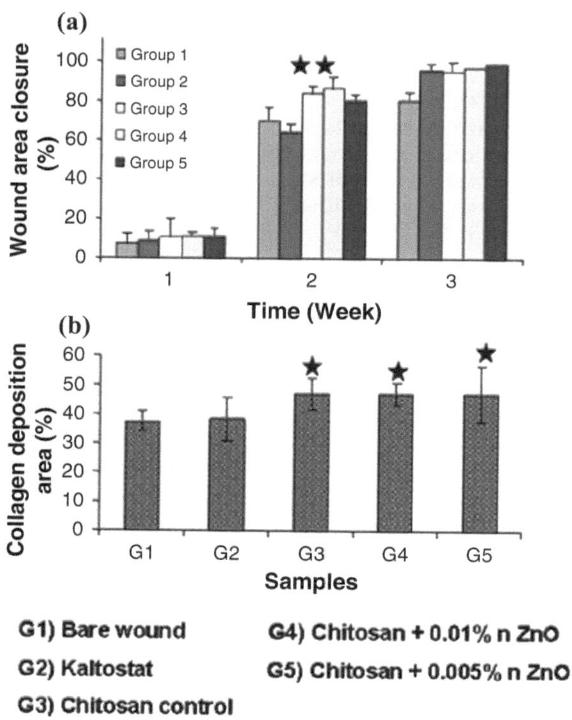

Fig. 10 **a** Evaluation of the wound area closure and **b** study of the collagen deposition area. (In both graphs, the *star symbols* represent the p 0.05 level, indicating that the means are significantly different, compared with the control) [100]

was enhanced collagen deposition in the wound healing assisted by the chitosan controls and CZBs after 4 weeks. CZBs showed controlled degradation, enhanced blood clotting, excellent platelet activation ability, cytocompatibility, and antibacterial activity. All these properties indicate that advanced CZBs can be used for burn, chronic, and diabetic wound infections. The development of a CS–hyaluronan/nano-chondroitin sulfate ternary composite sponge for medical use has been reported by Anisha et al. This nanocomposite sponges with enhanced swelling and blood clotting ability were cytocompatible and showed enhanced proliferation of human dermal fibroblast cells [101]. Nano-Ag incorporated chitosan films were developed which showed faster wound healing in Spraque Dawley rats [102].

Acknowledgments The author (A K. Mahanta) gratefully acknowledges the financial support from Council for Scientific and Industrial Research (CSIR-UGC), New Delhi in the form of fellowship. The authors also acknowledge the receipt of research funding from Council for Scientific and Industrial Research (CSIR), New Delhi, Government of India (Project No. 02(0074)/12/EMR-II).

References

1. Vacanti JP, Vacanti CA, Lanz RP, Langer R, Vacanti J (2000) Principles of tissue—engineering, 2nd edn. Academic Press, California 3
2. Hench LL, Polak JM (2002) Third-generation biomedical materials. Science 295:1014–1017
3. Ramires PA, Romito A, Cosentino F, Milella E (2001) The influence of titania/hydroxyapatite composite coatings on in vitro osteoblasts behavior. Biomaterials 22:1467–1474
4. Rinaudo M (2006) Chitin and chitosan: properties and applications. Prog Polym Sci 31:603–632
5. Aranaz I, Mengíbar M, Harris R, Panos I, Miralles B, Acosta N, Galed G, Heras A (2009) Functional characterization of chitosan. Curr Chem 3:203–230
6. Honarkar H, Barikani M, Honarkar H, Barikani M (2009) Applications of biopolymers I: Chitosan. Monatsh Chem 140:1403–1420
7. Hein S, Wang K, Stevens WF, Kjems J (2008) Chitosan composites for biomedical applications: status, challenges and perspectives. Mater Sci Technol 24:1053–1061
8. Shi C, Zhu Y, Ran X, Wang M, Su Y, Cheng T (2006) Therapeutic potential of chitosan and its derivatives in regenerative medicine. J Surg Res 133:185–192
9. Dash M, Chiellini F, Ottenbrite RM, Chiellini E (2011) Chitosan a versatile semi-synthetic polymer in biomedical applications. Prog Polym Sci 36:981–1014
10. Rudall K, Kenchington MW (1973) The chitin system. Biol Rev 48:597–633
11. Ifuku S, Morooka S, Morimoto M, Saimoto H (2010) Acetylation of chitin Nanofibers and their transparent nanocomposite films. Biomacromolecules 11:1326–1330
12. Onishi H, Machida Y (1999) Biodegradation and distribution of water-soluble Chitosan in mice. Biomaterials 20:175–182
13. Ravi Kumar MNV, Muzzarelli RAA, Muzzareli C, Sasfiwa H, Domb AJ (2004) Chitosan chemistry and pharmaceutical perspectives. Chem Rev 104:6017–6084
14. Honarkar H, Barikani M, Honarkar H, Barikani M (2009) Applications of biopolymers I: Chitosan. Monatsh Chem 140:1403–1420
15. Chitin Tasch P (1979) In: Fairbridge RW, Jablonski D (eds) Paleontology, encyclopedia of earth science. Springer, Berlin, pp 186–189
16. Jang MK, Kong G, Jeong Y, Lee CH, Nah JW (2004) Physicochemical characterization of α chitin, β chitin, and γ chitin separated from natural resources. J Polym Sci part A: PolymChem 42:3423–3432

17. Anitha A, Sowmya S, Sudheesh Kumar PT, Deepthi S, Chennazhi KP, Ehrlirch H, Tsurkan M, Jayakumar R (2014) Chitin and chitosan in selected biomedical applications. Prog Polym Sci 39:1644–1667
18. Gong X, Peng S, Wen W, Sheng P, Li W (2009) Design and fabrication of magnetically functionalized core/shell microspheres for smart drug delivery. Adv Funct Mater 19:292–297
19. Hsieha WC, Changb CP, Gaoc YL (2006) Controlled release properties of Chitosan encapsulated volatile Citronella Oil microcapsules by thermal treatments. Colloids Surf B 53:209–214
20. Yamamoto H, Amaike M (1997) Biodegradation of cross-linked chitosan gels by microorganisms. Macromolecules 30:3936–3937
21. Risbud MV, Bhonde RR (2000) Polyacrylamide-chitosan hydrogels: in vitro biocompatibility and sustained antibiotic release studies. Drug Deliv 7:69–75
22. Roughley P, Hoemann C, Desrosiers E, Mwale F, Antoniou J, Alini M (2006) The potential of chitosan-based gels containing invertebraldisc cells for nucleus pulposus supplementation. Biomaterials 27:388–396
23. Ehrlich H, Krajewska B, Hanke T, Born R, Heinemann S, KniebC Worch H (2006) Chitosan membrane as a template for hydroxyap-atite crystal growth in a model dual membrane diffusion system. J Membr Sci 273:124–128
24. Jayakumar R, Divya RVV, Shalumon KT, Sudheesh Kumar PT, Nair SV, Furuike T, Tamura H (2009) Bioactive and osteoblast cell attachmentstudies of novel α- and β- chitin membranes for tissue-engineering applications. Int J Biol Macromol 45:260–264
25. Jayakumar R, Prabaharan M, Nair SV, Tamura H (2010) Novel chitin and chitosan nanofibers in biomedical applications. Biotechnol Adv 28:142–150
26. Morganti P, Morganti G (2008) Chitin nanofibrils for advanced cosmeceuticals. Clin Dermatol 26:334–340
27. Jayakumar R, Reis RL, Mano JF (2007) Synthesis and characterization of pH-sensitive thiol-containing chitosan beads for controlled drugdelivery applications. Drug Deliv 14:9–17
28. Prabaharan M, Mano JF (2005) Chitosan-based particles as controlled drug delivery systems. Drug Deliv 12:41–57
29. Anitha A, Divya RVV, Krishna R, Sreeja V, Selvamurugan N, Nair SV, Tamura H, Jayakumar R (2009) Synthesis, characterization, cyto-toxicity and antibacterial studies of chitosan, O-carboxymethyl, N, O-carboxymethyl chitosan nanoparticles. Carbohydr Polym 78:672–677
30. Anitha A, Deepagan VG, Divya RVV, Deepthy M, Nair SV, Jayakumar R (2011) Preparation, characterization, in vitro drug release and biologicalstudies of curcumin loaded dextran sulphate-chitosan nanoparti-cles. Carbohydr Polym 84:1158–1164
31. Drury JL, Mooney DJ (2003) Hydrogels for tissue engineering: scaffold design variables and applications. Biomaterials 24:4337–4351
32. Jayakumar R, Prabaharan M, Sudheesh Kumar PT, Nair SV, Tamura H (2011) Biomaterials based on chitin and chitosan in wound dressing applications. Biotechnol Adv 29:322–337
33. Prabaharan M, Jayakumar R, Nair SV (2012) Electrospun nanofibrous scaffolds -current status and prospects in drug delivery. Adv Polym Sci 246:241–262
34. Muramatsu K, Masuda S, Yoshihara S, Fujisawa A (2003) In vitro degra-dation behavior of freeze-dried carboxymethyl-chitin sponges processed by vacuum-heating and gamma irradiation. Polym Degrad Stab 81:327–332
35. Vacanti CA (2006) The history of tissue engineering. J Cell Mol Med 10:569–576
36. Krajewska B (2005) Membrane-based processes performed with use of chitin/chitosan materials. Sep Purif Technol 41:305–312
37. Venkatesan J, Kim SK (2010) Chitosan composites for bone tissue engineering—an overview. Mar Drugs 8:2252–2266
38. Jayakumar R, Nwe N, Tokura S, Tamura H (2007) Sulfated chitin and chi-tosan as novel biomaterials. Int J Biol Macromol 40:175–181

39. Tigli RS, Gumusderelioglu M (2009) Evaluation of alginate-chitosan semi IPNs as cartilage scaffolds. J Mater Sci Mater Med 20:699–709
40. Ragetly GR, Slavik GJ, Cunningham BT, Schaeffer DJ, Griffon DJ (2010) Cartilage tissue engineering on fibrous chitosan scaffolds produced by a replica molding technique. J Biomed Mater Res, Part A 93:46–55
41. Ragetly GR, Griffon DJ, Lee HB, Fredericks LP, Gordon-Evans W, Chung YS (2010) Effect of chitosan scaffold microstructure on mesenchymal stem cells chondrogenesis. Acta Biomater 6:1430–1436
42. Mourya VK, Inamdar NN (2008) Chitosan-modifications and applications: opportunities galore. React Funct Polym 68:1013–1051
43. Suh JKF, Matthew HWT (2000) Application of chitosan-based polysaccharide biomaterials in cartilage tissue engineering: a review. Biomaterials 21:2589–2598
44. Muzzarelli RAA (2009) Chitins and chitosans for the repair of wounded skin, nerve, cartilage and bone. Carbohydr Polym 76:167–182
45. Dai M, Zheng XL, Xu X, Kong XY, Li XY, Guo G, Luo F, Zhao X, Wei YQ, Qian Z (2009) Chitosan–alginate sponge: preparation and application in curcumin delivery for dermal wound healing in rat. J Biomed Biotechnol 595126/1–8
46. Sudheesh Kumar PT, Abhilash S, Manzoor K, Nair SV, Tamura H, Jayakumar R (2010) Preparation and characterization of novel β-chitin/nano silver composite scaffolds for wound dressing applications. Carbohydr Polym 80:761–777
47. Kofuji K, Huang Y, Tsubaki K, Kokido F, Nishikaw K, Isobe T, Murata M (2010) Preparation and evaluation of a novel wound dressing sheet comprised of β-glucan-chitosan complex. React Funct Polym 70:784–789
48. Cai ZX, Mo XM, Zhang KH, Fan LP, Yin AL, He CL, Wang HS (2010) Fabrication of chitosan/silk fibroin composite nanofibers for wound-dressing applications. Int J Mol Sci 11:3529–3539
49. Kang YO, Yoon IS, Lee SY, Kim DD, Lee SJ, Park WH, Hudson SM (2010) Chitosan coated poly (vinyl alcohol) nanofibers for wound dressings. J Biomed Mater Res B 92:568–576
50. Dong Y, Liu HZ, Xu L, Li G, Ma ZN, Han F, Yao HM, Sun YH, Li SM (2010) A novel CHS/ALG bi-layer composite membrane with sustained antimicrobial efficacy used as wound dressing. Chin Chem Lett 21:1011–1014
51. Radhakumary C, Antonty M, Sreenivasan K (2011) Drug loaded thermo responsive and cytocompatible chitosan based hydrogel as a potential wound dressing. Carbohydr Polym 83:705–713
52. Dev A, Binulal NS, Anitha A, Nair SV, Furuike T, Tamura H, Jayakumar R (2010) Preparation of novel poly (lactic acid)/chitosan nanoparticles for anti HIV drug delivery applications. Carbohydr Polym 80:833–838
53. Dev A, Mohan JC, Sreeja V, Tamura H, Patzke GR, Hussain F, Weyeneth S, Nair SV, Jayakumar R (2010) Novel carboxymethyl chitin nanoparticles for cancer drug delivery applications. Carbohydr Polym 79:1073–1079
54. Ravi Kumar MNV (2000) A review of chitin and chitosan applications. React Funct Polym 46:1–27
55. Sashiwa H, Aiba S (2004) Chemically modified chitin and chitosan as bio-materials. Prog Polym Sci 29:887–888
56. Ravi Kumar MNV, Muzzarelli RAA, Muzzarelli C, Sashiwa H, Domb AJ (2004) Chitosan chemistry and pharmaceutical perspectives. Chem Rev 104:6017
57. Yi H, Wu LQ, Bentley WE, Ghodssi R, Rubloff GW, Culver JN, Payne GF (2005) Biofabrication with chitosan. Biomacromolecules 6:2881–2894
58. Goldstein AS, Juarez TM, Helmke CD, Gustin MC, Mikos AG (2001) Effect of convection on osteoblastic cell growth and function in biodegradable polymer foam scaffolds. Biomaterials 22:1279–1288
59. Dutta PK, Srivastava R, Dutta J (2013) Functionalized nanoparticles and chitosan-based functional nanomaterials. Adv Polym Sci 254:1–50

60. Regiel-Futyra A, Liskiewicz MK, Sebastian V, Irusta Silvia, Arruebo M, Stochel G, Kyzioł A (2015) Development of noncytotoxic chitosan−gold nanocomposites as efficient antibacterial materials. Appl Mater Interf 7:1087–1099
61. Hebeisha AA, Ramadana MA, Montasera AS, Farag AM (2014) Preparation, characterization and antibacterial activity ofchitosan-g-poly acrylonitrile/silver nanocomposite. Int J Biol Macromol 68:178–184
62. Naseri N, Algan C, Jacobs V, John M, Kristiin Oksman K, Mathew AP (2014) Electrospun chitosan-based nanocomposite mats reinforced withchitin nanocrystals for wound dressing. Carbohydr Polym 109:7–15
63. Liu M, Wu C, Jiao Y, Xiong S, Zhou C (2013) Chitosan–halloysite nanotubes nanocomposite scaffolds for tissue engineering. J Mater Chem B 1:2078–2089
64. Duan B, Liu F, He M, Zhang L (2014) Ag–Fe_3O_4 nanocomposites@chitin microspheres constructed by in situ one-pot synthesis for rapid hydrogenation catalysis. Green Chem 16:2835–2845
65. Huang H, Yuan Q, Yang X (2005) Morphology study of gold–chitosan nanocomposites. J Colloid Interf Sci 282:26–31
66. Wang SF, Shen L, Tong YJ, Chen L, Phang IY, Lim PQ, Liu TX (2005) Biopolymer chitosan/montmorillonite nanocomposites: preparation and characterization. Polym Degrad Stab 90:123–131
67. Darder M, Colilla M, Hitzky R (2003) Biopolymer—clay nanocomposites based on chitosan intercalated in montmorillonite. Chem Mater 15:3774–3780
68. Liu M, Zhang Y, Li J, Zhou C (2013) Chitin-natural clay nanotubes hybrid hydrogel. Int J Bio Macromol 58:23–30
69. Sudheesh Kumar PT, Srinivasan S, Lakshmanan VK, Tamura H, Nair SV, Jayakumar R (2011) b-Chitn hydrogel/nanohydroxyapatite composite scaffolds for tisuue engineering applications. Carbohydr Polym 85:584–591
70. Han YS, Lee SH, Choi KH, Park I (2010) Preparation and characterization of chitosan-clay nanocomposites with antimicrobial activity. J Phys Chem Solids 71:464–467
71. Anisha BS, Sankar D, Mohandas A, Chennazhi Nair, Jayakumar R (2013) chitosano-hyaluronan/nano chondroitin sulfate ternary composite sponges for medical use. Carbohydr Polym 92:1470–1476
72. Marroquin JB, Rhee KY, Park SJ (2013) Chitosan nanocomposite films: Enhanced electrical Conductivity, thermal stability, and mechanical properties. Carbohydr Polym 92:1783–1791
73. Silva RTD, Pasbakhsh P, Goh KL, Chai SP, Ismail H (2013) Physico-chemical characterization of chitosan/halloysite composite membranes. Polym Test 32:265–271
74. Azizi S, Ahmad MB, Ibrahim NA, Hussein MZ, Namvar F (2014) Preparation and properties of poly (vinyl alcohol)/chitosan Blend Bio-nanocomposites reinforced by cellulose nanocrystals. Chin J Polym Sci 12:1620–1627
75. Jayakumar R, Ramachandran R, Divyarani VV, Chennazhi KP, Tamura H, Nair SV (2011) Fabrication of chitin-chitosan/nano TiO_2-composite scaffolds for tissue engineering application applios. Int J Biol Macromol 48:336–344
76. Wang SF, Shen Lu, Zhang WD, Tong YJ (2005) Preparation and mechanical properties of chitosan/carbon nanotubes composites. Biomacromolecules 6:3067–3072
77. Yang X, Tu Y, Li L, Shang S, Tao X (2010) Well-disperesed chitosan/graphene oxide nanocomposites. Appl Mater Interf 6:1707–1713
78. Casariego A, Souza BWS, Cerqueira MA, Teixeira JA, Cruz L, Diaz R, Vicente AA (2009) Chitosan/clay films properties as affected by biopolymer and clay micro/nanoparticles' concentrations. Food Hydrocoll 23:1895–1902
79. Rubentheren V, Ward TA, Chee CY, Tang CK (2015) Processing and analysis of chitosan nanocomposites reinforced with chitin whiskers and tannic acid as a crosslinker. Carbohydr Polym 115:379–387
80. Huang D, Mu B, Wang A (2012) Preparation and properties of chitosan/poly (vinyl alcohol) nanocomposite film rein forced with rod-like sepiolite. Mater Lett 86:69–72

81. Liu M, Zhang Y, Li J, Zhou C (2013) Chitin-natural clay nanotubes hybrid hydrogel. Int J Biol Macromol 58:23–30
82. Tang C, Xiang L, Su J, Wang K, Yang C, Zhang Q, Fu Q (2008) Largely improved tensile properties of chitosan film via unique synergistic reinforcing effect of carbon nanotube and clay. J Phys Chem B 112:3876–3881
83. Liu M, Zhang Y, Wu C, Xiong S, Zhou C (2012) Chitosan/halloysite nanotubes bionanocomposite: structure, mechanical properties and biocompatibility. Int J Biol Macromol 51:566–575
84. Singh N, Koziol KKK, Chen J, Patil AJ, Gilman JW, Trulove PC, Kafienah W, Rahatekar SS (2013) Iionic liquids-based processing of electrically conducting chitin nanocomposite scaffolds for stem cell growth. Green Chem 15:1192–1202
85. Peter M, Sudheesh Kumar PT, Binulal NS, Nair SV, Tamura H (2009) Development of novel a-chitin/nanobioactive glass ceramic composite scaffolds for tissue engineering applications. Carbohydr Polym 78:926–931
86. Gaharwar AK, Schexnailder PJ, Jin Q, Wu CJ, Schmidt G (2010) Addition of chitosan to silicate cross-linked PEO for tuning osteoblast cell adhesion and mineralization. Appl Mater Interf 2:3119–3127
87. Chen R, Chen Q, Huo D, Ding Y, Hu Y, Jiang X (2012) In situ formation of chitosan-gold hybrid hydrogel and its application for drug delivery. Colloids Surf B 97:132–137
88. Sudheesh Kumar PT, Ramya C, Jayakumar R, Nair SV, Lakshmanan (2013) Drug delivery and tissue engineering applications of biocompatible pectin-chitin/nano $CaCO_3$ composite scaffolds. Colloids Surf B 106:109–116
89. Jayakumar R, Ramachandran R, Sudheesh Kumar PT, Divyarani VV, Srinivasan, Chennazhi KP, Tamura H, Nair SV (2011) Fabrication of chitin-chitosan/nano ZrO_2 composite scaffolds for tissue engineering applications. Int J Biol Macromol 49:274–280
90. Zhang JL, Misra RDK (2007) Magnetic drug-targeting carrier encapsulated with thermosensitive smart polymer: core–shell nanoparticle carrier and drug release response. Acta Biomater 3:838–850
91. Ardeshirzadesh B, Anaraki NA, Irani M, Rad LR, Shamshiri S (2015) Controlled release of doxorubicine from electrospun PEO/chitosan/grapheme oxide nanocomposite nanofiberous scaffolds. Mat Sci Eng C 48:384–390
92. Liu KH, Liu TY, Chen SY, Liu DM (2008) Drug release behavior of chitosan—montmorillonite nanocomposite hydrogels following electrostimulation. Acta Biomater 4:1038–1045
93. Nivethaa EAK, Dhanavel S, Narayanan V, Vasu CA, Stephen (2015) An in vitro cytotoxicity study of 5-fluorouracil encapsulated shitosan/gold nanocomposites towards MCF-7 cells. RSC Adv 5:1024–1032
94. Shariatinia Z, Nikfar Z, Gholivand K, Tarei SA (2015) Antibacterial activities of novel nanocomposite biofilms of chitosan/phosphoramide/Ag NPs. Polym Compos 454–466
95. Sharma S, Sanpui P, Chattopadhyay A, Ghosh SS (2012) fabrications of antibacterial silver nanoparticle-sodium alginate-chitosan composite films. RSC Adv 2:5837–5843
96. Rao KSV, Reddy PR, Lee YI, Kim C (2012) Synthesis and characterization of chitosan-PEG-Ag nancomposites for antimicrobial application. Carbohydr Polym 87:920–925
97. Youssef AM, Yousef HA, Sayed SME, Kamel S (2015) Mechanical and antibacterial properties of novel high performance chitosan/nanocomposite films. Int J Biol Macromol 76:25–32
98. Shariatinia Z, Fazli M (2015) Mechanical properties and antibacterial activities of novel nanobiocomposite films of chitosan and starch. Food Hydrocoll 46:112–124
99. Wysokowski M, Motylenko M, Stocker H, Bazhenov VV, Langer E, Dobrowolska A, Czaczyk K, Galli R, Stelling AL, Behm T, Klapiszewski L, Ambrozewicz D, Nowacka M, Molodtsov SL, Abendroth B, Meyer DC, Kurzydlowski KJ, Jesionowski T, Ehrlich H (2013) An extreme of biomemitic approach: hydrothermal synthesis of b-chitin/ZnO nanostrustured composites. J Mater Chem B 1:6469–6476

100. Sudheesh Kumar PT, Lakshmanan VK, Anilkumar TV, Ramya C, Reshmi P, Unnikrishnan AG, Nair SV, Jayakumar R (2012) Flexible and microporus chitosan hydrogel/nano ZnO composite bandages for wound dressing. in vitro and in vivo evaluation. Appl Mater Interf 4:2618–2629
101. Anisha BS, Deepthi S, Annapoorna M, Chennazhi KP, Nair SV, Jayakumar R (2013) Chitosan–hyaluronan/nano chondroitin sulfateternary composite sponges for medical use. Carbohydr Polym 92:1470–1476
102. Lu S, Gao W, Gu HY (2008) Construction, application and biosafety of silve nanocrystalline chitosan wound dressing. Burns 34:623–628

Chitin, Chitosan, and Silk Fibroin Electrospun Nanofibrous Scaffolds: A Prospective Approach for Regenerative Medicine

Brijesh K. Singh and Pradip Kumar Dutta

Abstract Intensive studies have been done to a wide range of natural and synthetic polymeric scaffolds which have been done for the use of implantable and temporal devices in tissue engineering. Biodegradable and biocompatible scaffolds having a highly open porous structure with compatible mechanical strength are needed to provide an optimal microenvironment for cell proliferation, migration, differentiation, and guidance for cellular in growth at host tissue. One of the most abundantly available biopolymer chitins and its deacetylated derivatives is chitosan which is non-toxic and biodegradable. It has potential biomedical applications such as tissue engineering scaffolds, wound dressings, separation membranes, antibacterial coatings, stent coatings, and biosensors. Recent literature shows the use of chitin and chitosan in electrospinning to produce scaffolds with improved cytocompatibility, which could mimic the native extra-cellular matrix (ECM). Similarly, silk from the *Bombyx mori* silkworm, a protein-based natural fiber, having superior machinability, biocompatibility, biodegradation, and bioresorbability, has evolved as an important candidate for biomedical porous material. This chapter focuses on recent advancements made in chitin, chitosan, and silk fibroin-based electrospun nanofibrous scaffolds, emphasizing on tissue engineering for regenerative medicine.

Keywords Chitin · Chitosan · Silk fibroin · Electrospinning · Nanofibers · Regenerative medicine

1 Introduction

The field of regenerative medicine is a ray of hope for those peoples whose conditions are beyond repair. It is known as game-changing area of medicine having good potential to heal damaged tissues and organs. However, to generate new

B.K. Singh · P.K. Dutta (✉)
Department of Chemistry, MN National Institute of Technology,
Allahabad 211004, India
e-mail: pkd_437@yahoo.com

P.K. Dutta (ed.), *Chitin and Chitosan for Regenerative Medicine*, Springer Series on Polymer and Composite Materials, DOI 10.1007/978-81-322-2511-9_7

tissue or organs, using stem cell needs a supporting network which should function as structural matrix. Tissue engineering and regenerative medicine mostly uses 3D scaffold matrix which mimics the natural extra-cellular matrix (ECM). This 3D matrix is used as structural template that supports cell adhesion, migration, differentiation, and proliferation, and provides guidance for neo-tissue formation. According to a survey, there is one death reported every 4 min on the streets of India, which is the highest number of deaths due to road accidents in whole world. Nearly, 5 lakh road accidents were reported in year 2013 in which more than 1 lakh people lost their lives. A large chunk of the victims were aged between 30 and 44 years. Up to the common middle age, human have very good capacity to regenerate the new cells. A lot of patients are still waiting for organ transplants and their growing demand is at an alarming stage, whereas in contrast, potential organ donors are not available. Hence, bioengineers are still working to find out high-tech solutions for regenerative medicine.

Due to trauma, accidents, and various infectious diseases, there is an extreme shortage of organs. According to a survey, only in U.S. approximately 28,000 transplants are made every year, whereas 120,000 people stuck on waiting lists. Tragically, most of the peoples waiting for organ transplant die before they ever get an organ by a suitable donor. It has been estimated that nearly 35 % of total deaths in United States could be prevented, or at least significantly delayed, by significant organ transplantation. According to WHO, currently organ transplantation is less than 10 % of the global demand. Considering kidneys alone, nearly 25,000 peoples died every year due to unavailability of potential donors and approximately two million peoples are in wait list in Europe, North America, and British Commonwealth. According to National Kidney Foundation India, nearly 100 peoples per million populations succumb to kidney diseases and approximately 90,000 kidney transplants are required every year in India and out of them 10,000 kidney transplants are required in Maharashtra only. It is interesting to know that 22.5 % of patients requiring dialysis receive treatment, whereas only 2.5 % of patients in kidney failure receive transplants. Due to medical, social, and financial reasons, only 25 % of willing donors are suitable for donation.

2 Materials for Regenerative Medicine

The main objective of regenerative medicine is to create therapies either by repairing or replacing the tissues and organs to restore their original functions which resulted due to congenital defects, disease, trauma, and aging [1]. The area of regenerative medicine is highly interdisciplinary and involves tissue engineering, drug delivery, multidisciplinary principles of materials science, medicine, and life science to generate tissues and organs of better biological structures and functions. Regenerative therapies aim to restore the functions of damaged or removed tissues and organs, and can be classified into three distinct approaches: somatic cell therapy (i.e., implantation of (stem) cells triggering the regeneration), tissue engineering (i.e., implantation of tissues grown outside the body), and induced endogenous regeneration (i.e., induction of regenerative processes without applying cells). The aim of regenerative medicine is to design systems with features that mimic the natural environment of the cells. Materials play very crucial role in a variety of medical applications encompassed by regenerative medicine and tissue engineering [2]. Historically, it is evident that wound healing materials were used for thousands of years, but from past few decades, the field of biomaterials has evolved greatly [3, 4]. Biocompatibility is one of the most important factors in selecting biomaterials for regenerative medicine. Besides biocompatibility, the designed material should ideally be inert to the tissue and such behavior could be defined by the ability to minimize cell–host interactions. As our understanding of biological phenomena and technologies becomes advanced, a new generation of biomaterials is emerging sharply, which are bio-instructive rather than passive.

An ideal scaffolding material should be biocompatible, porous with interconnected pores, desired mechanical strength, degradation rate as per required function, and ability to support cells in three-dimensional way [5, 6]. Food and drug administrations have issued a list of natural and synthetic polymers which can be used for human health. Biopolymers have attractive options because of their similarities with native ECM as well as their chemical and biological properties. Ideally, regenerative materials should be highly dynamic and responsive, and their physical and chemical properties should program to change over time or upon stimulation by cues in the environment. Biomaterials have been widely used in regenerative medicine for many applications including rapid expansion of stem cells, delivery of therapeutic cells or drugs, or replacement of damaged tissue. The process of regeneration involves implantation of cells onto the designed scaffolding materials for regenerating tissue and subsequent deposition of ECM. These artificial scaffolds play very important role because they act as an artificial ECM and provide a temporary environment for cells to infiltrate, adhere, proliferate, and differentiate [7–9]. They provide an initial structural support and retain cells in the defective area for cell growth, metabolism, and matrix production; hence, their role during the development of engineered tissues is of prime importance [10].

In regenerative medicine applications, natural as well as synthetic both types of polymers are used as scaffolding materials. Every material has its own good and bad part, i.e., if one has good biocompatibility then fails in mechanical strength or interconnected pore forming ability. So for a specific type of cell seeding, it is important to optimize every possible parameter according to the need of seeded cell for regeneration. At this place, a biomaterial temporarily provides a structural function and used as implant. The seeded cells settle on the material and replace it over time by functional neo-tissue while the implant degrades. Although synthetic biodegradable materials are the most popular sources utilized in regenerative medicine, they are mostly hydrophobic and showed poor interactions with cells and native tissue. Cells are surrounded by numerous ECM molecules that are assembled together to form complex 3D network. Interactions of cells with ECM are critical in the determination of cell fate processes such as adhesion, proliferation, differentiation, and survival. Therefore, many researchers are engaged to functionalize synthetic biomaterials to make them exhibit one or many of chemical, physical, biological, or structural properties of ECM that direct desirable cell and tissue responses.

2.1 Synthetic Polymers for Regenerative Medicine

Several types of polymers are used for electrospinning forming nano- to submicron range fibers and used for various applications. Several authors reported electrospinning of various types of polymers such as natural, synthetic, or blend of both, including proteins [11, 12], nucleic acids, and polysaccharides [13]. Up to now, more than 200 types of polymers, blends, or hybrids of natural and synthetic polymers have been electrospun successfully and characterized with respect to their respective applications [14]. Synthetic polymers are broadly divided into two categories: biodegradable and non-biodegradable. A number of synthetic biodegradable polymers have been subjected to electrospinning for providing a 3D nanofibrous scaffolds for better cell growth and tissue formation. For the purpose of regenerative medicine, poly (α-hydroxy acids) [15], poly (propylene fumarate) [16], poly (orthoester) [17], polycarbonate [18], polyurethanes [19, 20] poly-3-hydroxybutyrate [21], and polyphosphazenes [22] synthetic polymers were fabricated into nanofibrous scaffolds and characterized. Out of these synthetic biodegradable polymers, poly (α-hydroxy acids) including poly (lactic acid), poly (glycolic acid), and their copolymer poly(lactic acid-*co*-glycolic acid) have approved by FDA for certain regenerative medicine applications [23]. A schematic representation for the classification of biodegradable polymers is given in Fig. 1.

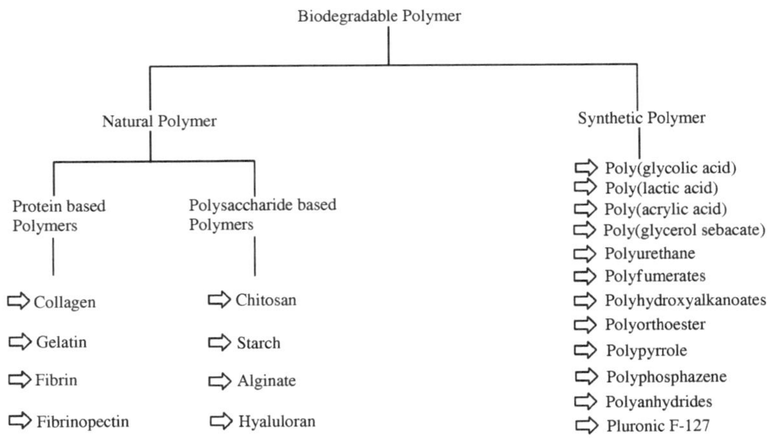

Fig. 1 Classification of biodegradable polymers

One of the important applications of biodegradable synthetic polymers includes drug delivery systems, resorbable sutures, and orthopedic fixation devices such as pins, rods, and screws [24, 25]. As and when these materials are used in vivo or in vitro, they starts encountering to protein-rich medium that provides nutritional support to cells by material surface [26], whereas non-biodegradable polymers viz. PVA, PMMA, and PNIPA are yet to be studied for their functional properties used in various applications. Synthetic polymers are used for scaffolding materials due to their processability, bulk biodegradation chemistry, good mechanical strength, cost effective, compatibility with sterilization techniques, and their greater shelf life. The above properties of synthetic scaffolding material do not necessarily ensure that the surface properties of the final scaffolds would be optimal for protein and cell interaction, and it is quite possible that the original material of synthetic polymers or their degradation products can induce inflammatory responses [27].

2.2 *Natural Polymers for Regenerative Medicine*

Biopolymers are preferred over synthetically derived polymers for regenerative medicine mainly because of their structural similarities with native ECM, chemical versatility, and typically good biological performance. Biopolymers normally exhibit better biocompatibility and low immunogenicity in comparison to synthetic polymers in biomedical applications [28]. The basic cause behind using biopolymers for electrospinning method is their inherent capacity for binding cells since they carry specific protein sequences, such as RGD (arginine/glycine/aspartic acid) [29]. Recently, proteins mainly from collagen, gelatin, elastin, and silk fibroin were used for electrospinning producing nanofibrous scaffolds for tissue regeneration [30]. Typical biopolymers which have been used for electrospinning purpose include chitin, chitosan, cellulose acetate, casein, collagen, gelatin, silk protein, fibrinogen, etc. Scaffolds of biopolymers show better clinical functionality. One of the most important benefits of using biopolymers instead of synthetic polymer is that they are structurally similar with the ECM and do not produce any chronic inflammation or immunological reactions and toxicity which is often detected in case of synthetic polymers. On the other hand, there are some limitations also for the use of biopolymers in place of synthetic polymers such as undesirable immune response which is observed often due to the presence of impurities and endotoxins. It is also notable that the properties of biopolymers may also differ from batch-to-batch large-scale production because of inability to control the processing parameters. In our present analysis, we are focused to polymers of natural origin and their electrospinning for regenerative medicine applications.

3 Electrospinning

Polymeric nanofibers are produced by various methods such as drawing, self-assembly, phase separation, template synthesis, and electrospinning [31]. Via electrospinning process, solid continuous and thinner fibers can be generated using electric fields [32]. Electrospinning is simpler method than template synthesis and self-assembly methods which can be used for different types of materials. Due to its unique advantages such as high surface-to-volume ratio, adjustable porosity of electrospun fibers, and most importantly the flexibility to electrospun into a variety of shapes and sizes, electrospun nanofibers have been widely used in combination with different materials [33] for several specific applications such as filtration [34], optical and chemical sensors [35], electrode materials [36], and biological scaffolding materials [37].

Basic electrospinning setup included three major components: high-voltage power supply, spinneret, and grounded plate collector. A polypropylene syringe pump is adjusted at the head of grounded plate collector and used for feeding polymeric solution at a constant and controllable rate. The polypropylene syringe contains a stainless steel needle of different holes as per the need. One electrode which is connected from the power supply is connected to the needle also which holds the spinning solution. To avoid the charging of polymer solution and electrical shock, an opposite polarity collector is attached to rotating drum collector which receives the electrospun nanofibers. A high voltage ranging from 0 to 30 kV is applied at the spinneret, and at the same time, flow rate of polymer was adjusted according to the needs for final product and then the surface of the fluid droplet at spinneret gets electrostatically charged at the tip, and these charged ions move in response to the applied electric field toward the electrode of opposite polarity; hence, they are collected on a rotating drum collector [38]. In the presence of high electric field, polymer solution at the tip of the capillary takes a cone-like projection and when the applied electrical field reaches a critical value required to overcome the surface tension of the liquid, a jet of liquid is ejected from the cone tip, resulting in fibers of micro- to nanorange [39].

The electrospinning process was first time patented by J.F. Cooley in February 1902 (U.S. Patent 692,631) and by W.J. Morton in July 1902 (U.S. Patent 705,691), whereas the first commercialization patent in year 1934 was credited to Anton Formhals for the fabrication of textile yarns (U.S. Patent 1,975,504). Biomedical applications of electrospun nanofibrous scaffolds are determined by their physical as well as biological properties such as biocompatibility, hydrophilicity, biodegradability, mechanical modulus, strength, and specific cell interactions, which in turn depend largely on the chemical compositions. Polymer blendings also known as polyelectrolyte solutions and hybrid solution are two most important methods for combining electrospinning solutions to get materials of novel properties. Thus, in our present investigation, we are analyzing the material properties of electrospun chitin, chitosan, silk fibroin, and their composite materials also.

3.1 Electrospinning of Natural Polymers

Synthetic polymers are preferred over biopolymers for electrospinning as they can be functionalised easily and they retain their lot-to-lot uniformity even at large-scale productions. Further, synthetic polymers synthesized in bulk, cost effective, and their raw materials are readily available. Synthetic hydrophobic biodegradable polymers such as polyesters, polyglycolide (PGA) [40], polylactide (PLA) [41], and poly(ε-caprolactone) (PCL) [42] blended or hybrid solutions were subjected to electrospinning for the production of electrospun nanofibrous scaffolds which are further used for different biomedical applications. Hydrophilic biodegradable polymers, such as polyurethane [43], poly(vinyl alcohol) (PVA) [44], PEO [45], polydioxanone [46], and polyphosphazene derivatives [47] blended or hybrid solutions were also subjected to electrospinning for the production of electrospun nanofibrous scaffolds for biomedical applications, whereas, in comparison to synthetic polymers, biopolymers always represent better biocompatibility and low immunogenicity, as and when used for biomedical applications. All types of naturally occurring polymers—proteins, polysaccharides, DNAs, and lipids—have been subjected to electrospinning process and successfully fabricated into electrospun fibrous scaffolds for respective applications. In our present investigation, we are focused to analyze and find out regenerative medicine applications of chitin, chitosan, and silk fibroin nanofibers.

3.2 Electrospinning of Chitin

Chitin is one of the most available naturally occurring polysaccharides (second to cellulose) and produced 10^{10}–10^{11} tons each year in nature and it is naturally synthesized by a number of living organisms of particular species. Chitin is the major structural component of cell walls of fungi and yeast, which is found in exoskeleton of arthropods and in nature as ordered microfibrils. Chitin is commercially prepared by crab and shrimp shells, which are abundantly present in sea, and after used by sea food industries, they are thrown out as waste products. Chitin is not soluble in common organic solvents; hence, it is chemically treated to give more deacetylated derivative, chitosan [48]. Every year about 300 billion tons of industrial and agricultural wastes are generated, deriving from processing of plant raw materials into intermediates or final products [49]. Only 3 % (13 billion tons/year) of world plant biomass is used for making goods, while 20 % of 154 billion tons/year of fishery and crustacean's processing are transformed into chitin, chitosan, and oligosaccharides, producing waste of 30 million/tons [50].

Sutures are probably most used material implants in human body which are used in case of trauma, accidents, severe diseases, and scissors made by surgeon, and its market is nearly $1.3 billions annually [51]. At the one end, chitin could be electrospun into fibers for the use of textile industry [52], whereas it could be prepared in the form of sutures which have remarkable properties for biomedical applications [53]. As chitin is insoluble, sutures resist attack in bile urine and pancreatic juices, whereas other biopolymers become absorbable sutures [54]. Unlike cellulose, chitin also has linear polymeric chain structure and is expected to give fiber and film-forming ability similar to those of cellulose [55]. Thus, microfibril of chitin with diameters from 2.5 to 2.8 nm in literature shows that chitin can be spun into fibers [56].

Chitin biopolymers have plenty of possibilities for chemical and mechanical modifications and produce a material with unique properties, functions, and applications mainly for biomedical applications. Chitin is available naturally in thousands of tons but its utilization limited due to its intractability and insolubility. Day-to-day scientists are working to address this drawback and overcome to those limitations which restrict the use of chin. The solvent system used for electrospinning technique is another big problem due to their corrosive and degradative nature. Chitin shows vigorous viscosity with most of the organic solvents which limits its use for electrospinning as it forms beads not fibers. Blending is one of most important options to overcome the mechanical and spinnability which are the major factors limiting use of chitin in electrospinning [57].

Chitin is mostly insoluble in organic solvents, so its applications are limited. Chitin is soluble only in specific type of solvents such as *N,N*-dimethylacetamide (DMAC), Lithium chloride (LiCl), Hexafluoroacetone, 1,1,1,3,3,3-hexafluoro-2-propanol (HFIP) [58], and saturated calcium solvent [59]. Min et al. performed electrospinning of chitin solution using HFIP solvent and found nanofibers of nearly 110 nm diameter. Before staring electrospinning process, authors depolymerized chitin by γ-radiation to improve its solubility. As-spun chitin nanofibers had the broad fiber diameter distribution showing that most of them are less than 100 nm [60]. Junkasem et al. reported electrospinning of PVA and α-chitin whiskers prepared from shells of *Penaeus merguiensis* shrimps and produces nanofibrous mat for tissue engineering applications [61]. This investigation was carried out since both PVA had previously been electrospun [62] in water and chitin/PVA nanocomposite films containing α-chitin whiskers have also been fabricated in organic solvent system [63]. Chitin/silk fibroin blended solution was electrospun in HFIP solvent and the nanofibers were obtained in the range of 920–340 nm [64]. Shalumon et al. reported electrospinning of water-soluble carboxymethyl chitin/PVA blend solution having the concentration of CMC (7 %) with PVA (8 %) in different ratios (0–100 %). The electrospun fibers were made water insoluble by crosslinking with glutaraldehyde vapors followed by thermal treatment [65].

Chitin is soluble in 1,1,1,3,3,3-hexafluoro-2-propanol (HFIP) and fabricated into nanofibers through self-assembly, microcontact printing, and electrospinning methods [66]. For electrospinning purpose, ionic liquids are known as a "green" solvent which can be used to dissolve cellulose, chitin, starch, and lignin [67]. Recently, neat chitin nanofibers of high molecular weight were successfully electrospun through a one-pot process in 1-ethyl-3-methylimidazolium acetate [68]. Various types of physical methods such as mechanical treatment and ultrasonication are also used to produce chitin nanofibers [69–71]. Due to chitin's less solubility in common organic/inorganic solvents, electrospinning of chitin-based nanofibers is still in developing stage. Here, freeze-drying technique was used to prepare nanofibers of chitin (20 nm diameter) size and morphology of electrospun nanofiber was tuned by adjusting freeze-drying temperature and heat transfer characteristics. It was observed that reducing the freezing temperature produces controlled nanostructures ranging from oriented sheets to 3D nanofibers of good interconnected pores. Authors assumed that these types of nanofibers with well-interconnected pores were formed probably due to slow rate of freeze drying which encapsulates and retains the network structure [72].

Single-spinneret electrospinning was used for chitin butyrate-coated nylon-6 blended solution, and TFAA/H_3PO_4-mediated *O*-acylation of chitin with butyric acid was used to improve the solubility of chitin for getting nanofibers. The electrospun fiber was characterized by FE-SEM, TEM, FTIR, and morphological analyses which revealed that both the polymeric solutions were immiscible in the as-spun nanofibers, and it becomes more pronounced when chitin butyrate amount increases in blend solution. The cell behavior and cytocompatibility of nanofibers were investigated by incubating in biomimetic-simulated body fluid and results were found within the biomaterials range. The bone formation ability of blended electrospun fibers was also evaluated and authors concluded that the hydrophilicity, bone formation ability, and cell biocompatibility of nylon-6 nanofibrous scaffolds were strongly influenced by the incorporation of chitin butyrate, and hence this material can be used for hard tissue engineering [73].

Nanosize fillers such as nanocrystals and nanofibers of chitin are of great importance and were added in thermoplastic starch matrix by melt-mixing method. Nanocrystals and nanofibers of chitin were incorporated to thermoplastic starch-based nano-biocomposites and characterized, compared in terms of morphology, chemical and crystal structure, thermal and mechanical properties, and water resistance. Mostly, all starch-based thermoplastics show good thermal stability, mechanical properties, and storage modulus than thermoplastic starch matrix without fillers. It was assumed that nanosize fillers get dispersed very well in the matrix, and hence due to their chemical similarity, they get attached by hydrogen bonding interactions. Due to these interactions and dispersion of nanofillers within the matrix, they show very good thermal as well as mechanical property but it is greatly influenced by the concentration and type of chitin nanosize fillers [74]. Morphologically electrospun nanofibrous scaffolds resemble with human native extra-cellular matrix [75], and hence these artificial architectures are promising material for cell culture and tissue engineering applications. Using electrospinning technique, it is possible to produce complex, seamless, and three-dimensional nanofibrous scaffolds that support diverse types of cells to grow into the artificial tissues that can be used for regenerative medicine.

Chitin and chitosan have very poor mechanical strength naturally, and therefore they could not be used for bone repair and reconstruction but addition of biomaterials like hydroxyapatite, bioactive glass ceramic, etc. make them suitable candidate for bone regeneration applications [76]. Park et al. [77] reported electrospinning of chitin/silk fibroin blended solutions using HFIP solvent and found fibers in range of diameters of 920–340 nm and used for cell attachment and spreading for NHEK and NHEF. Carboxymethyl chitin (CMC)/PVA chains were subjected to intercalation into the interlayer of organic rectorite (OREC), and this composite solution was electrospun to get nanofibrous mat and results revealed that interlayer distance of OREC increased from 3.68 to 4.08 nm. It was also observed that the composite solution electrospun nanofibrous mat shows greater thermal stability but after intercalation, the crystalline effect was greatly reduced. The electrospun material was investigated for its MTT assay and found that nanofibrous mats containing OREC show its slightly toxic nature. Mouse lung fibroblasts were used for the evaluation of cell cytocompatibility and cell viability, and results indicate that the CMC-OREC/PVA nanofibrous mats are biocompatible. Authors concluded that the prepared electrospun nanofibrous materials could be used for catalysis, sensors, tissue engineering, food packaging, and antimicrobial wound dressing applications [78].

Using ultrasonic irradiation method, a clear and transparent nanofibril solution was obtained by treating purified squid pen powder in aqueous solution. The analysis of nanofibril solution revealed the presence of β-chitin nanofibrils having 3–10 nm width and several micrometers in length. Chitin nanofibril suspension could be transformed into a durable 3-D hydrogels by simply heating to 180 °C for 1–4 h in an autoclave and concluded that since chitin is a biodegradable and cytocompatible biopolymer, the chitin hydrogels molded into many shapes and prepared directly by elevating hydrothermal temperature without employing any chemical cross linkers which should have much potential biomedical applications such as for the wound dressing and scaffolds in tissue engineering [79].

A deep eutectic solvent mostly resembles with ionic liquids with special properties and composed of a mixture which forms a eutectic solvent with a melting point much lower than either of the individual components. Most primitive and significant eutectic activity was observed with a mixture of choline chloride and urea in the molar ratio of 1:2. Here, choline halide (chloride/bromide)–urea and choline chloride–thiourea were used as deep eutectic solvents for electrospinning purpose. The electrospinning parameters were optimized mostly concerning to concentration of used deep eutectic solvents. The electrospun nanofibers were thus compared a couple of imidazolium-based ionic liquids, viz. 1-butyl-3-methylimidazolium hydrogen sulfate and 1-methylimidazolium hydrogen sulfate and with choline-based ionic liquids, choline hydrogen sulfate, and choline acrylate. Authors concluded that chitin nanofibers obtained by using deep eutectic solvents show greater elasticity than calcium alginate bio-nanocomposite gel beads, and hence used the prepared matrix to observe the release profile of anticancer drug 5-fluorouracil and found that it gives very slow release which may be very beneficial in chemotherapy [80].

The basic aim of any tissue engineer is to fabricate a model scaffolding system which could work as native extra-cellular matrix for a seeded cell. In light of this view, two scaffolding systems of chitin/silk fibroin nanofibrous scaffolds (blend scaffolds and hybrid scaffolds) were suitably designed by either electrospinning or simultaneous electrospinning of chitin/silk fibroin solutions. The average diameter of electrospun nanofiber was found in the range of 340–920 nm, whereas pure chitin and silk fibroin show 140 and 1260 nm diameter, respectively. The chitin/SF bicomponent scaffolds were after-treated with water vapor, and their nanofibrous structures were almost maintained. It was observed that best electrospun fibers were found when the ratios of chitin and silk fibroin were taken as 25 and 75 %, respectively. As chitin and Silk fibroin are immiscible in solutions, hence, bimodal-type distribution of fiber diameters was obtained. Water vapor treatment was given to electrospun scaffolds because they get stabilized due to solvent-induced crystallization. From the cytocompatibility and cell behavior on the chitin/SF blend or hybrid nanofibrous scaffolds, the hybrid matrix with 25 % chitin and 75 % SF as well as the chitin/SF blend nanofibers could be a potential candidate for tissue engineering scaffolds. Human epidermal keratinocytes were taken for cytocompatibility and cell viability investigation, and it was observed that the chitin/silk fibroin electrospun nanofibrous scaffold was found as very prominent scaffold system for the attachment and spreading between cells and scaffolds [81].

Degree of deacetylation of chitin and chitosan greatly affects the electrospinning process. Chitin and chitosan solutions having different degrees of deacetylation were subjected to electrospinning using spinning solvent HFIP. The electrospinning fibers were then deacetylated at 100 °C for 0–150 min with 40 % aqueous sodium hydroxide solution. To evaluate the thermal behavior of electrospun chitin/chitosan nanofibers, TGA and DSC analyses were performed, and results revealed that thermal stability of fibers decreases with increasing the degree of deacetylation. Chitin nanofiber decomposes in single step, whereas chitosan with degree of deacetylation >50 % represents two-step degradation mechanism. It was also observed that activation energy and maximum decomposition temperature of the chitin nanofibers were higher than that of the chitosan, and authors assumed that reason behind this is possibly due to the higher thermal stability of the *N*-acetyl glucosamine units. In DSC thermogram, chitosan shows two decomposition patterns probably one is for amino glucosamine and another one is for *N*-acetyl glucosamine units [82].

Polycaprolactone is biodegradable polyester which has been approved by Food and Drug Administration for specific applications in human body. Chitin nanofibrils and polycaprolactone were blended and electrospun with 2,2,2-trifluoroethanol solvent. The electrospun nanofiber mat was characterized for their physicochemical and biological properties. As concentration of polycaprolactone increases in chitin/polycaprolactone blended solution, the mechanical property of electrospun nanofibers becomes enhanced. Interestingly, it was observed that electrospun nanofibers change their nature from hydrophobic to hydrophilic when chitin nanofibril content in blending material becomes more than 25 wt%. On the basis of in vitro cell culture results, authors concluded that the chitin nanofibril is a potential reinforcing and bioactive filler for PCL [83].

Silver nanoparticles are highly antimicrobial in nature and used in various medications. Ifuku et al. reported preparation of silver nanoparticle on the surfaces of chitin nanofibers by UV light reduction of silver ions. The mixing ratio of silver nitrate and UV irradiation time greatly affects the synthesis of silver nanoparticles. Chitin nanofibril behaves like a substrate to disperse silver nanoparticles stably without any coagulation. Due to high fungal activity, chitin nanofibrils could be used for a range of commercial applications such as medical, food, and cosmetics [84].

Chitin nanofibrils, also known as nanowhiskers, are the purest crystal form of chitin. Chitin nanofibrils have been used in making innovative cosmetics, drug delivery systems, and advanced medications [85–87]. Chitin nanofibril-based biomedical nanocomposites can be used for drug/gene delivery, for tissue engineering as scaffolds and cosmetic orthodontics [88], because they are able to support the growth of cells inducing tissue regeneration. Best results are obtained when a scaffold or non-woven tissue has a proper architecture, which is designed in such a way that the cellular response desirable for biological function of specific organs is triggered [89].

3.3 Electrospinning of Chitosan

Chitin is insoluble in most of the solvents; hence, chitosan is preferred over chitin for electrospinning purpose. However, both are biocompatible, biodegradable, non-toxic, antimicrobial, and hydrating agents, and are easily processed into gels [90, 91] membranes [92, 93], nanofibers [94, 95], beads [96], microparticles [97], nanoparticles [98], scaffolds [99], and sponge [100] forms. They have several promising applications in the form of thin films and fibers [101–103].

Chitosan is deacetylated derivative of natural biopolymer chitin which is one of the most abundant organic materials after cellulose. Unlike cellulose, chitin is a linear polymer having analogous structure except the amino group which replaces the hydroxyl group present on the C-2 position. The amine groups present on C-2 position of chitosan backbone provide several unique properties such as solubility in acidic aqueous solvents, antifungal, and antimicrobial properties [104]. A schematic illustration is represented in Fig. 2 for the preparation of chitin and chitosan. Figure 3 shows method for preparation of silk fibre from silk cocoons.

Chitosan has lots of challenges to electrospinning in its original state, viz. rigidity of *d*-glucosamine repeating units, its highly crystalline nature, and hydrogen

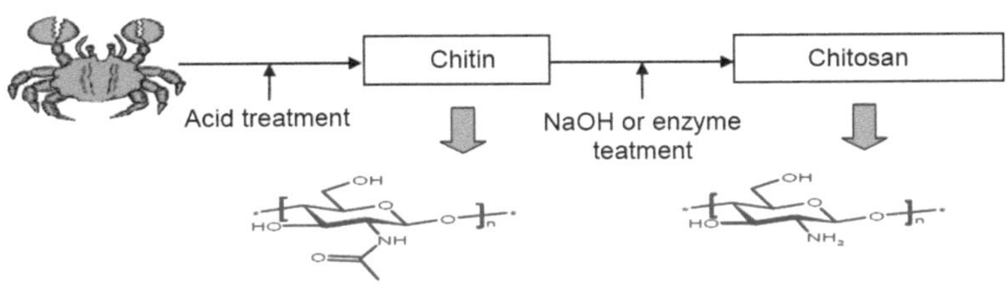

Fig. 2 Preparation of chitosan from crustacean cell

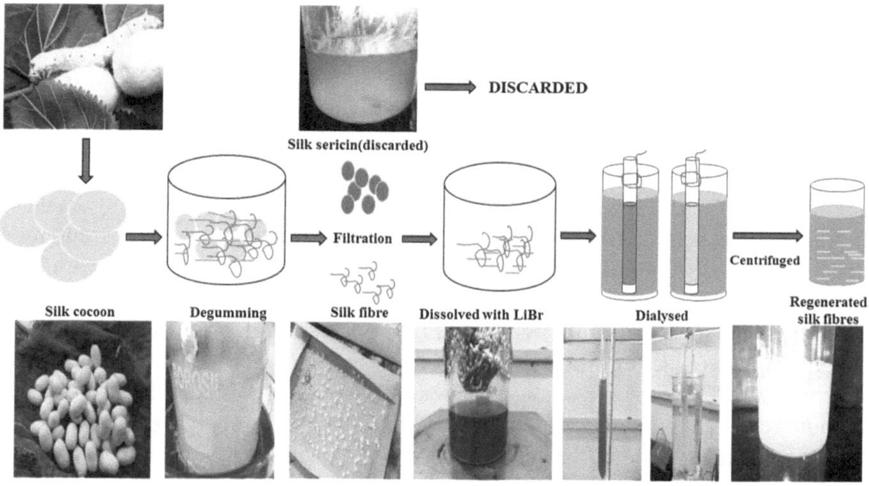

Fig. 3 Preparation of Silk fibroin from silk cocoon

bond forming ability which leads to poor solubility in common organic solvents. When polymeric chain segments were exposed to the electrical field, they form strong hydrogen bonds within them and prevent free movement of polymeric chains which leads to jet break up during the process [105]. Nanofibers of chitosan have been prepared by dissolving it in TFA [106] and their mixtures with DCM and TCM [107]. Amino group of chitosan at C-2 position forms stable salts with TFA, which hinders the intermolecular interactions between chitosan chains and facilitate electrospinning [106]. Some authors have reported electrospinning of neat chitosan with degrees of deacetylation 5419 in highly concentrated aqueous acetic acid solution (80–90 %) [108]. It was also observed that as surface tension of the solution decreases by increasing the acetic acid quantity, it helps in the electrospinnability of chitosan solution [109]. However, use of TFA-based solvents is not preferred because electrospun nanofibrous membranes are dissolved in neutral and weak basic aqueous solvents producing TFA-chitosan salt residues [110].

McKee et al. [111] reported that in order to get defect-free electrospun nanofibers of polymeric solutions, the concentration of the polymer must be at least 2–2.5 times higher than that of entanglement concentration. However, chitosan solutions at these concentrations are often very difficult to electrospun nanofibers because of the high viscosity in solution. As it is evident that chitosan is a cationic biopolymer and it also affects the rheology of the solutions, optimization of the chitosan solution viscosity is one of the important tasks to electrospun nanofibers successfully. The chitosan solution becomes too viscous and fiber-forming ability is hindered at upper threshold of entanglement, and electric field is not strong enough to overcome to this situation, whereas below the lower threshold limit, the polymer chains do not entangled. Hence, the fiber formation is not possible and polymer beads oftenly created. So it can be concluded that the polymer chain entanglement concentration plays a crucial role in electrospinning of polymeric solutions. Probably, this is one of the prime reasons that most of the research groups are using blended chitosan with other synthetic/biopolymers in an attempt to improve the electrospinnability [112]. The importance of polymer blending technology is that it is versatile, simple, and cost effective.

Blending of chitosan with other polymeric materials facilitates its processing and makes chitosan electrospinnable. The basic property of blending material should have excellent fiber-forming characteristics in order to create entanglements and physical bonds with chitosan, and act as a carrier in the electrospinning process. Synthetic polymers such as polyethylene oxide (PEO) [113], PVA [114], polylactic acid (PLA) [115], nylon-6 [116], polycaprolactone (PCL) [117], and proteins such as silk fibroin [118], zein [119], and collagen [120] have been successfully blended with chitosan to produce electrospun nanofibers. It was observed through research publications that author-to-author quantity of the co-spinning agent ranges from 20 to 80 wt% [121].

Pakravan et al. reported electrospinning of defect-free nanofibers of 60–120 nm diameter by highly deacetylated (97.5 %) chitosan in 50 % acetic acid at moderate temperatures (25–70 °C) in the presence of a low content of polyethylene oxide (10 wt% PEO) to beadless nanofibers. Rheological analysis of chitosan and PEO solutions was performed in order to explain how PEO improves the electrospinnability of chitosan. They assumed that positive charges on the chitosan molecule and its chain stiffness are the main limiting factors for electrospinnability of neat chitosan as compared to PEO, even though surface tension and viscosity of the respective solutions were similar. The authors concluded that the success of chitosan PEO-assisted electrospinning is believed to be the consequence of strong hydrogen bonds formed between ether groups in PEO and hydroxyl and amino groups in chitosan and speculated that PEO may act as a “carrier” of chitosan in the electrospinning process via those physical bonds [122].

A similar study was also performed by Kriegel et al. [123] and reported electrospinning of chitosan and poly(ethylene oxide) at a ratio of 3:1 using aqueous acetic acid solvent system. Electrospinning conditions such as flow rate (0.02 ml/min), applied voltage (20 kV), distance between spinneret and collector (10 cm), and temperature (25 °C) were adjusted. Optimum results were obtained only after addition of surfactants of different charges at concentrations well above their critical micellar concentrations. After getting electrospun nanofibers, they were investigated for the influence of viscosity, conductivity, and surface tension on the morphology and physicochemical properties.

Zhang et al. reported [124] coaxial electrospinning of chitosan/PEO blended solution in DMSO solvent system and found that with increase of chitosan content in blended soliton, the thickness of shell layer enlarges which vanishes at higher temperature (70 °C). In similar way, Desai et al. reported electrospinning of chitosan and poly(ethyleneoxide) blended solutions. Electrospinning of neat chitosan is hindered due to its poor solubility in aqueous acids and high degree of inter- and intra-chain hydrogen bonding. Electrospun nanofibers of 80 ± 35 nm size without bead defects were obtained using high molecular weight chitosan/PEO in 95:5

ratios. Fiber formation was characterized by fiber shape and size, and was found to be strongly affected by the polymer molecular weight, blend ratios, polymer concentration, choice of solvent, and degree of deacetylation of chitosan. Surface morphology of electrospun fiber was revealed by quantifying binding efficiency of toxic heavy metal ions like chromium, and they resembled with fiber composition and structure [125].

Jia et al. reported the electrospinning of chitosan/PVA blends and observed that the average diameter of the electrospun fiber gradually decreases with increasing chitosan content ranging from 10 up to 30 %. Beyond this limit, blended material could not form nanofibers. In the similar way, morphology and diameter of the electrospun fiber were mostly affected by the concentration of solution from 3 to 9 wt%. Below 3 wt% concentration, only beads are formed whereas gradually increasing the concentration up to 9 wt% produces fine fibers. Above this concentration, blended solution becomes too much viscous and electrospinning process was hard to maintain [126]. Paipitak et al. [127] prepared nanocomposite of PVA/Chitosan in weight ratio of 80:20 in 2 % acetic acid and electrospun to get fibers. The concentration of the PVA/Chitosan solution was varied from 3 to 5 wt% and authors got most suitable fibers of nearly 100 nm at the concentration of 5 wt%.

In another study, researchers synthesized the water-soluble *N*-carboxyethyl chitosan by Michael addition reaction of chitosan and acrylic acid. The synthesized water-soluble *N*-carboxyethyl chitosan was then subjected to electrospinning process but failed to get electrospun nanofiber. So blending technology was used to get electrospun nanofibrous mat along with two co-spinning agents such as silk fibroin and PVA because of its good fiber-forming, biocompatibility, and chemical resistance properties [128]. The electrospinning was performed with water which is considered as an ideal solvent system and electrospinning was carried out at neutral pH. The morphological characterization and crystallization were investigated. Cytotoxicity tests were done to check the biocompatibility of the scaffolds in vitro. The results revealed that the prepared composite materials are found biocompatible with mouse fibroblast L929, and in blend composite material, nanofibers had smaller diameters and narrower diameter distributions with increasing SF nanoparticles content [129].

In similar way, a chitosan derivative, *N*-[(2-hydroxy-3-trimethylammonium) propyl] chitosan chloride (HTCC) blended with PVA in aqueous solution, was electrospun and get fibers of average diameter of 200–600 nm. The authors concluded that nanofibers were mainly affected by weight ratio of the blend and applied voltage. The results revealed that increasing HTCC content in the blends decreases the average fiber diameter. Antimicrobiological evaluation showed that the PVA–HTCC mats have a good antibacterial activity against Gram-positive bacteria, *Staphylococcus aureus*, and Gram-negative bacteria, *Escherichia coli* [130].

Several reports have been published for the fabrication of either neat chitosan or chitosan-based blend nanofibers via electrospinning [131–133]. Out of them, a very few reports are found which describe the preparation of electrospun nanofibers in aqueous environment that demonstrate structural stability. It is evident by literature that as the fraction of the chitosan component in the nanofiber increases, maintaining the nanofiber uniformity and structural integrity becomes increasingly difficult. This happens mostly due to the lack of appropriate solvents to prepare chitosan solutions that are used for electrospinning process. Acetic acid is most versatile solvent used for the preparation of chitosan solution. However, it can dissolve a very small amount of chitosan (depending upon degree of deacetylation), which does not provide a sufficient viscosity or polymer chain entanglement required for electrospinning [134]. Chitosan forms very viscous solution which cannot be electrospun; hence, a significant fraction of a non-ionic polymer, such as polyethylene glycol or polyethylene oxide, is required to render the solution electrospinnable by decreasing the solution viscosity.

Trifluoroacetic acid is another solvent used for electrospinning of chitosan to yield a relatively high output but it produces nanofibers of relatively inferior morphology such as non-uniformity, beaded fibers and interconnectivity, and embedded ammonium salt residue that dissolves upon exposure to aqueous media. To stabilize chitosan electrospun nanofibers, several stabilizer/neutralization agents of a base or basic salt solution [135], or chemical crosslinking agents [136] are used. The alkali treatment hydrolyzes chitosan chains and so decreases its molecular weight. The neutralization of chitosan nanofibers in aqueous media usually results in severe fiber contraction leading to partial or even complete loss of the nanofeatures. Chemical crosslinking also raises concerns of the potential toxicity of the crosslinking agents.

Cooper et al. reported electrospinning of chitosan derivative nanofibers with structural stability in an aqueous environment. For their study, they used lactic acid-modified chitosan as a substrate material. The resulting chitosan salt was dissolved in TFA and methylene chloride for electrospinning. The rationale behind the use of lactic acid-modified chitosan for electrospinning was that lactic acid is an organic acid commonly present in the human body, and chitosan lactate salt has been demonstrated to be biocompatible and biodegradable in drug and gene delivery systems [137].

Geng et al. reported electrospinning of chitosan dissolved in concentrated acetic acid solution in which authors tried to optimize the effect of acetic acid concentration, chitosan concentration and molecular weight, electric field and concluded that acetic acid concentration is one of the most important parameters which decreases surface tension of the chitosan solution and at the same time increases charge density of jet without significant effect on viscosity. The morphology of electrospun fiber was also changed from beaded fiber to uniform fiber with increasing acetic acid concentration, implying significant influence of the concentrated acetic acid on reducing the applied field required for electrospinning. Decreasing the acetic acid concentration in the solvent increases the mean diameter of the nanofibers. The best electrospun fiber was obtained with chitosan solution of a molecular weight of 106,000 g/mol at 7 % concentration, dissolved in 90 % aqueous acetic acid solution at electric field higher than 3 kV/cm but lower than 5 kV/cm [138].

Different chitosan derivatives have also been successfully electrospun into fibers. Hexanoyl derivative of chitosan solution was electrospun with chloroform solvent system. The concentration of prepared spinning solutions ranging between 4 and 14 % w/v produces electrospun fibers of ribbonlike structure having average diameters of 0.64–3.93 μm. With increasing hexanoyl chitosan concentration, the average fiber diameter was found to increase, while the bead density was found to decrease. An increase in the applied electrical potential was responsible for increasing the average diameter of the as-spun fibers. An organic salt of pyridinium formate was used as alkali stabilizer to increase the conductivity of the spinning solution, which increases the average diameter and a general decrease in the bead density of the resulting fibers [139].

For electrospinning purpose, both natural and synthetic types of polymers are used as co-spinning agents. Here, bioblends of zein/chitosan solution was electrospun to get fibers of highly antimicrobial activity which further can be used for various applications such as food packaging material and in pharmaceutical and biomedical applications. However, in case of bioblends, a relatively low amount of chitosan was able to provide an efficient biocide effect. Controlled release of protonated glucosamine units and of entrapped TFA can thus be used to provide antimicrobial properties to zein ultrathin fiber-based structures. Because both the blending materials are biomaterials, hence, they could therefore be of great interest [140].

In an attempt, chitosan was blended with polyamide-6 and electrospun to get fibers which shows high degree of electrical conductivity. The blend for electrospinning was prepared by dissolving polyamide-6 pellets and chitosan powder in 85 % formic acid. Before spinning, 18 wt% of polyamide-6 with varying concentrations of chitosan (0, 1, 1.5, and 2 wt%) was used to get electrospun nanofiber mats. Blended polymer solution was taken into a 5-ml plastic syringe equipped with a polystyrene microtip (0.3 mm inner diameter and 10 mm length), which was subjected to a high-voltage power supply of about 22 kV. The distance between microtip and rotating drum collector was kept at 15 cm. The electrospinning of prepared blended solution produced nanofibers with diameters of nearly 20–40 nm. The electrical conductivity of blended solution increased with increasing chitosan content which was attributed to the formation of ultra-fine nanofibers [141].

The chitosan derivative, quaternized chitosan, exhibits broader range of antimicrobial and antimycotic activity as compared to chitosan. Hence, Ignatova et al. prepared nanofibers of polylactide-based materials loaded with a natural polyphenolic compound gossypol which is well known for its antitumor properties. After preparation of electrospun fibrous mats of blended material, it was coated with a thin film of crosslinked quaternized chitosan. The in vitro release of gossypol depends upon the mat composition and is of diffusion controlled in nature. The electrospun nanofibers exhibit high degree of cytotoxicity toward HeLa tumor cells and strong antiproliferative effect mainly due to the induction of cell apoptosis [142].

A new solvent system of formic acid and acetone mixture was used for electrospinning of chitosan and poly(Ɛ-caprolactone). The concentration of chitosan solution was varied from 0.5 to 2 % by fixing poly(Ɛ-caprolactone) (PCL) concentration at 6 %. Chitosan (1 %) was again blended with 4–10 % PCL to get smooth nanofibers. The mixing composition was taken in the ratio of 1:3, 1:1, and 3:1 of chitosan and PCL. Lower concentrations of PCL produce beaded fibers, whereas 8 and 10 % of PCL in lower compositions of chitosan resulted in fine nanofibers. Viscosity and conductivity measurements revealed the optimum values for the spinnability (1 % CS and 8 % PCL) in 1:3 compositions produced fine nanofibers of 102 $\pm$ 24 nm in diameter. The prepared electrospun nanofibrous scaffold could be used as an excellent matrix for biomedical applications [143].

Quaternized derivative of chitosan and aqueous solution of poly(vinyl pyrrolidone) blend was subjected to electrospinning to get continuous defect-free fibers, and it was observed that on increasing the polyelectrolyte content, the average fiber diameter significantly decreases from 2800 to 1500 nm. In order to impart stability of electrospun fibers to water and water vapor, the fibers were crosslinked by a photo-crosslinking additive before spinning and subsequently UV irradiation to the electrospun fibers which shows high degree of antibacterial activity against *S. aureus* and *E. coli* [144].

3.4 Electrospinning of Silk Fibroin

Compared with other fiber fabrication techniques, electrospinning is most versatile method for preparation of continuous ultra-fine fibers in the range of micro- to nanometer range with the help of electrical forces. Besides traditional two-dimensional nanofibrous structures, electrospinning is powerful technique in fabrication of three-dimensional fibrous macrostructures, especially for tissue engineering and regenerative medicine applications [145].

Among various natural polymers [146], chitosan is one of the most promising biopolymers for antibacterial drug delivery [147] and cartilage regeneration [148]. One of the major concerns with chitosan limiting its use for regenerative medicine is poor mechanical strength, swelling property, and biodegradability in aqueous solution [149] which can be overcome to an extent by combining it with other natural or synthetic polymers. In recent years, considerable attention has been focused into materials for steered stem cell regeneration and drug-release kinetics. To obtain this goal, blending of silk fibroin is one of the most prominent choices. Silk has bouquet of unique properties like low immunogenicity [150], slow degradation rates [151], and impressive mechanical properties. After blending with chitosan, silk further gets ameliorated, showing increase in its mechanical strength with palpable effect on reduction of degradation rate [152], and thus controlling the drug-release kinetics [153]. A diagrammatic representation for the preparation of regenerated silk fibers is shown in Fig. 2.

Silk is very well known in the textile industry for its luster and mechanical properties. It is produced by cultured silkworms and is a fibrous protein synthesized in specialized epithelial cells that line glands in the class of Arachnida and in several worms of the Lepidoptera larvae such as silkworms, spiders, scorpions, mites, and flies. Silk fibroin polymers contain protein series and are responsible for cocoon and web formation, nest building, etc. As they have mostly β-sheet structures consisting of short side-chain amino acids in the primary sequence, they permit tight packing of piled sheets of hydrogen bonded anti-parallel protein chains. Because of both large hydrophobic and smaller hydrophilic areas, they allow to promote the assembly of silk which is very good in strength and resiliency. In the literature, there are several researches on *B. mori* to study the properties of these silk proteins for biomaterial purposes. Silk fibers from *B. mori* are generally 10–20 μm in thickness, and each fiber is in fact duplet of two individual fibers. Each of these fibers has its own silk coating (sericin) and an inner core (fibroin) [154].

Silk is an ancient and the only natural filament fiber which is used for thousands of years and for the year 2011, China (126 ktons), India (20.4 ktons), Viet Nam (7.05 ktons), Romania (2.1 ktons), Thailand (1.6 ktons), and Uzbekistan (1.2 ktons) are the six major silk producers in the world [155]. Silk is a fine, strong continuous fibroin filament produced by the larva of certain insects especially the cultured silkworms constructing their cocoons. Raw silk fiber has natural impurities such as wax, carbohydrates, inorganic matter, and pigment less than 3 wt% of a fiber. In a composition of a silk fiber, fibroin content is about 70–80 % and the content of sericin, which is silk gum and usually removed in processing, is about 20–30 % [156].

There are two main types of silk fiber, cultivated and wild; they differ in diameter, cross-sectional shape, and in fine structure. Its length is about 300–600 m [157]. Because of its beauty, handling, and high cost, silk is also known as a luxury fiber, and hence it remains its use both in various textiles, i.e., fabrics, underwear, socks, and leggings, and has a great place for the fashion designers' collection. But more recently, silk has started to fascinate nanotechnologists in their studies which consist of fibroin [158] in biomaterial applications. Before discussing some of these studies and their applications which are mentioned in the following sections, here is a method discussed on processing for silk fibroin.

As it is already mentioned that silk fibroin (SF) is a natural protein fiber and it has been investigated thoroughly for biomaterial applications. Briefly, silk process involves the selection of a proper silk source, e.g., *B. mori* or a wild silk, silk degumming to isolate sericin from fibroin fibers, fibroin dissolution, and fabrication of new structures with different protein conformations and different matrices starting from regenerated fibroin water solution.

SF fibers are about 10–25 μm in diameter, and a single cocoon may provide over 1000 m of SF fibers when subjected to a degumming process either in boiling water or alkaline solutions [159]. Dissolution can be performed between 40 and 65 °C by preparing from a regenerated protein obtained by the dissolution of natural fibers. Natural silk fibers can be dissolved in concentrated aqueous organic solutions of salts as NH_4OH, $Ca(NO_3)_2$, or in concentrated aqueous solutions of acids, i.e., phosphoric, sulfuric, etc. [160]. Removal of the solvent is achieved by dialyzing the solution against distilled water for 3–4 days. Solvent with a lower boiling point such as formic acid was seen to be effective in dissolving SF and to obtain a suitable fibroin solution for the electrospinning process (see in Sect. 2.1 of processing).

SF fibers are biocompatible, biodegradable, and highly crystalline, with a highly organized β-sheet structure. They present high resistance to tension, good elasticity, and resilience [161]. The studies on the silk fibroin from silkworm cocoons were first reported in 1994. A hexafluoroisopropanol solvent is used and nanoscaled fibers can be produced by the electrospinning process. This process is similar to the spinning process of a silkworm where the polymer solution is subjected to a high electric field and this field overcomes the surface tension so that polymer ejects, stretches, and deposits a kind of non-woven mat. Therefore, with this technique, the polymer fibers of micrometers produce fibers in nanometer scales. Electrospun fibers have a high specific surface area and high porosity which is good for cell attachment and good biocompatibility. Fibroin concentration and pH of the solution have also significant role in the morphology of the SF nanofibers, i.e., diameter and shape [162]. It is as well to be noted that electrospun SF fibers require a post-treatment with organic solutions, for example, methanol or ethanol, to change the random coil conformation to β-sheet which is more stable and insoluble in water. Early researchers in silk fibroin were faced some difficulties with the polymer solubilization during the electrospinning processing. Some [163] have produced different blends of silk fibroin nanofibers, which are 800–1000 nm in diameter, and polyethylene oxide (PEO) in hexafluoroisopropanol; they have studied the structural properties of these fibers with FTIR spectroscopy.

Aligned nanofibrous scaffold of chitosan/silk fibroin blended solution was prepared by electrospinning technique and was investigated for growth and osteogenic differentiation of human bone marrow mesenchymal stem cells (hMSCs). The authors evaluated morphological as well as physicochemical properties of prepared nanofibrous scaffolds. The effect of chitosan and silk fibroin on cell proliferation was assessed by the MTS assay, whereas osteogenic differentiation was investigated through Alizarin Red staining, alkaline phosphatase activity, and expression of osteogenic marker genes. It was observed that osteogenic differentiation and proliferation of hMSCs were enhanced by chitosan and silk fibroin nanofibers, respectively. It was observed that blending of chitosan with silk fibroin retained the osteogenesis nature of chitosan without negatively influencing the cell proliferative effect of silk fibroin [164].

Two different species of silk viz. eri and tasar was blended in the ratio of 70:30 and the solution was subjected to electrospinning and got nanoscale scaffolds. The electrospun nanoscale scaffolds showed better hydrophilicity and mechanical strength in comparison to *B. Mori*. The hMSCs showed enhanced attachments and better proliferation which ensures retention of biological superiority even after the fabrication of fibers. The prepared scaffold shows good number of cell adhesion and higher metabolic activity. The authors concluded that on the basis of result findings, electrospun scaffolds have better surface and cell supportive property so it could be applicable for bone tissue engineering [165].

Silk fibroin solution was electrospun in a mixture of formic acid and calcium chloride using needleless electrospinning method. The calcium chloride works as a stabilizer for electrospun nanofibers and its concentration plays an important role in spinnability and diameter of the obtained fibers. As concentration of $Cacl_2$ increases, the fiber diameter ranges from 200 to 2300 nm. Authors concluded that the most suitable concentration of calcium chloride for improved solubility of silk fibroin in formic acid is of 2–3 wt% and get best results [166].

Electrospun nanofibrous scaffolds play very important role in regenerative medicine but due to small pore size, their potential use is restricted. To overcome this disadvantage, scaffold matrix could be significantly enhanced by combining two or more nano- and microfibrous structures. To achieve the target, hybrid electrospinning, combining solution electrospinning with melt electrospinning, could be of great importance. Authors prepared silk fibroin/poly(ε-caprolactone) blended nano-/microfibrous composite scaffolds and their morphological as well as physicochemical characterizations were done. Cell proliferation, cell viability, and infiltration ability were performed on hMSCs and results were found to closely associate with the nanofiber content in the composite scaffolds [167].

Kang et al. [168] reported an electrospinning method for *B. mori* fibroin solution producing silk non-woven membranes consisting of nanofibers of diameter 460 ± 40 nm, and these membranes along with nanofibers were again subjected to the aqueous solution of multi-walled carbon nanotubes (MWCNTs) which adsorb on their surface enhancing its electrically conducting property. Triton X-100 was used as the surfactant, and electrospun silk fibroin membranes were dipped into aqueous solution of MWCNT dispersion bath for 60 s and then rinsed by deionized water and then dried under vacuum oven. The electrical conductivity of the electrospun membranes was measured by a four-probe method using a picometer with an internal voltage source (487, Keithley, USA) and an impedance analyzer (4284 A, HP, USA). The adsorbed MWCNTs on the surface of silk fibroin nanofibrous membranes show a large increase in their electrical conductivity (from $\sim 10^{-15}$ to $\sim 10^{-4}$ S cm^{-1}) with electrical conductivity of 2.4×10^{-4} S/cm. Authors observed strong interaction between the MWCNTs and electrospun nanofibers and cannot be separated even after ultrasonication. This unique method could also be used for those fibers and membranes which are nonconducting in nature, and their electrical conductivity can be enhanced significantly and this method could explore a new era in the development of new materials, such as electromagnetic interference shielding or electrostatic dissipation membranes.

The composite materials of water-soluble *N*-carboxyethyl chitosan/PVA/silk fibroin were successfully electrospun by electrospinning method. The common organic solvents or organic acid solvents used for silk electrospinning include 1,1,1,3,3,3-hexafluoro-2-propanol (HFIP), trifluoroacetic acid, dichloromethane, and formic acid. When these electrospun fibers are applied to wounded human skin or tissue, the traces of solvents or acids present in electrospun products show harmful effects. Due to this reason, water-soluble *N*-carboxyethyl chitosan/PVA/silk fibroin blended material was electrospun with aqueous solvent system. Authors found that when CECS/PVA was taken in 40/60 ratio, smooth and homogeneous fibers were obtained and to this electrospun fibrous mat, 0–8 % silk fibroin nanoparticle was added resulting in the average diameter of the blend nanofibers gradually decreased from 643 to 126 nm. The in vitro cytotoxicity assessment of blended electrospun CECS/PVA/SF nanofiber mats with mouse fibroblasts (L929) showed good biocompatibility and could be used as wound dressing material [169].

The regenerated silk fibroin was electrospun with formic acid solvent to produce nanofiber non-wovens for cell culture of normal human keratinocytes and fibroblasts. To investigate insolubilization of as-spun silk fibroin nanofiber, non-wovens were chemically treated with aqueous solution of 50 % methanol. Morphological and physic-chemical properties of as-spun and chemically treated SF nanofibers were also investigated and revealed that nanofibers exhibited an average diameter of 80 nm and their diameters ranged from 30 to 120 nm. After treatment with 50 % methanol, the porosity of non-wovens nanofibers decreased from 76.1 % up to 68.1 %. Cytocompatibility and cell behavior onto the electrospun SF nanofibers, cell attachment and spreading of normal human keratinocytes, and fibroblasts seeded on the SF nanofibers and interaction between cells and SF nanofibers were studied. Our results indicate that the SF nanofibers may be a good candidate for the biomedical applications, such as wound dressing and scaffolds for tissue engineering [170].

Paşcu et al. reported electrospinning of polyhydroxybutyrate-*co*-(3-hydroxyvalerate) with 2 % valerate fraction (PHBV), nanohydroxyapatite (nHAp), and *B. mori* silk fibroin and investigated by SEM/EDX, FTIR, uniaxial tensile and compressive mechanical testing, degradation tests, and in vitro bioactivity tests. Results showed that smooth, uniform, and continuous fibers were obtained without any bead formation having fiber diameter of 10–15 μm. To investigate biodegradation of electrospun fibers, it was subjected to deionised water for one month and results showed less than 2 % weight loss and retaining their fibrous morphology which confirms slow biodegradability of matrix. To check the bioactivity of electrospun matrix, it was immersed in simulated body fluid and results reveal that after 28 days of immersion, an apatite layer was visible on the surface of the fibers, which confirms bioactivity of the electrospun matrix. Cell cytotoxicity was found within the range of biomaterials. Tensile strength of electrospun fiber increases with increase in nHAP and silk fibroin concentration [171].

In United States, Food and Drug Administration issued a list of natural polymers such as collagen, silk fibroin, chitosan, and synthetic biopolymers such as polylactic acid, polycaprolactone, polyglycolic acid, and their copolymers for regenerative applications. Here, eri silk fibroin (ESF) and hydroxyapatite (Hap) blended electrospun nanofibrous mat were formed and soaked in calcium chloride and then in sodium diammonium phosphate solution to get scaffold. The average tensile strength and water uptake in water as well as phosphate buffer saline of the pure ESF and hydroxyapatite-coated ESF scaffold (ESF-Hap) were found to be 1.84 and 0.378 MPa and 69 and 340 %, respectively. The hemolysis assay and cytocompatibility studies for both the scaffolds show good blood compatibility and cytocompatibility. Cell attachment and growth on the scaffold were performed on hMSCs, and authors concluded that the ESF-Hap scaffold is better suited for cell growth than the pure ESF scaffold [172].

Silk fibroin/cellulose acetate blended solution was electrospun into nanofibers using 98 % formic acid as a spinning solvent, and electrospun nanofibers were investigated by SEM, FTIR, XRD, and DSC. The spinnability of silk fibroin solution was remarkably improved by blending it with cellulose acetate solution (<10 %), and SEM image analysis revealed that without cellulose acetate addition, fine and uniform fibers were in the diameter range of 50–300 nm, whereas blending of silk fibroin solution by 10 % (by weight) cellulose acetate shows excellent spinnability and found nanofibers without any bead formation with average diameter of 142.1 nm. However, blending with more than 30 % (by weight) of cellulose acetate, the spinning process was seriously deteriorated producing discontinuous and tacky fibers with beads and dendritic structures. It was also observed that the thermal and mechanical properties were improved by the addition of cellulose acetate up to 10 % [173].

Electrospinning technique was used to produce nanofibrous scaffolds of silk fibroin/hydroxybutyl chitosan (HBC) blended solution using 1,1,1,3,3,3-hexafluoro-2-propanol and trifluoroacetic acid solvents to biomimic the native ECM. SEM analysis clearly indicates that the average diameter increases as the content of HBC was raised from 20 to 100 %. Water contact angle measurements confirmed that blended nanofibrous scaffolds have good hydrophilicity and the tensile strength and elongation-at-break were improved when the weight ratio of silk fibroin and HBC reached 20:80. Here, genipin vapor was used for induced conformation of silk fibroin to convert from random coil to β-sheet structure and it also act as a crosslinking agent for both the solutions. Cell viability results showed good cellular compatibility [174].

Air-impedance electrospinning was used to fabricate poly (L-lactic acid-co-ε-caprolactone) P(LLA-CL)/silk fibroin blended solution using perforated mandrel subjected to various intraluminal air pressures (0–300 kPa) and solid mandrel. A mechanical property as well as in vitro cellular infiltration of scaffolds was evaluated. Smooth muscle cells (SMCs) were seeded to vascular scaffolds for the evaluation of cellular infiltration at 1, 7, and 14 days. The result of the output was analyzed and found that air-impedance scaffolds allowed significantly more cell infiltration in comparison to the scaffolds fabricated with solid mandrel. It was also observed that both electrospinning models determined the interfiber distance and the alignment of fibers in the greater porosity regions which directly influences cell infiltration. Hence, it can be concluded that air-impedance electrospun scaffolds could be a choice of material for various biomedical applications [175].

Neat silk fibroin does not represent any strong adhesion site, so it was assumed that coating or blending it with any biodegradable polymer could enhance its biocompatibility toward regenerative medicine applications. Hence, collagen type I was selected for this purpose which also mimic the native extra-cellular matrix of most of the species. Silk fibroin and collagen were taken in equal amount 7 %

and blended the solutions at ratios of 100:0 (pure SF), 95:5, 90:10, and 85:15 (SF: collagen, v/v) before electrospinning. Morphological and physicochemical investigations were performed to characterize the electrospun scaffolds. It was observed that the blending of silk fibroin with collagen decreases porosity. Mammary epithelial cell MCF10A was investigated for its related biological studies such as adhesion, viability, and infiltration onto blended or coated silk fibroin/collagen scaffolds, and results revealed that the coated or blended materials showed significantly better cell adhesion, viability, and infiltration. On the basis of results, it could be concluded that the silk fibroin/collagen scaffold may serve as better matrix for tissue regeneration applications [176].

4 Chitin, Chitosan, and Silk Fibroin-Based Nanofibers: Applications in Regenerative Medicine

The preparation of an efficient extra-cellular matrix analog which is similar to native one is a tough challenge for bioengineers up to now. A cell competent ECM must not only imitate the architecture and compositional properties of the natural ECM, but it should also have the ability to connect with cells in 3D environment and promote communication between them [177]. Electrospinning technique is used for the fabrication of polymeric nanofiber non-woven materials which show extremely high surface-to-mass/volume ratio and excellent pore-interconnected porous scaffolding material. Electrospun nanofibrous mats are considered as one of the most promising scaffolds for cells which offer an artificial environment that mimics the native ECM, and electrospinning technique is relatively new for preparation of fibers and is capable of producing scaffolds as per the needs of the tissues to be repaired [178].

There are some specific needs for a cell to regenerate on electrospun fibrous scaffolds. One of the main drawbacks with polymers electrospun into scaffolds is that they do not have any specific group which could promote communication with the cells. Therefore, sometimes biomolecules are incorporated into electrospun nanofibrous scaffolds which could now work as biofunctional scaffold and determine the efficiency of fibers for the regeneration of new tissues [179]. As scaffolding material needs sufficient amount of water for the regeneration of cells, it should also be kept in mind that the electrospun matrix should be hydrophilic. To overcome this problem, electrospun fibers are subjected to functionalization such as plasma, UV, γ-radiation, acid or basic solutions treatment, and high-temperature treatments but they can damage the original structure of the fibers, whereas polymer blending is one of the mostly used techniques nowadays in which a hydrophilic biodegradable polymer attached by adsorption or covalently bonded to electrospun fibers [180].

It is evident from literature that neither silk fibroin nor chitin and chitosan in their neat solutions could be electrospun in nanofibrous mats which mimic the extra-cellular matrix. Hence, polymers approved by FDA could be blended with chitin, chitosan, and silk fibroin and after optimizing different electrospinning parameters such as concentration of amount of blending materials, viscosity, applied voltage, flow of blended solution by syringe, distance between collector and syringe, rotary speed of drum collector, etc. After successful spinning of blended solution, they could be further functionalized as per the needs.

If we talk about broad area of application of the electrospun fibers, they could be used for filtrations, affinity membranes and recovery of metal ions, tissue engineering scaffolds, wound healing, release control, catalyst and enzyme carriers, sensors, and energy storage applications [181]. But in our present analysis, we are focused on applications of nanofibrous materials used for regenerative medicine only. For regenerative medicine, only biocompatible and biodegradable polymers are used. The selection of scaffolding material is based on the properties of the tissue to be regenerated, as well as its time of regeneration. Synthetic polymers used for regenerative medicine include polyesters [182], such as poly(lactide), poly(glycolyde), and poly(caprolactone). But sometimes as per the need for regeneration of cells, two or more polymers are used simultaneously. For this type of use, polyelectrolyte complexes or blends or hybrids of polymers are selected for electrospinning [183].

Due to very weak mechanical strength of electrospun chitin and chitosan nanofibers, they are used with different blended biodegradable polymers. Presently, bioengineers are working on different fields of regenerative medicine such as bone regeneration, cartilage regeneration, ligament and tendon regeneration, soft to hard tissue interface regeneration, nerve regeneration, skeletal regeneration, etc.

Electrospun nanofibers of chitin and chitosan are potential candidate for many biomedical applications due to structural similarity to glycosaminoglycans present in the extra-cellular matrix and their morphological similarity to fibrous collagen matrices which are present in the ECM on nanometer scale [184]. Noh et al. reported electrospinning of chitin nanofibrous matrices (163 nm), and their cellular behavior as well as biodegradability was compared with commercially available chitin microfiber (8.77 μm) and it was found that chitin nanofibrous matrices promote cell attachment and spreading of normal human epidermal keratinocytes and normal human epidermal fibroblasts more better than chitin microfiber; hence, chitin nanofibrous matrices are known as important candidate for wound healing and tissue regeneration of oral mucosa and skin [185].

Simultaneous electrospinning was used to fabricate nanostructured composite matrix of PLGA/chitin using HFIP solvent and obtained nanofibers of 310 nm diameter at the ratio of 80/20 (w/w) [186]. It was observed that when normal human epidermal keratinocytes were seeded on the electrospun nanofibrous matrix, cell attachment and proliferation substantially improved in comparison to neat PLGA matrices. In a similar study, NHEKs and NHEFs were seeded to the chitin/silk fibroin blended electrospun nanofibrous matrices to observe the cellular responses and results revealed that it significantly improved [187].

Chemically modified hyaluronic acid/chitin blended electrospun nanofibrous mat was used for bone regeneration as it is osteoinductive and exhibited rapid degradation and neovascularization in vivo [188]. Chitin nanofibrous scaffolds were also found very suitable for skin regeneration system.

A hybrid matrix of PLGA–chitosan/PVA was electrospun by simultaneous electrospinning method, and obtained nanofibrous matrix was considered as promising candidate for culture of human embryo skin fibroblasts [189]. Bhattarai et al. [190] reported electrospinning of chitosan/PEO (90/10, w/w) blended solution with Triton X-100 surfactant. The obtained fibrous matrix retained its structural integrity in aqueous solution and promoted the adhesion of chondrocyte cells. Altman et al. [191] reported electrospinning of bilayered chitosan tube which is composed of an outer layer of chitosan film and an inner layer of chitosan nonwoven nano/micromesh. In the inner layer of electrospun chitosan tube, laminin-1 was applied to the nano/microfiber mesh surface of chitosan by covalent bonding. The nerve regeneration using chitosan tubes was found to be efficient and similar to that of the isograft, indicating a promising scaffold for peripheral nerve repair.

Electrospinning of chitosan/poly(vinyl alcohol) and *N*-carboxyethyl chitosan/PVA electrospun membranes was performed using an composite scaffold of hydroxyapatite which is formed in supersaturated $CaCl_2$ and KH_2PO_4 solution [192]. To improve hydroxyapatite formation having uniform distribution on electrospun membranes, poly(acrylic acid) was added to the incubation solution in different proportions. Mouse fibroblasts (L929) were seeded on the surface of HAp-CECS/PVA electrospun nanofibrous membrane, and results revealed that the cell viability and morphology were found to be well maintained and thus it suggests potential applications in bone tissue regeneration.

Jiang et al. [193] reported electrospinning of PLGA/PEG-g-chitosan blended solution followed by ibuprofen loaded solution, and results revealed that PEG-g-chitosan significantly reduces initial burst release of loaded ibuprofen drug from the electrospun PLGA membranes. It was also observed that conjugating ibuprofen to the side chains of PEG-g-chitosan gives sustained release of ibuprofen for more than 2 weeks.

Electrospun nanofibrous membranes are proved as one of the promising candidates for immobilization of enzymes due to their high specific surface area and porous structure. Huang et al. [194] reported immobilization of lipase enzyme in an electrospun nanofibrous chitosan/PVA membrane, whereas glutaraldehyde was used as a coupling agent. The loading of lipase enzyme to prepared nanofibrous membrane was up to 63.6 mg/g, and the residual activities of the immobilized lipase were more than 50 % after 30 days, which indicates excellent reusability and storage stability.

Wang et al. reported [195] electrospinning of chitosan solution and modified the surfaces of electrospun fiber using a peptide known as RGD. In comparison to the pure chitosan fibers, the cell adhesion activity of NIH3T3 cells on modified-RGD nanofibers was found to increase during first three days. Therefore, surface-modified nanofibers could be a powerful strategy for the enhancement of cell attachment which is an important factor required of regenerative medicine.

Chitosan/PCL electrospun nanofibers show 48 % higher rat Schwann cell proliferation compared to neat PCL nanofibers after eight days of culture [196].

Jung et al. [197] reported electrospinning of chitosan/PET blended solution and fabricated nanofibers using TFA/HFIP solvent system. The fabricated electrospun fibers were subjected for their antimicrobial activity, and results showed that blended nanofibers were found to be more effective against *S. aureus* and *Klebsiella pneumoniae* than pure PET matrices. In a similar study, Spasova et al. [198] reported that electrospinning of chitosan/PEO blended solution was fabricated into nanofiber matrices and potassium 5-nitro-8-quinolinolate was incorporated into them to investigate the antimicrobial and antimycotic effect against gram-negative and gram-positive bacteria and fungi.

Silk fibroin and its derivatives show excellent biocompatibility, slow and controlled degradation, minimal inflammatory response, and very good mechanical properties which made them a unique material for regenerative medicine [199]. Demands for blood vessels having very small diameter are needed for vessel transplants and bioengineers are continuously making their efforts to develop them, but up to now all efforts have encountered failure. Electrospun silk nanofiber could be potential candidate for vascular graft as they support the cell adhesion, proliferation, and differentiation of vascular cells on electrospun matrix which resist shear stress and pressure from simulated blood flow. Bondar et al. prepared silk fibroin-based nanofibers and investigated endothelial cell responses such as cell adhesion, proliferation, formation of intercellular contact, and expression of adhesion molecules [200]. It was observed that cells grown on nanofibrous scaffold show better results in comparison to microfibrous scaffolds which were confirmed by real-time PCR analysis which represents significant upregulation of integrin-β1 in endothelial cells. In another investigation, Soffer et al. [201] reported electrospinning of silk nanofibers into tubular structures showing inner diameter of ~3 mm with average wall thickness of 0.15 mm having mechanical strength of 2.42 ± 0.48 with linear modulus 2.45 ± 0.47 MPa. The average burst strength of the tubular scaffolds was measured as 811 mmHg which supposed to be excellent result in comparison to burst strength of collagen (71 mmHg) or any other mostly prepared nanofibers [202]. However, the mile stone is so far away to reach the gold standard of the saphenous vein whose burst strength is 1800 mmHg [203]. For near future work, different blends of silk fibroin with silk will need to focus for the preparation of electrospun small-diameter vascular grafts of endothelial and smooth muscle cells in a tubular and perfusion environment which may be very close to mimicking the in vivo environment.

Skin works as a protecting layer of human body against the direct exposure to environment, dehydration, and infectious agents. Skin cells are found in two main layers: a keratinized, stratified epidermis and an underlying thick layer of collagen-rich dermal connective tissue. Due to burns, wounds, or disease, skin comes to direct contact with infections resulting severe disability or sometimes even death. However, up to now, skin replacement is a challenging task for surgeons because of transplant limitations. Min et al. reported electrospinning of silk fibroin and found that fabricated matrix after treatment with water vapor followed by collagen I coating and this material have very potential use for accelerating early stages wound healing [204].

Yoo et al. fabricated chitin/silk blend nanofibrous matrices and hybrid matrices both in ratio of 75:25 and concluded that chitin/silk blend matrices represent better cell adhesion and spreading than that of chitin/silk hybrid matrices, which confirms that chitin/silk blend matrices show potential use as skin regeneration substitutes [205]. Jin et al. [206] demonstrated electrospinning of silk fibroin–PEO-extracted matrix and showed electrospun silk matrices as potential platform for mesenchymal stem cells attachment, proliferation, and extra-cellular deposition which results in improved cell attachment for PEO-extracted matrices in comparison to non-extracted matrices. Kim et al. [207] observed that osteocalcin protein content significantly increases in cell culture medium on day 14 in comparison to days 1 and 7. Comparing with standard tissue culture plastic, electrospun silk nanofibrous matrices showed comparable cellular response in terms of alkaline phosphatase activity and mineralization.

Modified electrospinning technique was used to fabricate three-dimensional silk matrices for bone generation using MC3T3-E1 cells [208]. In comparison to two-dimensional matrices, three-dimensional matrices significantly improve cell spreading and proliferation. It was also observed that the higher porosity of three-dimensional matrices was attributed to the improved cell adhesion and spreading. Electrospinning technique could be used for easily incorporating growth factors into electrospun nanofibers which are responsible for enhanced bone regeneration. Li et al. reported encapsulation of BMP-2 growth factors within hydroxyapatite nanoparticles and silk nanofibers alone or in combination. Growth factor loaded silk matrices were seeded with hMSCs and assessed for bone regeneration [209]. It was evident from results that matrices which contain BMP-2 growth factor showed increased calcium deposition, apatite crystallinity, and transcript levels of bone-specific markers compared to those matrices without growth factors.

In our body, cartilage is found at different places such as rib cage, ear, intervertebral disks, and articular surfaces between bones. The main function of cartilages is to provide structure and support to the body tissues without the rigidity of bone, and provide cushioning in joints. Due to aging, joint injury, and developmental disorders, cartilage defects/disorders are found in humans who generally felt by joint pain and loss of mobility. Baek et al. reported electrospinning of silk fibroin solution to fabricate nanofibrous silk fibroin scaffold which was subjected to microwave-induced argon plasma treatment [210] and found that plasma treatment nanofibrous scaffold significantly increases cell adhesion and proliferation of neonatal human knee articular chondrocyte cells. It was also notable that glycosaminoglycan synthesis was significantly high on plasma-induced silk matrices within a day after cell seeding. Surprisingly, there was no difference observed after one week of cell culture which suggested that plasma treatment is only effective in the early stages of culture in cartilage tissue engineering and repair.

5 Concluding Remarks

Developing scaffold matrix that potentially mimics as natural extra-cellular matrix for expanded cells at the nanoscale is still one of challenging areas of regenerative medicine. Presently, three types of techniques are used for the preparation of nanofibers: electrospinning, self-assembly, and phase separation. Out of these techniques, electrospinning is considered as one of the most versatile techniques, which demonstrates the most promising results in terms of regenerative medicine applications. Biomedical applications of electrospun nanofibrous scaffolds are determined by their physical as well as biological properties such as mechanical modulus, strength, biocompatibility, hydrophilicity, biodegradability, and specific cell interactions, which in turn depend largely on the material compositions of electrospinning solutions. For regenerative medicine applications, natural as well as synthetic both types of polymers are used as scaffolding materials but every material has its own good and bad part. Synthetic polymers are preferred over biopolymers for electrospinning as they can be functionalised easily and they retain their lot-to-lot uniformity even at large-scale productions. Further, synthetic polymers are synthesized in bulk and cost effective, and their raw materials are readily available but they are mostly hydrophobic and show poor interactions with cells and native tissue. Hence, nowadays, biopolymers are preferred over synthetically derived polymers mainly because of their structural similarities with native ECM, chemical versatility, and typically good biological performance such as better biocompatibility and low immunogenicity. But due to weak mechanical strength of biopolymers, they are used as blends, polyelectrolyte complexes, or hybrids with biodegradable synthetic polymers as co-spinning agents. There are several factors which affect the formation of three-dimensional electrospun nanofibrous scaffolds such as concentration of base material and co-spinning material, solvent, viscosity, applied voltage, distance between spinneret and collector, rotational speed of collector, etc.

Solubility and viscosity are the two important physical parameters for electrospinnability of chitin, chitosan, and silk fibroin biomaterials. To improve the solubility, their derivatives are synthesized and mixed solvent systems are commonly used. A high viscosity is caused mainly due to chain entanglement of chitin and chitosan layers and the applied electrical charges could not overcome to produce nanofibers. The surface chemistry, microstructure, and architecture of nanofibrous matrices significantly influence cellular adhesion, proliferation, and differentiation. It was observed that aligned nanofibers significantly improve cell spreading and proliferation in comparison to randomly oriented nanofibers. Before applying electrospun nanofibrous scaffolds for in vivo models, it is important to solve the critical issues such as porosity, mechanical strength and structural integrity, and low cell infiltration rates. Moreover, it was realized that the incorporated growth factors/nutrients were abundantly present at the surface of the scaffold; hence, cells will likely prefer to stay on the surface without migrating into the bulk. The standard upper limit diameter of native extra-cellular matrix is below 100 nm which is a challenging task for tissue engineering scientists. For the fabrication of electrospun nanofibrous scaffold which could be used as ECM, the preferable fiber diameter should be in range of 10–50 nm but at this stage, the porosity would automatically be decreased. Hence, there is an urgent need for the fabrication of a nanofibrous scaffold with desired diameter and porosity which should be identical to native ECM fibers and allow high rates for cell infiltration and mass transport. In our present review, we have discussed almost every factor which affects the electrospinning of chitin, chitosan, and silk fibroin in reference to regenerative medicine applications.

Acknowledgments One of the authors (PKD) thankfully acknowledged the financial support from CSIR New Delhi in the form of a research project and other author (BKS) gratefully acknowledged the financial support in the form of research associateship from CSIR New Delhi.

References

1. Chien KR (2008) Regenerative medicine and human models of human disease. Nature 453:302–305
2. Stevens MM, George JH (2005) Exploring and engineering the cell surface interface. Science 310:1135–1138
3. Khademhosseini A, Vacanti JP, Langer R (2009) Progress in tissue engineering. Sci Am 300:64–71
4. Place ES, Evans ND, Stevens MM (2009) Complexity in biomaterials for tissue engineering. Nat Mater 8:457–470
5. Senni K, Pereira J, Gueniche F, Delbarre-Ladrat C, Sinquin C, Ratiskol J, Godeau G, Fischer AM, Helley D, Colliec-Jouault S (2011) Marine polysaccharides: a source of bioactive molecules for cell therapy and tissue engineering Mar. Drugs 9:1664–1681
6. Wu J, Tan H, Li L, Gao C (2009) Covalently immobilized gelatin gradients within three-dimensional porous scaffolds. Chin Sci Bull 54:3174–3180
7. Lee KY, Mooney DJ (2012) Alginate: Properties and biomedical applications. Prog Polym Sci 37:106–126
8. Lee KY, Yuk SH (2007) Polymeric protein delivery systems. Progr Polym Sci 32:669–697
9. Pawar SN, Edgar KJ (2012) Alginate derivatization: a review of chemistry, properties and applications. Biomaterials 33:3279–3305
10. Tan H, Gong Y, Lao L, Mao Z, Gao C (2007) Gelatin/chitosan/hyaluronan ternary complex scaffold containing basic fibroblast growth factor for cartilage tissue engineering. J Mater Sci Mater Med 18:1961–1968
11. Ohgo K, Zhao C, Kobayashi M, Asakura T (2003) Preparation of non- woven nanofibers for Bombyx mori silk, Samia cythia ricini silk and recombinant hybrid silk with electrospinnning method. Polymer 44:841–846
12. Wnek GE, Carr ME, Simpson DG, Bowlin GL (2003) Electrospinning of nanofibers fibrinogen structures. Nano Lett 3:213–216
13. Fang X, Reneker DH (1997) DNA fibers by electrospinning. J Macromol Sci Phys B 36:169–173
14. Jiang HL, Fang DF, Hsiao BS, Chu B, Chen WL (2004) Optimization and characterization of dextran membranes prepared by electrospinning. Biomacromolecules 5:326–333
15. Kulkarni RK, Pani KC, Neuman C, Leonard F (1966) Polylactic acid for surgical implants. Arch Surg 93:839–843
16. Fisher JP, Vehof JW, Dean D, van der Waerden JP, Holland TA, Mikos AG, Jansen JA (2002) Soft and hard tissue response to photocrosslinked poly(propylene fumarate) scaffolds in a rabbit model. J Biomed Mater Res 59:547–556

17. Heller J, Barr J, Ng SY, Abdellauoi KS, Gurny R (2002) Poly(ortho esters): synthesis, characterization, properties and uses. Adv Drug Deliv Rev 54:1015–1039
18. Lee SJ, Choi JS, Park KS, Khang G, Lee YM, Lee HB (2004) Response of MG63 osteoblast-like cells onto polycarbonate membrane surfaces with different micropore sizes. Biomaterials 25:4699–4707
19. Santerre JP, Woodhouse K, Laroche G, Labow RS (2005) Understanding the biodegradation of polyurethanes: From classical implants to tissue engineering materials. Biomaterials 26:7457–7470
20. GuanJJ Fujimoto KL, Sacks MS, Wagner WR (2005) Preparation and characterization of highly porous, biodegradable polyurethane scaffolds for soft tissue applications. Biomaterials 26:3961–3971
21. Chen GQ, Wu Q (2005) The application of polyhydroxyalkanoates as tissue engineering materials. Biomaterials 26:6565–6578
22. Conconi MT, Lora S, Menti AM, Carampin P, Parnigotto PP (2006) In vitro evaluation of poly bis(ethyl alanato)phosphazene as a scaffold for bone tissue engineering. Tissue Eng 12:811–819
23. Rashidi H, Yang J, Kevin M, Shakesheff (2014) Surface engineering of synthetic polymer materials for tissue engineering and regenerative medicine applications. Biomater Sci 2:1318–1331
24. Behravesh E, Yasko AW, Engle PS, Mikos AG (1999) Synthetic biodegradable polymers for orthopaedic applications. Clin Orthop 367S:118–185
25. Middleton JC, Tipton AJ (2000) Synthetic biodegradable polymers as orthopaedic devices. Biomaterials 21:2335–2346
26. Vroman L, Adams AL, Fischer GC, Munoz PC (1980) Interaction of high molecular-weight kininogen, factor-XII, and fibrinogen in plasma at interfaces. Blood 55:156–159
27. Kim Moon Suk, Kim Jae Ho, Min Byoung Hyun, Chun Heung Jae, Han Dong Keun, Lee Hai Bang (2011) Polymeric scaffolds for regenerative medicine. Polym Rev 51:23–52
28. Bhardwaj N, Kundu SC (2010) Electrospinning: a fascinating fiber fabrication technique. Biotechnol Adv 28:325–347
29. Pierschbacher MD, Ruoslahti E (1984) Cell attachment activity of fibronectin can be duplicated by small synthetic fragments of the molecule. Nature 309:30–33
30. Li M, Mondrinos MJ, Gandhi MR, Ko FK, Weiss AS, Lelkes PI (2005) Electrospun protein fibers as matrices for tissue engineering. Biomaterials 26:5999–6008
31. Ramakrishna S, Fujihara K, Teo WE, Lim TC, Ma Z (2005) An introduction to electrospinning and nanofibers. World Scientific, Singapore
32. Li D, Xia Y (2004) Electrospinning of nanofibers: reinventing the wheel? Adv Mater 16:1151–1170
33. Zhang CX, Yuan XY, Wu LL, Han Y, Sheng J (2005) Study on morphology of electrospun poly(vinyl alcohol) mats. Eur Polym J 41:423–432
34. Schreuder-Gibson HL, Gibson P, Tsai P, Gupta P, Wilkes G (2004) Cooperative charging effects of fibers from electrospinning of electrically dissimilar polymers. Int Nonwovens J 13:39–45
35. Wannatong L, Sirivat A (2004) Electrospun fibers of polypyrrole/polystyrene blend for gas sensing applications. PMSE Prepr 91:692–693
36. Kim C, Park SH, Lee WJ, Yang KS (2004) Characteristics of supercapacitor electrodes of PBI-based carbon nanofiber web prepared by electrospinning. Electrochim Acta 50:877–881
37. Khil MS, Bhattarai SR, Kim HY, Kim SZ, Lee KH (2005) Novel fabricated matrix via electrospinning for tissue engineering. J Biomed Mater Res B Appl Biomater 72B:117–124
38. Reneker DH, Chun I (1996) Nanometre diameter fibres of polymer, produced by electrospinning. Nanotechnol 7:216–223
39. Taylor GI (1969) Electrically driven jets. Proc Roy Soc London 313:453–475
40. Jun Z, Hou HQ, Schaper A, Wendorff JH, Greiner A (2003) Poly-L-lactide nanofibers by electrospinning-influence of solution viscosity and electrical conductivity on fiber diameter and fiber morphology. e-Polymers 3:102–110

41. Boland ED, Wnek GE, Simpson DG, Pawlowski KJ, Bowlin GL (2001) Tailoring tissue engineering scaffolds using electrostatic processing techniques: a study of poly(glycolic acid) electrospinning. J Macromol Sci Pure Appl Chem 38:1231–1243
42. Hsu CM, Shivkumar S (2004) N, N-dimethylformamide additions to the solution for the electrospinning of poly(e-caprolactone) nanofibers. Macromol Mater Eng 289:334–340
43. Rockwood D, Fromstein J, Woodhouse K, Chase B, Rabolt JF (2004) Electrospinning of a biodegradable polyurethane for use in tissue engineering. Polym Prepr (Am Chem Soc, Div Polym Chem) 45:406
44. Ding B, Kim HY, Lee SC, Shao CL, Lee DR, Park SJ, Kwag GB, Choi KJ (2002) Preparation and characterization of a nanoscale poly(vinyl alcohol) fiber aggregate produced by an electrospinning method. J Polym Sci, Part B: Polym Phys 40:1261–1268
45. Son WK, Youk JH, Lee TS, Park WH (2004) The effects of solution properties and polyelectrolyte on electrospinning of ultrafine poly (ethylene oxide) fibers. Polymer 45:2959–2966
46. Boland Eugene D, Coleman Branch D, Barnes Catherine P, Simpson David G, Wnek Gary E, Bowlin Gary L (2005) Electrospinning polydioxanone for biomedical applications. Acta Biomater 1:115–123
47. Nair LS, Bhattacharyya S, Bender JD, Greish YE, Brown PW, Allcock HR, Laurencin CT (2004) Fabrication and optimization of methylphenoxy substituted polyphosphazene nanofibers for biomedical applications. Biomacromolecules 5:2212–2220
48. Kurita K (2001) Controlled functionalization of the polysaccharide chitin. Prog Polym Sci 269:1921–1971
49. Morganti P (2013) Saving the environment by nanotechnology and waste raw material: use of chitin nanofibril by eu research projects. J Appl Cosmetol 31:89–96
50. Dtie U (2009) Converting waste agricultural biomass into a resource. Compendium of technologies. Osaka, United Nations Environment Programme
51. Ajmeri JR, Ajemri CJ (2006) Surgical sutures: the largest textile implant material. In: Anand SC, Kennedy JF, Rajendran S (eds) Medical textiles and biomaterials for health care. CRC Press/Woodhead Publishing Ltd., Boca Raton, Cambridge, pp 432–440
52. Rajendran S, Anand SC (2002) Developments inmedical textiles. Text Progr 32:10–13
53. Muzzarelli RAA, Muzzarelli C (2005) Chitosan chemistry: relevance to the biomedical sciences. Adv Polym Sci 186:151–209
54. Austin PR, Brine CJ, Castle JE, Zikakis JP (1981) Chitin: new facets of research. Science 212:749–753
55. Rathke TD, Hudson SM (1994) Review of chitin and chitosan as fiber and film formers. Polym Rev 34:375–437
56. Samuels RJ (1981) Solid state characterization of the structure of chitosan films. J Polym Sci A-2 Polym Phys 19:1081–1105
57. Pillai CKS, Paul W, Sharma CP (2009) Chitin and chitosan polymers: chemistry, solubility and fiber formation. Prog Polym Sci 34:641–678
58. Kurita K (2001) Controlled functionalization pf the polysaccharide chitin. Prog Polym Sci 269:1921–1971
59. Jayakumar R, Tamura H (2008) Synthesis, characterization and thermal properties of chitin-g-poly(caprolactone) copolymers using chitin hydrogel. Int J Biol Macromol 43:32–36
60. Min BM, Lee SW, Lim JN, You Y, Lee TS, Kang PH, Park WH (2004) Chitin and chitosan nanofibers: electrospinning of chitin and deacetylation of chitin nanofibers. Polymer 45:7137–7142
61. Junkasem J, Rujiravanit R, Supaphol P (2006) Fabrication of α-chitin whiskerreinforced poly(vinyl alcohol) nanocomposite nanofibres by electrospinning. Nanotechnology 17:4519–4528
62. Ding B, Kim HY, Lee SC, Shao CL, Lee DR, Park SJ, Kwag GB, Choi KJ (2002) Preparation and characterization of a nanoscale poly(vinyl alcohol) fiber aggregate produced by an electrospinning method. J Polym Sci, Part B: Polym Phys 40:1261–1268
63. Sriupayo J, Supaphol P, Blackwell J, Rujiravanit R (2005) Preparation and characterization of α-chitin whisker-reinforced chitosan nanocomposite films with or without heat treatment. Carbohydr Polym 62:130–136

64. Park KE, Jung SY, Lee SJ, Min BM, Park WH (2006) Biomimetic nanofibrous scaffolds: preparation and characterization of chitin/silk fibroin blend nanofibers. Int J Biol Macromol 38:165–173
65. Shalumon KT, Binulal NS, Selvamurugan N, Nair SV, Menon D, Furuike T, Tamura H, Jayakumar R (2009) Electrospinning of carboxymethyl chitin/poly(vinyl alcohol) nanofibrous scaffolds for tissue engineering applications. Carbohydr Polym 77:863–869
66. Zhong C, Kapetanovic A, Deng Y, Rolandi M (2011) Adv Mater 23:4776–4781
67. Viswanathan G, Murugesan S, Pushparaj V, Nalamasu O, Ajayan PM, Linhardt RJ (2006) Biomacromolecules 7:415–418
68. Barber PS, Griggs CS, Bonner JR, Rogers RD (2013) Electrospinning of chitin nanofibers directly from an ionic liquid extract of shrimp shells. Green Chem 15:601–607
69. Ifuku S, Saimoto H (2012) Chitin nanofibers: preparations, modifications, and applications. Nanoscale 4:3308–3318
70. Ifuku S, Nogi M, Yoshioka M, Morimoto M, Yano H, Saimoto H (2010) Fibrillation of dried chitin into 10–20 nm nanofibers by a simple grinding method under acidic conditions. Carbohydr Polym 81:134–139
71. Ifuku S, Nogi M, Abe K, Yoshioka M, Morimoto M, Saimoto H, Yano H (2011) Simple preparation method of chitin nanofibers with a uniform width of 10–20 nm from prawn shell under neutral conditions. Carbohydr Polym 84:762–764
72. Wu J, Meredith JC (2014) Assembly of chitin nanofibers into porous biomimetic structures via freeze drying. ACS Macro Lett 3:185–190
73. Pant HR, Kim HJ, Bhatt LR, Joshi MK, Kim EK, Kim JI, Abdal-hay A, Hui KS, Kim CS (2013) Chitin butyrate coated electrospun nylon-6 fibers for biomedical applications. Appl Surf Sci 285:538–544
74. Salaberria AM, Labidi J, Fernandes SCM (2014) Chitin nanocrystals and nanofibers as nano-sized fillers into thermoplastic starch-based biocomposites processed by melt-mixing. Biochem Eng J 256:356–364
75. Jayakumar R, Prabaharan M, Nair SV, Tamura H (2010) Novel chitin and chitosan nanofibers in biomedical applications. Biotechnol Adv 28:142–150
76. Jayakumar R, Menon D, Manzoor K, Nair SV, Tamura H (2010) Biomedical applications of chitin and chitosan based nanomaterials-a short review. Carbohydr Polym 82:227–232
77. Park KE, Jung SY, Lee SJ, Min BM, Park WH (2006) Biomimetic nanofibrous scaffolds: Preparation and characterization of chitin/silk fibroin blend nanofibers. Int J Biol Macromol 38:165–173
78. Xin S, Li Y, Li W, Du J, Huang R, Du Y, Deng H (2012) Carboxymethyl chitin/organic rectorite composites based nanofibrous mats and their cell compatibility. Carbohydr Polym 90:1069–1074
79. Nata IF, Wang SSS, Wu TM, Lee CK (2012) β-Chitin nanofibrils for self-sustaining hydrogels preparation via hydrothermal treatment. Carbohydr Polym 90:1509–1514
80. Mukesh C, Mondal D, Sharma M, Prasad K (2014) Choline chloride–thiourea, a deep eutectic solvent for the production of chitin nanofibers. Carbohydr Polym 103:466–471
81. Yoo CR, Yeo IS, Park KE, Park JH, Lee SJ, Park WH, Min YM (2008) Effect of chitin/silk fibroin nanofibrous bicomponent structures on interaction with human epidermal keratinocytes. Int J Biol Macromol 42:324–334
82. Nama YS, Park WH, Ihm D, Hudson SM (2010) Effect of the degree of deacetylation on the thermal decomposition of chitin and chitosan nanofibers. Carbohydr Polym 80:291–295
83. Jia Y, Liang K, Shen X, Bowlin GL (2014) Electrospinning and characterization of chitinna nofibril/polycaprolactone nanocomposite fiber mats. Carbohydr Polym 101:68–74
84. Ifuku S, Tsukiyama Y, Yukawa T, Egusa M, Kaminaka H, Izawa H, Morimoto M, Saimoto H (2015) Facile preparation of silver nanoparticles immobilized on chitin nanofiber surfaces to endow antifungal activities. Carbohydr Polym 117:813–817
85. Morganti P (2009) Chitin nanofibrils for cosmetic delivery. Cosmet Toiletries 125:36–393
86. Mincea M, Negrulescu A, Ostafe V (2012) Preparation, modification, and application of chitin nanowiskers: a review. Rev Adv Mater Sci 30:225–242

87. Muzzarelli RAA, Morganti P, Morganti G, Palombo P, Palombo M, Biagini G, Mattioli-Belmonte M, Giantomassi F, Orlandi F, Muzzarelli C (2007) Chitin nanofibril/chitosan composites as wound medicaments. Carbohydr Polym 70:274–284
88. Rosen Y, Elman N (2012) Biomaterial science. an integrated clinical and engineering approach. CRC-Press, New York
89. Palsson BO, Bathia SN (2004) Tailoring biomaterials. Tissue Engineering. Pearson Prendice Hall, Upper Saddle River, pp 270–287
90. Nagahama H, Kashiki T, Nwe N, Jayakumar R, Furuike T, Tamura H (2008) Preparation of biodegradable chitin/gelatin membranes with GlcNAc for tissue engineering applications. Carbohydr Polym 73:456–463
91. Nagahama H, Nwe N, Jayakumar R, Koiwa S, Furuike T, Tamura H (2008) Novel biodegradable chitin membranes for tissue engineering applications. Carbohydr Polym 73:295–302
92. Jayakumar R, Prabaharan M, Reis RL, Mano JF (2005) Graft copolymerized chitosan-Present status and applications. Carbohydr Polym 62:142–158
93. Jayakumar R, Nwe N, Tokura S, Tamura H Sulfated chitin and chitosan as novel biomaterials. Int J Biol Macromol 40:175–181
94. Schiffman JD, Schauer CL (2007) Crosslinking chitosan nanofibers. Biomacromolecules 8:594–601
95. Shalumon KT, Binulal NS, Selvamurugan N, Nair SV, Menon D, Furuike T, Tamura H, Jayakumar R (2009) Electrospinning of carboxymethyl chitin/poly (vinyl alcohol) nanofibrous scaffolds for tissue engineering applications. Carbohydr Polym 77:863–869
96. Jayakumar R, Reis RL, Mano JF (2006) Phosphorous containing chitosan beads for controlled oral drug delivery. J Bioact Compat Polym 21:327–340
97. Prabaharan M, Mano JF (2005) Chitosan-based particles as controlled drug delivery systems. Drug Deliv 12:41–57
98. Anitha A, Divya Rani VV, Krishna R, Sreeja V, Selvamurugan N, Nair SV, Tamura H, Jayakumar R (2009) Synthesis, characterization, cytotoxicity and antibacterial studies of chitosan, O-carboxymethyl, N, O-carboxymethyl chitosan nanoparticles. Carbohydr Polym 78:672–677
99. Madhumathi K, Sudhessh Kumar PT, Kavya KC, Furuike T, Tamura H, Nair SV, Jayakumar R (2009) Novel chitin/nanosilica composite scaffolds for bone tissue engineering applications. Int J Biol Macromol 45:289–292
100. Muramatsu K, Masuda S, Yoshihara Y, Fujisawa A (2003) In vitro degradation behavior of freeze-dried carboxymethyl-chitin sponges processed by vacuum-heating and gamma irradiation. Polym Degrad Stab 81:327–332
101. Wang JW, Hon MH (2003) Preparation and characterization of pH sensitive sugar mediated (polyethylene glycol/chitosan) membrane. J Mater Sci: Mater Med 14:1079–1088
102. Rinaudo M (2006) Chitin and chitosan: Properties and applications. Prog Polym Sci 31:603–632
103. Pillai CKS, Paul W, Sharma CP (2009) Chitin and chitosan polymers: chemistry, solubility and fiber formation. Prog Polym Sci 34:641–678
104. Rinaudo M (2006) Chitin and chitosan: properties and applications. Prog Polym Sci 31:603–632
105. Li L, Hsieh YL (2006) Chitosan bicomponent nanofibers and nanoporous fibers. Carbohydr Res 341:374–381
106. Ohkawa K, Minato KI, Kumagai G, Hayashi S, Yamamoto H (2006) Chitosan nanofiber. Biomacromolecules 7:3291–3294
107. Schiffman JD, Schauer CL (2007) Cross—linking chitosan nanofibers. Biomacromolecules 8:594–601
108. Homayoni H, Ravandi SAH, Valizadeh M (2009) Electrospinning of chitosan nanofibers: processing optimization. Carbohydr Polym 77:656–661

109. Geng XY, Kwon OH, Jang JH (2005) Electrospinning of chitosan dissolved in concentrated acetic acid solution. Biomaterials 26:5427–5432
110. Sangsanoh P, Supaphol P (2006) Stability improvement of electrospun chitosan nanofibrous membranes in neutral or weak basic aqueous solutions. Biomacromolecules 7:2710–2714
111. McKee MG, Wilkes GL, Colby RH, Long TE (2004) Correlations of solution rheology with electrospun fiber formation of linear and branched polyesters. Biomacromolecules 37:1760–1767
112. Schiffman JD, Schauer CL (2008) A review: Electrospinning of biopolymer nanofibers and their applications. Polym Rev 48:317–352
113. Desai K, Kit K, Li J, Zivanovic S (2008) Morphological and surface properties of electrospun chitosan nanofibers. Biomacromolecules 9:1000–1006
114. Zhou YS, Yang DZ, Chen XM, Xu Q, Lu FM, Nie J (2008) Electrospun water-soluble carboxyethyl chitosan/poly(vinyl alcohol) nanofibrous membrane as potential wound dressing for skin regeneration. Biomacromolecules 9:349–354
115. Ignatova M, Manolova N, Markova N, Rashkov I (2009) electrospun non-woven nanofibrous hybrid mats based on chitosan and PLA for wound-dressing applications. Macromol Biosci 9:102–111
116. Zhang HT, Wu CY, Zhang YL, White CJB, Xue Y, Nie HL, Zhu LM (2010) Elaboration, characterization and study of a novel affinity membrane made from electrospun hybrid chitosan/nylon-6 nanofibers for papain purification. J Mater Sci 45:2296–2304
117. Shalumon KT, Anulekha KH, Girish CM, Prasanth R, Nair SV, Jayakumar R (2010) Single step electrospinning of chitosan/poly(caprolactone) nanofibers using formic acid/acetone solvent mixture. Carbohydr Polym 80:413–419
118. Park WH, Jeong L, Yoo DI, Hudson S (2004) Effect of chitosan on morphology and conformation of electrospun silk fibroin nanofibers. Polymer 45:7151–7157
119. Torres-Giner S, Ocio MJ, Lagaron JM (2009) Novel antimicrobial ultrathin structures of zeinlchitosan blends obtained by electrospinning. Carbohydr Polym 77:261–266
120. Chen ZG, Mo XM, Qing FL (2007) Electrospinning of collagen–chitosan complex. Mater Lett 61:3490–3494
121. Pakravan M, Heuzey MC, Ajji A (2012) Core–shell structured peo-chitosan nanofibers by coaxial electrospinning. Biomacromolecules 13:412–421
122. Pakravan M, Heuzey MC, Ajji A (2011) A fundamental study of chitosan/PEO electrospinning. Polymer 52:4813–4824
123. Kriegel C, Kit KM, McClements DJ, Weiss J (2009) Electrospinning of chitosan–poly(ethylene oxide) blend nanofibers in the presence of micellar surfactant solutions. Polymer 50:189–200
124. Zhang JF, Yang DZ, Xu F, Zhang ZP, Yin RX, Jun Nie J (2009) Electrospun core-shell structure nanofibers from homogeneous solution of poly(ethylene oxide)/chitosan. Macromolecules 42:5278–5284
125. Desai K, Kit K, Li J, Zivanovic S (2008) Morphological and surface properties of electrospun chitosan nanofibers. Biomacromolecules 9:1000–1006
126. Jia YT, Gong J, Gu XH, Kim HY, Dong J, Shen XY (2007) Fabrication and characterization of poly (vinyl alcohol)/chitosan blend nanoWbers produced by electrospinning method. Carbohydr Polym 67:403–409
127. Paipitak K, Pornpra T, Mongkontalang P, Techitdheera W, Pecharapa W (2011) Characterization of PVA-chitosan nanofibers prepared by electrospinning. Procedia Eng 101–105
128. Gimenez V, Mantecon A, Cadiz VJ (1996) Modification of poly(vinyl alcohol) with acid chlorides and crosslinking with difunctional hardeners. J Polym Sci Part A: Polym Chem 34:925–934
129. Zhou Y, Yang H, Liu X, Mao J, Gu S, Xu W (2013) Electrospinning of carboxyethyl chitosan/poly(vinyl alcohol)/silk fibroin nanoparticles for wound dressings. Int J Biol Macromol 53:88–92

130. Alipour SM, Nouri M, Mokhtari J, Bahrami SH (2009) Electrospinning of poly(vinyl alcohol)–water-soluble quaternized chitosan derivative blend. Carbohydr Res 344:2496–2501
131. Jayakumar R, Prabaharan M, Nair SV, Tamura H (2010) Novel chitin and chitosan nanofibers in biomedical applications. Biotechnol Adv 28:142–150
132. Bhattarai N, Edmondson D, Veiseh O, Matsen FA, Zhang M (2005) Electrospun chitosan-based nanofibers and their cellular compatibility Biomaterials. 26:6176–6184
133. Desai K, Kit K, Li J, Zivanovic S (2008) Morphological and surface properties of electrospun chitosan nanofibers. Biomacromolecules 9:1000–1006
134. Jayakumar R, Prabaharan M, Nair SV, Tamura H (2010) Novel chitin and chitosan nanofibers in biomedical applications. Biotechnol Adv 28:142–150
135. Sangsanoh P, Supaphol P (2006) Stability improvement of electrospun chitosan nanofibrous membranes in neutral or weak basic aqueous solutions. Biomacromolecules 7:2710–2714
136. Schiffman J, Schauer C (2007) One-step electrospinning of cross-linked chitosan fibers. Biomacromolecules 8:2665–2667
137. Cooper A, Bhattarai N, Kievit FM, Rossol M, Zhang M (2011) Electrospinning of chitosan derivative nanofibers with structural stability in an aqueous environment. Phys Chem Chem Phys 13:9969–9972
138. Geng X, Kwon OH, Jang J (2005) Electrospinning of chitosan dissolved in concentrated acetic acid solution. Biomaterials 26:5427–5432
139. Neamnark A, Rujiravanit R, Supaphol P (2006) Electrospinning of hexanoyl chitosan. Carbohydr Polym 66:298–305
140. Torres-Giner S, Ocio MJ, Lagaron JM (2009) Novel antimicrobial ultrathin structures of zein/chitosan blends obtained by electrospinning. Carbohydr Polym 77:261–266
141. Nirmala R, Navamathavan R, El-Newehy MH, Kim HY (2011) Preparation and electrical characterization of polyamide-6/chitosan composite nanofibers via electrospinning. Mater Lett 65:493–496
142. Ignatova M, Manolova N, Toshkova R, Rashkov I, Gardeva E, Yossifova L, Alexandrov M (2012) Quaternized chitosan-coated nanofibrous materials containing gossypol: Preparation by electrospinning, characterization and antiproliferative activity towards HeLa cells. Int J Pharm 436:10–24
143. Shalumon KT, Anulekha KH, Girish CM, Prasanth R, Nair SV, Jayakumar R (2010) Single step electrospinning of chitosan/poly(caprolactone) nanofibers using formic acid/acetone solvent mixture. Carbohydr Polym 80:413–419
144. Alipour SM, Nouri M, Mokhtari J, Bahrami SH (2009) Electrospinning of poly(vinyl alcohol)-water-soluble quaternized chitosan derivative blend. Carbohydr Res 344:2496–2501
145. Sun B, Long YZ, Zhang HD, Li MM, Duvail JL, Jiang XY, Yin HL (2014) Advances in three-dimensional nanofibrous macrostructures via electrospinning. Prog Polym Sci 39:862–890
146. Cao H, Kuboyama N (2010) A biodegradable porous composite scaffold of PGA/ß TCP for bone tissue engineering. Bone 46:386–395
147. Lai HL, Abulkhalil A, Craig DQ (2003) The preparation and characterization of drug loaded alginate and chitosan sponges. Int J Pharm 251:175–181
148. Kumar S, Koh J, Tiwari DK, Dutta PK (2011) Optical study of chitosan-ofloxacin complex for biomedical applications. J Macromol Sci Part A 48:789–795
149. Alves da Silva ML, Crawford A, Mundy JM, Correlo VM, Sol P, Bhattacharya M, Hatton PV, Reis RL, Neves NM (2010) Chitosan/polyester-based scaffolds for cartilage tissue engineering: assessment of extracellular matrix formation. Acta Biomaterialia 6:1149–1157
150. Lu GY, Kong LJ, Sheng BY, Wang G, Gong Y, Zhang X (2007) Degradation of covalently cross-linked carboxymethyl chitosan and its potential application for peripheral nerve regeneration. Eur Polym J 43:3807–3818
151. Meinel L, Hofmann S, Karageorgiou V, Kirker-Head C, McCool J, Gronowicz G, Zichner L, Langer R, Vunjak-Novakovic G, Kaplan DL (2005) The inflammatory responses to silk films in vitro and in vivo. Biomaterials 26:147–155

152. Mandal BB, Kundu SC (2008) Non-bioengineered silk fibroin protein 3D scaffolds for potential biotechnological and tissue engineering applications. Macromol Biosci 8:807–818
153. Mehta AS, Singh BK, Singh N, Archana D, Snigdha K, Harniman R, Rahatekar SS, Tewari RP, Dutta PK (2015) Chitosan silk-based three-dimensional scaffolds containing gentamicin encapsulated calcium alginate beads for drug administration and blood compatibility. J Biomater Appl 29:1314–1325
154. Mandal BB, Kundu SC (2009) Calcium alginate beads embedded in silk fibroin as 3D dual drug releasing scaffolds. Biomaterials 30:5170–5177
155. Yukseloglu SM, Sokmen N, Canoglu S (2015) Biomaterial applications of silk fibroin electrospun nanofibres. Microelectron Eng, Article in Press. doi: http://dx.doi.org/10.1016/j.mee.2015.04.008
156. FAO http://faostat3.fao.org/search/*/E
157. Gulrajani M (1992) Degumming of silk. Rev Prog Color Relat Top 22:79–89
158. Li Y, Dai XQ (2006) Biomechanical Engineering of Textiles and Clothing. Wood head publishing in textiles, Cambridge 164
159. Chutipakdeevong J, Ruktanonchai UR, Supaphol P (2013) Process optimization of electrospun silk fibroin fiber mat for accelerated wound healing. J Appl Polym Sci 130:3634–3644
160. Vepari C, Kaplan DL (2007) Silk as a Biomaterial. Prog Polym Sci 32:991–1007
161. Ki CS, Lee KH, Baek DH, Hattori M, Um IC, Ihm DW, Park YH (2007) Dissolution and wet spinning of silk fibroin using phosphoric acid/formic acid mixture solvent system. J Appl Polym Sci 105:1605–1610
162. Park KE, Jung SY, Lee SJ, Min BM, Park WH (2006) Biomimetic nanofibrous scaffolds: Preparation and characterization of chitin/silk fibroin blend nanofibers. Int J Biol Macromol 38:165–173
163. Zhu J, Zhang Y, Saho H, Hu X (2008) Electrospinning and rheology of regenerated Bombyx mori silk fibroin aqueous solutions: The effects of pH and concentration. Polymer 49:2880–2885
164. Jin HJ, Fridrikh SV, Rutledge GC, Kaplan DL (2002) Electrospinning Bombyx mori silk with poly(ethylene oxide). Biomacromolecules 3:1233–1239
165. Lai GJ, Shalumon KT, Chen SH, Chen JP (2014) Composite chitosan/silk fibroin nanofibers for modulation of osteogenic differentiation and proliferation of human mesenchymal stem cells. Carbohydr Polym 111:288–297
166. Panda N, Bissoyi A, Pramanik K, Biswas A (2015) Development of novel electrospun nanofibrous scaffold from P. ricini and A. mylitta silk fibroin blend with improved surface and biological properties. Mater Sci Eng C Mater Biol Appl 48:521–532
167. Sasithorn N, Martinová L (2012) Effect of calcium chloride on electrospinning of silk fibroin nanofibres. In: International conference on textiles and fashion, 3–4 July 2012
168. Kim BS, Park KE, Kim MH, You HK, Lee J, Park WH (2015) Effect of nanofiber content on bone regeneration of silk fibroin/poly(ε-caprolactone) nano/microfibrous composite scaffolds. Int J Nanomedicine 10:485–502
169. Kang M, Jin HJ (2007) Electrically conducting electrospun silk membranes fabricated by adsorption of carbon nanotubes. Colloid Polym Sci 285:1163–1167
170. Zhou Y, Yang H, Liu X, Mao J, Gu S, Xu W (2013) Electrospinning of carboxyethyl chitosan/poly(vinyl alcohol)/silk fibroin nanoparticles for wound dressings. Int J Biol Macromol 53:88–92
171. Min BM, Lee G, Kim SH, Nam YS, Lee TS, Park WH (2004) Electrospinning of silk fibroin nanofibers and its effect on the adhesion and spreading of normal human keratinocytes and fibroblasts in vitro. Biomaterials 25:1289–1297
172. Paşcu EI, Stokes J, McGuinness GB (2013) Electrospun composites of PHBV, silk fibroin and nano-hydroxyapatite for bone tissue engineering. Mater Sci Eng C Mater Biol Appl 33:4905–4916
173. Andiappan M, Sundaramoorthy S, Panda N, Meiyazhaban G, Winfred SB, Venkataraman G, Krishna P (2013) Electrospun eri silk fibroin scaffold coated with hydroxyapatite for bone tissue engineering applications. Prog biomater 2:6

174. Zhou Weitao, He Jianxin, Shan Du, Cui Shizhong, Gao Weidong (2011) Electrospun silk fibroin/cellulose acetate blend nanofibres: structure and properties. Iran polym j 20:389–397
175. Zhang K, Qian Y, Wang H, Fan L, Huang C, Mo X (2011) Electrospun silk fibroin-hydroxybutyl chitosan nanofibrous scaffolds to biomimic extracellular matrix. J Biomater Sci 22:1069–1082
176. Yin A, Li J, Bowlin GL, Li D, Rodriguez IA, Wang J, Wu T, El-Hamshary HA, Al-Deyab SS, Mo X (2014) Fabrication of cell penetration enhanced poly (l-lactic acid-co-ε-caprolactone)/silk vascular scaffolds utilizing air-impedance electrospinning. Colloids Surf B: Biointerfaces 120:47–54
177. Maghdouri-White Y, Bowlin GL, Lemmon CA, Dréau D (2014) Mammary epithelial cell adhesion, viability, and infiltration on blended or coated silk fibroin-collagen type I electrospun scaffolds. Mat Sci Eng C 43:37–44
178. Barnes CP, Sell SA, Boland ED, Simpson DG, Bowling GL (2007) Nanofibers technology: designing the next generation of tissue engineering scaffolds. Adv Drug Deliv Rev 59:1413–1433
179. Braghirolli DI, Steffens D, Pranke P (2014) Electrospinning for regenerative medicine: a review of the main topics. Drug Discov Today 19:743–753
180. Agarwal S, Wendorff JH, Greiner A (2008) Use of electrospinning technique for biomedical applications. Polymer 49:5603–5621
181. Agarwal S, Wendorff JH, Greiner A (2009) Progress in the field of electrospinning for tissue engineering applications. Adv Mater 21:3343–3351
182. Jian F, HaiTao H, Tong L, XunGai W (2008) Applications of electrospun nanofibers. Chin. Sci. Bull 53:2265–2286
183. Ramakrishna S, Fujihara K, Teo WE, Lim TC, Ma Z (eds) (2005) An introduction to electrospinning and nanofibers. World Scientific, Singapore
184. Jiang T, Carbone EJ, Lo KWH, Laurencin CT (2015) Electrospinning of polymer nanofibers for tissue regeneration. Prog Polym Sci, Article in Press, http://dx.doi.org/10.1016/j.progpolymsci.2014.12.001
185. Noh HK, Lee SW, Kim JM, Oh JE, Kim KW, Chung CP, Choi SC, Park WH, Min B (2006) Electrospinning of chitin nanofibers: degradation behavior and cellular response to normal human keratinocytes and fibroblasts. Biomaterials 27:3934–3944
186. Min B, You Y, Kim JM, Lee SJ, Park WH (2004) Formation of nanostructured poly(lactic-co-glycolic acid)/chitin matrix and its cellular response to normal human keratinocytes and fibroblasts. Carbohydr Polym 57:285–292
187. Park KE, Jung SY, Lee SJ, Min B, Park WH (2006) Biomimetic nanofibrous scaffolds: preparation and characterization of chitin/silk fibroin blend nanofibers. Int J Biol Macromo 38:165–173
188. Frenot A, Henriksson MW, Walkenstrom P (2007) Electrospinning of cellulose-based nanofibers. J Appl Polym Sci 103:1473–1482
189. Duan B, Yuan X, Zhu Y, Zhang Y, Li X, Zhang Y, Yao K (2006) A nanofibrous composite membrane of PLGA-chitosan/PVA prepared by electrospinning. Eur Polym J 42:2013–2022
190. Bhattarai N, Edmondson D, Veiseh O, Matsen FA, Zhang M (2005) Electrospun chitosan-based nanofibers and their cellular compatibility. Biomaterials 26:6176–6184
191. Altman GH, Horan RL, Lu HH, Moreau J, Martin I, Richmond JC, Kaplan DL (2002) Silk matrix for tissue engineered anterior cruciate ligaments. Biomaterials 23:4131–4141
192. Yang D, Jin Y, Zhou Y, Ma G, Chen X, Lu F, Nie J (2008) In situ mineralization of hydroxyapatite on electrospun chitosan-based nanofibrous scaffolds. Macromol Biosci 8:239–246
193. Jiang H, Fang D, Hsiao B, Chu B, Chen W (2004) Preparation and characterization of ibuprofen-loaded poly(lactide-co-glycolide)/poly(ethylene glycol)-g-chitosan electrospun membranes. J Biomat Sci Polym Ed 15:279–296
194. Huang X, Ge D, Xu Z (2007) Preparation and characterization of stable chitosan nanofibrous membrane for lipase immobilization. Eur Polym J 43:3710–3718

195. Wang YY, Lu LX, Feng ZQ, Xiao ZD, Huang NP (2010) Cellular compatibility of RGD-modified chitosan nanofibers with aligned or random orientation. Biomed Mater 5:054112
196. Prabhakaran MP, Venugopal JR, Chyan TT, Hai LB, Chan CK, Lim AY, Ramakrishna S (2008) Electrospun biocomposite nanofibrous scaffolds for neural tissue engineering. Tissue Eng. Part A 14:1787–1797
197. Jung K, Huh M, Meng W, Yuan J, Hyun SH, Bae J, Hudson SM, Kang I (2007) Preparation and antibacterial activity of PET/chitosan nanofibrous mats using an electrospinning technique. J Appl Polym Sci 105:2816–2823
198. Spasova M, Manolova N, Paneva D, Rashkov I (2004) Preparation of chitosan containing nanofibers by electrospinning of chitosan/poly(ethylene oxide) blend solution. e-Polymers 56:1–12
199. Horan RL, Antle K, Collette AL, Wang Y, Huang J, Moreau JE, Volloch V, Kaplan DL, Altman GH (2005) In vitro degradation of silk fibroin. Biomaterials 26:3385–3393
200. Bondar B, Fuchs S, Motta A, Migliaresi C, Kirkpatrick CJ (2008) Functionality of endothelial cells on silk fibroin nets: comparative study of micro- and nanometric fibre size. Biomaterials 29:561–572
201. Soffer L, Wang X, Zhang X, Kluge J, Dorfmann L, Kaplan DL, Leisk G (2008) Silk-based electrospun tubular scaffolds for tissue-engineered vascular grafts. J Biomater Sci Polymer Edn 19:653–664
202. Orban JM, Wilson LB, Kofroth JA, El-Kurdi MS, Maul TM, Vorp DA (2004) Crosslinking of collagen gels by transglutaminase. J Biomed Mater Res. A 68(68):756–762
203. Nishibe T, Kondo Y, Muto A, Dardik A (2007) Optimal prosthetic graft design for small diameter vascular grafts. Vascular 15:356–360
204. Min BM, Lee G, Kim SH, Nam YS, Lee TS, Park WH (2004) Electrospinning of silk fibroin nanofibers and its effect on the adhesion and spreading of normal human keratinocytes and fibroblasts in vitro. Biomaterials 25:1289–1297
205. Yoo CR, Yeo IS, Park KE, Park JH, Lee SJ, Park WH, Min BM (2008) Effect of chitin/silk fibroin nanofibrous bicomponent structures on interaction with human epidermal keratinocytes. Int J Biol Macromol 42:324–334
206. Jin HJ, Chen J, Karageorgiou V, Altman GH, Kaplan DL (2004) Human bone marrow stromal cell responses on electrospun silk fibroin mats. Biomaterials 25(25):1039–1047
207. Kim KH, Jeong L, Park HN, Shin SY, Park WH, Lee SC, Kim TI, Park YJ, Seol YJ, Lee YM, Ku Y, Rhyu IC, Han SB, Chung CP (2005) Biological efficacy of silk fibroin nanofiber membranes for guided bone regeneration. J Biotechnol 120:120327–120339
208. Ki CS, Park SY, Kim HJ, Jung HM, Woo KM, Lee JW, Park YH (2008) Development of 3-D nanofibrous fibroin scaffold with high porosity by electrospinning: implications for bone regeneration. Biotechnol Lett 30:405–410
209. Li C, Vepari C, Jin HJ, Kim HJ, Keplan DL (2006) Electrospun silk-BMP-2 scaffolds for bone tissue engineering. Biomaterials. 27:3115–3124
210. Baek HS, Park YH, Ki CS, Park JC, Rahd DK (2008) Enhanced chondrogenic responses of articular chondrocytes onto porous silk fibroin scaffolds treated with microwave-induced argon plasma. Surf Coat Technol 202:5794–5797

Part II
Focus on Therapeutics, Functionalization and Computer Aided Techniques

Chitosan: A Potential Therapeutic Dressing Material for Wound Healing

D. Archana, Pradip Kumar Dutta and Joydeep Dutta

Abstract A wide variety of polymers have been used over decades for the preparation of dressing materials for wound healing applications. But the dressing materials based on polysaccharides such as chitosan (CS) have received tremendous attention of the worldwide researchers as a consequence of its important properties like anti-infectional activity, biocompatibility, biodegradability, nontoxicity to mention a few. CS helps in every phase of wound healing such as acting as barrier against microbes, absorbing exudates, accelerates the infiltration of inflammatory cells like neutrophils and helps in healing without scar formation. A reason behind the popularity of CS is that not only it can easily be processed as gels, films, fibers, and scaffolds but also can be blended with natural as well as synthetic polymers to reduce price and improve properties like mechanical, wettability, gas permeability, and handling. Apart from natural and synthetic polymers, CS is also blended with nanoparticles and growth factors to which it shows better antibacterial activity and reduce time span for wound healing. The present chapter aims to focus on feasibility of combining natural polymers, synthetic polymers, nanoparticles, and growth factors with CS for the preparation of wound dressings as basic healthcare materials for regenerative medicine.

Keywords Chitosan · Synthetic polymers · Natural polymers · Nanoparticles · Growth factors · Wound healing applications

D. Archana · P.K. Dutta
Department of Chemistry, MN National Institute of Technology, Allahabad, U.P 211004, India

P.K. Dutta
Centre for Medical Diagnostics and Research, MN National Institute of Technology, Allahabad, U.P 211004, India

J. Dutta (✉)
Department of Chemistry, Amity School of Applied Sciences,
Amity University Haryana, Pachgaon, Near IMT, Manesar, Haryana, 122413, India
e-mail: dutta_joy@yahoo.co.in

P.K. Dutta (ed.), *Chitin and Chitosan for Regenerative Medicine*, Springer Series on Polymer and Composite Materials, DOI 10.1007/978-81-322-2511-9_8

Abbreviations

BC	Bacterial cellulose
BC–Ch	Bacterial cellulose–chitosan
BDE	Bilayer dermal equivalent
CMCS	Carboxymethylated chitosan
CS	Chitosan
ChL	Chitosan lactate
DA	Degree of acetylation
DD	Degree of deacetylation
EGF	Epidermal growth factor
EDC	Ethylcarbodiimide hydrochloride
EDTA	Ethylenediamine tetraacetic acid
ECM	Extracellular matrix
FGF	Fibroblast growth factor
MMP	Induces matrix metalloproteinase
Mw	Molecular weight
OCMC	Oxidation of carboxymethyl cellulose
PDGF	Platelet-derived growth factor
PGA	Poly(L-glutamic acid)
PEO	Poly(ethylene oxide)
PVA	Poly(vinyl alcohol)
PVP	Poly(vinyl pyrrolidone)
PEC	Polyelectrolyte complex
rhGM-CSF	Recombinant human granulocyte-macrophage colony-stimulating factor
AgNPs	Silver nanoparticles
TiO_2	Titanium dioxide
β (TGF-β).	Transforming growth factor
VEGF	Vascular endothelial growth factor

1 Introduction

Wound is caused due to injury when skin is torn, cut, or punctured. According to wound healing society, the definition of wound is 'disruption of normal anatomic structure and function' [1]. In general, the body responds via a complex process when the skin is damaged, known as wound healing [2]. Wounds can be classified on the basis of appearance as necrotic (typically dry wounds of black color), sloughy (yellow color, may be contaminated with unpleasant smell), granulating (red in color, may be contaminated with unpleasant smell), and epithelializing (pink in color, less exudate, near to healing) wounds [3].

As time is an important factor for wound management, that is why wounds can be clinically categorized as acute or chronic according to their time frame of healing. In general, acute wounds are tissue injuries which have complete healing, less scarring, less span of time (usually 8–12 weeks) [4]. In chronic wounds, inflammatory phase is prolonged which results as delayed healing (may be more than 3 months) which includes arterial, venous, diabetic and pressure ulcers.

1.1 Phases of Wound Healing

The wound healing progresses through three important overlapping and complex phases, namely inflammation, proliferation, and remodeling.

1.1.1 Inflammatory Phase

As soon as injury occurs hemostasis takes place to protect from exsanguination. Platelet releases wound healing mediators which include platelet-derived growth factor (PDGF) and transforming growth factor α (TGF-α), transforming growth factor β (TGF-β), epidermal growth factor (EGF), and vascular endothelial growth factor (VEGF). After hemostasis phase, inflammatory phase begins within the first 24 h of an injury. This phase is signed by redness, pain, swelling and heat, which results from release of vasoactive amines and histamine-rich granules from the mast cells. The stage can last up to 2 weeks in patients whose wounds are healing appropriately but can last longer in those patients with chronic nonhealing wounds.

Neutrophils and macrophages also play very important roles in the inflammatory phase. Neutrophils at first enter wound site, provides defense against infection by phagocytosing bacteria, damaged extracellular components, and foreign materials which were remained after initial bleeding. The wound macrophages function to remove any residual bacteria, foreign bodies, and remaining necrotic tissues. Therefore, the function of these macrophages is similar to that of neutrophils, but macrophages better regulate proteolytic destruction of wound tissue by secreting protease inhibitors.

1.1.2 Proliferative Phase

Proliferative phase starts just after inflammatory phase. In this proliferative phase, fibroblast proliferation, collagen synthesis, granulation tissue formation, and epithelization occur between 4 and 21 days. In addition, angiogenesis occurs such that new blood vessels replace the previously damaged capillaries and provide nourishment for the matrix.

In response to mediators which are released from platelets and macrophages, fibroblast migrates into the wound and move through the extracellular matrix.

Fibroblast activity is predominately regulated by PDGF and TGF-β. Factors like FGF and PDGF stimulate fibroblasts which produce extracellular matrix (ECM) components like collagen, elastin, and glycosaminoglycan to generate granulation tissue. Granulation tissue fills in wound defects.

Collectively, FGF and VEGF as angiogenic factors stimulate endothelial cells. The production of collagen results in increase in wound tensile strength and forms a matrix that permits myofibroblasts to start wound contraction, which then leads to decrease in wound size by closing wound margins [5].

1.1.3 Remodeling Phase

This is the last phase of wound healing, which begins after 21 days of injury. Remodeling is carried out by collagenases that were secreted in the previous phase. During this phase collagen turnover allows the randomly deposited scar tissue to be arranged in both linear and lateral orientations. Wound contraction and gradual increase in strength are seen as this phase progresses. Amount of collagen synthesis and their cross-link network determine the extent of healing.

1.2 Wound Healing: Ancient and Modern Concepts

The first documentation of wound care was found in the ancient Egyptian Edwin Smith Papyrus of 1600 BC, with a description of the removal of devitalised skin and pus following war injuries [6]. After that they used bandage soaked with grease with little thought of wound management. Wound care in ancient was followed by treatment with astringents, herbs, and honey. After the usage of bandage as a dressing astringent, herbs and honey were used in the treatment of wounds.

Though sterile gauze dressings can protect from infection, but because of absorbing wound exudates, it loses its property. The problem associated with these dressings is that they get adhered to the wound bed, which causes pain to patients and later leads to trauma.

Winter [7] introduced concept of moist wound healing in 1962 after that wound care took a new direction. After that some antibacterial agents like silver, povidone-iodine were incorporated into dressings to prevent from bacterial infection. In a modern concept, brief description of available dressings up-to-date is semipermeable film dressings, non-adherent contact layer dressings, alginate dressings, hydrocolloid dressings, hydrofiber dressings, foam dressings, hydrogel dressings, and antimicrobial dressings [8].

Semipermeable dressings transmit moisture vapor but do not absorb sufficient exudates. These types of dressings are impermeable to microbes and protect the wound bed from contamination. Opsite^{TM}, Cutifilm^{TM}, Biooclusive^{TM}, and Tegaderm are the examples of semipermeable dressings available in the market.

Hydrocolloid dressings are used both for acute and chronic wounds because removing of dressing is not painful to the patients. Further, it allows gas and water to pass. Granuflex™, AQUACEL™, Comfeel™, Tegasorb™ are some examples of these types of dressings.

Hydrogel dressings are prepared with polymers of hydrophilic nature and water. These dressings almost fulfill the properties of an ideal dressing and suitable for sloughy or necrotic wounds. 3M™ Tegaderm and Derma GeL® are the examples of hydrogels.

Non-adherent contact layer dressings are directly applied onto the wound to provide an interface with the secondary dressing or pad. Some non-adherent dressings are impregnated with paraffin or silicone to improve non-adherence.

Foam dressings are mostly made of polyurethane or silicone, which enable them to handle large volumes of wound exudates. Examples of foam dressings include: Lyofoam® and ALLEVYN® to name a few.

1.3 Factors Affecting Wound Healing and Important Factors for Dressing Choice

Wound healing is a complex process and hence there are many factors on which proper wound healing depends. The most important factor is adequate supply of blood and tissue perfusion. Tissue perfusion is mainly influx of white blood cells which play very important role in debridement and decontamination. The factors affecting wound repair can be divided as local factor and systemic factor. Wound healing affected due to wound site is local factor, whereas affected due to overall health without direct relation to wound site is systematic factor.

Local factors which influence healing oxygenation play very important role. Oxygen tension is a predetermined factor for collagen synthesis and angiogenesis, two events. Various events indicate that increased oxygen tension at wound site increases the healing rate and decreases the incident of infection [9].

The second important local factor largely affecting wound healing process is infection. Presence of bacterial infection adversely affects angiogenesis, the formation of granulation tissue and epithelialization. Infection is an important local factor due to which delayed inflammatory phase occurs.

In systematic factors, age is one of the factors as advanced age chances for delayed wound healing increases. Gosain et al. [10] observed the changes due to age include increased aggregation of platelet, slow angiogenesis, and re-epithelization. Diabetes which is very common among people plays an important role. The impaired healing may be due to lack of cellular and molecular signals required for normal wound repair process such as angiogenesis, granulation tissue formation, epithelialization, and remodeling. One of the systematic factors is psychological stress that impairs normal cell-mediated immunity at the wound site and thus causes an extensive delay in the healing process [11].

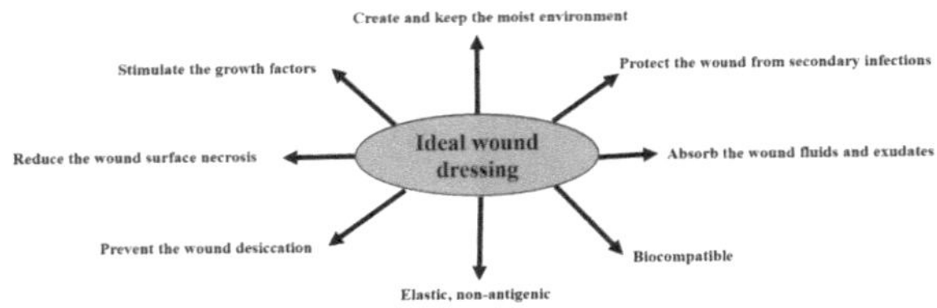

Fig. 1 Properties of ideal wound dressing [65–67]

Wound dressing provides an immediate protection layer to the skin so as to minimize the loss of its function. Choosing the right material for wound dressing is important and apart from its healing activity, antimicrobial activity, nontoxicity, it should be cost-effective, readily available, and can be fabricated into new products according to wound conditions.

The primary principle of wound healing is to maintain the wound in a moist environment. Traditional dry healing allows wounds to dehydrate, whereas moist wound healing increases healing rates via complex mechanisms, such as prevention of scab formation and drying of exudates containing nutrients, oxygen, growth factors, and leukocytes [12].

According to the needs, different types of wounds require separate types of dressings. As necrotic wound requires detachment of dead tissues from normal tissues, so dressing requires to maintain moist environment to prevent dehydration. As sloughy wounds like burn and ulcer wounds must be cleaned, so these types of wounds require dressings which absorb exudates. Granulating wound requires a dressing according to the shape of wound and exudate produced. The epithelializing wounds require dressings, which can provide protection to the wounds, allow gaseous exchange, and finally can be removed without ripping the skin (Fig. 1). On the other hand, an infected wound requires a dressing, which absorbs exudates and protects against microbes [3].

1.4 Polymers in Wound Healing Dressings

The most of the dressings are made up of polymeric materials (synthetic and natural). Natural polymers such as alginates, chitin, CS, gelatin, heparin, collagen, chondroitin, fibrin, keratin, silk fibroin, and eggshell membrane are biocompatible, biodegradable, and similar to macromolecules recognized by the human body and extensively used in wounds and burns management [13]. In comparison to synthetic polymers, natural macromolecules show relatively low mechanical strength. So, to enhance the mechanical property of biopolymers, blending or cross-linking with synthetic polymers is used in regenerative medicine for cartilage, vascular, bone, nerve repair, and restoration. For the regeneration of full-thickness wounds, polysaccharides and proteins are the most common natural polymers used in the field of tissue engineering, because of their biocompatibility, biodegradability, and similarity with ECM [14, 15].

Yang et al. [16] reported the reactive groups including free radicals, –C=C–, –COOH, $–NH_2$, –OH, –SH, –Si–O, S–S, –C=O, and/or formation of cyclic structures for facilitating self-healing.

1.5 Chitosan

CS, a mixture of β-(1,4)-D-*N*-acetylglucosamine and β-(1,4)-D glucosamine, is a polysaccharide derived from deacetylation of chitin, which is found in the cell walls of lower plants and in the exoskeleton tissues of lower animals including arthropods, crustaceans, and molluscs [17]. On the basis of source and preparation method, its molecular weight ranges from 300 to over 1000 kDa with a degree of deacetylation from 30 to 95 % [18]. CS is normally soluble in acidic pH and solubility depends upon distribution of free amino and *N*-acetyl groups. In dilute acids ($pH < 6$), the free amino group gets protonated and the molecule becomes soluble [19].

1.5.1 Structural Analysis and Its Importance

CS has three types of reactive functional groups, an amino group as well as both primary and secondary hydroxyl groups at the C-2, C-3, and C-6 positions respectively. CS is crystalline in nature and shows polymorphism depending on its physical state. It can be fabricated into films, scaffolds, hydrogels, fibers, etc. These are shown in Fig. 2.

Nowadays, CS is receiving a positive response in medical and pharmaceutical applications because of its promising biocompatibility effects, possible topical ocular applications used in implantation or injection. CS shows a broad spectrum of antimicrobial activity against both gram-positive and gram-negative bacteria and fungi [20]. The interaction between positively charged CS molecule and negatively charged microbial cell membrane components leads to change in cell permeability [21] and cellular contents [22] by forming an impermeable layer around cell, which prevents the transport of essential solutes. The permeabilizing effect in protonated CS observed is slightly acidic in nature, but this permeabilizing effect of CS is reversible [21]. In another reason, CS may enter the nuclei of bacteria and fungi and inhibit mRNA and protein synthesis by binding to microbial DNA [23, 24].

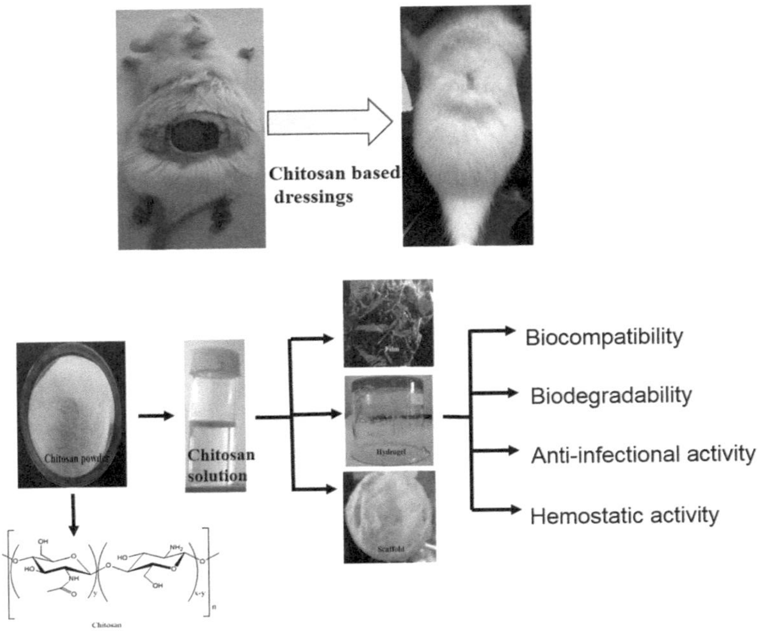

Fig. 2 CS powder, film, hydrogel, and scaffold

1.5.2 Effect of Degree of Deacetylation, Crystallinity, Degradation, and Molecular Weight

The proportion of *N*-acetylglucosamine units is described by the degree of acetylation (DA). Thus, chitin has a degree of acetylation of 100 %. Chitin is insoluble in most solvents. Therefore, usually chemical deacetylation is performed to produce the most common derivative termed as CS [25]. In acidic conditions, CS becomes positively charged due to NH_2 protonation and soluble in aqueous medium having n number of applications. The properties of CS, such as solubility, viscosity, and biocompatibility are inversely proportional to DA. Biodegradability is directly related to DA, whereas it is inversely related to the molecular weight (Mw) [26–32]. The biological properties such as antimicrobial, analgesic, hemostatic, mucoadhesion, permeation-enhancing effects of CS-based systems are inversely related to the DA; also, the properties, permeation enhancing effect and mucoadhesion, increase with the increase in molecular weight [33, 34]. Thus Anitha et al. [35] concluded that the DA, crystallinity, degradation, and molecular weight affect the physicochemical and biological properties of CS membranes and scaffolds.

As per the reports, degree of deacetylation (DD) and Mw could affect the activity of CS. Claire et al. [36] described in his study the role of DD effect on biological properties of CS films and concluded that DD does not affect its compatibility and plays a crucial part in cell adhesion and proliferation.

Minagawa et al. [37] researched on the effect of DD and molecular weight for wound healing. They have taken different chitins varied in deacetylation, i.e., 14, 33, 63, and 96 % having no change in molecular size (50 kD) for comparison of DD, monomers and oligomers of chitin and CS for comparison of Mw on wound healing. The collagenase activity, wound break strength, and activated fibroblast in CS group were found higher than chitin group. From histological study, they found that collagen fibers approached vertically, parallel with incisional line in the oligosaccharide group and control, monomers respectively. In higher DD found that strong break strength with more activated fibroblast.

2 Chitosan in Wound Dressings

From clinical studies, it is proved that CS shows positive effects in all phases of wound repair and potent hemostatic ability [38, 39]. The reason behind hemostasis is due to involvement of combined blood proteins and platelet activation to encourage clot formation [40]. Acting on repairing processes such as fibroplasia and re-epithelization, CS regulates the migration of neutrophils and macrophages [33]. To clean the wound site from foreign agents, CS accelerates infiltration of inflammatory cells in inflammatory stage. CS enhances activity of inflammatory cells, macrophages, and fibroblasts which helps in wound healing activity [34, 41].

At the time of new tissue formation, granulation tissue forms in wound area along with fibrous tissue formation. In open wound, healing remains the scar due to imbalance of collagen, while open wounds treated with CS healed without scar [42]. The clinical studies of CS-based wound healing effects are discussed below.

In a study, CS membrane was prepared for wound healing applications. The CS membrane, Bactigras was applied on the skin graft donor site. They found that membrane adhered uniformly to wound surface can reduce pain. The high rate of healing with abundant fibroblast was observed in wounds treated with CS membrane [43].

In another study, influences of the DA of 3D CS scaffolds were reported on the inflammatory response. In this study, male BALB/c mice were treated with DA 4, 15 % and untreated animals. The high density of adherent cells was found in CS with degree of acetylation at 15 %, whereas it is low at 4 % degree of acetylation. The thick fibrous capsule and a high infiltration of inflammatory cells were observed in CS scaffold with DA of 15 % [44].

Bio-inspired bilayered physical hydrogel was prepared using CS and water. The rigid layer provides good mechanical property and gas exchange, while second flexible layer allows the material to superficial contact. In vivo experiments were carried on female pigs (burn wounds) and Tulle Gras^{TM}, monolayered and bilayered CS were applied. All the results reveal that CS material actively stimulates tissue regeneration. It induces inflammatory cell migration and favors high vascularization of the newer tissues. Study indicates particularly enhanced deposition of type I and type IV collagen. Type I collagen provides mechanical strength to the skin, while type IV collagen contains the major part of the basal epidermal membrane. Physical hydrogel stimulate rapid reconstruction of the dermal epidermal junction as compared to viscous solution of CS [45].

A sponge-like asymmetric CS membrane is fabricated with immersion-precipitation phase-inversion method. The thickness and porosity of design membrane could be controlled by phase separation method. Asymmetric membrane showed good oxygen, permeability controlled water loss, and facilitated fluid drainage ability. The prepared membrane was able to inhibit invasion of microorganisms because of dense skin-like upper layer and antimicrobial activity of parent CS. Wistar rats were used in this study. In vivo studies indicate asymmetric CS membrane acted as hemostatic agent due to stopping bleeding at the wound site. Rapid epithelization and deposition of collagen in dermis were observed in asymmetric membrane treated wounds [46].

A water-insoluble CS patch scaffold is prepared with chitosan, glycerol, and applied on the right perforated eardrums of Sprague Dawley rats. The left perforated eardrum remained untreated. They found that CS and glycerol (3 %, 3 %) have optimum CS patch scaffold. From in vivo study, they found that regenerated eardrums showed thinner thickness and smoother surface, and compact tissues compared to spontaneously healed eardrums so they are effective in repair traumatic tympanic membranes perforations [5].

In another study, hydrogel was prepared with CS for wound healing applications. In vivo studies were compared between CS covered with elastic bag and PBS covered with elastic bag on Wistar rats. They found fast decreasing of wound area as compared to control and neither specific inflammation nor reactive granulomas in wounds treated with hydrogel was observed. In wounds treated with hydrogel, pathological abnormalities in heart, brain, liver, kidney, lung, and spleen obtained during necropsy were not observed. They supported the role of CS in skin repair [47].

A CS nonwoven membrane was prepared and its effects studied on periodontal healing. One-wall intrabony defects were created surgically on 2-year-old beagle dogs and resorbable membrane, CS nonwoven membrane was applied. In one-wall intrabony defects, new bone formation and new cementum were observed in CS nonwoven membrane treated groups. In this study, CS nonwoven membrane showed the potency in inducing periodontal tissue regeneration [48].

In another study, CS hydrogel was prepared with different Mw and DD ranges for wound healing applications. The burn wounds were created on Wistar rats and animals were treated with 2 % (w/v) low Mw, medium Mw, high Mw, 2 % Fucidin^R ointment, and no treatment. Complete re-epithelization with more granulation is observed in high molecular weight treated groups. They concluded that high Mw with high DD CS are more suitable for dermal burns [49].

Okamoto et al. [50] studied the importance of CS on experimental abscess with *Staphylococcus aureus* in dogs. The healing was compared between granulated CS (again three subgroups 0.01, 0.1, and 1.0 mg/subgroups), ampicillin, and saline (control group). The 0.1 and 1.0 mg CS treated groups showed quicker wound healing, the formed granulation tissue had more vascularization.

A chitosan film evaluated the healing at split skin graft donor (human) site. Healing was compared between CS and conventional dressing. Rapid wound re-epithelialization, smaller dermal nerve fibers, looser connective tissue stroma in the papillary dermis, and changing skin color to normal skin color were found in CS film treated wounds [51].

Freeze-dried *N*-carboxybutyl CS was prepared for wound healing applications. Control (phytostimuline gauze), *N*-carboxybutyl CS were used for comparison on human (Skin graft donor sites). At dermal level better histoarchitectural order and vascularization, regular distribution of collagen, absence of inflammatory cells were found in *N*-carboxybutyl CS treated groups. Organized epithelium was found in *N*-carboxybutyl CS treated groups [52].

N-carboxymethyl CS membrane was prepared for second-degree burn wounds. Male Wistar rats were treated with sterilized saline (control) and *N*-carboxymethyl CS membrane. The wound healing time is 25 and 35 for wounds treated with *N*-carboxymethyl CS membrane and control respectively. Reduction in wound size, ability to inducing the expression of inflammatory cytokines, activating the wound healing cascade, and targeting downstream protein such as Smad3 was observed in *N*-carboxymethyl CS membrane treated wounds. They concluded that *N*-carboxymethyl CS membrane as promising material for second-degree burn wounds [53].

In a very recent study CS-based derivatives 5-methyl pyrrolidinone CS, *N*-carboxymethyl chitosan, and *N*-succinyl CS, for the sustained delivery of neurotensin (NT) were prepared with the aim of wound healing applications. The order of the hydrophilicity changes is SC > 5-methyl pyrrolidinone CS > *N*-carboxymethyl CS and reduced glutathione release per mpolymer (g) is found the order of *N*-carboxymethyl CS (32.33 ± 0.72 %) < 5-methyl pyrrolidinone CS (67.65 ± 6.77 %) < *N*-succinyl CS (287.18 ± 14.92 %). C57BL/6 mice were used and neurotensin alone, 5-methyl pyrrolidinone CS foam alone and neurotensin-loaded 5-methyl pyrrolidinone CS foam in both control and diabetic mice. Neurotensin-loaded 5-methyl pyrrolidinone CS treatment was more effective than 5-methyl pyrrolidinone CS alone treated groups. Neurotensin-loaded 5-methyl pyrrolidinone CS can help healing in three phases by reducing the inflammatory response, higher collagen formation, and stimulating re-epithelialization [54].

3 Wound Dressings with Chitosan and Natural Polymers

A lot of CS-based blends are investigated for the purpose of wound dressing in which composite blends play an important role. Properties like absorbent, gel forming, and biocompatibility of alginate are used as a carrier for drug delivery and hemostatic wound dressing [55]. The formation of PEC on the combination of CS and alginate was prepared and found effective in wound healing material.

A PEC consists of CS and sodium alginate prepared for wound dressing applications. Briefly, at first they prepared freeze-dried CS sponge and kept in alginate solution to form PEC complex and freeze-dried and then silver sulfadiazine and dihydroepiandrosterone. The efficacy of prepared dressing was compared by an in vivo study on mice with vaseline gauze, CS sponge, silver sulfadiazine-impregnated PEC, and silver sulfadiazine and dihydroepiandrosterone-impregnated PEC wound dressing and results showed that complete wound contraction in silver sulfadiazine-impregnated PEC, and silver sulfadiazine and dihydroepiandrosterone-impregnated PEC wound dressing, whereas irregularly thickened epidermis is observed in vaseline gauze treated wounds and infiltration of leukocytes and proliferation of fibroblasts are more in the upper dermis. Well-organized epidermis, and the thickness is thinner in silver sulfadiazine-impregnated PEC than vaseline treated groups. In silver sulfadiazine and DHEA-impregnated PEC wound dressing almost normal thickness of epidermis was observed. Kim et al. [56] concluded that dihydroepiandrosterone may prevent the systemic change of the immune function after burn, as well as progressive destruction of the thermally injured skin tissues which makes silver sulfadiazine and dihydroepiandrosterone-impregnated PEC wound dressing as plays effective role in wound healing than other treatment groups.

Wang et al. [57] prepared a transparent, flexible membrane with CS–alginate PEC membranes with calcium chloride for in vivo test inbred male Sprague Dawley rats. In PEC membranes treated wounds, accelerated healing of incision wounds were observed as compared to conventional gauze. Histological studies showed mature epidermal structure with normal thickness of keratinized surface and reduced inflammation in the dermis. They found that the use of CS–alginate membrane facilitated the remodeling of scar tissue by increasing the rate of collagen synthesis and mature fibroblast on the day 21. In control wounds it still shows the signs of an active inflammatory phase under scab on day 21. The results proved the effectiveness of the membrane for wound healing.

HemCon® dressing was tested by in vitro and in vivo studies for its efficacy. The dressing was prepared by using CS acetate and mice were treated with no treatment control, HemCon bandage, sodium alginate bandage, and silver sulfadiazine for comparison. Male BALB/c mice were used for the Gram-negative experiments, and three groups (no antimicrobial cream) for *S. aureus*. The bacteria were applied on the wound for contaminated wounds. They found that adhesion time was more for HemCon® dressing compared to alginate dressing. According to Kaplan–Meier survival curves, mice treated with HemCon, alginate, silver sulfadiazine cream, untreated wounds, demonstrated 100, 65, 50, and 25 % survival [58]. The reason for the effectiveness of HemCon® dressing is due to primary amino groups of the polyglucosamine backbone which contains more number of positive charges at acidic pH values and interacts with the outer membrane of Gram-negative bacteria, destabilizes the lipopolysaccharide and permeabilizes both the outer and inner membranes [20] leading to leakage of cellular contents [21].

A Smad3 antisense oligonucleotide-impregnated CS and sodium alginate polyelectrolyte complex prepared by lyophilisation technique for wound healing applications. Cytotoxicity was evaluated in NIH3T3 cells using MTT assay. The cell viability of prepared scaffolds was increased by increasing CS content. The increased area of cells and faster proliferation rate of prepared scaffolds were more than PEC. In vivo test was performed on C57BL6 mice and their histopathology finding shows that Smad3 antisense oligonucleotide–PEC scaffolds facilitate collagen formation at early phase of wound healing reduced the inflammatory cells and exhibit fast wound closure [59].

In a study, curcumin-loaded CS–alginate sponges were prepared having CS–alginate ratio (3:1 (C3A1), 1:1(C2A2), and 1:3(C1A3)). To evaluate its efficacy in wound healing process, in vivo study was performed on Sprague Dawley (SD) with full-thickness wounds. The water uptake ability of sponges was found to be C3A1 (4290)>C2A2 (3806)>C1A3 (1218) with weight loss order C1A3>C3A1>C2A2. Observing the results, authors found that C2A2 sponge, C2A2-curcumin sponge, and control group treated wounds contracted 88, 90, and 74 % respectively. An immatured granulation tissue with many inflammatory cells and congested vessels were observed in gauze treated groups, granulation tissue and collagen alignment were more advanced in C2A2, C2A2-curcumin sponge treated wounds. Arrangement of collagen is loose in control wounds and compact in C2A2 treated wounds [60]. In the previous reported in vitro and in vivo studies, curcumin was exposed as a potential anti-inflammatory, antimicrobial, antioxidant, and accelerated wound healing activities [61–63].

A hydrosheet by using alginate, chitin/CS, and fucoidan. In vivo analysis was performed on mitomycin C-treated wounds on Sprague Dawley rats. The researchers first tested the wound closure difference between treated and non-treated wounds with mitomycin C and achieved 50 % wound closure at 7th day and 18 days treated and non-treated wounds with mitomycin C. Effectiveness of hydrosheet for repairing mitomycin C-treated wounds were compared with alginate (Kaltostat®), DuoACTIVE®, and non-treated wounds. High wound closure rate, significantly enhanced re-epithelization were observed in hydrosheet treated wounds as compared to alginate (Kaltostat®), DuoACTIVE®, and non-treated wounds. But no statistical differences between all the groups were observed on non-mitomycin C-treated wounds. This concludes that hydrosheet is active to impaired wounds and had little effect on non-mitomycin C-treated wounds [64].

The sodium alginate/CS and glycerol were used to prepare the film and studied the effect of low-level laser therapy for dermal burn wound healing. For in vivo study, adult male Rattus norvegicus albinus, Wistar strain were used. Laser therapy was applied for 7 days. The results were compared between untreated, cellulose films, sodium alginate/CS-based films, laser-irradiated undressed wounds, laser-irradiated wounds with cellulose and sodium alginate/CS-based films. Among all these, sodium alginate/CS-based films showed highest epithelisation rate, more blood vessels, and grossly interlaced type I collagen bundles [65].

In a study, hydrogel prepared by using curcumin, *N,O*-carboxymethyl CS, and oxidized alginate for wound healing applications and the results were compared with control group, blank CS/oxidized alginate hydrogel group, unmodified curcumin loaded CS/oxidized alginate hydrogel group, and nanocurcumin/CS/oxidized alginate hydrogel group. Due to larger surface area of curcumin, nanocurcumin-loaded CS/oxidized alginate hydrogel showed complete wound closure rate as compared to other groups. So, they concluded that nanocurcumin played an important role by enhancing full re-epithelialization on 14th day and greatly increasing the collagen deposition in the wound tissue on 7th day and nanocurcumin/*N*,Ocarboxymethyl CS/oxidized alginate hydrogel treatment significantly increased the DNA and protein content in wound tissue results in significantly enhanced healing process in mouse wound models [66].

Mohanty et al. [67] prepared curcumin-loaded oleic acid-based polymeric bandage with different blend ratios of alginate:CS (1:1, 1:2, and 1:3) for wound healing applications. The in vivo activity tested between prepared bandage and control (cotton gauze) in a Sprague Dawley male rat model. The authors observed that the collagen content and wound contraction rate on 10th day found high in curcumin-loaded oleic acid polymeric bandage treated wounds (94 %) than control (70 %) and void oleic acid based polymeric bandage (without curcumin)(80 %). The better healing of curcumin-loaded oleic acid-based polymeric bandage than void oleic acid-based polymeric bandage may be due to anti-inflammatory activity of curcumin. In DNA fragmentation assay analyzed that greater amount of apoptosis (DNA ladder) in an earlier phase (4th day) leads to quicker progress from inflammatory to the proliferative phase in curcumin-loaded oleic acid based polymeric bandage treated wounds than other groups. In Western blotting and semi-quantitative PCR analysis clearly gives an idea that curcumin-loaded oleic acid-based polymeric bandage can efficiently quench free radicals leading to reduced antioxidative enzyme activity.

In a recent study, CS and alginate bilayer composite membrane containing ciprofloxacin hydrochloride was developed. They observed that the membrane plays important role as reservoir because the amount of drug added was directly proportional to antibacterial activity. In drug release study they observed that (i) increasing cross-linking time slows drug release and (ii) lower concentration of calcium chloride prevents erosion of alginate from film and delays kinetics of drug release. So they selected 10 % calcium chloride for further studies. While New Zealand rabbits were used to study primary skin irritation test, in vivo pharmacodynamics efficacy, and histopathological studies. The drug-loaded composite membrane was compared with saline, Vaseline gauze containing mupirocin ointment, and blank composite film. The healing was more in drug-loaded composite membrane compared to other groups. No rotten flesh outside the skin and newly formed epithelium were observed in bilayer treated groups [68].

In a study, Ac2-26 peptide-loaded gels were prepared by using CS, alginate, and ANXA1 N-terminal peptide Ac2-26. In peptide release study they observed that CS hydrogel showed prolonged release and alginate/peptide gels exhibited burst release. C57BL/6 mice were used to compare healing activity among controlgroup (saline), 5.0 % alginate gel, Ac2-26 100 nM/alginate hydrogel group, Ac2-26 500 nM/alginate hydrogel group, and Ac2-261 μM/alginate hydrogel group. They observed that high M alginate/peptide Ac2-26 formulations showed 97 % wound closure so they concluded that prepared hydrogels had great potential for dermal wound healing applications [69].

In addition to biopolymeric materials used for tissue engineering and wound healing applications, collagen is one of the most important biopolymer due to its biocompatibility, nontoxicity, low antigenicity [70], and it absorbs excess of wound exudate along with maintaining a moist environment [71].

A durable sandwich wound dressing system was designed by Wang et al. [72] using different weight ratios of collagen to CS and immobilized on the polyproline fabric which were pre-grafted with acrylic acid or *N*-isopropyl acrolamide. First layer consists of porous polypropylene and second layer consists of *N*-isopropyl acrylamide acrylic acid. First layer provides mechanical support, whereas second layer provides hydrophilic surface for the adhesion of third layer, which consists of CS and collagen. Extend of immobilization increased with percent of CS because the amine group of CS has greater reactivity with glutaraldehyde. Similarly, antibacterial activity increased with CS percentage. The water uptake and water diffusion coefficient for the prepared film is also quite good and decrease with increasing immobilization percentage. In vivo study performed on rat using three sets of testing samples which are as follows: porous polypropylene nonwoven as control, porous polypropylene + acrylic acid + collagen + CS, porous polypropylene + *N*-isopropyl acrolamide + collagen and CS. The result shows the healing ability of third sample is high. This is because of collagen which is biocompatible material and has the ability to introduce fibroblast into wound area. In addition, the wound cover with poly-*N*-isopropyl acrylamide + collagen + CS film was automatically separated on day 21 post-injury, hence it does not create any scar onto the wound area.

A novel dermal substitute of combining recombinant human granulocyte-macrophage colony-stimulating factor (rhGM-CSF) with a porous heparinized collagen/CS scaffolds. At first collagen/CS scaffolds (9:1) were prepared and then heparinized with *N*-(3-dimethylamino-propyl)-N1-ethylcarbodiimide hydrochloride (EDC) and *N*-hydroxysuccinimide (NHS). After that without and with heparin scaffolds were washed with 0.1 M Na2HPO4 (2 h), 4 M NaCl (four times for 24 h) and denoted as H0E, H1E respectively. Sprague Dawley (SD) rats were used and tested the prepared scaffolds by subcutaneous implantation experiment. The research groups studied to prove efficacy of scaffolds by H1E/rhGMCSF (heparinized scaffold loaded with rhGM-CSF), H0E/rhGM-CSF (unheparinized scaffold loaded with rhGM-CSF), H1E (heparinized scaffold without rhGM-CSF), and H0E (unheparinized scaffold without rhGM-CSF). Their gross observation initially confirmed that H1E/rhGM-CSF scaffolds were surrounded by many blood vessels and by histological analysis they found that extensive neovascularization was observed and secretion of extracellular matrix into the scaffold. But very few vessels were observed in H0E/rhGM-CSF, H1E, and H0E control implantation surroundings. The number of vessels were observed highest in H1E/rhGM-CSF group (28.46 $\pm$ 7.6) and other groups found in the order of H0E/rhGM-CSF (21.06 $\pm$ 4.5), H1E (14.66 $\pm$ 3.2), and H0E (14.26 $\pm$ 4.4) at 14th day. The VEGF and TGF-β1 expressions of the rhGM-CSF groups were found high. This study shows that H1E/rhGMCSF helps in all phases of wound healing [73]. TGF-β1 helps to initiate granulation by increasing the expression of genes associated with extracellular matrix formation [74]. VEGF by acting on VEGFR-2 stimulates endothelial cells and then induces matrix metalloproteinase (MMP) production and endothelial cell proliferation and migration [75].

To prepare RNAi functionalized collagen–CS/silicone membrane bilayer at first siRNA solution (dissolved siRNA in RNase-free water) was mixed with trimethyl CS (TMC) and then porous collagen/CS scaffold was prepared and cross-linked with glutaraldehyde solution. Collagen–CS scaffold and silicone membrane were combined with a silicone adhesive to regenerate the bilayer dermal equivalent (BDE). TMC/siRNA particles were dropped onto the BDE and incubated for overnight. Bama miniature pigs were used for in vivo study. The RNAi-BDE worked as a reservoir for the incorporated TMC/siRNA complexes, enabling a prolonged siRNA release. From histochemical analysis, in vivo inhibition of TGF-β1 expression with RNAi-BDE treated wounds. The factors related to scar were regulated and the regenerated skin found similar like normal skin in RNAi-BDE treated wounds [76].

Gelatin is water soluble macromolecule obtained from animal collagen and has been widely used in pharmaceutics due to its biodegradability and low immunogenicity [77]. It is used in ointments and capsules [77, 78]. Gelatin is used as a nano carrier for drug delivery orally and directly to tumor has been studied a lot [79, 80]. An amino acid found in gelatin helps liver for removal of toxins in our body.

In a study, CS and gelatin were mixed with cross-linking agent and then freeze-dried to prepare porous membrane (CGSWD) structure. Big New Zealand rabbits were used in animal experiments and wounds were made on the back of rabbits. The first group of wounds was treated with 0.2 % (v/v) ethacridinee at a dose of 0.2 mL, second and third groups were treated with CGSWD and Vaseline respectively. Less healing time, absence of infection, collagen fasciculi of new tissues were found in CGSWD treated wounds. The reason for good healing was explained with CS could induce hyperplasia of partial macrophage and thus build up its activity, gelatin as it takes part in the movement, differentiation, and multiplication of cell [81].

Hydrogel sheets (HS) consist of CS, honey and gelatin are prepared in different weights (Gram) HS (0.5:20:20), HS-1 (0.25:20:20), HS-2 (0.5:10:20), HS-3 (0:20:20), and HS-4 (0.5:0:20) respectively. HS-4 showed highest modulus, HS-2 showed lowest modulus. The HS showed high swelling ratio and good antibacterial activity with nontoxicity. For testing of dressing, burn wounds were created on Kunming mice and New Zealand rabbits. For comparison HS and MEBO® ointment were used. HS group, MEBO® ointment, and untreated groups showed healing in 12, 14, 17 days respectively, so HS treated wounds were showed highest wound closure rate. In histological findings untreated wounds' hair follicles basically disappeared, scab fallen off naturally in MEBO-treated burns but small ulcers were observed at the center of burn and in HS-treated wounds healed perfectly with well-organized dermis, which look like a normal skin. The researchers also concluded that feasible effects played important role in wound healing [82]. Gelatin could take part in the movement, differentiation, and multiplication of cell [83]. The solutions with high viscosity like honey, sugar glucose inhibits microbial growth thus encourages healing [84]. Similarly in honey, sugar due to osmotic is responsible for antibacterial activity [85–88].

A hybrid hydrogel with gelatin and CM–CS hybrid hydrogel were prepared with different ratios (gelatin to CM–CS) at 10/0, 8/2, 6/4, 4/6, 2/8, and 0/10. They observed that by increasing the CM–CS content in the hybrid hydrogel, swelling property and degradation rate increased, whereas compressive modulus decreased. The prepared hydrogel was compatible to NIH3T3 fibroblast cells and cells could grow actively on the hydrogels [89].

Huang et al. [90] extended this work and checked the efficacy of CM–CS/gelatin for cutaneous wound healing. After optimization of properties, CS–gelatin hydrogel of 4:6 ratios was used for further wound healing test. Male young adult Sprague Dawley rats were used for in vivo study and transparent polyurethane was used as control. Radiation cross-linked carboxymethyl–CS/gelatin hydrogel group showed a higher wound healing rate ($p < 0.05$) on all phases of wound healing. In wounds treated with hydrogel regeneration of granulation tissues was observed as compared to control. In proliferative phase, VEGF-positive granulation tissues were evenly spread from the surface to the basal layer in hydrogel-treated groups, whereas the mainly located at base of wound in blank control group. In remodeling phase, thick and well-organized collagen fibers/bands were observed in regenerated epithelium in the hydrogel-treated group, whereas thinner collagen fibers were seen in the control group.

Microbial cellulose (MC), is a polymer synthesized by Acetobacter xylinum. Although it is potent and becomes a high-value product of biotechnology has not attracted much more [91]. In another study, oxidation of carboxymethyl cellulose (OCMC) and carboxymethylated CS (CMCS) prepared with volume ratios 15/2, 15/5 and 15/8, the hydrogels were represented as CSOC-2, CSOC-5, and CSOC-8. As OCMC weight percent was increased in the hydrogel, gelation time was decreased so they selected CSOC-5 hydrogel for further testing. On adult male Wistar rat wounds CSOC-5 hydrogel was introduced and for control kept constant. At 7th day 69.8, 65.8 and on 14th day 99.8 %, only 79.9 % was observed by wounds treated with testing group and control respectively ($p < 0.05$). Large number of inflammatory cells in dermis was observed in control animal groups, whereas small amount of inflammatory cells with new epithelium on surface of the defect was observed in CSOC-5 treated wounds which show the potency of used hydrogel as wound healing material [92].

Bacterial cellulose (BC) and bacterial cellulose–CS (BC–Ch) membranes are considered to be potent wound healing materials. Lin et al. [93] synthesized BC by Acetobacter xylinum. Sprague Dawley rats were used for in vivo testing and BC and BC–Ch membranes, 3M Tegaderm hydrocolloid and transparent films were used. In vivo study reveals that BC–Ch membranes treated wounds not only healed more rapidly but also epithelialized and regenerated faster than those treated with BC or Tegaderm. The overall results reveal that BC–Ch membranes are to be considered as very important wound treatment materials.

Huang et al. [9694] prepared biomimetic layer-by-layer nanofibrous matrices with CS, poly(ε-caprolactone), cellulose acetate, and collagen. First template was prepared with poly(ε-caprolactone), cellulose acetate and then these templates were soaked with CS and collagen respectively for 5, 5.5, 10, 10.5, 15, 15.5, 20, 20.5 times to prepare layer-by-layer nanofibrous matrices. Physical properties such as stress at break, strain at break, and loaded protein were found high in mats coated 10 times with CS and collagen. The higher degradation was found in larger number of coated groups. SD rats with thickness cutaneous wound used for in vivo testing. In 5, 10, 15, 20 times coated layer-by-layer mats, full wound closure was observed after 4, 4, 3, 2 weeks after treatment. Other important tests like α-tubulin, collagen IV, integrin β1, and phosphorylated FAK at Tyr 397 showed highest expression in larger number coated mats.

4 Wound Dressings with Chitosan and Synthetic Polymers

CS blends with synthetic polymers including poly(vinyl alcohol) (PVA), poly(vinyl pyrrolidone) (PVP), poly(ethylene oxide) (PEO), and polyglutamic acid were investigated for wound healing applications. PVA with its well-known properties such as nontoxic which has a little cell adhesion and protein absorption is widely used as wound healing material [95].

A hydroactive dressing was prepared by mixing poly(vinyl alcohol)/CS lactate (PVA/ChL) and nitrofurazone (0.1–0.2-mm thickness) using phase inversion technique. The results of hemocompatibility shows that increase in ChL content in PVA/ChL membrane reduce the fibrinogen adsorption, superior blood compatibility than that of each polymers. In their further study they used PVA/ChL membrane with $x_{\mathrm{PVA}} = 0.5$ (mole fraction of PVA) and this membrane showed higher water content and low interfacial tension, which helps in enhancing biocompatibility of blends. Wistar rats were used in their study and PVA, ChL, and PVA/ChL were compared. After 7 postoperative days wound treated with ChL and PVA membranes showed discrete inflammation, whereas no inflammation was observed in PVA/ChL treated wounds. Vascular proliferation and fibroblastic scar tissue formation were clearly observed which strongly prove the use of PVA/ChL and nitrofurazone membrane as dressing material [96].

Semi-interpenetrating polymer networks (SIPNs) composed of CS (CS), poloxame and dehydroepiandrosterone (DHEA) are used for wound dressing applications. Poloxamer macromer, having acrylate-terminated PEOs is used in biomedical fields [97]. BALB/c mice were used and prepared dressings were compared with gauze (control), Vaseline gauze, CS/poloxamer SIPNs, and DHEA-loaded CS/poloxamer SIPNs. DHEA (2 wt% of polymer weight). The results supported that CS/poloxamer SIPNs were capable in providing suitable circumstances for granulation tissue formation, new epithelium in wounded place and mature collagen in dermis. So, the researchers concluded that the reason for wound healing of prepared SIPNs may be affected by released DHEA which prevent progressive destruction of the thermally injured skin tissues and systemic change of immune function after burn [98].

The acrylic acid-grafted and gamma-polyglutamic acid/CS-immobilized polypropylene nonwoven fabric (PP–AAg–PGCi) were prepared and in vivo study was performed on male Wistar rats. Healing of PP-nonwoven fabric (PP), AAgrafted and gamma-PGA immobilized PP nonwoven fabric (PP–AAg–PGi), AA grafted and CS-immobilized PP nonwoven fabric (PP–AAg–Ci), and AA grafted and gamma-PGA/CS-immobilized PP nonwoven fabric (PP–AAg–PGCi) were compared accordingly. Results showed that increase of gamma-PGA in the immobilizing increases water diffusion and bacteria inhibition zone of PP–AAg–PGCi found highest. Wound treated with PP–AAg–PGCi membrane was found to be healed quickly as removal of this membrane was easy. The healing of the wound area for the various membranes was found in the order of PP–AAg–PGCi>PP–AAg–Ci>PPAAg–PGCi. The dense and intact hair follicles cells with complete epithelialization were observed in PP–AAg–PGCi membrane treated wounds and delay of the follicles formation in PP–AAg–Ci membrane treated wounds [99].

CS–hyaluronic acid (CS–HA) hybrid films were reported by solution casting method. The CS and hyaluronic acid solutions were prepared in the ratio of 5:1, 4:2 to 3:3, corresponding to a weight ratio of HA of 0.1, 0.25, and 0.5 % represented as CS5HA1, CS4HA2, and CS3HA3, respectively. Researchers found that by addition of hyaluronic acid water uptake capacity increased in comparison with CS film. That is, CS3HA3 showed highest swelling ratio. Based on the results they decided to select CS4HA2 for testing in vivo and SD mice were used for the same. Vaseline gauge and CS4HA2 were compared and result reveals that CS/HA film is more easily peeled off without causing any harm and shows faster healing rate [100].

Cross-linked CS hydrogel films were prepared by PVA and CS loaded with minocycline by freeze-thawing method for wound healing applications by Sung et al. [101]. Results indicate that amount of CS affects cross-linking strength of hydrogel fraction and also flexibility of prepared gel. On increasing CS concentration, gel fraction and cross-linking ability decreases. It also decreases thermal stability of PVA hydrogel, increases swelling ability up to 400 %, water vapor transmission rate, and porosity of hydrogel. Data analysis reveals that loading of minocycline does not affect physical property of hydrogel. In wound healing testing minocycline-loaded PVA hydrogel shows faster rate of healing in rat dorsum than the control (sterile gauze). Hydrogel loaded with drugs were found to be more favorable for re-epithelization microvessel formation, collagen tissue and reconstruction of skin tissue than conventional product.

The CS-coated PVA nanofiber prepared by electrospinning method. Heat-treated PVA nanofibers were then coated by immersing in CS solution. The wound healing ability of CS-coated PVA and heat treated PVA (H-PVA) nanofibers matrices were compared. The result indicates that CS-coated PVA (C-PVA) shows less hydrophilic and better tensile strength as compared to H-PVA nanofibrous matrix. Wound healing ability of coated matrix was examined on a mouse. The nanomatrices (C-PVA and H-PVA) accelerate wound healing compared to control. The enhanced wound healing activity may be due to high surface area and microporous structure of matrices which facilitate cell attachment and proliferation. Moreover, the collagen density was found to be high in C-PVA than H-PVA. The epidermis of C-PVA treated wound was thicker than that of H-PVA treated wound [102].

Hydrogel prepared by using PVA, water-soluble CS and glycerol for wound healing applications. The wound healing ability was evaluated on Sprague Dawley rats. They observed that gauze dressing adhered to the healing tissues resulted in bleeding and hydrogel could be easily removed from wound surface. The hydrogel-treated wounds were healed almost at 11th day but control showed only 90 % wound contraction on 14th day. The histology of wounds at 16th day supported the efficacy of hydrogel as good wound dressing. Mature collagen in dermis, well-structured epidermis found in hydrogel used dressings. Moderate number of inflammatory cells in dermis, incomplete covered with epithelium was observed in control treated wounds [103].

Tsao et al. prepared CS/γ-poly(glutamic acid) (PGA) PEC for wound dressing materials in the molar ratio of 3:1, 1:1, and 1:3. They observed that water uptake capacity was found in order to C25P75 (959 ± 97 %), C75P25 (675 ± 55 %), C50P50 (232 ± 44 %) because of higher water absorption capabilities of free –COOH groups in γ-PGA than that of the free $-NH_2$ groups in CS existing in the CS/γ-PGA PECs. In vivo experiment was conducted on ICR mice and found that CS/γ-PGA PECs treated wounds healed almost but not in control. It was also observed that the suppression of inflammatory increased as CS content in prepared wound dressing but not in neat CS treated wounds which showed that γ-PGA was important factor in reduction of inflammatory. As γ-PGA content increases in CS/γ-PGA PECs, amount of epithelialization increases. New hair follicle formation, regeneration of new tissues were observed in CS/γ-PGA PECs [104].

In a study, rutin-conjugated CS–poly(ethylene glycol)–tyramine (RCPT) hydrogel was prepared. Male Sprague Dawley rats were used for in vivo analysis and PBS, CS–poly(ethylene glycol)–tyramine (CPT), commercialized wound dressing Duoderm, and RCPT were used for in vivo study. The formation of extracellular proteins and epithelization of incisions increased in order of PBS<CPT<Duoderm≈RCPT. The formation of new blood vessels, well-proliferated fibroblast, and high granulation tissue formation was due to collagen enhanced in RCPT hydrogel for the use in dermal healing [105].

A new type of layered hydrogel prepared by using alginate, CS, poly(L-glutamic acid) (PGA). For comparison, dry sodium alginate film was soaked in either 10 wt% $CaCl_2$ (alginate–$CaCl_2$) or 3 wt% Ca–PGA (alginate–PGA) aqueous solution. The swelling ratio of alginate–PGA and alginate–CS–PGA were higher than that of alginate–$CaCl_2$. Calcium ions released from hydrogels in DMEM were found in order alginate–PGA>alginate–CS–PGA>alginate–$CaCl_2$. The animal experiment was performed over diabetic female Wistar rats by creating full-thickness rectangular wounds. The healing effectiveness was compared with commercial product, alginate–$CaCl_2$, alginate–PGA, and alginate–CS–PGA. They found that wound healing was more in alginate–CS–PGA compared to conventional alginate hydrogels. Suppressed inflammatory phenomena, abundant mature collagen were observed in CS/PGA PECs treated wounds [106].

In a study, curcumin–MPEG–CS film prepared by using curcumin, methoxy poly(ethylene glycol) and CS by using casting/solvent evaporation method. For in vivo testing SD rats were used and healing rate was compared curcumin–MPEG–CS film (film area: 3 cm^2) and a control group with treatment of MPEG–CS film. Researchers observed that curcumin–MPEG–CS films treated group showed 90 and 60 % wound reduction in MPEG–CS film treatment proves the strong role of curcumin in wound healing. They concluded that complete epithelialization of the wound, compact as well as aligned collagen and actual collagen deposition in curcumin–MPEG–CS films treated wounds showed the great potency for use in the wound healing applications [107].

In another study, ciprofloxacin hydrochloride-loaded CS/poly(ethylene glycol) (PEG) composite scaffold was prepared and antibacterial results indicated that antibiotic loaded CS–PEG composite showed highest inhibition zone over antibiotic loaded pure CS scaffold. Albino male mice were used and effective wound healing was observed in antibiotic loaded CS–PEG composite scaffolds, antibiotic-loaded pure CS scaffold and control (gauze) in 9, 14, and 21 days respectively. Well-organized epithelial cells, hair follicle, and hair growth were observed in antibiotic-loaded CS–PEG composite scaffolds which show the effectiveness of this scaffold for wound healing applications [108].

Charernsriwilaiwat et al. [109] prepared nanofiber mats using CS–ethylenediamine tetraacetic acid (EDTA) solution, PVA solution at a weight ratio of 30/70 and lysozyme (LZ) (10, 20 and 30 wt% to polymer), and then electrospun to produce fibrous mats. Male Wistar rats wound were covered with a nanofiber mat, gauze, and SOFRA-TULLE® (antibacterial). Researchers found that 30 % LZ loaded CS–EDTA/PVA nanofiber mats were faster than bare-CS–EDTA/PVA and bare-CS–acetate/PVA nanofiber. The two main reasons proposed for this healing were (i) assists in the acceleration of the healing process by deactivating bacterial and depolymerizing the CS [110] and (ii) By products from CS depolymerisation such as EDTA and Nacetyl-*d*-glucosamine enhances the healing rate because of their antibacterial properties and aid in fibroblast proliferation [111].

A CS–poly(ethylene glycol)-tyramine (CPT) conjugate was evaluated for wound healing applications. Hemorrhaging liver mouse model was employed (C57BL/6 mouse) for hemocompatibility test. The total blood loss measured for 3 min after the liver was pricked with a needle and for control liver (no treatment), prepared dressing was 154 mg and 59 mg respectively. The healing ability of CPT hydrogels compared with suture, fibrin glue- and cyanoacrylate-treated models, as well as normal skins. In suture treated groups the epidermal layer has penetrated into sutured incision and large gap still present, together with a suture hole. Better healing was observed in cyanoacrylate-treated models than suture treated groups, although healing process was incomplete, demonstrating fibrosis around the incision. The wound treated with fibrin glue found better than suture, and cyanoacrylate-treated models but fibrosis was still observed, with incomplete dermis recovery. In hydrogel-treated wounds incision was completely recovered, with a newly formed dermis and no visible fibrosis after 2 weeks, and found better than other research groups [112].

In a study, polypropylene (PP) nonwoven fabric (NWF) was prepared by using CS, poly(*N*-isopropylacrylamide) (PNIPAAm). Cytotoxicity test result revealed that original and acrylic acid (AAc) coated NWFs showed some cytotoxicity, but grafting of CS and PNIPAAm into temperature-responsive PP–g–CS–g–PNIPAAm NWF reduced cytotoxicity and supported proliferation. The wound areas remain after 17 days for gauze, PP–g–CS, and PP–g–CS–g–PNIPAAm NWFs was 25.6, 11.9, and 4.9 % respectively. From histological images of the newborn skin tissues at the wound site confirms that no infection and inflammation were observed in wound site covered with PP–g–CS–g–PNIPAAm NWF throughout the wound healing process. Hence it confirms that the PPg–CS–g–PNIPAAm is useful and potential wound dressing material [113].

Chen et al. [114] prepared reinforced poly(ethylene glycol) (PEG)-CS (RPC) hydrogels with different Mw (600, 1000 and 2000 Da) of PEG and different molar ratios of carboxylic acid to amine group (5:95, 10:90, and 15:85, neat CS). 1000-RP10C90 has good mechanical strength and proper degradation rates. Adult BALB/c mice were used and healing activity among all prepared dressing material and control (received no dressing) were compared. In macroscopic evaluation control wounds (86.6 ± 3.8 %), as well as the C100 (88.6 ± 3.3 %) and 600-RP15C85 (90.9 ± 2.6 %) dressings showed less healing as compared with 2000-RP15C85 (91.7 ± 2.3 %) and 1000-RP10C90 (93.3 ± 2.8 %). They observed that as increasing amount of cross-linked PEG the number of inflammatory cells were increased (i.e., the number of inflammatory cells was 1000-RP15C85>1000-RP10C90>1000-RP5C95). 1000-RP10C90 dressings showed continuous increase in thick epidermal layer due to accelerated epithelial cell repair site and superior wound healing effects were found as compared to other dressing materials.

A wound healing hydrogel for deep second-degree burn was prepared by adding CS, poly(vinyl alcohol), polyethylene glycol (PEG 400), and agar. Male Albino Wistar rats were used for in vivo study and copper punch was used to create burn wound. Cotton gauze, paraffin gauze and PVA/CS hydrogel dressings were applied on rats. Hydrogel PVA/CS was peeled off very easily as compared to other control groups. In cotton and paraffin treated groups between 2 and 6 days fragments of scab, light coat of scab which persists in the wound observed in between day 6 and 9 and only 90 % wound was observed after 16 days. Homogeneous formation of the fibrin between 2 and 6 days was observed in PVA/CS hydrogel dressings and good quality of wound healing was observed in between day 6 and 9. The main advantage of hydrogel healing is providing moist environment so that epidermal cells migrates easily over a wet wound area and easy removal of dressing without causing additional damage. In hydrogel-treated wounds on day 9 observed positive signs such as fleshy buds constituted with blood vessels, disappeared inflammatory cells, quick epithelisation, whereas presence of inflammatory signs were observed in cotton gauze, paraffin gauze treated wounds [115].

5 Wound Dressings with Chitosan and Growth Factors

Growth factors play an important role in wound healing by playing important role in each phase of wound healing. Growth factors are polypeptides which control the growth, differentiation, and metabolism of cells and regulate the process of tissue repair [116–118]. Several defined peptide growth factors, including EGF, PDGF, FGF and TGF-β, have been shown to stimulate cellular proliferation and synthesis of the extracellular matrix [119, 120].

In a study, photocross-linkable CS (Az–CH–LA) aqueous solution including fibroblast growth factor-2 (FGF-2) was prepared for wound healing applications. The rate of wound closure rate was evaluated by creating full-thickness skin incisions on healing-impaired diabetic (db/db) mice and their normal (db/+) littermates. The wound closure in db/db mice accelerated in growth factor added hydrogel but not in db/+mice. Histological findings revealed that wounds treated with growth factor added CS hydrogels in db/db mice shows formation of advanced granulation tissue, formation of capillary and epithelialization [121].

A prepared CS film (freeze-drying hydroxyl propyl CS acetate buffer solution) by incorporating different concentrations of bFGF (0.6, 2 and 6 μg/film) was evaluated for wound healing applications in genetically diabetic mice. Genetically diabetic (C57BLKs/J-m+/db, db/db) female mice were used in this study and results were compared with different concentrations of bFGF and without bFGF in the CS film. They found that for stimulating wound the dose of 0.6 μg/film was not sufficient after 15 days. More capillaries, proliferation of fibroblast were observed in both groups but more granulation tissue was found in the bFGF–CS film group. Hence, they concluded bFGF with CS film is effective for wound healing [122].

A flexible hydrogel was prepared by irradiating ultraviolet light to a photo-cross-linkable CS (Az–CH–LA) aqueous solution and FGF-2. Healing-impaired diabetic (db/db) mice and their normal (db/þ) littermates were used in in vivo study. In their results addition of FGF-2 into the CS hydrogel had small effect on wound healing in db/þ mice and accelerated the adjusted tissue filling rate in db/db mice. The CS hydrogel containing FGF-2 noticeably increased the number of CD-34-positive vessels in the wound which proves the applicability of the hydrogel to further use in wound healing in healing-impaired db/db mice [123].

Alemdaroglu et al. [124] prepared a CS gel with EGF for second-degree burn wounds in rats. The efficacy of prepared gel was compared with silver sulfadiazine, CS gel with EGF, CS gel without EGF and without EGF, EGF solution and untreated control groups unburned and untreated female Sprague Dawley rats. In case of CS gel without EGF treated wounds some degeneration of mitochondria, loss of crista, and for healing indication active fibroblasts in dermis were observed. In CS gel with EGF treated wounds, some vacuolization of dermal cells with healing, infiltration of inflammatory cells, epithelial cells with normal structure and some active fibroblasts and inflamed cells at connective tissue layer were observed. In the case of EGF solution treated groups continuation of fibroblastic activity, wound scab, invasive inflammatory cell infiltrations underneath, degeneration of cells at epithelial surface were noticed. Positive indication observations for wound healing such as myofibroblasts were observed. Significant increase in cell proliferation is clearly observed in CS gel with EGF treated groups from their experiment ($p < 0.001$). Although the wound healing rate was faster in CS gel with EGF and EGF solution treated groups as when compared to other groups highest wound healing rate and epidermis thickness were observed in CS gel with EGF treated groups ($p < 0.001$). These results prove the importance of CS as well as EGF in wound healing applications.

A scaffold was developed by using CS, bFGF, and gelatin microparticles to evaluate wound closure of pressure ulcers in aged mice. C57BL/6NHsd female mice were used to compare the wound healing effect of control (cotton gauze), CS, and CS loaded with bFGF. CS accelerated healing with or without bFGF by attracting more neutrophils. In vivo elastase protein levels were reduced on day 10 for CS loaded with bFGF. CS potentiates the bFGF activity and results accelerated of pressure ulcers in aged mice [125].

In a study, PDGF-containing collagen–CS gel (CCP) was used as wound healing material. The control (without treatment), CS with PGDF and CCP wound healing efficacy was tested on rat. Complete epithelisation in control, CS with PGDF and CCP occurred in 20, 14 and 10 days respectively. The wound contraction rate was also observed in same order. In control groups loosely packed collagen fibers and undifferentiated keratinocytes under basal lamina layer. In CS with PDGF treated wounds after 10th day, epidermis consists of more number of capillaries were observed. In PDGF-containing collagen–CS group hair follicle structure, no inflammatory cells and moderate number of fibroblasts were observed with all signs of complete wound healing on 10th day. Researchers suggested that PDGF recruits additional macrophages into the wound, which produce endogenous growth factors, such as TGF-β, and increased collagen synthesis [126].

Further Takei et al. [127] modified the above prepared gel with PVA and prepared CS–gluconic acid conjugate/poly(vinyl alcohol) cryogels. Male Wistar rats were induced diabetes with streptozotocin. The results were compared with CG/PVA cryogel, bFGF-containing CG/PVA cryogel, Viewgel® and dry cotton gauze. Time period required for the healing of 50 % of the initial area of wounds treated with gauze (n = 5), Viewgel® (n = 6), 1 % CG/5 % PVA cryogels (n = 7) and 1 % CG/5 % PVA cryogels containing bFGF (n = 7) were 6.5 ± 0.3, 5.7 ± 0.3, 4.0 ± 1.1 and 2.7 ± 1.0 days respectively. Their results indicated that healing rate of the wounds covered with the bFGF incorporated cryogels was 1.5 times higher than that of wounds covered without bFGF cryogels. Histopathological study reveals that no formation of fibrotic capsules as a consequence of the inflammatory process in the 1 % CG/5 % PVA cryogel-treated wounds were found after complete closure.

Takei et al. [12] extended their work and prepared aqueous solution soluble (neutral pH) CS-gluconic acid conjugate and formed gel by freeze-thawing (cryogelation) without using additives. Male Wistar rats were used for in vivo study. They observed that the prepared cryogel promoted increase of inflammatory cells such as PMN which release chemical mediators effective for wound healing. Hence they concluded that CS-gluconic acid conjugate was found efficient than cotton gauze and Viewgel® in wound healing.

CS-based microspheres prepared by using different quantities of VEGF, EGF, and EGF+VEGF in CS solutions. The microspheres were prepared by electrospinning technique. By using Wistar rats wound healing efficacy was compared between six groups; namely EGF+VEGF dissolved in PBS; oxidized dextran loaded with CS microparticles without Growth factors; oxidized dextran loaded with microparticles with VEGF incorporated; oxidized dextran with microparticles loaded with EGF; PBS (control); oxidized dextran loaded with microparticles containing EGF+VEGF. The treatment group treated with oxidized dextran loaded with microparticles containing EGF+VEGF wound closure was observed earlier. They concluded that no specific inflammation or reactive granulomas were found because of the presence of oxidized dextran CS and growth factors in all groups and both growth factors were contributed in wound healing [128]. The slow wound healing was observed in growth factors absent groups, single growth factor added groups compared with two growth factors added groups that improve the establishment of the regenerative cascade in order to produce new extracellular matrix and promote angiogenesis that are fundamental for skin regeneration [129].

6 Wound Dressings with Chitosan and Nanoparticles

Due to nanoparticles small size (often 100 nm or smaller) are well merged in CS-based dressing material which enhance tensile strength. Nanoparticle can interact at molecular and cellular level, due to which it shows better antibacterial activity and reduce time span for wound healing.

In a study, CS dressing developed by incorporating a polyphosphate (procoagulant) and silver (antimicrobial). Silver-loaded CS–polyphosphate exhibited higher bactericidal activity than CS–polyphosphate in vitro, by 100 and 90 % kill of *Pseudomonas aeruginosa* and *S. aureus* consistently. Male BALB/c mice were used to test the efficacy. *P. aeruginosa* was incorporated on the wound surface. The effect of gauze and silver-loaded CS–polyphosphate on bacteria treated wounds were compared on day 1 and 2 only because gauze treated mice were died on 2nd day. The more inflammatory cell infiltrates and less infiltration in wounds treated with silver-loaded CS–polyphosphate. Gram stain of wounds treated with silver-loaded CS–polyphosphate revealed that bacteria in inflammatory cell infiltrates on 7th day but not at day 14. On 14th day dermis showed organized with thicker collagen fibers, more fibroblasts, and in epidermis maturation of keratinocyte with silver-loaded CS–polyphosphate treated wounds [130].

The mechanism of antibacterial activity of silver is studied by many researchers as silver nanoparticles could interact with the sulfur containing, phosphorus containing compounds such as DNA in bacteria [131]. The silver ions inactivate cellular proteins and DNA loses replication ability [132]. Kim et al. reported that silver ions help in cell death by enable formation of free radicals which induce damage to the bacterial membrane [133].

CS and nanosilver was prepared for wound healing applications. They found negative result with sterility and pyrogen tests with silver nanocrystalline CS dressing. For in vivo study, Sprague Dawley rats were used and healing activity was compared with silver sulfadiazine, CS film, and composite dressing. The wound healing rates were 98.98 $\pm$ 6.09 %, 81.67 $\pm$ 6.30 % and healing time was 13.51 $\pm$ 4.56, 17.45 $\pm$ 6.23 days for silver nanocrystalline CS dressing treated group and silver sulfadiazine dressing treated control groups respectively. The antibacterial activity of silver nano found more than same amount at ionic level and finally they concluded that the prepared dressing has more efficacy in wound healing applications [134].

Arockianathan et al. [135] prepared composite films with CS, sago starch, silver nanoparticles, and gentamycin. To see the effect of gentamycin some films were without it. They found that addition of 2 mL of ethylene glycol as plasticizer increased tensile strength. Male Albino Wister rats were used to compare activity between cotton gauze dipped with gentamicin (control), CS–sago–silver nanoparticles and CS–sago–silver nanoparticles with gentamicin. Complete healing time for control, CS–sago–silver nanoparticles and CS–sago–silver nanoparticles with gentamicin were 24, 18 and 16 days respectively. Same proved by histopathological study. Increased tensile strength thus increase in collagen matrix in CS–sago–silver nanoparticles with gentamicin treated wounds reflects faster healing. They concluded that this prepared dressing may be used as wound healing material.

In a study, a fibrous mats using PVA, CS oligosaccharides (COS) and silver nanoparticles (AgNPs) by electrospun for wound healing applications. The antibacterial activity was more for PVA/COS-AgNP nanofiber. The gauze, PVA/COS/$AgNO_3$ (silver nitrate) nanofiber, commercial wound paste and PVA/COS-AgNP were applied on the wound created on Sprague Dawley rats. PVA/COS-AgNP treated groups showed better wound healing than other groups. 7 days after operation in PVA/COS/$AgNO_3$ nanofiber and gauze treated groups granulation tissue with infiltration of inflammatory cells were observed which was disappeared without capillary hyperplasia in PVA/COS-AgNP treated group. In each group, on 14th day newly synthesized fibrous tissue and sparse inflammatory cells in the dermis were observed and subcutis were covered by completely re-epithelialized epidermis. They concluded that PVA/COS-AgNP nanofibrous membrane could be used for wound healing [136].

In recent study, Archana et al. [137] prepared CS–PVP–silver oxide loaded film for wound healing applications. They found that prepared films had higher antibacterial activity and nontoxic nature. Adult male albino rats (140–180 g) were used for wound healing study. They found that prepared film has more wound healing property than that of cotton gauge and 100 % CS.

Kumar et al. [138] prepared a CS/nano ZnO composite bandage for wound healing applications. Sprague Dawley (SD) rats were used to compare healing effect of CS/nano ZnO (nZnO (0.005 % and 0.01 %), CS control, Kaltostat, and bare wound (negative control). After 2 weeks, wound closure ~90 % achieved by

CS, CS/nano ZnO and ~70 % achieved in Kaltostat-treated wound and untreated wounds. More collagen deposition, complete re-epithelialization observed in CS controls and the CS/nano ZnO treated wounds, compared to the Kaltostat and the untreated wounds. Swabs collected from the rat wound cultured with bacteria showed reduced viability in the first week and further reduced in the second week with CS/nano ZnO treated wounds than other treatment groups. They concluded that CS/nano ZnO dressings were effective.

The effect of CS-based copper nanocomposite on healing was evaluated on excision wound model in Wistar rats. They found that the upregulation of VEGF and transforming growth factor-β1, more fibroblast proliferation, collagen deposition, and intact re-epithelialization in CS-based copper nanocomposite treated wounds as compared to 0.1 % acetic acid-treated groups and 1 % CS-treated group [139].

Nano–TiO_2–CS with collagen artificial skin (NTCAS) was prepared for wound healing applications. The formation dried for homogeneous membranes. Animals used in in vivo experiment were Sprague Dawley rats and results were compared between uncovered group (control), Duoderm (positive control) and NTCAS. Low levels of IL-6 interleukin 6 and tumor necrosis factor alpha in rat serum, better healing (photographic view) in NTCAS-treated group when compared to other control groups [140].

In one study titanium dioxide nanoparticle loaded CS–pectin dressing was prepared to investigate wound healing efficacy. The incorporation of nanoparticle increased the mechanical strength of dressing. The feasibility of two antibacterial materials CS and TiO_2 increased antibacterial efficacy as well as wound healing property. The in vivo open excision type wounds were created on adult male albino rats and healing rate was compared between nano dressing, CS, and conventional gauze. Wound closure rate of the gauze, CS, and CS–pectin–TiO_2 matrices treated groups were found 91.22, 94.98, and 99.01 % respectively. In nano dressing treated wounds, well-developed dermis and epidermis, formation of matured fibrous tissues were observed. Necrotic patches, loose granulation, and sparse blood capillary formation were observed in CS treated wounds clearly indicate that the involvement of nanoparticle as well as pectin in wound healing applications [141].

In another study, CS–PVP–titanium dioxide nanocomposites were prepared for wound healing applications. The dressing contains antibacterial activity, hemocompatible nature, and nontoxic nature. The dressing was compared with gauze, soframycin skin ointment and CS, they found the contraction rate 90.42, 93.34, 94.21, and 99.09 % respectively. Photographic appearance was supported the results and adult male albino rats were used for in vivo testing [142].

7 Conclusions

A number of publications and patents in this field indicate the necessity of wound dressings. The selection of material plays an important role in wound healing point of view. The main advantage of CS is it can be easily blended with simple process and easily fabricated into gels, films, scaffolds, and fibers. This chapter mainly focuses on advantages of CS-based dressings in the field of wound healing applications. The CS-based dressings were prepared by blending technology using synthetic, natural, growth factors and nanoparticles. This technology is an effective way to prepare new classical materials for improving its versatile property. Even though deep rooted literature provided that a deep study is still required to find the exact mechanism of CS activity in each phase of healing.

Acknowledgments One of the authors (DA) gratefully acknowledged the full cooperation as obtained from MNNIT Allahabad during her post doctoral research work.

References

1. Lazarus GS, Cooper DM, Knighton DR, Margolis DJ, Percoraro ER, Rodeheaver G, Robson MC (1994) Definitions and guidelines for assessment of wounds and evaluation of healing. Arch Dermatol 130:489–493
2. Gurtner GC, Werner S, Barrandon Y, Longaker MT (2008) Wound repair and regeneration. Nature 453:314–321
3. Qin Y (2001) Advanced wound dressings. J Text Inst 92:127–138
4. Percival JN (2002) Classification of wounds and their management. Surgery 20:114–117
5. Kim JH, Bae JH, Lim KT, Choung PH, Park JS, Choi SJ, Im AL, Lee ET, Choung YH, Chung JH (2009) Development of water-insoluble chitosan patch scaffold to repair traumatic tympanic membrane perforations. J Biomed Mater Res, Part A 90:446–455
6. Majuno G (1975) The healing hand: man and wound in the ancient world. Harvard University Press, Cambridge
7. Winter GD (1962) Formation of the scab and the rate of epithelization of superficial wounds in the skin of the young domestic pig. Nature 193:293–294
8. Boateng JS, Matthews KH, Stevens HNE, Eccleston GM (2008) Wound healing dressings and drug delivery systems: a review. J Pharm Sci 97:2892–2923
9. Guo S, DiPietro LA (2010) Factors affecting wound healing. J Dent Res 89:219–229
10. Gosain A, DiPietro LA (2004) Aging and wound healing. World J Surg 28:321–326
11. Godbout JP, Glaser R (2001) Stress-induced immune dysregulation: implications for wound healing, infectious disease and cancer. J Neuroimmune Pharmacol 1:421–427
12. Takei T, Nakahara H, Ijima H, Kawakami K (2012) Synthesis of a chitosan derivative soluble at neutral pH and gellable by freeze–thawing, and its application in wound care. Acta Biomater 8:686–693
13. Thomas S, Fisher B, Fram PJ, Waring MJ (1998) Odour-absorbing Dressings. Journal of Wound Care 7:246–250
14. Malafaya PB, Silva GA, Reis RL (2007) Natural-origin polymers as carriers and scaffolds for biomolecules and cell delivery in tissue engineering applications. Adv Drug Deliv Rev 59:207–233
15. Wiegand C, Hipler UC (2010) Polymer-based biomaterials as dressings for chronic stagnating wounds. Macromol Symp 294:1–13
16. Yang Y, Marek W (2013) Urban self-healing polymeric materials. Chem Soc Rev 42:7446–7467
17. Muzzarelli RAA, Peter MG (1997) Chitin handbook; Atec, Grottammare, Italy
18. VandeVord PJ, Matthew HW, DeSilva SP, Mayton L, Wu B, Wooley PH (2002) Evaluation of the biocompatibility of a chitosan scaffold in mice. J Biomed Mater Res 59:585–590
19. Madihally SV, Matthew HWT (1999) Porous chitosan scaffolds for tissue engineering. Biomaterials 20:1133–1142

20. Kumar ABV, Varadaraj MC, Gowda LR, Tharanathan RN (2005) Characterization of chitooligosaccharides prepared by chitosanolysis with the aid of papain and pronase, and their bactericidal action. Biochem J 391:167–175
21. Helander IM, Nurmiaho-Lassila EL, Ahvenainen R, Rhoades J, Roller S (2001) Chitosan disrupts the barrier properties of the outer membrane of Gram-negative bacteria. Int J Food Microbiol 71:235–244
22. Tsai GJ, Su WH (1999) Antibacterial activity of shrimp chitosan against Escherichia coli. J Food Prot 62:239–243
23. Qi L, Xu Z, Jiang X (2005) Preparation and antibacterial activity of chitosan nanoparticles. Carbohydr Res 339:2693–2700
24. Ma Y, Zhou T, Zhao C (2008) Preparation of chitosan-nylon-6 blended membranes containing silver ions as antibacterial materials. Carbohydr Res 343:230–237
25. Roberts GAF (1992) Chitin chemistry, 1st edn. Macmillan Press, London
26. Kumirska J, Weinhold MX, Czerwicka M, Kaczyński Z, Bychowska A, Brzozowski K, Thoming J, Stepnowski P (2011) Influence of the chemical structure and physicochemical properties of chitin and chitosan based materials on their biomedical activity. In: Laskovski AN (ed) Biomedical engineering, trends in materials science. Rijeka Croatia, InTech, pp 25–64
27. Kumirska J, Weinhold MX, Thoming J, Stepnowski P (2011) Biomedical activity of chitin/chitosan based materials influence of physicochemical properties apart from molecular weight and degree of N-acetylation. Polymers 3:1801–1875
28. Weinhold MX, Sauvageau JCM, Keddig N, Matzke M, Tartsch B, Grunwald I, Kubel C, Jastorff B, Thoming J (2009) Strategy to improve the characterization of chitosan for sustainable biomedical applications: SAR guided multi-dimensional analysis. Green Chem 11:498–509
29. Domard A (2011) A perspective on 30 years research on chitin and chitosan. Carbohydrate Polymer 84:696–703
30. No HK, Park NY, Lee SH, Meyers SP (2002) Antibacterial activity of chitosans and chitosan oligomers with different molecular weights. Int J Food Microbiol 74:65–72
31. Tsao CT, Chang CH, Lin YY, Wu MF, Wang JL, Han JL, Hsieh KH (2010) Antibacterial activity and biocompatibility of a chitosan-γ-poly (glutamic acid) polyelectrolyte complex hydrogel. Carbohydr Res 345:1774–1780
32. Kong M, Chen XG, Xing K, Park HJ (2010) Antimicrobial properties of chitosan and mode of action: a state of the art review. Int J Food Microbiol 144:51–63
33. Ishihara M, Nakanishi K, Ono K, Sato M, Kikuchi M, Saito Y (2002) Photocrosslinkable chitosan as a dressing for wound occlusion and accelerator in healing process. Biomaterials 23:833–840
34. Adamson R (2009) Role of macrophages in normal wound healing: an overview. J Wound Care 18:349–351
35. Anitha A, Sowmya S, Sudheesh Kumar PT, Deepthi S, Chennazhi KP, Ehrlich H, Tsurkan EM, Jayakumar R (2014) Chitin and chitosan in selected biomedical applications. Prog Polym Sci 39:1644–1667
36. Chatelet C, Damour O, Domard A (2001) Inuence of the degree of acetylation on some biological properties of chitosan films. Biomaterials 22:261–268
37. Minagawa T, Okamura Y, Shigemasa Y, Minami S, Okamoto Y (2007) Effects of molecular weight and deacetylation degree of chitin/chitosan on wound healing. Carbohydr Polym 67:640–644
38. Howling GI, Dettmar PW, Goddard PA, Hampson FC, Dornish M, Wood EJ (2001) The effect of chitin and chitosan on the proliferation of human skin fibroblasts and keratinocytes in vitro. Biomaterials 22:2959–2966
39. Hirano S, Zhang M, Nakagawa M, Miyata T (2000) Wet spun chitosan-collagen fibers, their chemical N-modifications, and blood compatibility. Biomaterials 21:997–1003
40. Rao SB, Sharma CP (1997) Use of chitosan as a biomaterial: studies on its safety and hemostatic potential. J Biomed Mater Res 34:21–28

41. Lawrence WT (1998) Physiology of the acute in wound. Clin Plast Surg 25:321–340
42. Ueno H, Yamada H, Tanaka I, Kaba N, Matsuura M, Okumura M, Kadosawa T, Fujinaga T (1999) Accelerating effects of chitosan for healing at early phase of experimental open wound in dogs. Biomaterials 20:1407–1414
43. Azad AK, Sermsintham N, Chandrkrachang S, Stevens WF (2002) Chitosan membrane as a wound-healing dressing: characterization and clinical application. J Biomed Mater Res B Appl Biomater 69B:216–222
44. Barbosa JN, Amaral IF, Águas AP, Barbosa MA (2010) Evaluation of the effect of the degree of acetylation on the inflammatory response to 3D porous chitosan scaffolds. J Biomed Mater Res, Part A 93A:20–28
45. Boucarda N, Vitona C, Agayb D, Maric E, Rogerc T, Chancerelleb Y, Domard A (2007) The use of physical hydrogels of chitosan for skin regeneration following third-degree burns. Biomaterials 28:3478–3488
46. Mi FL, Shyu SS, Wu YB, Lee ST, Shyong JY, Huang RN (2001) Fabrication and characterization of a sponge-like asymmetric chitosan membrane as a wound dressing. Biomaterials 22:163–173
47. Ribeiro MP, Espiga A, Silva D, Baptista P, Henriques J, Ferreira C, Silva JC, Borges JP, Pires E, Chaves P, Correia IJ (2009) Development of a new chitosan hydrogel for wound dressing. Wound Repair Regeneration 17:817–824
48. Yeo YJ, Jeon DW, Kim CS, Choi SH, Cho KS, Lee YK, Kim CK (2005) Effects of chitosan nonwoven membrane on periodontal healing of surgically created one-wall intrabony defects in beagle dogs. J Biomed Mater Res B Appl Biomater 72:86–93
49. Alsarra IA (2009) Chitosan topical gel formulation in the management of burn wounds. Int J Biol Macromol 45:16–21
50. Okamoto Y, Tomita T, Minami S, Matsuhashi A, Kumazawa NH, Tanioka S, Shigemasa Y (1995) Effects of chitosan on experimental abscess with Staphylococcus aureus in dogs. J Vet Med Sci 57:765–767
51. Stone CA, Wright H, Clarke T, Powell R, Devaraj VS (2000) Healing at skin graft donor sites dressed with chitosan. Br J Plast Surg 53:601–606
52. Biagini G, Bertani A, Muzzarelli R, Damadei A, DiBenedetto G, Belligolli A, Riccotti G, Zucchini C, Rizzoli C (1991) Wound management with N-carboxybutyl chitosan. Biomaterials 12:281–286
53. Chang J, Liu W, Han B, Peng S, He B, Gu Z (2013) Investigation of the skin repair and healing mechanism of N-carboxymethyl chitosan in second-degree burn. Wound Repair Regeneration 21:113–121
54. Moura LIF, Dias AMA, Leal EC, Carvalho L, De Sousa HC, Carvalho E (2014) Chitosan-based dressings loaded with neurotensin an efficient strategy to improve early diabetic wound healing. Acta Biomater 10:843–857
55. Kim JO, Park JK, Kim JH, Jin SG, Yong CS, Li DX, Choi JY, Woo JS, Yoo BK, Lyoo WS, Kim JA, Choi HG (2008) Development of polyvinyl alcohol-sodium alginate gel-matrix-based wound dressing system containing nitrofurazone. Int J Pharm 359:79–86
56. Kim HJ, Lee HC, Oh JS, Shin BA, Oh CS, Park RD, Yang KS, Cho CS (1999) Polyelectrolyte complex composed of chitosan and sodium alginate for wound dressing application. J Biomater Sci-Polym Ed 10:543–556
57. Wang L, Khor E, Wee A, Lim LY (2002) Chitosan-Alginate PEC membrane as a wound dressing. Appl Biomater 63:610–618
58. Burkatovskaya M, Tegos GP, Swietlik E, Demidova TN, Castano AP, Hamblin MR (2006) Hamblin use of chitosan bandage to prevent fatal infections developing from highly contaminated wounds in mice. Biomaterials 27:4157–4164
59. Hong HJ, Jin SE, Park JS, Ahn WS, Kim CK (2008) Accelerated wound healing by smad3 antisense oligonucleotides-impregnated chitosan/alginate polyelectrolyte complex. Biomaterials 29:4831–4837

60. Dai M, Zheng XL, Xu X, Kong XY, Li XY, Guo G, Luo F, Zhao X, Wei YQ, Qian Z (2009) Chitosan-Alginate sponge: preparation and application in curcumin delivery for dermal wound healing in rat. J Biomed Biotechnol Article ID 595126:1–8
61. Brouet I, Ohshima H (1995) Curcumin, an anti-tumour promoter and anti-inflammatory agent, inhibits induction of nitric oxide synthase in activated macrophages. Biochem Biophys Res Commun 206:533–540
62. Han S, Yang Y (2005) Antimicrobial activity of wool fabric treated with curcumin. Dyes Pigm 64:157–161
63. Panchatcharam M, Miriyala S, Gayathri VS, Suguna L (2006) Curcumin improves wound healing by modulating collagen and decreasing reactive oxygen species. Mol Cell Biochem 290:87–96
64. Murakami K, Ishihara M, Aoki H, Nakamura S, Nakamura S, Yanagibayashi S, Takikawa M, Kishimoto S, Yokoe H, Kiyosawa T, Sato Y (2010) Enhanced healing of mitomycin C-treated healing-impaired wounds in rats with hydrosheets composed of chitin/chitosan, fucoidan, and alginate as wound dressings. Wound Repair Regeneration 18:478–485
65. Dantas MD, Cavalcante DR, Araújo FE, Barretto SR, Aciole GT, Pinheiro AL, Ribeiro MA, Lima-Verde IB, Melo CM, Cardoso JC, Albuquerque Júnior RL (2011) Improvement of dermal burn healing by combining sodium alginate/chitosan-based films and low level laser therapy. J Photochem Photobiol, B 105:51–59
66. Li X, Chen S, Zhang B, Li M, Diao K, Zhang Z, Li J, Xu Y, Wang X, Chen H (2012) In situ injectable nano-composite hydrogel composed of curcumin, N, O-carboxymethyl chitosan and oxidized alginate for wound healing application. Int J Pharm 437:110–119
67. Mohanty C, Das M, Sahoo SK (2012) Sustained wound healing activity of curcumin loaded oleic acid based polymeric bandage in a rat model. Mol Pharm 9:2801–2811
68. Han F, Dong Y, Song A, Yin R, Li S (2014) Alginate/chitosan based bi-layer composite membrane as potential sustained-release wound dressing containing Ciprofloxacin hydrochloride. Appl Surf Sci 311:626–634
69. Gaudio PD, Cicco FD, Aquino RP, Picerno P, Russo P, Piaz FD, Bizzarro V, Belvedere R, Parente L, Petrella A (2015) Evaluation of in situ injectable hydrogels as controlled release devicefor ANXA1 derived peptide in wound healing. Carbohydr Polym 115:629–635
70. Cen L, Liu W, Cui L, Zhang W, Cao Y (2008) Collagen tissue engineering: development of novel biomaterials and applications. Pediatr Res 63:492–496
71. Antonio F, Guillem R, Sonia T, Clara M, Piergiorgio G, Valeria C, Gianluca C, Tzanov T (2011) Cross-linked collagen sponges loadedwith plant polyphenols with inhibitory activity towards chronic wound enzymes. Biotechnol J 6:1208–1218
72. Wang CC, Su CH, Chen CC (2008) Water absorbing and antibacterial properties of N-isopropyl acrylamide grafted and collagen/chitosan immobilized polypropylene nonwoven fabric and its application on wound healing enhancement. J Biomed Mater Res 84:1006–1017
73. Sun H, Wang X, Hu X, Yu W, You C, Hu H, Han C (2012) Promotion of angiogenesis by sustained release of rhGM-CSF from heparinized collagen/chitosan scaffolds. Appl Biomater 100B:788–798
74. Goldberg MT, Han YP, Yan C, Shaw MC, Garner WL (2007) TNF-alpha suppresses alpha-smooth muscle actin expression in human dermal fibroblasts: an implication for abnormal wound healing. J Invest Dermatol 127:2645–2655
75. Nillesen ST, Geutjes PJ, Wismans R, Schalkwijk J, Daamen WF, Kuppevelt TH (2007) Increased angiogenesis and blood vessel maturation in acellular collagen-heparin scaffolds containing both FGF2 and VEGF. Biomaterials 28:1123–1131
76. Liu X, Ma L, Liang J, Zhang B, Teng J, Gao C (2013) RNAi functionalized collagen-chitosan/silicone membrane bilayer dermal equivalent for full-thickness skin regeneration with inhibited scarring. Biomaterials 34:2038–2048
77. Djagny VB, Wang Z, Xu S (2001) Gelatin: a valuable protein for food and pharmaceutical industries: review. Crit Rev Food Sci Nutr 41:481–492

78. Schwick HG, Heide K (1969) Immunochemistry and immunology of collagen and gelatin. Bibliotheca Haematologica 33:111–125
79. Goldie K, Mansoor A (2005) Tumor-targeted gene delivery using Poly(ethyleneglycol)-modified gelatin nanoparticles: in vitro and in vivo studies. Pharm Res 22:951–961
80. Rakesh K, Ramesh CN, Meenakshi D, Jayanta KP (2011) In-vitro and in-vivo study of indomethacin loaded gelatin nanoparticles. J Biomed Nanotechnol 7:325–333
81. Deng CM, He LZ, Zhao M, Yang D, Liu Y (2007) Biological properties of the chitosan-gelatin sponge wound dressing. Carbohydr Polym 69:583–589
82. Wang T, Zhu XK, Xue XT, Wu DY (2012) Hydrogel sheets of chitosan, honey and gelatin as burn wound dressings. Carbohydr Polym 88:75–83
83. Chirife J, Scarmato G (1982) Scientific basis for use of granulated sugar in treatment of infected wounds. Lancet 1:560–561
84. Archer HG, Barnett S, Irving S, Middleton KR, Seal DV (1990) A controlled model of moist wound healing: comparison between semi-permeable film, antiseptics and sugar paste. J Exp Pathol 71:155–170
85. Somerfield S (1991) Honey & healing. J R Soc Med 84:179
86. Green A (1988) Wound healing properties of honey. Br J Surg 75:1278
87. Condon RE (1993) Curious interaction of bugs and bees. Surgery 113:234–235
88. Seymour F, West K (1951) Honey—its role in medicine. Med Times 79:104–107
89. Yang C, Xu L, Zhou Y, Zhang X, Huang X, Wang M, Han Y, Zhai M, Wei S, Li J (2010) A green fabrication approach of gelatin/CM-chitosan hybrid hydrogel for wound healing. Carbohydr Polym 82:1297–1305
90. Huang X, Zhang Y, Zhang X, Xu L, Chen X, Wei S (2013) Influence of radiation crosslinked carboxymethyl-chitosan/gelatin hydrogel on cutaneous wound healing. Mater Sci Eng, C 33:4816–4824
91. Bielecki S, Krystynowicz A, Turkiewicz M, Kalinowska H (2002) Bacterial cellulose. In: Steinbuchel A (ed) Biopolymers, vol 5, Polysaccharides I. Munster, Gremany: Wiley-VCH, Verlag GmbH, pp 37–90
92. Fan L, Tan C, Wang L, Pan X, Cao M, Wen F, Xie W, Nie M (2013) Preparation, characterization and the effect of carboxymethylated chitosan–cellulose derivatives hydrogels on wound healing. J Appl Polym Sci 128:2789–2796
93. Lin WC, Lien CC, Yeh HJ, Yu CM, Hsu SH (2013) Bacterial cellulose and bacterial cellulose–chitosan membranes for wound dressing applications. Carbohydr Polym 94:603–611
94. Huang R, Li W, Lv X, Lei Z, Bian Y, Deng H, Wang H, Li J, Xueyong Li X (2015) Biomimetic LBL structured nanofibrous matrices assembled by chitosan/collagen for promoting wound healing. Biomaterials 53:58–75
95. Chuang WY, Young TH, Yao CH, Chiu WY (1999) Properties of the poly(vinyl alcohol)/chitosan blend and its effect on the culture off fibroblast in vitro. Biomaterials 20:1479–1487
96. De Queiroz AA, Ferraz HG, Abraham GA, del Mar Fernández M, Bravo AL, Román JS (2003) Development of new hydroactive dressings based on chitosan membranes: Characterization and in vivo behaviour. J Biomed Mater Res 64A:147–154
97. Ruel-Gariépy E, Leroux JC (2004) In situ-forming hydrogels—review of temperature-sensitive systems. Eur J Pharm Biopharm 58:409–426
89. Kim IY, Yoo MK, Seo JH, Park SS, Na HS, Lee HC, Kim SK, Cho CS (2007) Evaluation of semi-interpenetrating polymer networks composed of chitosan and poloxamer for wound dressing application. Int J Pharm 341:35–43
99 Wang CC, Su CH, Chen JP, Chen CC (2009) An enhancement on healing effect of wound dressing: Acrylic acid grafted and gamma-polyglutamic acid/chitosan immobilized polypropylene non-woven. Mater Sci Eng, C 29:1715–1724
100. Xu H, Ma L, Shi H, Gao C, Han C (2007) Chitosan-hyaluronic acid hybrid film as a novel wound dressing: in vitro and in vivo studies. Polym Adv Technol 18:869–875

101. Sung JH, Hwang MR, Kim JO, Lee JH, Kim YI, Kim JH, Chang SW, Jin SG, Kim JA, Lyoo WS, Han SS, Ku SK, Yong CS, Choi HG (2010) Gel characterisation and in vivo evaluation of minocycline-loaded wound dressing with enhanced wound healing using polyvinyl alcohol and chitosan. Int J Pharm 39:2232–2240
102. Kang YO, Yoon IS, Lee SY, Kim DD, Lee SJ, Park WH, Hudson SM (2010) Chitosan-Coated Poly(vinyl alcohol) nanofibers for wound dressings. J Biomed Mater Res B Appl Biomater 92B:568–576
103. Yang X, Yang K, Wu S, Chen X, FengYu Li J, Ma M, Zhu Z (2010) Cytotoxicity and wound healing properties of PVA/ws-chitosan/glycerol hydrogels made by irradiation followed by freeze–thawing. Radiat Phys Chem 79:606–611
104. Tsao CT, Chang CH, Lin YY, Wud MF, Wang JL, Young TH, Hane JL, Hsieh KH (2011) Evaluation of chitosan/γ-poly(glutamic acid) polyelectrolyte complex for wound dressing materials. Carbohydr Polym 84:812–819
105. Tran NQ, Joung YK, Lih E, Park KD (2011) In situ forming and rutin-releasing chitosan hydrogels as injectable dressings for dermal wound healing. Biomacromolecules 12:2872–2880
106. Lee YH, Chang JJ, Yang MC, Chien CT, Lai WF (2012) Acceleration of wound healing in diabetic rats by layered hydrogel dressing. Carbohydr Polym 88:809–819
107. Li X, Nan K, Li L, Zhang Z, Chen H (2012) In vivo evaluation of curcumin nanoformulation loaded methoxy poly (ethylene glycol)-graft-chitosan composite film for wound healing application. Carbohydr Polym 88:84–90
108. Sinha M, Banik RM, Haldar C, Maiti P (2013) Development of ciprofloxacin hydrochloride loaded poly (ethylene glycol)/chitosan scaffold as wound dressing. J Porous Mater 20:799–807
109. Charernsriwilaiwat N, Opanasopit P, Rojanarata T, Ngawhirunpat T (2012) Lysozyme-loaded, electrospun chitosan-based nanofiber mats for wound healing. Int J Pharm 427:379–384
110. Mecitoflu Ç, Yemenicioflu A, Arslanoflu A, Elmacâ ZS, Korel F, Çetin AE (2006) Incorporation of partially purified hen egg white lysozyme into zein films for antimicrobial food packaging. Food Res Int 39:12–21
111. Reshetov IV, Yudanova TN, Matorin OV, Morozov DS (2004) A coating material containing chlorhexidine and lysozyme for wound treatment. Pharm Chem J 38:388–390
112. Lih E, Lee JS, Park KM, Park KD (2012) Rapidly curable chitosan–PEG hydrogels as tissue adhesives for hemostasis and wound healing. Acta Biomater 8:3261–3269
113. Chen JP, Kuo CY, Lee WL (2012) Thermo-responsive wound dressings by grafting chitosan and poly(N isopropylacrylamide) to plasma-induced graft polymerization modified non-woven fabrics. Appl Surf Sci 262:95–101
114. Chen SH, Tsao CT, Chang CH, Lai YT, Wu MF, Chuang CN, Chou HC, Wang CK, Hsieh KH (2013) Assessment of reinforced poly (ethylene glycol) chitosan hydrogels as dressings in a mouse skin wound defect model. Mater Sci Eng, C 33:2584–2594
115. Nacer Khodja A, Mahlous M, Tahtat D, Benamer S, Larbi Youcef S, Chader H, Mouhoub L, Sedgelmaci M, Ammi N, Mansouri MB, Mameri S (2013) Evaluation of healing activity of PVA/chitosan hydrogels on deep second degree burn: pharmacological and toxicological tests. Burns 39:98–104
116. Fu X, Shen Z, Chen Y, Xie J, Guo Z, Zhang M, Sheng Z (1998) Randomised placebo-controlled trial of use of topical recombinant bovine basic fibroblast growth factor for second-degree burns. Lancet 352:1661–1664
117. McCarthy DW, Downing MT, Brigstock DR, Luquette MH, Brown KD, Abad MS, Besner GE (1996) Production of heparin-binding epidermal growth factor-like growth-factor (Hb-EGF) at sites of thermal injury in pediatric patients. J Invest Dermatol 106:49–56
118. Brown GL, Curtsinger LJ, White M, Mitchell RO, Pietsch J, Nordquist R, Fraunhofer AV, Schultz GS (1988) Acceleration of tensile strength of incisions treated with EGF and TGF-b. Ann Surg 208:788–794

119. Robson MC, Mustoe TA, Hunt TK (1998) The future of recombinant growth factors in wound healing. Am J Surg Pathol 176:80S–82S
120. Steed DL (1998) Modifying the wound healing response with exogenous growth factors. Clin Plast Surg 25:397–405
121. Obara K, Ishihara M, Ishizuka T, Fujita M, Ozeki Y, Maehara T, Saito Y, Yura H, Matsui T, Hattori H, Kikuchi M, Kurita A (2003) Photocrosslinkable chitosan hydrogel containing fibroblast growth factor-2 stimulates wound healing in healing-impaired db/db mice. Biomaterials 24:3437–3444
122. Mizuno K, Yamamura K, Yano K, Osada T, Saeki S, Takimoto N, Sakurai T, Nimura Y (2003) Effect of chitosan film containing basic fibroblast growth factor on wound healing in genetically diabetic mice. J Biomed Mater Res 64A:177–181
123. Obara K, Ishihara M, Fujita M, Kanatani Y, Hattori H, Matsui T, Takase B, Ozeki Y, Nakamura S, Ishizuka T, Tominaga S, Hiroi S, Kawai T, Maehara T (2005) Acceleration of wound healing in healing-impaired db/db mice with a photo cross linkable chitosan hydrogel containing fibroblast growth factor-2. Wound Repair Regeneration 13:390–397
124. Alemdaroğlu C, Değim Z, Celebi N, Zor F, Oztürk S, Erdogan D (2006) An investigation on burn wound healing in rats with chitosan gel formulation containing epidermal growth factor. Burns 32:319–327
125. Park CJ, Clark SG, Lichtensteiger CA, Jamison RD, Johnson AJ (2009) Accelerated wound closure of pressure ulcers in aged mice by chitosan scaffolds with and without bFGF. Acta Biomater 5:1926–1936
126. Pierce GF, Mustoe TA, Lingelbach J, Masakowski VR, Griffin GL, Senior RM, Duel TF (1989) Platelet-derived growth factor and transforming growth factor-β enhance tissue repair activities by unique mechanisms. J Cell Biol 109:429–440
127. Takei T, Nakahara H, Tanaka S, Nishimata H, Yoshida M, Kawakami K (2013) Effect of chitosan-gluconic acid conjugate/poly (vinyl alcohol) cryogels as wound dressing on partial-thickness wounds in diabetic rats. J Mater Sci 24:2479–2487
128. Ribeiro MP, Morgado PI, Miguel SP, Coutinho P, Correia IJ (2013) Dextran-based hydrogel containing chitosan microparticles loaded with growth factors to be used in wound healing. Mater Sci Eng C 33:2958–2966
129. Sheridan MH, Shea LD, Peters MC, Mooney DJ (2000) Bioabsorbable polymer scaffolds for tissue engineering capable of sustained growth factor delivery. J Control Release 64:91–102
130. Ong SY, Wu J, Moochhala SM, Tan MH, Lu J (2008) Development of a chitosan-based wound dressing with improved hemostatic and antimicrobial properties. Biomaterials 29:4323–4332
131. Morones JR, Elechiguerra JL, Camacho A, Holt K, Kouri JB, Ramírez JT, Yacaman MJ (2005) The bactericidal effect of silver nanoparticles. Nanotechnology 16:2346–2353
132. Gupta P, Bajpai M, Bajpai SK (2008) Investigation of antibacterial properties of silver nanoparticle-loaded Poly (acrylamide-co-itaconic acid)-grafted cotton fabric. J Cotton Sci 12:280–286
133. Kim JS, Kuk E, Yu KN, Kim JH, Park SJ, Lee HJ, Kim SH, Park YK, Park YH, Hwang CY, Kim YK, Lee YS, Jeong DH, Cho MH (2007) Antimicrobial effects of silver nanoparticles. Nanomed: Nanotechnol Biol Med 3:95–101
134. Lu S, Gao W, Gu HY (2008) Construction, application and biosafety of silver nanocrystalline chitosan wound dressing. Burns 34:623–628
135. Arockianathan PM, Sekar S, Kumaran B, Sastry TP (2012) Preparation, characterization and evaluation of biocomposite films containing chitosan and sago starch impregnated with silver nanoparticles. Int J Biol Macromol 50:939–946
136. Li C, Fu R, Yu C, Li Z, Guan H, Hu D, Zhao D, Lu L (2013) Silver nanoparticle/chitosan oligosaccharide/poly (vinyl alcohol) nanofibers as wound dressings: a preclinical study. Int J Nanomed 8:4131–4145
137. Archana D, Singh Brijesh K, Dutta Joydeep, Dutta PK (2015) Chitosan-PVP-nano silver oxide wound dressing: In vitro and in vivo evaluation. Int J Biol Macromol 73:49–57

138. Kumar PT, Lakshmanan VK, Anilkumar TV, Ramya C, Reshmi P, Unnikrishnan AG, Nair SV, Jayakumar R (2012) Flexible and microporous chitosan hydrogel/nano ZnO composite bandages for wound dressing: in vitro and in vivo evaluation. ACS Appl Mater Interfaces 4:2618–2629
139. Gopal A, Kant V, Gopalakrishnan A, Tandan SK, Kumar Dinesh (2014) Chitosan-based copper nanocomposite accelerates healing in excision wound model in rats. Eur J Pharmacol 731:8–19
140. Peng CC, Yang MH, Chiu WT, Chiu CH, Yang CS, Chen YW, Chen KC, Peng RY (2008) Composite nano-titanium oxide–chitosan artificial skin exhibits strong wound-healing effect-an approach with anti-inflammatory and bactericidal kinetics. Macromol Biosci 8:316–327
141. Archana D, Dutta J, Dutta PK (2013) Evaluation of chitosan nano dressing for wound healing: Characterization, in vitro and in vivo studies. Int J Biol Macromol 57:193–203
142. Archana D, Singh BK, Dutta J, Dutta PK (2013) In vivo evaluation of chitosan–PVP–titanium dioxide nanocomposite as wound dressing material. Carbohydr Polym 95:530–539

Recent Advances in Chitosan-Based Nanomedicines for Cancer Chemotherapy

Ankit Saneja, Chetan Nehate, Noor Alam and Prem N. Gupta

Abstract Chitosan, a cationic polysaccharide, has prompted the continuous impetus for the development of tumor targeted drug delivery systems, thanks to the polymer's biocompatibility, low toxicity, and biodegradability. The presence of primary hydroxyl and amine groups on its backbone allows it for chemical modifications to control its physical properties. The nanomedicines prepared from chitosan and its derivatives can be delivered through different routes, such as oral, intravenous, and intraperitoneal. Chitosan-based nanomedicines including nanoparticles, microspheres, drug conjugates, micelles, hydrogels, etc. are in various stages of development. This polymer is being currently investigated for simultaneous delivery of two chemotherapeutic agents or chemotherapeutic agent with a gene carrier to produce synergistic effects. This chapter summarizes the recent advances in application of chitosan and its derivatives as a carrier for chemotherapeutic agents as well as gene carriers for cancer chemotherapeutics.

Keywords Chitosan · Delivery systems · Nanoparticles · Targeted delivery

1 Introduction

Cancer is regarded as one of the most dreadful diseases and chemotherapy remains to be the necessary approach for treatment mostly for patients who do not respond to local excision or radiation treatment. However, numerous limitations are associated with chemotherapeutic agents that hinder the effectiveness of chemotherapy, e.g., poor aqueous solubility (making them difficult to administer), narrow therapeutic index (which limits the dose and frequency of treatment), non-specific

A. Saneja · C. Nehate · N. Alam · P.N. Gupta (✉)
Formulation & Drug Delivery Division,
CSIR-Indian Institute of Integrative Medicine Canal Road, Jammu 180001, India
e-mail: pngupta@iiim.ac.in; pngupta10@gmail.com

P.K. Dutta (ed.), *Chitin and Chitosan for Regenerative Medicine*, Springer Series on Polymer and Composite Materials, DOI 10.1007/978-81-322-2511-9_9

distribution throughout the body (which leads to insufficient penetration into cancer cells) and efflux transporter specificity (which confers cancer cell drug resistance) [1, 2]. To overcome these limitations over the past several decades, drug delivery researchers, and formulation scientists have focused their attention toward the development of drug delivery systems that specifically target the site of action and significantly affect the pharmacokinetics and pharmacodynamics of therapeutic agents [3, 4]. A wide range of carriers including natural or synthetic polymers, lipids, surfactants, and dendrimers, have been employed to design such drug delivery systems [4–6]. Among these carriers, polysaccharides have attracted more and more attentions due to their outstanding merits such as easy availability, nontoxicity, biocompatibility, biodegradation, and easy modification [7, 8]. Chitosan, a linear aminopolysaccharide is one of the most abundant polysaccharides found in nature, in fact second only to cellulose [9]. It is a copolymer of *D*-glucosamine (deacetylated unit) and *N*-acetyl-*D*-glucosamine (acetylated unit), linked via (1 → 4) glycosidic bonds and obtained by partial *N*-deacetylation and hydrolysis of chitin, which is the main component of the exoskeleton of crustaceans such as crab and shrimp [4, 10, 11]. Generally, deacetylation of chitin is not completely achieved and usually ranges from 60 to 98 %. Therefore, chitosan is available with various molecular weights (ranging from 50 to 2000 kDa) and degree of deacetylation [12]. Every deacetylated subunit of chitosan contains a primary−NH_2 group, having pKa value of ~6.5 and makes it pH sensitive polymer. Therefore, chitosan is insoluble in water, alkali, and organic solvents and soluble in acidic media, such as acetic acid, glutamic acid, citric acid, aspartic acid, hydrochloric acid, lactic acid when the pH of the solution is below 6. Other than pH, chitosan solubility is significantly influenced by the extent of degree of deacetylation, molecular weight, and ionic strength of the solution [13–15].

Chitosan has remarkable physicochemical properties (Fig. 1) such as biodegradability [16, 17], biocompatibility [18], nontoxicity [19, 20], mucoadhesivity [21], low immunogenicity [22, 23], antibacterial [24, 25], and anticancer activity [26].

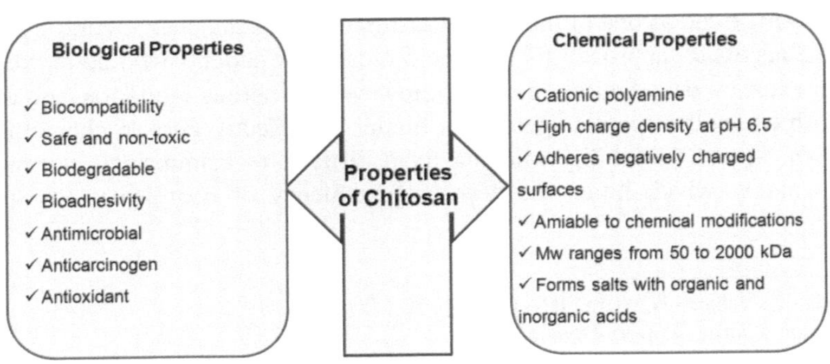

Fig. 1 Physicochemical properties of chitosan

Chitosan is biodegraded by a number of enzymes, such as lysozyme, *N*-acetyl-beta-*D*-glucosaminidase, di-*N*-acetylchitobiase, and chitiotriosidase, which can be produced by the normal flora in the human intestine or exist in the blood [27–29]. Further, there are certain reports which states that chitosan and its degradation products activate human macrophages and lymphocyte proliferation without leading to inflammatory symptoms [29, 30]. With regard to the in vivo toxicity profile, it is practically nontoxic in animals as well as humans, with an LD_{50} in rats of 16 g/kg [15, 31]. Chitosan has also gained generally recognized as safe (GRAS) status by FDA for wound dressing applications [32].

Chitosan is polycationic due to the occurrence of an amino functionality at the 2-position and its polycationic nature has attracted the drug delivery researchers for gene delivery due to their negative charge [15, 33]. The presence of amino, acetamido, and/or hydroxyl group in the chitosan presents great opportunity for its chemical modification (Fig. 2) [11]. Several derivatives of chitosan has been designed in order to improve its pH dependent solubility, enhance mucoadhesive properties, transfection efficiency, such as succinyl chitosan [34], quaternized chitosan [35, 36], carboxymethylated chitosan [37], PEG-chitosan [38], thiolate chitosan [39], etc. It is reported, for example, thiolation of chitosan remarkably improves its mucoadhesive properties due to the formation of disulfide bonds with cysteine-rich subdomains of mucus glycoproteins [40, 41]; *N*-acyl and *N*-alkyl derivatives of chitosan having carboxy or sulfate group are soluble at basic pH region [42]. Various synthetic strategies for the modification of the chitosan and their advantages are well described in several comprehensive reviews [13, 43–45]. Chitosan has also been conjugated to chemotherapeutic molecules via cleavable linkers and these conjugates form self assembled nanoparticles in aqueous solutions. Several researchers have also conjugated chitosan to other polymers to form self-assembled polymeric micelles of chemotherapeutic agents. The interesting biopharmaceutical and toxicological profile of chitosan, attracted drug delivery researchers to design nanocarriers for cancer chemotherapy. In this chapter, we present the application of various nanomedicine approaches based on chitosan and its derivatives explored for cancer drug delivery during the last few years.

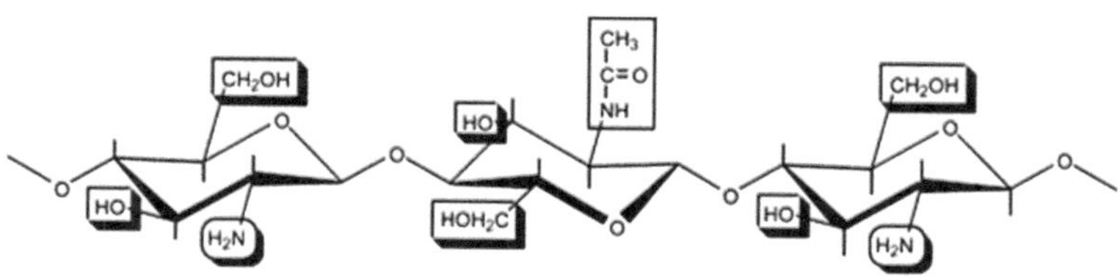

Fig. 2 Illustration of possible reaction sites in chitin and chitosan (Reprinted from Ref. [11] Copyright 2009, with permission from Elsevier)

2 Chitosan-Based Nanomedicines for Delivery of Chemotherapeutic Agents

Chitosan-based nanomedicines have been successfully employed for delivery of chemotherapeutic agents due to their nanoscale size, high surface-to-volume ratio, and favorable physicochemical characteristics [10, 15, 46, 47]. These nanomedicines exhibit relatively long circulation times and allow for controlled drug release of chemotherapeutic agents. They have the potential to modulate pharmacokinetic as well as pharmacodynamic profiles of chemotherapeutic agents due to their ability to inhibit P-glycoprotein, uptake by microfold cells and ability to bypass-tight junctions (Tj) in epithelial cells (Fig. 3) [1, 2, 48, 49]. Chitosan is considered to possess a direct effect on the tight junctions of epithelial cells by acting on the expression of claudin-4 protein and its redistribution from the cell membrane to the cytosol which is associated with its degradation in lysosomes and a decrease in tight junction strength, thus enhancing the paracellular permeability [49, 50].

Further, due to their smaller size they have lower uptake by reticuloendothelial system (RES) and accumulate preferably at the tumor site due to leaky vasculature of tumor tissue (≥200 nm), the phenomena is known as "enhanced permeability and retention (EPR)" effect [51]. This is referred as passive tumor targeting and often associated with some limitations such as inability to achieve sufficiently high level of drug concentration in tumor resulting in lower therapeutic efficacy and may elicit adverse systemic effects. To overcome these limitations, in recent years, there has been focus in developing these nanomedicines with active tumor

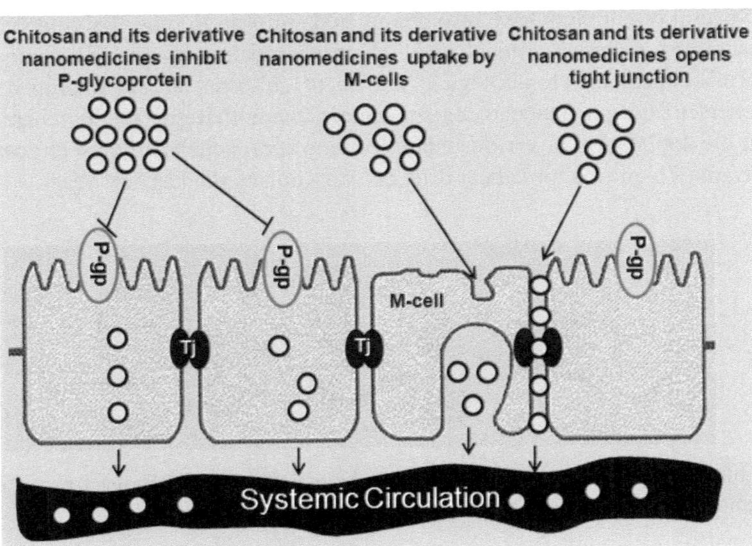

Fig. 3 Proposed mechanisms for modulating the pharmacokinetics and pharmacodynamics of chemotherapeutic agents by chitosan nanomedicines

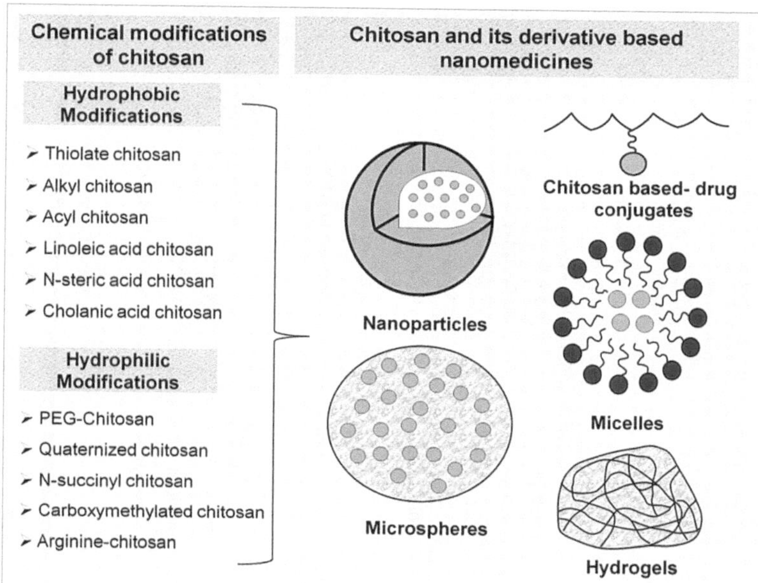

Fig. 4 Various chemical modifications of chitosan and its application in designing different nanomedicines

targeting ligands, such as folic acid, hyaluronic acid, antibodies, peptides, biotin, and avidin, which can recognize and bind to specific receptors that are unique to cancer cells [10, 15, 52–54]. Various types of nanomedicines have been designed based on chitosan and its derivatives, such as nanoparticles, microspheres, drug conjugates, micelles, hydrogels (Fig. 4) and are described in the following sections.

2.1 Nanoparticles

Chitosan has ability to promote cross-linkage with various cross-linking agents, such as sodium tripolyphosphate (TPP), glutaraldehyde, geneipin, etc. due to the presence of amine groups in backbone and it provides an efficient network to entrap the chemotherapeutic agent into the NPs [10]. Chitosan NPs can be prepared by a number of methods such as emulsion cross-linking, thermal cross-linking, spray drying, emulsion droplet coalescence method, etc. Detailed methods of preparation of chitosan-based nanoparticles are well described in several comprehensive reviews [55–57] and detailed discussion in this area is beyond the scope of this chapter. Some of the recent examples of chitosan-based nanoparticles for the delivery of chemotherapeutic agents are described in Table 1.

Table 1 Summary of some recent studies on nanoparticles using chitosan for delivery of chemotherapeutic agents along with cell lines studied for in vitro cytotoxicity

Therapeutic agent	Method of preparation	Size (nm)	Targeting moiety	Cell lines	References
DMC	Mixing chitosan in water with DMC in methanol followed by conjugation with EGFR antibody	<200	anti-EGFR antibody	A549	[58]
Selenocysteine	Ionic gelation followed by conjugation with folic acid	200–210	Folic acid	MCF-7, A375, HeLa, L02	[59]
EGCG	Ionic gelation followed by conjugation with folic acid	400	Folic acid	H1299, HeLa, Capan-1	[60]
Doxorubicin	Double emulsion	220–229	iRGD (CRGDKGPDC) peptide	HUVECs, MB-MDA-231, 4T1	[61]
Paclitaxel	Simple dialysis	302	–	MDA-MB231	[63]
Doxorubicin	Ionic cross linking	12–50	Folic acid	MCF-7 Cells, SMMC-7721	[70]
Methotrexate	Ionic gelation combined with chemical cross-linking	190–213	Folic acid	HeLa and T3-E1 cells	[71]
Metformin	Ionic gelation	210–290	–	MiaPaCa-2	[62]
Taxol	Synthesizing PLA-PCL-TPGS nanoparticles by solvent extraction followed by surface modification by thiolated chitosan	200–220	–	Caco-2 and A549	[64]
Docetaxel	Ionic gelation followed by cetuximab conjugation	200–250	Cetuximab	A549 and NIH3T3	[65]
Bufalin	Solvent-dialysis	150–200	Biotin	MCF-7	[66]
SN38	Ionic gelation followed by MUC-1 conjugation	200 nm	MUC-1 DNA aptamer	HT-29	
Paclitaxel	Synthesizing folate-conjugated chitosan–polylactide followed by ionic gelation	129–142	Folic acid	MCF-7, MCF-10A	[68]
Oxaliplatin	Solvent evaporation and emulsification cross-linking	89–98	–	MCF-7	[69]
Curcumin	Ionic gelation	146–150	–	HEK293	[72]

Table Abbreviations DMC Demethoxycurcumin; *CMC* Carboxymethyl-hexanoyl chitosan; *EGFR* epidermal growth factor receptor; *EGCG* epigallocatechin gallate; *PLA* polylactic acid; *PCL* polycaprolactone; *TPGS* D-α tocopheryl polyethylene glycol 1000 succinate; *MUC*-1 (mucin 1)

In a recent study, Huang et al. designed demethoxycurcumin (DMC) loaded amphiphilic chitosan (carboxymethyl-hexanoyl chitosan) nanoparticles coated with an anti-EGFR (epidermal growth factor receptor) antibody layer to overcome multidrug resistant (MDR) in lung cancer [58]. In vitro cell culture studies demonstrated EGFR targeted nanocarriers improved the effect of DMC under both normoxic and hypoxic conditions, with IC_{50} of 10 and 5 μg/mL for hypoxic and normoxic lung cancer (A549) cells, which were significantly lower than for free DMC. In vivo studies in A549 xenograft lung tumor mice model demonstrated DMC loaded core-shell nanocarriers achieved about eightfold reduction in tumor volume as compare to control group over the 8 weeks of the investigation. Yu et al. developed folate decorated chitosan nanoparticles of selenocystine (FA-SeC-CSNPs) and demonstrated FA-SeC-CSNPs exhibited broad-spectrum growth inhibition on the MCF-7 (breast cancer), A375 (human malignant melanoma), and HeLa (human epithelial carcinoma) cell lines, with IC_{50} values ranging from 1.2 to 3.12 μM [59]. Further, these folate decorated NPs significantly suppressed the migration and invasion of HeLa cells in a dose dependent manner. In mechanistic investigation the study revealed intracellular nanosystem triggered the overproduction of reactive oxygen species (ROS), which activated various downstream signaling pathways, such as p53, AKT, and MAPKs, to regulate the cell fate. In another investigation, Liang and co-workers designed folate-conjugated chitosan NPs of EGCG (FCS-EGCG-NPs) and evaluated in vitro cytotoxicity in HeLa, H1299 (human non-small-lung-cancer) and Capan-1 (human pancreatic adenocarcinoma) cells [60]. The maximum inhibition rate of FCS-EGCG-NPs in the HeLa, H1299 and Capan-1 cancer cells was 85, 56, and 64 %, respectively at concentration of 1 mg/mL. There was stronger tumor inhibition effect in HeLa cancer cells due to a larger expression of folic acid receptors on the surface, than in other cell types with lower expressions at the same concentration.

Doxorubicin loaded in amphiphilic chitosan-*co*-(D,L-lactide)/1,2-dipalmitoyl-sn-glycero-3 phosphoethanolamine copolymer (CS-*co*-PLA/DPPE) nanoparticles with the modification of an iRGD (CRGDKGPDC) peptide as the targeting agent were developed by Nie et al. [61]. In vitro studies in HUVEC (human umbilical vein endothelial cells, abundant expression of $\alpha_v\beta_3$ integrin receptors) and two breast carcinoma cell lines, 4T1 (which has a moderate expression of $\alpha_v\beta_3$ integrin receptors) and MB-MDA-231 (which has minor expression of $\alpha_v\beta_3$ integrin receptors) demonstrated a marginal higher toxicity of DOX-iRGD-NPs in HUVECs and MB-MDA-231, but lower toxicity in 4T1 as compared to free DOX which may be due to the difference in the capacities of endocytosis-mediated cellular uptake of the nanoparticles. The IC_{50} values of DOX-iRGD-NPs was lower than DOX-NPs, approximately twofold less for HUVECs (2.82 vs. 5.97), 4T1 (0.79 vs. 1.31) and slightly less in MB-MDA-231 (1.77 vs. 1.98). Further, increased cellular uptake of DOX-iRGD-NPs was observed using laser confocal scanning microscope and confirmed to be a consequence of a specific endocytosis pathway mediated by ligand–receptor interactions.

Snima et al. studied *O*-Carboxymethyl chitosan encapsulated metformin nanoparticles (O-CMC-metformin NPs) for pancreatic cancer therapy and investigated reduction in colony formation ability of the cancer cells in case of O-CMC-metformin NPs using in vitro clonogenic assay [62]. The cell migration assay results revealed the marginal reduction in an area covered by the cancer cells after 20 h incubation with O-CMC-metformin NPs (36 %) as compare to untreated sample (45 %) showed the less effect of NPs on migration property of the cancer cells on time dependent manner, whereas scratch assay results demonstrated these NPs could elicit pronounced cell growth inhibitory effect on pancreatic cells. Further, in vivo biocompatibility evaluation by histopathology of O-CMC-metformin NPs demonstrated normal cellular architecture of mice organs, such as spleen, liver, kidney, brain, lung, and heart [62]. Koo et al. developed the glycol chitosan nanoparticles with hydrotropic oligomers (HO-CNPs) of paclitaxel (PTX) and demonstrated rapid cellular uptake and lower cytotoxicity in MDA-MB231 (human breast cancer) cells, compared to Cremophor EL/ethanol formulation of PTX [63]. Intravenously injection of PTX-HO-CNPs (20 mg/kg) in mice showed superior tumor-targeting ability which may be due to EPR effect of nanoparticles through fenestrated vessels in angiogenic tumor tissue. Thirty-four days post-injection, the tumor volume of mice showed smaller tumor size treated with Abraxane® (commercial formulation of PTX) and PTX-HO-CNPs, and there was only negligible tumor growth especially in the case of PTXHO-CNPs. While tumor volume in mice treated with saline grew to about 2700 mm^3 and in the case of PTX formulation, it reached to about 1300 mm^3 showing insufficient suppression of tumor growth.

In another investigation, Jiang et al. synethesized PTX loaded in different formulations, i.e., 5 % thiolated chitosan-modified PCL nanoparticles (CNP), unmodified PLA-PCL-TPGS nanoparticles (UNP), 5 % thiolated chitosan-modified PLA PCL-TPGS nanoparticles (TNP), and 20 % thiolated chitosan-modified PLA-PCL-TPGS nanoparticles (DNP) [64]. In vitro cytotoxicity studies in A549 cells demonstrated cell viability after 24-h incubation at the 10 μg/mL drug concentration was 44.41 % for Taxol®, and 28.65 % (i.e., a 28.39 % increase in cytotoxicity) for TNP. Moreover, it increased by 37.65 % ($p < 0.05$, $n = 6$) and 18.72 % ($p < 0.05$, $n = 6$) for TNP after 48 and 72-h incubation at 10 μg/mL PTX concentration. These findings might be due to thiolated chitosan could greatly increase its mucoadhesiveness, increase PTX transport by opening tight junctions and bypassing the efflux pump of P-gp. Maya et al. designed chitosan cross-linked γ-poly(glutamic acid) (γ-PGA) nanoparticles loaded with docetaxel (DTXL) and conjugating the nanoparticles with Cetuximab (CET) for targeting the EGFR over-expressing non-small-cell-lung-cancer (NSCLC) cells (A549) [65]. In vitro cytotoxicity in A549 cells ($EGFR^{+ve}$) demonstrated the IC_{50} value of CET-DTXL-γ-PGA NPs (50 μg/mL) as 5 times less than that of DTXL-γ-PGA NPs (250 μg/mL). However, there was no significant difference in the IC_{50} values in NIH3T3 cells (250 μg/mL) for both targeted and nontargeted NPs. Further, flow cytometry investigation demonstrated A549 cells treated with CET-DTXL-γ-PGA NPs underwent a G2/M phase cell cycle arrest followed by reduction in mitochondrial membrane potential, inducing apoptosis and necrosis resulting in enhanced cancer cell death. Tian et al. designed bufalin loaded biotinylated chitosan nanoparticles (Bu-BCS-NPs) for targeting breast cancer [66]. In vitro cytotoxicity studies in MCF-7 cells demonstrated Bu-BCS-NPs exhibited a lower cytotoxicity (IC_{50}; 0.582 μg/mL) vs native bufalin (IC_{50}; 1.896 μg/mL). Further, Bu-BCS-NPs induced increase in intracellular ROS which was about 2.5 times that of free bufalin and triggered more mitochondria dysfunction than free bufalin which ultimately increased numbers of apoptotic cells. Treatment with Bu-BCS-NPs down-regulated expression of anti-apoptotic molecules Bcl-2 and survivin while up-regulating expression of pro-apoptotic molecules Bax and caspase-3. In vivo studies using MCF-7 tumor models in nude mice demonstrated significantly decreased body weight in case of free bufalin (10 mg/kg) while there was no apparent loss of body weight in case of Bu-BCS-NPs treatment.

In another chemical modification of chitosan, Sayari et al. synthesized *N*-carboxyethyl chitosan-SN38 conjugate (CS-SN38) and forming it into the nanoaparticles by ionotropic gelation method using tripolyphosphate [67]. The nanoparticles were conjugated by MUC1 DNA aptamer as a targeting agent. In vitro cytotoxicity studies in HT-29 (human colon adenocarcinoma, $MUC1^{+}$ cancerous cells) cell line demonstrated diminished efficacy of CS-SN38 NPs with respect to free drug which can be related to covalent conjugation of SN38 to chitosan. However, the enhanced and selective uptake of CS-SN38-Apt as compared to CS-SN38 nontargeted NPs using confocal microscopy compensate for reduced efficacy of conjugate. Huang et al. synthesized FA-conjugated chitosan–polylactide (FA-CH-PLA) copolymers and encapsulated PTX using ionic gelation method [68]. In vitro cytotoxicity in MCF-7 cells was twofold or even threefold higher in case of FA-CH-PLA nanoparticles as compared to free PTX at an equivalent PTX dose of 1000 or 10,000 ng/mL. Further, the uptake of the nanoparticles in MCF-7 cells was more than sixfold higher as compared to native PTX during 24-h incubation. Recently, Vivek et al. designed oxaliplatin loaded chitosan nanoparticles by solvent evaporation and emulsification cross-linking method [69]. In vitro cell cytotoxicity in MCF-7 cells demonstrated twofold increase in cytotoxicity of oxaliplatin loaded chitosan nanoparticle when compared to native oxaliplatin ($p \leq 0.05$) after 48 h. Further, IC_{50} value decreased 13-times in case of oxaliplatin loaded chitosan nanoparticle (10 μg/mL) as compare to native oxaliplatin after 48 h incubation. Cellular uptake using confocal microscopy demonstrated enhanced cellular uptake of Oxaliplatin loaded chitosan nanoparticle in the MCF-7 cells after 4 h, compared with native oxaliplatin. In the case of oxaliplatin loaded chitosan nanoparticle, the flow cytometry analysis showed that the number of apoptotic cells distribution increased to 21.3 ± 1.57, 54.4 ± 5.79, and 93.6 ± 11.52 % when the concentration of oxaliplatin was 10, 15, and 20 μg/mL, while in case of native oxaliplatin it was 11.74 ± 0.13, 21.57 ± 0.47, and 70.68 ± 7.26 %. The study demonstrates that oxaliplatin chitosan NPs may be a potential drug delivery system for breast carcinoma.

2.2 Microspheres

Microspheres are spherical particles having the diameter in micrometer range. The microspheres have been prepared by the same techniques used for preparation of nanoparticles. Peng et al. designed composite microspheres based on gelatin (Gel) and chitosan (Cs) loaded with 5-fluorouracil (5-FU) using glutaraldehyde (GA) as a crosslinker [73]. In vitro cytotoxicity in the MCF-7 cell line demonstrated that as the amount of glutaraldehyde increased, cell viability was decreased in case of the empty Gel/Cs microspheres; however, the cell viability increased with increasing glutaraldehyde content for the Gel/Cs microspheres loaded with 5-FU. This might be because 5-FU has much higher cytotoxicity than the crosslinker (GA); when the Gel/Cs microspheres with 5 FU crosslinked by higher GA then the release of 5-FU was slower and the cell viability was higher. Zhao et al. designed doxorubicin-loaded multilayer polyelectrolyte hollow microspheres fabricated via a layer-by-layer (LbL) assembly of fluorescein isothiocyanate (FITC)-modified chitosan (CSFITC) and sodium hyaluronate (HA) (as the polycation and polyanion, respectively) on polystyrene sulfonate (PSS) templates with galactosylated chitosan (GC) as the outermost layer to target hepatocytes specifically [74]. The cytocompatibility studies in HepG2 (liver hepatocellular carcinoma) cells demonstrated that viability of the HepG2 cells was near 100 % (96.57−104.70 %) at the tested concentrations (0 −100 μg/mL) after 48 h of incubation, indicating that hollow microspheres have excellent nontoxicity toward HepG2 cells. Park et al. designed doxorubicin-loaded chitosan microspheres by emulsification and crosslinking methods [75]. In vivo study demonstrated blank chitosan microspheres led to a reduction in the tumour size; the size decreased by approximately 25 % and in case of doxorubicin-loaded chitosan microspheres tumour sizes were shrunk to one-third of the initial size. Therefore, synergistic anticancer activities of doxorubicin-releasing chitosan microspheres were observed to VX2 cells in the rabbit auricle model.

2.3 Chitosan-Drug Conjugates

A polymer–drug conjugate is an area of polymer therapeutics where the chemotherapeutic agent is not encapsulated but covalently linked to a polymeric macromolecular carrier via biodegradable spacer [3, 76, 77]. The spacer is usually stable in the systemic circulation but cleaved at the target site in response to specific stimuli, such as temperature, pH, redox potential, light, and enzymes. Such drug conjugates can be selectively accumulated at the tumor site by the phenomena of EPR, followed by release of the chemotherapeutic drug by cleavage of the spacer [4, 78]. Owing to presence of reactive amino moieties which are present in the backbone of chitosan, it and its derivatives has been successfully conjugated to various chemotherapeutic agents (Fig. 5).

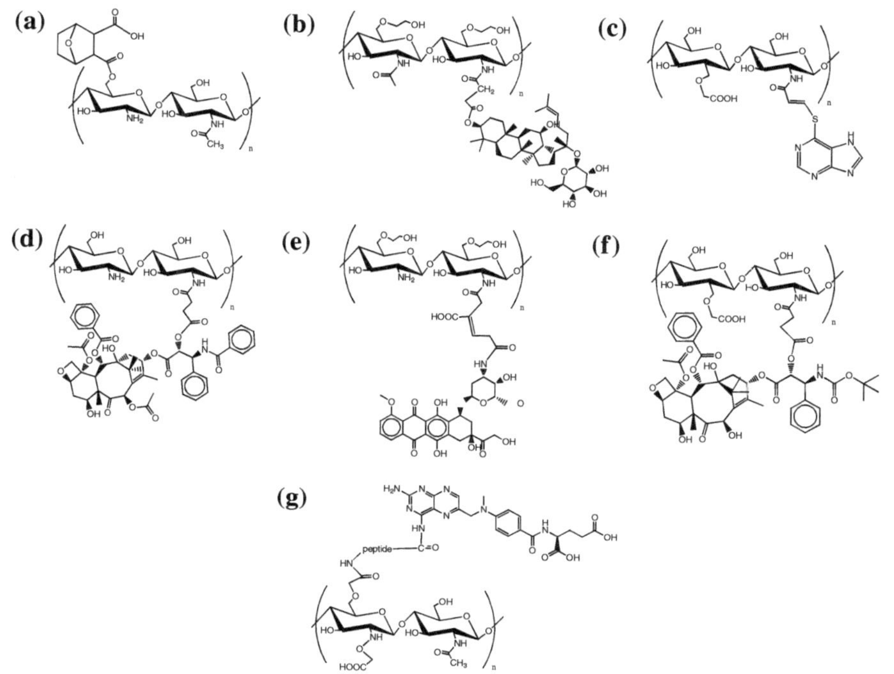

Fig. 5 Illustration of chitosan and its derivative conjugates with chemotherapeutic drugs. **a** Norcantharidin-chitosan conjugate. **b** Glycol chitosan ginsenoside compound K conjugate. **c** Carboxymethyl chitosan-6-mercaptopurine conjugate. **d** Chitosan paclitaxel conjugate. **e** Glycol chitosan-doxorubicin conjugate. **f** Carboxymethyl chitosan-docetaxel conjuaget. **g** *O*, *N*-carboxymethyl chitosan-methotrexate conjugate (peptide used were Gly-Gly; Gly-Phe, and Gly-Tyr)

Recently Li et al. synthesized *O*, *N*-carboxymethyl chitosan methotrexate (MTX) conjugate via dipeptide spacers [glycylglycine (Gly–Gly), glycyl–*L*-phenylalanine (Gly–Phe), glycyl–*L*–tyrosine (Gly–Tyr)] [79]. In vitro cytotoxicity in RPMI 8226 (human myeloma) cells demonstrated free MTX had the maximum cell inhibition ratio, while prodrugs had the lower inhibition ratio at doses of 50, 36, 24, 12, and 4 μg/mL after 48 h. However, MTX is released, when these prodrugs are internalized in tumor cells via the endocytosis pathway and entrapped in the acidic endosomal/lysosomal compartments. Kim et al. synthesized PTX conjugate to low molecular weight chitosan via succinate linker in three different molar ratios, yielding LMWC L (840 mg, 0.14 mmol), LMWC M (504 mg, 0.084 mmol), and LMWC H (315 mg, 0.053 mmol) [80]. In vitro cytotoxicity in human non-small-cell-lung cancer cells (NCIH358) demonstrated IC_{50} values were similar to that of parent compound with all LMWC-PTX. However, LMWC PTX L and M exhibited higher cytotoxicity (IC_{50} 0.26 and 2.71 ng/mL, respectively) against murine breast cancer cells (4T1) than did free PTX (IC_{50} 12.4 ng/mL). In vivo pharmacokinetics demonstrated improved half-life of PTX in the blood (at least sixfold) in all of the LMWC-PTX formulations as compared to that of the free PTX formulation (10 mg/kg). Mathiyalagan et al. synthesized ginsenoside compound K covalently conjugated to the backbone of hydrophilic glycol chitosan (GC) through succinic acid linker [81]. In vitro cytotoxicity demonstrated GC-CK exhibited significantly higher or similar cytotoxicity compared as compare to CK in HT29 and HepG2cells. However, GC-CK conjugates demonstrated slightly lower toxicity than CK in case of HT22 cells. Further, GC-CK conjugates also inhibited LPS induced NO production in RAW264.7 (murine macrophage) cells to a higher degree than CK. Li et al. synthesized norcantharidin-chitosan conjugate (NCTD-CS) via alcoholysis reaction and examined the IC_{50} against human esophageal carcinoma ECA-109 cells and murine breast cancer EMT6 cells (168.8 ± 8.9 and 90.7 ± 8.1 μg/mL, respectively) [82]. The IC_{50} values were much higher than that of NCTD, i.e., 9.4 ± 0.9 and 3.1 ± 0.3 mg/mL for ECA-109 (human esophageal squamous carcinoma) and EMT6 (breast cancer) cells, respectively. However, due to prolonged release of the drug from the conjugate these IC_{50} values were understandable. Flow cytometric analysis indicated that NCTD-CS arrested ECA-109 cell cycle at the S phase and reduced the number of cells in the G0/G1 phase. In vivo antitumour efficacies in a breast tumor-bearing Kunming mouse model demonstrated control group had an average tumor burden of 1.02 g while the NCTD and NCTD-CS groups had average tumor burdens of 0.66 and 0.55 g, respectively, a 35.87, and 45.82 % decrease in tumor growth compared with the control group.

In another study, Liu et al. synthesized carboxymethyl chitosan–docetaxel (CMCS-DTX) conjugates which could spontaneously self-assemble into nanoparticles in aqueous buffer, with uniform size of 127.2 ± 3.58 nm [83]. In vitro cytotoxicity in HepG2 and B16 (mouse melanoma) cell line demonstrate that when administrated at low doses (≤0.1 μg/mL), free DTX exhibited higher cytotoxicity than CMCS-DTX conjugates, but when treated with relative high concentration of drug, CMCS-DTX conjugates showed comparable (1–10 μg/mL) or even higher (50 μg/mL) cytotoxicity than free DTX. Apoptosis analysis on HepG2 cells using Hoechst staining of nucleus showed the nucleuses of the control cells were homogeneously stained, while the nucleuses of the cells treated with the DTX or CMCS-DTX conjugates exhibited chromatin condensation and further distributed into apoptotic bodies. In vivo antitumor efficacy in murine carcinoma xenograft models in mice showed strong antitumor effect not only by prolonging the median survival time of the mice, but also by inhibiting the tumor growth. During the administration of CMCS–DTX conjugates (5 mg/kg equivalent of DTX), 37.5 % of the mice died during 30 days of treatment while in case of Duopafei® (commercial formulation of DTX) 50 % of the mice died within 21 days. On the other hand, mice in control group (administered with normal saline) and CMCS group exhibited rapid death with 50 % dying within 15 days and 100 % dying within 21 days. Moreover, the median survival time for control, Duopafei® and CMCS-DTX conjugates were 12, 24, and >30 days, respectively. There was an enhancement up to fourfold in maximum tolerated dose (MTD) of CMCS-DTX conjugates (>250 mg/kg) compared to Duopafei® (approximately 50 mg/kg), which was most likely attributed to the excellent biocompatibility of CMCS and alleviation of the toxic solubilizer.

In other study by Gong et al. amphiphilic conjugates of 6-mercaptopurine (6-MP) and carboxymethyl chitosan (CMCS) was synthesized via a glutathione (GSH) sensitive α, β unsaturated linker for intracellular delivery [84]. GSH linker was chosen to design the conjugate because the concentration of GSH in cancer cells is fourfold higher than that in normal cells and will cause cleavage of α, β unsaturated bonds in cancer cells. In vitro cytotoxicity in acute promyelocytic leukemia cell line (HL-60) showed that free 6-MP had the maximum inhibition ratio while conjugate had the minimum inhibition ratio at each concentration varying from 50 to 250 μg/mL. This is due to interaction of free 6-MP with the plasma protein and conjugate protected 6-MP from interaction with other proteins and cells. However, in vitro cytotoxicity of conjugate on mouse fibroblast cell line L929 was much lower than that of 6-MP. Son et al. synthesized doxorubicin conjugated to glycol chitosan (GC-DOX) via cis-aconityl linkage due to its pH-sensitive behavior which results in hydrolysis in the acid environment of endosomes/lysosomes and the release of the drug to the cytoplasm [85]. In vivo antitumor activity of free DOX and GC-DOX nanoaggregates (10 mg/kg) in male Fisher 344 rats (implanted with solid tumor induced by II45 mesothelioma cells) demonstrated size of the tumor was suppressed over 10 days after i.v. injection of the free DOX and GC-DOX. Further, body weight of free DOX-treated rat significantly decreased while that of the GC-DOX-treated rat was maintained constantly, over 10 days.From the examples evident above it is noteworthy that generally these chitosan-drug conjugates show lower or comparable in vitro cytotoxicity in comparison with the corresponding free drug but their superiority is the significant performance in enzymatic physiological condition, where in vivo distribution will affect efficacy and safety of the treatment [7].

2.4 Micelles

Polymeric micelles are another form of nanomedicines formed by the self-aggregation of copolymeric amphiphiles above a certain concentration called critical micellar concentration (CMC) [86]. These micelles usually comprises of copolymers such as X–Y diblocks and X–Y–X triblocks, where X and Y represent the hydrophilic and the hydrophobic blocks, respectively. During their synthesis hydrophilic blocks accommodate at the interface between the inner hydrophobic domain composed of the hydrophobic segments, known as core and the external medium, forming the micellar corona which control the release of the drug payload [86–88]. Ideally, micelles developed for chemotherapeutic agents must be (i) small enough (10–200 nm) to effectively penetrate into cancer cells; (ii) unrecognizable by the mononuclear phagocyte system (MPS) for a sufficient time in order to accumulate in target tissue; (iii) able to interact with the target cells; (iv) eliminated from the organism either after degradation or dissolution; (v) have tunable stability; (vi) improve the pharmacokinetic (PK) profile of the chemotherapeutic agent; (vii) possess high loading capacity; and (viii) synthesized in a reproducible, facile method which is reasonably inexpensive [89–91]. To improve the delivery of chemotherapeutic agents most of the chitosan micelles has been developed with modification of chitosan with hydrophobic moieties such as oleic acid [92], stearic acid [93–97], deoxycholic acid [98], linoleic acid [99], cholesterol [100], glycyrrhetinic acid [101], polycaprolactone [102] etc. These chitosan-modified hydrophobic moieties form self-assembled micelles when inserted into aqueous solutions.

Recently, Su et al. synthesized doxorubicin (DOX) conjugated with a stearic acid-grafted chitosan oligosaccharide (CSO-SA) via disulfide linkers (DOX-SS-CSO-SA) and these conjugates self assembled into nano-sized micelles of 62.8 nm [96]. In vitro cytotoxicity demonstrated IC_{50} values of CSO-SA micelles for BEL-7402, MCF-7, and MCF-7/Adr cells were approximately 347, 307, and 488 μg/mL, respectively. Further, there was only 3.4-fold difference between IC_{50} values of DOX-SS-CSO-SA in MCF-7/Adr and MCF-7 cells while in case of native DOX it was 118.6-fold, thus demonstrating reversal ability up to 34.8-fold. In vivo biodistribution in various tissues were calculated as %ID/g (the percentage of the injected dose per gram of tissue) and the results demonstrated that there was no accumulation in the heart (2.0 %, ID/g) thus relieved the DOX-induced cardiac injury in case of DOX-SS-CSO-SA micelles. There was accumulation of DOX in the liver (41.4 %, ID/g), spleen (35.5 %, ID/g), tumors (14.7 %, ID/g), and lungs (18.1 %, ID/g). In vivo tumor study in BEL-7402 xenograft tumor bearing nude mice showed tumor inhibition rate (IR) as 82.7 % for adriamycin treatment group and 70.8 % for DOX-SS-CSO-SA treatment group.

Jin et al. synthesized amphiphilic graft copolymer, *N*-octyl-*O*-sulfate chitosan (NOSC), and its PTX-encapsulated micelles (PTX-M) in order to overcome multidrug resistance [103]. In vitro cytotoxicity on both the human hepatocellular liver carcinoma (HepG2) cells and the multidrug resistance HepG2 (HepG2-P) cells demonstrated PTX-M presented the highest cellular uptake and the lowest efflux rate of PTX, which was due to a combination of the inhibiting P-gp effect of NOSC and the bypassing P-gp action of the intact PTX-M. The cellular uptake of PTX in HepG2-P cells with the NOSC concentrations of 0.008, 0.04, and 0.08 % (w/v) were detected to be 583.25 ± 138.33, 605.16 ± 192.11, and 598.46 ± 139.02 ng/mg, respectively, which corresponds to 6.0-fold, 6.3-fold, and 6.2-fold that of free PTX control, respectively. In vivo tumor efficacy in the Heps-P tumor-bearing mice demonstrated tumor inhibition rate was in order of PTX-M > Taxol + NOSC > Taxol. These results were mainly explained by a combination effect: (1) enhanced accumulation of PTX at the tumor site due to the EPR effect of PTX-M; (2) improved cellular uptake of PTX via endocytosis of PTX-M; (3) sustained release of PTX from PTX-M inside the cells and (4) persistent P-gp inhibition effect by NOSC polymer and its micelles.

Yang et al. designed self assembled micelles (~100 nm) using lactobionic acid (LA)-modified chitosan–stearic acid (CSS-LA) for the delivery of doxorubicin (Fig. 6) [97]. In vitro cytotoxicity in HepG2 and BEL-7402 cells demonstrated

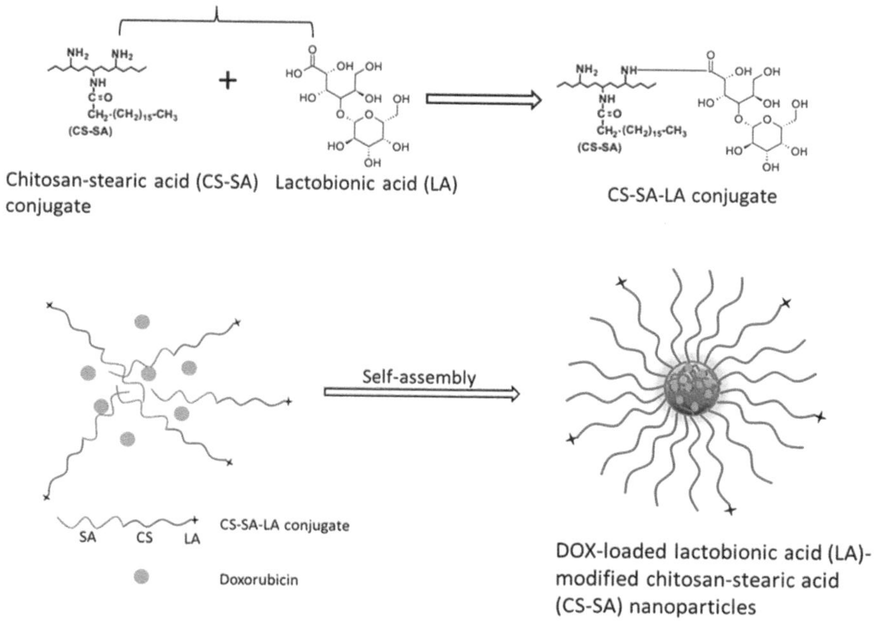

Fig. 6 Schematic illustration of conjugation of chitosan (*CS*), stearic acid (*SA*), and lactobionic acid (*LA*). Formation of polymeric micelles via self-assembly process has been depicted (Reprinted from Ref. [97], Copyright 2015, with permission from American Chemical Society)

reduction in IC_{50} value of DOX by ~85 and ~90 % respectively in case of CSS-LA/DOX, while it was by ~35 and ~32 %, respectively in case of CSS/DOX. Cellular internalization capacity of DOX in HepG2 cell line at the end of 6 h was ~450 and ~300 ng/mL for CSS-LA and CSS micelles, respectively. Further, in vivo studies in HepG2 cancer cell bearing xenograft nude mice demonstrated greater reduction in tumor volumes i.e. ~2400, 1500, 1100, and 650 mm^3 for saline, free DOX, CSS/DOX, and CSS-LA/DOX treated mice, respectively. The study demonstrated that active targeted ligand conjugated micelles is a promising strategy in the treatment of hepatic cancer.

2.5 Hydrogels

Hydrogels are comprised of cross-linked polymer networks that are hydrophilic and are able to provide sustained, local delivery of a variety of therapeutic agents [104]. These polymer networks have a high affinity for water, but are prevented from dissolving due to either by non-covalent attractions (such as hydrogen bonds, π–π interactions, and charge interactions) and physical entanglements, or by covalent cross-linkages [105, 106]. Hydrogels usually have advantages than other drug delivery system in terms of higher drug loading content (more than 10 %) and encapsulation of chemotherapeutic agents into the hydrogel or formation of the hydrogel by a drug derivative itself can eliminate the unexpected adverse effects from excipients. However, the most important feature of hydrogels is that they can be designed to undergo controllable and sustained release of chemotherapeutic agents by incorporating enzyme cleavable sites in the gelators [107]. These hydrogels can also be made thermoreversible, i.e., form transient gel or liquid states depending upon the environmental temperature. Chitosan itself have ability to form a gel, without using any additive by neutralization of $-NH_2$ groups on chitosan. Further due to its polycationic nature in acidic medium, it forms hydrogels with polyanions via electrostatic interactions. Several types of cross-linkers such as small molecule crosslinkers, polymer–polymer crosslinkers, photo-cross-linkers, enzymatic cross-linkers have been used to design chitosan hydrogels (Table 2) [104, 106, 107].

Li et al. developed a thermosensitive chitosan/β-glycerophosphate (C/GP) hydrogel loaded with docetaxel DTX for intratumoral delivery [108]. In vivo study of DTX-loaded C/GP hydrogel (DTX-C/GP) at 20 mg/kg in H22 tumor-bearing mice demonstrated significant reduction in tumor volume. As compare to saline group, the tumor inhibition rate of blank gel, intravenous DTX solution, intratumoral DTX solution, and DTX-C/GP was 2.3, 29.8, 41.9, and 58.1 %, respectively. Further, in vivo pharmacokinetic studies in H22 tumor-bearing mice demonstrated that DTX concentration of the gel was maintained above the detection limit until 21 days due to the sustained release nature of the DTX-C/GP and it had a long MRT of 9 days and a AUC_{plasma} of 219.43 μg/L day.

Alexander et al. synthesized PEGylated melphalan conjugates and these conjugates were incorporated into the medium molecular weight chitosan (CS)-based smart thermoreversible in situ forming injectable hydrogel (MLPEG-CP gel) in order to improve dose frequency and local effect [109]. in vitro release of MLPEG-CP gel demonstrated that there was a controlled drug release of melphalan approximately 80 % up to 100 h in hydrogel comprising CS (3.22 %, w/v) and glycerophosphate disodium salt (GP) (16 %, w/v).Taleb et al. synthesized sodium alginate/chitosan/hydroxyapatite (SA/CS/HAP) nanocomposite hydrogel of doxorubicin (DOX) containing different amounts, 0.6, 2.0, 3.5, and 5.0 % wt/v, of HAP using gamma radiation as cross-linker for liver cancer [110]. In vitro release study behavior of the hydrogel showed Fickian profile and DOX release was pH dependent, where samples showed higher release up to 95 % at pH 5, but 60 % at pH 7.4.

Ju et al. designed thermosensitive micelles–hydrogel hybrid system based on poloxamer 407 (P407) in order to resolve the fast erosion and low loading capability for local delivery of palcitaxel (PTX) (PTX-M-MG) [111]. For designing of hydrogels glutaraldehyde (GA) was used to generate cross-linked networks with carboxymethyl chitosan (CMCS) interpenetrated in P407 gels, in which PTX-loaded *N*-octyl-*O*-sulfate chitosan micelles (PTX-M) were dispersed uniformly. The tissue distribution profiles after a single intratumoral administration (20 mg/kg) in Heps tumor bearing mice demonstrated Taxol® and PTX-M exhibited higher concentrations of PTX in plasma or liver than PTX-M-P407 and PTXM-MG in the first 24 h after administration; however, PTX in Taxol® and PTX-M

Table 2 Examples of various small linker molecules and their network structures in covalently cross-linked chitosan hydrogels (Reprinted from Ref. [104], Copyright 2013, with permission from Elsevier)

Agent	Target functional groups	Reaction conditions	Cross-linkage	Comments
 Glutaraldehyde	Primary amines and aldehydes	Reaction favors basic and neutral pH	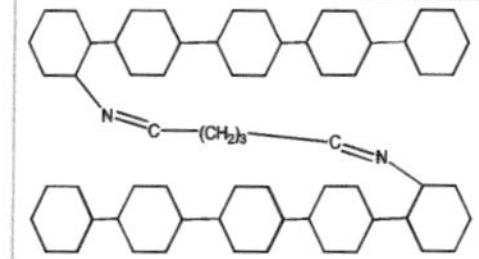	Reaction completes within 1 h; difficult to remove trace gluteraldehyde
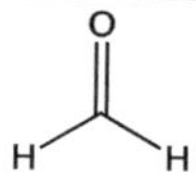 Formaldehyde	Primary amines and aldehydes	Basic and neutral pH	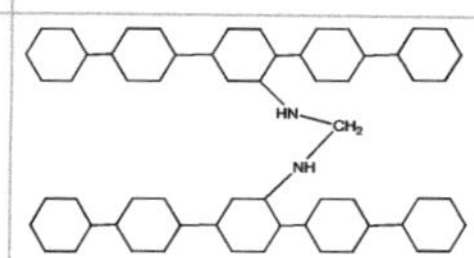	Reaction completes within 1 h; difficult to remove trace formaldehyde
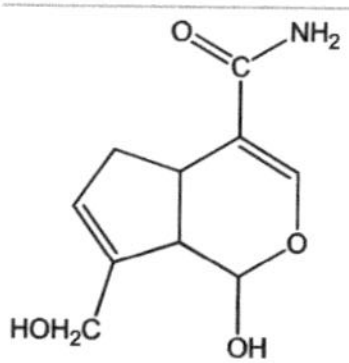 Genepin	Primary amines and aldehydes	Independent of pH	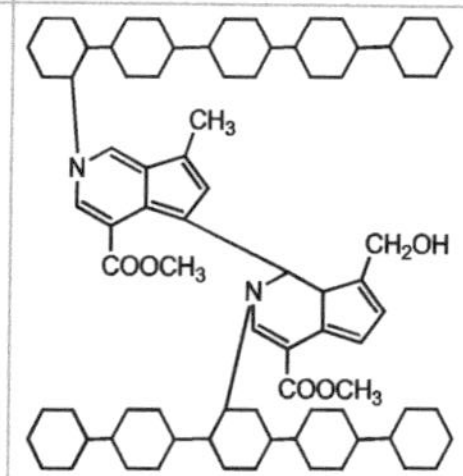	Nontoxic linker; can undergo self polymerization
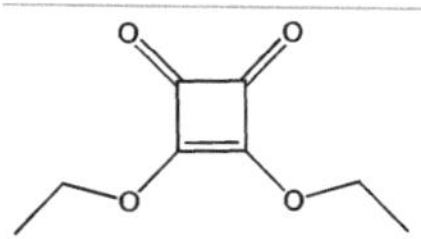 Diethyl squarate	Primary amines	pH 4.5–5.5; solution precipitates at higher pH; reaction favors elevated temperatures	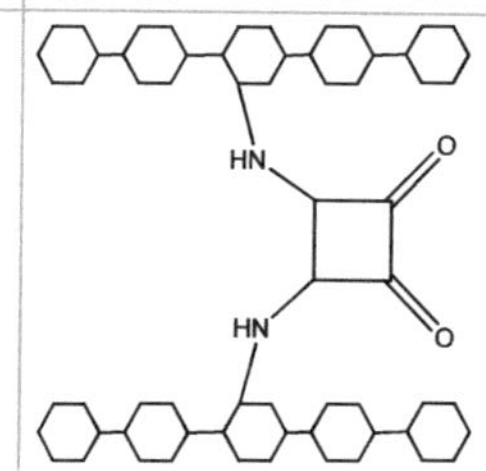	Reacts under mild conditions and is nontoxic; long reaction time precludes use for in situ gelation

(continued)

Table 2 (continued)

Agent	Target functional groups	Reaction conditions	Cross-linkage	Comments
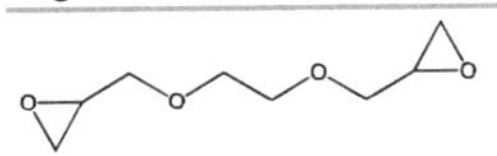 Ethyleneglycol diglycidylether	Primary amines and oxiranes	Basic pH; reaction favors elevated temperatures	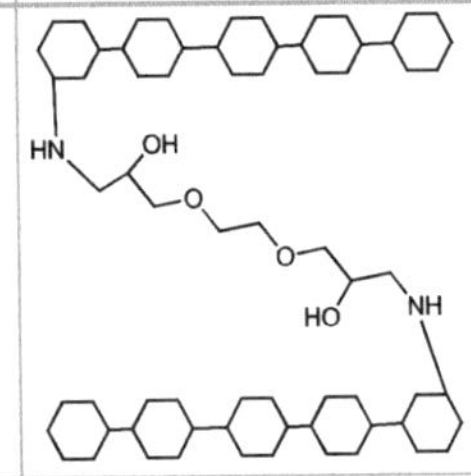	Difficult to remove EDGE traces; long reaction times and basic pH can yield hydrogel beads
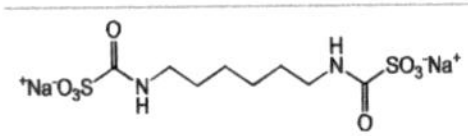 Blocked diisocyanate	Primary amines	Basic pH; reaction requires elevated temperatures	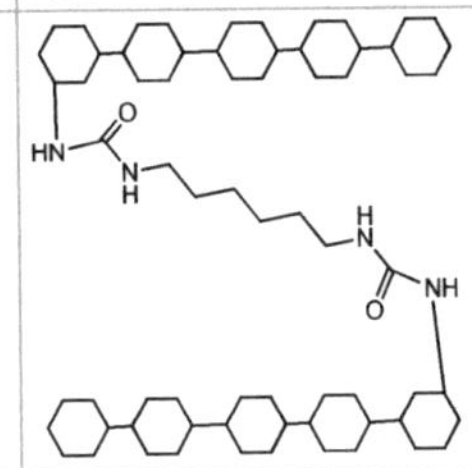	Long reaction times and basic pH requirements preclude in situ gelation; can produce hydrogel beads

were beyond the detection limit after 3 days whereas PTX in PTX-M-P407 and PTXM-MG hydrogels were above the detection limit till 7 and 15 days, respectively. The study also revealed CMCS-modified P407 gels (PTXa-M-MG) prolonged retention at tumor sites, lasting for 20 days, and a superior tumor inhibition rate (64.27 %) with reduced toxicity compared with Taxol®, PTX-M, and PTX-M loaded unmodified P407 gels (PTX-M-P407).

Zhang et al. designed magnetic thermosensitive hydrogel for sustained and prolonged delivery of Bacillus Calmettee Guérin (BCG), which was formulated with chitosan (CS), β-glycerophosphate (GP) and Fe_3O_4 magnetic nanoparticle (Fe_3O_4-MNP) [112]. In vivo antitumor studies revealed that both Fe_3O_4-BCG-CS/GP gel and BCG inhibited rat bladder carcinogenesis induced by nitrosamine (BBN) administration. Moreover, intravesical Fe_3O_4-BCG-CS/GP gel was superior to BCG alone in reducing the volume of tumors. Further, immunohistochemical staining and urinary cytokines analysis revealed the ability of Fe_3O_4-BCG-CS/GP gel to induce intenser $CD4^+$ lymphocytic infiltration in bladder submucosa and greater levels of urinary cytokines (a stronger Th1 immune response) than BCG alone correlates with its superior antitumor activity (Fig. 7). These results support the potential application of hydrogels as nanocarriers for cancer chemotherapy.

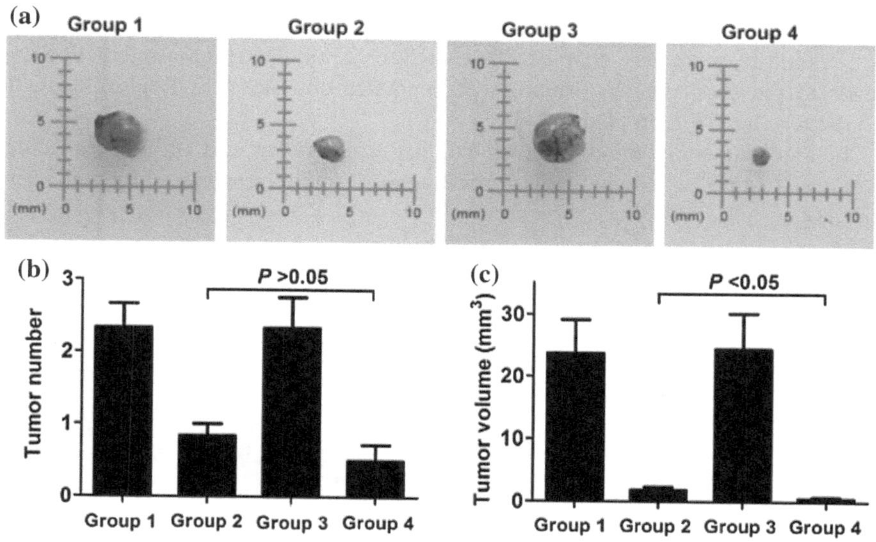

Fig. 7 In vivo antitumor effect on bladder cancer orthotopic model in different groups*. **a** Representative images of bladder tumors from four groups. **b** Summary data of bladder tumor number. **c** Summary data of bladder tumor volume. *This study was comprised of 24 female Wistar rats. After 10 weeks, these rats were divided into four groups (six rats each) according to the treatment administered intravesically: group 1, a control group given 0.1 ml phosphate-buffered saline (PBS); group 2, a treatment group given 1 mg/0.1 ml BCG; group 3, a CS/GP solution control group given 0.1 ml; group 4, a treatment group given 0.1 ml Fe_3O_4-BCG-CS/GP mixture (containing 1 mg BCG). All the groups received intravesical treatment weekly for six times (Reprinted from Ref. [112], Copyright 2013, with permission from Elsevier)

3 Chitosan-Based Gene Delivery for Cancer Chemotherapy

In recent decades, many studies have reported the potential of gene delivery for cancer chemotherapy by silencing a gene (siRNA or shRNA), replacing defective genes or substituting missing genes [15, 113]. However, gene delivery is confronted by many hurdles such as rapid degradation by nucleases, poor cellular uptake, non-specificity to the target cells, and low transfection efficiency [114]. For this purpose, viral systems have been widely used due to their efficient in vivo transfection, however, their undesirable side effects including possible toxicity, immunogenicity, and inflammatory effects have limited their clinical applications [15, 115]. In contrast, most nonviral vectors, such as chitosan, could be used as potential gene carriers for plasmid DNA (pDNA), oligonucleotides, and siRNA due to their biodegradability, biocompatibility, low toxicity and strong DNA binding ability. However, chitosan has low transfection efficiency and low cell specificity which has to be taken into consideration [114, 116]. Various factors which affect the transfection efficiency of chitosan are described in (Fig. 8). Further to overcome these limitations chitosan, has been modified by hydrophobic, hydrophilic, thermosensitive, pH-sensitive, and cell-specific ligand groups [114]. Chitosan has a potential for gene delivery due to its cationic property by which it can bind to negatively charged DNA/siRNA via an electrostatic interaction and leads to the spontaneous formation of nano-size complexes (polyplexes) in the aqueous milieu (Fig. 9) [15].

In a recent study Santosh et al. coupled primary amines of low molecular weight (LMW) chitosan with 2-acrylamido-2-methylpropane sulphonic acid

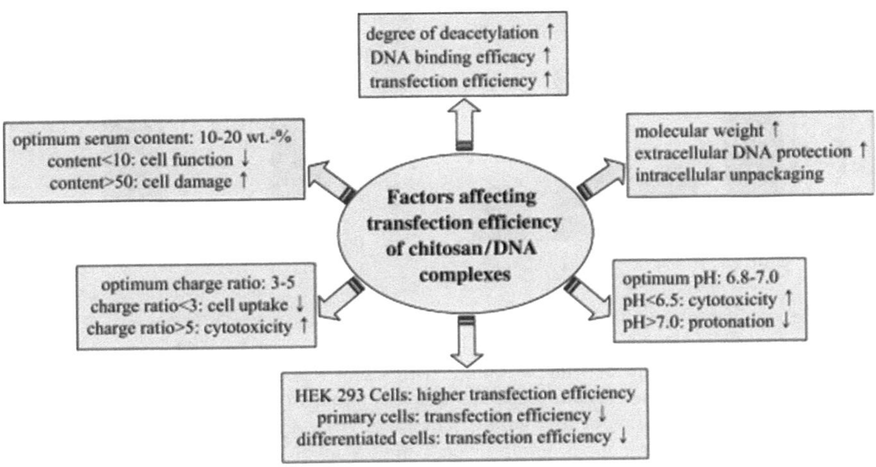

Fig. 8 Schematic illustration of the factors affecting transfection efficiency of chitosan/DNA complexes (Reprinted from Ref. [114], Copyright 2007, with permission from Elsevier)

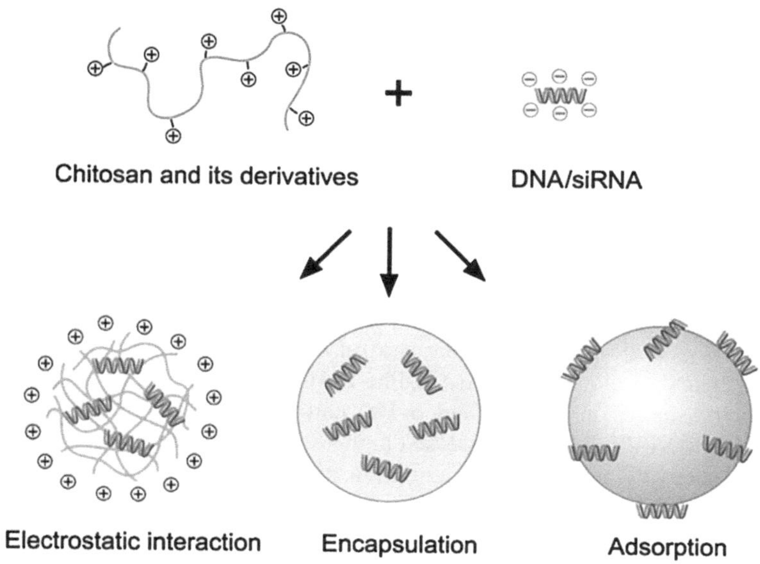

Fig. 9 Preparation of chitosan-based DNA/siRNA nanoparticles based on different mechanisms (Reprinted from Ref. [15], Copyright 2010, with permission from Elsevier)

(AMP) for its application in gene delivery [117]. Circular dichroism spectra (CD) and molecular docking studies showed AMP modified chitosan (CSAMP) has good interaction with DNA. In vitro cytotoxicity of DNA/CSAMP polyplexes showed 98 % cell viability in HepG2, HeLa and A549 mammalian carcinoma cell lines. Further, flow cytometric studies demonstrated transfection efficiency of CSAMP as 33–37 % compared to the 12–13 % transfection efficiency of native chitosan. In vivo biodistribution of luciferase reporter pDNA (pGL3)/CSAMP complexes in a Balb/c mouse model showed higher gene expression in spleen, lungs, liver, and kidney as comparison to native chitosan.

Ki et al. designed chitosan-based hybrid nanocomplex (GP-L-CT) of Survivin (SVN) siRNA for prostate cancer [118]. Survivin is overexpressed in the many cancer cells and plays a very important role in inhibiting the apoptosis process of tumor cells [119]. In vitro gene silencing efficiency in prostate cancer (PC-3) cells demonstrated reduction in SVN expression rates to about 21.9 % in case of GP-L-CT and the gene silencing efficacy was 79.2 %. In vivo antitumor efficacy of GP-L-CT nano-complexes in PC-3 tumor xenograft mouse model demonstrated naked siRNA did not have any antitumor efficacy while tumor volume of the SVN siRNA GP-L-CT group was significantly smaller than those of the control (treated with PBS).

4 Combination Therapy Using Chitosan-Based Nanomedicines

Combination therapy has been recommended for the treatment of cancer because it generates synergistic anticancer effects, reduces individual drug-related toxicity and suppresses drug resistance through distinct mechanisms of action [120, 121]. Chitosan-based NPs has also been prepared to deliver more than one chemotherapeutic agents with different mechanism of actions in a single formulation. Shen et al. combined doxorubicin and verapamil (an antiangiocardiopathy drug) in with chitosan shell coated on magnetic nanoparticles (MNPs), followed by entrapping into the poly (lactic acid-co-glycolic acid) nanoparticles conjugated with tumor-targeting ligand, cyclo(Arg-Gly-Asp-D-Phe-Lys) (cRGDfK) peptide [122]. In vitro cytotoxicity studies in HepG2 and S-180 cell lines demonstrated that cRGD-DOX/VER-MNP-PLGA NPs possessed higher cell cytotoxicity in order of 75 and 79 %, respectively at a DOX concentration of 15 μg/mL than free drugs due to cRGD-mediated targeting to cancer cells. Biodistribution and whole-mouse optical imaging studies demonstrated consistently preferential accumulation capability of the co-formulated NPs in the tumor tissue under magnetic guidance and DOX concentration was in order of tumor > liver > blood > kidney > heart > lung > spleen. Further, cardiac architecture of mice when treated with co-formulation showed regular cell distribution and normal myocardium histology without any damage.

A study by Anitha et al. [123] evaluated combinatorial anticancer effect of curcumin/5-fluorouracil loaded thiolated chitosan nanoparticles in separate formulations on colon cancer cells. In vitro cytotoxicity using MTT assay in HT29 cells, combination of both nanoparticles enhances the percentage of cell death (~86 %) after 48 h incubation. The mitochondrial membrane potential demonstrated highest number of apoptotic cells or dead mitochondria (~73 %) as compare to free curcumin and 5-fluorouracil or in combination in native form. Further, the in vivo pharmacokinetics demonstrated improved plasma concentrations of 5-FU and curcumin up to 72 h, unlike bare curcumin and 5-FU.

In a recent study Jia et al. developed both methotrexate (MTX) and mitomycin C (MMC) loaded PEGylated chitosan nanoparticles (PEG-CS-NPs) prepared by ionic gelation combined with a chemical cross-linking method [124]. In this study, methotrexate was used as a targeting agent due to its structural resemblance with folic acid. In vitro cytotoxicity on HeLa cells demonstrated (MTX+MMC)-PEG-CS-NPs induced a low cytotoxicity as compare to the free MTX+MMC at an equivalent MTX or MMC concentration which might be due to prolonged drug release from the multiple drugs-loaded NPs. However, in vitro cellular uptake was higher in case of FITC conjugated (MTX+MMC)-PEG-CS-NPs as compare to FITC conjugated PEG-CS-NPs, MMC-PEG-CSNPs, which demonstrated MTX modification improves the cellular uptake efficacy. In vivo fluorescent images demonstrated that Cy5.5 NHS (near-infrared fluorescent dye) conjugated (MTX+MMC)-PEG-CS-NPs clearly accumulated in the tumor, which was induced by inoculating H_{22} cells in the right axillary region of BALB/c nude mice. In case of Cy5.5 NHS conjugated (MTX+MMC)-PEG-CS-NPs, the fluorescent signal was highest at 1 h and maintained up to 6 h while less fluorescent signal were observed in case of Cy5.5 NHS conjugated MMCPEG-CS-NPs and PEG-CS-NPs. Similar results were also obtained in ex vivo images and there was decreased accumulation of the fluorescence in the major organs, on contrary increased accumulation was observed in the tumor, indicating the less RES uptake in case of Cy5.5 NHS conjugated (MTX+MMC)-PEG-CS-NPs (Fig. 10). The study concluded that targeted drug co-delivery systems might have potential clinical implications for combination cancer chemotherapy.

Besides delivery of two drugs in a same formulation, it has also been proposed that co-delivery of chemotherapeutic agent and gene carrier would be efficient in overcoming multidrug cancer resistance [125]. Wei et al. [126] used *N*-((2-hydroxy-3-trimethylammonium) propyl) chitosan chloride (HTCC) nanoparticles (HNPs) to deliver siRNA and paclitaxel (PTX). The co-delivery system (HNP/siRNA/PTX) at very low drug concentration (3 nmol/L of siRNA) significantly improved the in vitro cytotoxicity against lung carcinoma cells (LLC) and demonstrated no significant side effects. The co-delivery system was capable of improving the intestinal absorption, enhancing cellular uptake, and avoiding lysosomal entrapment. Further, this co-formulation achieved synergistic effects by simultaneously transporting siRNA and PTX into the same cancer cells and enhancing the intratumoral drug concentrations.

Recently, Deng et al. designed hyaluronic acid-chitosan nanoparticles for co-delivery of MiR-34a and doxorubicin by iontropic gelation technique [127]. It is reported that inclusion of hyaluronic acid enhances the colloidal stability and transfection efficiency of siRNA-loaded nanoparticles [128]. In vitro cytotoxicity in human breast cancer (MDA-MB-231) cells demonstrated that miR-34a/doxorubicin dual loaded nanoparticles decreased BCL-2 expression and resulted in greater cytotoxicity. The IC_{50} of the co-formulation (0.12 μM) was approximately 4.58 times lower than the IC_{50} of free DOX (0.55 μM). Flow cytometric analysis also revealed co-delivery of miR-34 with DOX enhances DOX mediated cell apoptosis, which in turn increases chemotherapeutic efficacy of DOX. Furthermore, co-delivered formulation also inhibited cell migration which may be potentially contributed by not only the intracellular delivery of miR-34a that down-regulated the expression of Notch-1, but also the increased killing effects of the co-delivery of DOX and miR-34a. In vivo studies in mice bearing MDA-MB-231 subcutaneous xenografts following intravenous co-delivery in nanoparticles demonstrated significant reduction in tumor volume as compare to free doxorubicin. This study revealed co-delivery of miRNAs and chemotherapeutic agents in nano-system has a great potential as combined therapeutic strategy in anticancer treatment.

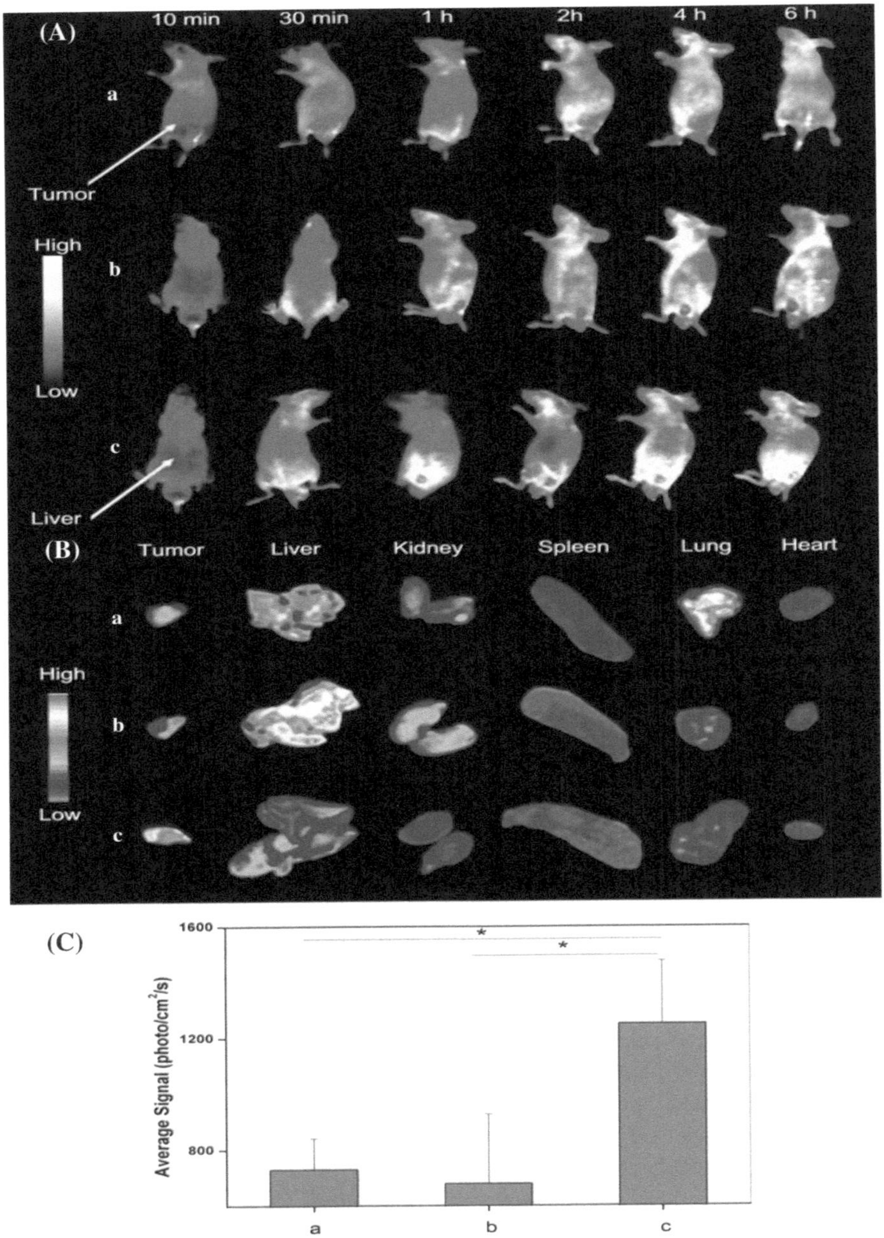
(A)
10 min
30 min
1 h
2h
4 h
6 h
a
b
c
Tumor
High
Low
Liver
(B)
Tumor
Liver
Kidney
Spleen
Lung
Heart
a
b
c
High
Low
(C)
1600
1200
800
Average Signal (photo/cm²/s)
*
*
a
b
c

◀ **Fig. 10** In vivo tumor targeted imaging of the (MTX+MMC)-PEGCS-NPs in the BALB/c mice bearing H22 tumor. (*a*) Cy5.5 NHS conjugated PEG-CS-NPs, (*b*) Cy5.5 NHS conjugated MMC-PEG-CSNPs, (*c*) Cy5.5 NHS conjugated (MTX+MMC)-PEG-CS-NPs. **A** In vivo fluorescence imaging of H22 tumor-bearing mice treated with different formulations at different time points. **B** Ex vivo fluorescence imaging of main organs and tumor excised from the tumor-bearing mice at 6 h post injection. **C** Quantitative tumor target characteristics of different formulations (mean ± SD, $n = 3$). * $P < 0.05$ (Reprinted from Ref. [124], Copyright 2014, with permission from Americal Chemical Society)

5 Conclusion

In cancer chemotherapy there must be a balance between potential benefit and side-effects of the treatment. Nanomedicines are being developed in order to shift this balance in favor of the benefits. As reviewed in this chapter, a great deal of research has been dedicated to design, synthesis and evaluation of chitosan-based nanomedicines in cancer chemotherapy. Though an exponential rise in chitosan-based nanomedicines have been witnesssed, questions arise around the absence of commercially available chitosan-based therapeutic formulations. One of the possible reasons may be different molecular weight and deacetylation degree of chitosan alters its biological characteristics and poses many difficulties to regulatory agencies which are required to perform rigorous evaluations. The co-formulation of chemotherapeutic agent with other pharmacological agent or siRNA using chitosan nanomedicines have shown promising results by reducing adverse effects related with the chemotherapeutic agents. The exponential improvements in chitosan-based nanomedicines indicated that in a near future, a chitosan-based commercial nano-formulation could be a reality.

References

1. Saneja A, Dubey RD, Alam N, Khare V, Gupta PN (2014) Co-formulation of P-glycoprotein substrate and inhibitor in nanocarriers: an emerging strategy for cancer chemotherapy. Curr Cancer Drug Targets 14(5):419–433
2. Saneja A, Khare V, Alam N, Dubey RD, Gupta PN (2014) Advances in P-glycoprotein-based approaches for delivering anticancer drugs: pharmacokinetic perspective and clinical relevance. Expert Opin Drug Deliv 11(1):121–138
3. Duncan R (2003) The dawning era of polymer therapeutics. Nat Rev Drug Discov 2(5):347–360
4. Park JH, Saravanakumar G, Kim K, Kwon IC (2010) Targeted delivery of low molecular drugs using chitosan and its derivatives. Adv Drug Deliv Rev 62(1):28–41
5. Alam N, Khare V, Dubey R, Saneja A, Kushwaha M, Singh G, Sharma N, Chandan B, Gupta PN (2014) Biodegradable polymeric system for cisplatin delivery: development, in vitro characterization and investigation of toxicity profile. Mater Sci Eng C Mater Biol Appl 38:85–93
6. Gupta PN, Jain S, Nehate C, Alam N, Khare V, Dubey RD, Saneja A, Kour S, Singh SK (2014) Development and evaluation of paclitaxel loaded PLGA:poloxamer blend nanoparticles for cancer chemotherapy. Int J Biol Macromol 69:393–399

7. Goodarzi N, Varshochian R, Kamalinia G, Atyabi F, Dinarvand R (2013) A review of polysaccharide cytotoxic drug conjugates for cancer therapy. Carbohydr Polym 92(2):1280–1293
8. Yang J, Han S, Zheng H, Dong H, Liu J (2015) Preparation and application of micro/nanoparticles based on natural polysaccharides. Carbohydr Polym 123:53–66
9. Gulbake A, Jain SK (2012) Chitosan: a potential polymer for colon-specific drug delivery system. Expert Opin Drug Deliv 9(6):713–729
10. Prabaharan M (2015) Chitosan-based nanoparticles for tumor-targeted drug delivery. Int J Biol Macromol 72:1313–1322
11. Pillai CKS, Paul W, Sharma CP (2009) Chitin and chitosan polymers: chemistry, solubility and fiber formation. Prog Polym Sci 34(7):641–678
12. Chenite A, Buschmann M, Wang D, Chaput C, Kandani N (2001) Rheological characterisation of thermogelling chitosan/glycerol-phosphate solutions. Carbohydr Polym 46(1):39–47
13. Thakur VK, Thakur MK (2014) Recent advances in graft copolymerization and applications of chitosan: a review. ACS Sustain Chem Eng 2(12):2637–2652
14. Upadhyaya L, Singh J, Agarwal V, Tewari RP (2014) The implications of recent advances in carboxymethyl chitosan based targeted drug delivery and tissue engineering applications. J Controlled Release 186:54–87
15. Mao S, Sun W, Kissel T (2010) Chitosan-based formulations for delivery of DNA and siRNA. Adv Drug Deliv Rev 62(1):12–27
16. Onishi H, Machida Y (1999) Biodegradation and distribution of water-soluble chitosan in mice. Biomaterials 20(2):175–182
17. Freier T, Koh HS, Kazazian K, Shoichet MS (2005) Controlling cell adhesion and degradation of chitosan films by N-acetylation. Biomaterials 26(29):5872–5878
18. Zhang H, Mardyani S, Chan WC, Kumacheva E (2006) Design of biocompatible chitosan microgels for targeted pH-mediated intracellular release of cancer therapeutics. Biomacromolecules 7(5):1568–1572
19. Rao SB, Sharma CP (1997) Use of chitosan as a biomaterial: studies on its safety and hemostatic potential. J Biomed Mater Res 34(1):21–28
20. Illum L (1998) Chitosan and its use as a pharmaceutical excipient. Pharm Res 15(9):1326–1331
21. Perioli L, Ambrogi V, Venezia L, Pagano C, Ricci M, Rossi C (2008) Chitosan and a modified chitosan as agents to improve performances of mucoadhesive vaginal gels. Colloids Surf B Biointerfaces 66(1):141–145
22. Baldrick P (2011) The safety of chitosan as a pharmaceutical excipient. Regul Toxicol Pharmacol 56(3):290–299
23. Zhang J, Xia W, Liu P, Cheng Q, Tahi T, Gu W, Li B (2010) Chitosan modification and pharmaceutical/biomedical applications. Marine drugs 8(7):1962–1987
24. Sadeghi AM, Dorkoosh FA, Avadi MR, Saadat P, Rafiee-Tehrani M, Junginger HE (2008) Preparation, characterization and antibacterial activities of chitosan, N-trimethyl chitosan (TMC) and N-diethylmethyl chitosan (DEMC) nanoparticles loaded with insulin using both the ionotropic gelation and polyelectrolyte complexation methods. Int J Pharm 355(1–2):299–306
25. Huang J, Jiang H, Qiu M, Geng X, Yang R, Li J, Zhang C (2013) Antibacterial activity evaluation of quaternary chitin against Escherichia coli and Staphylococcus aureus. Int J Biol Macromol 52:85–91
26. Hasegawa M, Yagi K, Iwakawa S, Hirai M (2001) Chitosan induces apoptosis via caspase-3 activation in bladder tumor cells. Jpn J Cancer Res 92(4):459–466
27. Kean T, Thanou M (2010) Biodegradation, biodistribution and toxicity of chitosan. Adv Drug Deliv Rev 62(1):3–11
28. Varum KM, Myhr MM, Hjerde RJ, Smidsrod O (1997) In vitro degradation rates of partially N-acetylated chitosans in human serum. Carbohydr Res 299(1–2):99–101
29. Garcia-Fuentes M, Alonso MJ (2012) Chitosan-based drug nanocarriers: where do we stand? J Controlled Release 161(2):496–504

30. Gorzelanny C, Poppelmann B, Pappelbaum K, Moerschbacher BM, Schneider SW (2010) Human macrophage activation triggered by chitotriosidase-mediated chitin and chitosan degradation. Biomaterials 31(33):8556–8563
31. Chandy T, Sharma CP (1990) Chitosan-as a biomaterial. Artif Cells Blood Substit Biotechnol 18(1):1–24
32. Boateng JS, Matthews KH, Stevens HN, Eccleston GM (2008) Wound healing dressings and drug delivery systems: a review. J Pharm Sci 97(8):2892–2923
33. Dash M, Chiellini F, Ottenbrite RM, Chiellini E (2011) Chitosan—a versatile semi-synthetic polymer in biomedical applications. Prog Polym Sci 36(8):981–1014
34. Kato Y, Onishi H, Machida Y (2004) N-succinyl-chitosan as a drug carrier: water-insoluble and water-soluble conjugates. Biomaterials 25(5):907–915
35. Thanou MM, Kotze AF, Scharringhausen T, Luessen HL, de Boer AG, Verhoef JC, Junginger HE (2000) Effect of degree of quaternization of N-trimethyl chitosan chloride for enhanced transport of hydrophilic compounds across intestinal caco-2 cell monolayers. J Controlled Release 64(1–3):15–25
36. Jia Z, shen D, Xu W (2001) Synthesis and antibacterial activities of quaternary ammonium salt of chitosan. Carbohydr Res 333(1):1–6
37. Xu T, Xin M, Li M, Huang H, Zhou S (2010) Synthesis, characteristic and antibacterial activity of N, N, N-trimethyl chitosan and its carboxymethyl derivatives. Carbohydr Polym 81(4):931–936
38. Prego C, Torres D, Fernandez-Megia E, Novoa-Carballal R, QuiÃ ± oÃ¡ E, Alonso MJ (2006) Chitosan-PEG nanocapsules as new carriers for oral peptide delivery: effect of chitosan pegylation degree. J Controlled Release 111(3):299–308
39. Bernkop-Schnurch A, Hornof M, Guggi D (2004) Thiolated chitosans. Eur J Pharm Biopharm 57(1):9–17
40. Leitner VM, Walker GF, Bernkop-Schnurch A (2003) Thiolated polymers: evidence for the formation of disulphide bonds with mucus glycoproteins. Eur J Pharm Biopharm 56(2):207–214
41. Roldo M, Hornof M, Caliceti P, Bernkop-Schnurch A (2004) Mucoadhesive thiolated chitosans as platforms for oral controlled drug delivery: synthesis and in vitro evaluation. Eur J Pharm Biopharm 57(1):115–121
42. Sashiwa H, Shigemasa Y (1999) Chemical modification of chitin and chitosan 2: preparation and water soluble property of N-acylated or N-alkylated partially deacetylated chitins. Carbohydr Polym 39(2):127–138
43. Alves NM, Mano JF (2008) Chitosan derivatives obtained by chemical modifications for biomedical and environmental applications. Int J Biol Macromol 43(5):401–414
44. Sashiwa H, S-i Aiba (2004) Chemically modified chitin and chitosan as biomaterials. Prog Polym Sci 29(9):887–908
45. Ji J, Wang L, Yu H, Chen Y, Zhao Y, Zhang H, Amer WA, Sun Y, Huang L, Saleem M (2014) Chemical modifications of chitosan and its applications. Polym Plast Technol Eng 53(14):1494–1505
46. Khawar IA, Kim JH, Kuh HJ (2015) Improving drug delivery to solid tumors: priming the tumor microenvironment. J Control Release 201C:78–89
47. Wicki A, Witzigmann D, Balasubramanian V, Huwyler J (2015) Nanomedicine in cancer therapy: challenges, opportunities, and clinical applications. J Controlled Release 200:138–157
48. Thanki K, Gangwal RP, Sangamwar AT, Jain S (2013) Oral delivery of anticancer drugs: challenges and opportunities. J Controlled Release 170(1):15–40
49. Chen MC, Mi FL, Liao ZX, Hsiao CW, Sonaje K, Chung MF, Hsu LW, Sung HW (2013) Recent advances in chitosan-based nanoparticles for oral delivery of macromolecules. Adv Drug Deliv Rev 65(6):865–879
50. Yeh TH, Hsu LW, Tseng MT, Lee PL, Sonjae K, Ho YC, Sung HW (2011) Mechanism and consequence of chitosan-mediated reversible epithelial tight junction opening. Biomaterials 32(26):6164–6173

51. Matsumura Y, Maeda H (1986) A new concept for macromolecular therapeutics in cancer chemotherapy: mechanism of tumoritropic accumulation of proteins and the antitumor agent smancs. Cancer Res 46(12 Pt 1):6387–6392
52. Yang Y, Wang S, Wang Y, Wang X, Wang Q, Chen M (2014) Advances in self-assembled chitosan nanomaterials for drug delivery. Biotechnol Adv 32(7):1301–1316
53. Kanapathipillai M, Brock A, Ingber DE (2014) Nanoparticle targeting of anti-cancer drugs that alter intracellular signaling or influence the tumor microenvironment. Adv Drug Deliv Rev 79–80:107–118
54. Khare V, Alam N, Saneja A, Dubey RD, Gupta PN (2014) Targeted drug delivery systems for pancreatic cancer. J Biomed Nanotechnol 10(12):3462–3482
55. Grenha A (2012) Chitosan nanoparticles: a survey of preparation methods. J Drug Target 20(4):291–300
56. Sarvaiya J, Agrawal YK (2015) Chitosan as a suitable nanocarrier material for anti-Alzheimer drug delivery. Int J Biol Macromol 72:454–465
57. Mitra A, Dey B (2011) Chitosan microspheres in novel drug delivery systems. Indian J Pharm Sci 73(4):355–366
58. Huang WT, Larsson M, Wang YR, Chiou SH, Lin HY, Liu DM (2015) Demethoxycurcumin-carrying chitosan-antibody core-shell nanoparticles with multi-therapeutic efficacy toward malignant A549 lung tumor-from in-vitro characterization to in-vivo evaluation. Mol Pharm 12(4):1242–1249
59. Yu B, Li H, Zhang J, Zheng W, Chen T (2015) Rational design and fabrication of a cancer-targeted chitosan nanocarrier to enhance selective cellular uptake and anticancer efficacy of selenocystine. J Mater Chem B 3(12):2497–2504
60. Liang J, Cao L, Zhang L, Wan X-C (2014) Preparation, characterization, and in vitro antitumor activity of folate conjugated chitosan coated EGCG nanoparticles. Food Sci Biotechnol 23(2):569–575
61. Nie X, Zhang J, Xu Q, Liu X, Li Y, Wu Y, Chen C (2014) Targeting peptide iRGD-conjugated amphiphilic chitosan-co-PLA/DPPE drug delivery system for enhanced tumor therapy. J. Mater Chem B 2(21):3232–3242
62. Snima KS, Jayakumar R, Lakshmanan VK (2014) In vitro and in vivo biological evaluation of O-carboxymethyl chitosan encapsulated metformin nanoparticles for pancreatic cancer therapy. Pharm Res 31(12):3361–3370
63. Koo H, Min KH, Lee SC, Park JH, Park K, Jeong SY, Choi K, Kwon IC, Kim K (2013) Enhanced drug-loading and therapeutic efficacy of hydrotropic oligomer-conjugated glycol chitosan nanoparticles for tumor-targeted paclitaxel delivery. J Controlled Release 172(3):823–831
64. Jiang L, Li X, Liu L, Zhang Q (2013) Thiolated chitosan-modified PLA-PCL-TPGS nanoparticles for oral chemotherapy of lung cancer. Nanoscale Res Lett 8(1):1–11
65. Maya S, Sarmento B, Lakshmanan VK, Menon D, Seabra V, Jayakumar R (2014) Chitosan cross-linked docetaxel loaded EGF receptor targeted nanoparticles for lung cancer cells. Int J Biol Macromol 69:532–541
66. Tian X, Yin H, Zhang S, Luo Y, Xu K, Ma P, Sui C, Meng F, Liu Y, Jiang Y, Fang J (2014) Bufalin loaded biotinylated chitosan nanoparticles: an efficient drug delivery system for targeted chemotherapy against breast carcinoma. Eur J Pharm Biopharm 87(3):445–453
67. Sayari E, Dinarvand M, Amini M, Azhdarzadeh M, Mollarazi E, Ghasemi Z, Atyabi F (2014) MUC1 aptamer conjugated to chitosan nanoparticles, an efficient targeted carrier designed for anticancer SN38 delivery. Int J Pharm 473(1–2):304–315
68. Huang S, Wan Y, Wang Z, Wu J (2013) Folate-conjugated chitosan-polylactide nanoparticles for enhanced intracellular uptake of anticancer drug. J Nanopart Res 15(12):1–15
69. Vivek R, Thangam R, Nipunbabu V, Ponraj T, Kannan S (2014) Oxaliplatin-chitosan nanoparticles induced intrinsic apoptotic signaling pathway: a "smart" drug delivery system to breast cancer cell therapy. Int J Biol Macromol 65:289–297
70. Song H, Su C, Cui W, Zhu B, Liu L, Chen Z, Zhao L (2013) Folic acid-chitosan conjugated nanoparticles for improving tumor-targeted drug delivery. Biomed Res Int 2013:723158

71. Luo F, Li Y, Jia M, Cui F, Wu H, Yu F, Lin J, Yang X, Hou Z, Zhang Q (2014) Validation of a Janus role of methotrexate-based PEGylated chitosan nanoparticles in vitro. Nanoscale Res Lett 9(1):1–13
72. Le TMP, Pham VP, Dang TML, La TH, Le TH, Le QH (2013) Preparation of curcumin-loaded pluronic F127/chitosan nanoparticles for cancer therapy. Adv Nat Sci Nanosci Nanotechnol 4(2):025001
73. Peng Z, Li Z, Zhang F, Peng X (2014) In-vitro degradation and cytotoxicity of gelatin/chitosan microspheres for drug controlled release. J Macromol Sci Part A 51(8):646–652
74. Zhao X, Liu P (2015) pH-sensitive fluorescent hepatocyte-targeting multilayer polyelectrolyte hollow microspheres as a smart drug delivery system. Mol Pharm 11(5):1599–1610
75. Park JM, Lee SY, Lee GH, Chung EY, Chang KM, Kwak BK, Kuh HJ, Lee J (2012) Design and characterisation of doxorubicin-releasing chitosan microspheres for anti-cancer chemoembolisation. J Microencapsul 29(7):695–705
76. Pang X, Du HL, Zhang HQ, Zhai YJ, Zhai GX (2013) Polymer-drug conjugates: present state of play and future perspectives. Drug Discovery Today 18(23–24):1316–1322
77. Haag R, Kratz F (2006) Polymer therapeutics: concepts and applications. Angew Chem Int Ed Engl 45(8):1198–1215
78. Maeda H (2010) Tumor-selective delivery of macromolecular drugs via the EPR effect: background and future prospects. Bioconjug Chem 21(5):797–802
79. Li D, Lu B, Zhang H, Huang Z, Xu P, Zheng H, Yin Y, Xu H, Liu X, Lou Y (2014) Synthesis and in vitro evaluation of methotrexate conjugated O, N-carboxymethyl chitosan via peptidyl spacers. J Nanopart Res 16(9):1–12
80. Kim H, Lee E, Lee I-H, Lee J, Kim J, Kim S, Lee Y, Kim D, Choi M, Kim Y-C (2014) Preparation and therapeutic evaluation of paclitaxel-conjugated low-molecular-weight chitosan nanoparticles. Macromol Res 22(8):805–808
81. Mathiyalagan R, Subramaniyam S, Kim YJ, Kim YC, Yang DC (2014) Ginsenoside compound K-bearing glycol chitosan conjugates: synthesis, physicochemical characterization, and in vitro biological studies. Carbohydr Polym 112:359–366
82. Li M, Xu X, Lu F, Guo S (2014) Primary in vitro and in vivo evaluation of norcantharidin-chitosan/poly (vinyl alcohol) for cancer treatment. Drug Deliv 21(4):293–301
83. Liu F, Feng L, Zhang L, Zhang X, Zhang N (2013) Synthesis, characterization and antitumor evaluation of CMCS-DTX conjugates as novel delivery platform for docetaxel. Int J Pharm 451(1–2):41–49
84. Gong XY, Yin YH, Huang ZJ, Lu B, Xu PH, Zheng H, Xiong FL, Xu HX, Xiong X, Gu XB (2012) Preparation, characterization and in vitro release study of a glutathione-dependent polymeric prodrug Cis-3-(9H-purin-6-ylthio)-acrylic acid-graft-carboxymethyl chitosan. Int J Pharm 436(1–2):240–247
85. Son YJ, Jang JS, Cho YW, Chung H, Park RW, Kwon IC, Kim IS, Park JY, Seo SB, Park CR, Jeong SY (2003) Biodistribution and anti-tumor efficacy of doxorubicin loaded glycol-chitosan nanoaggregates by EPR effect. J Controlled Release 91(1–2):135–145
86. Sosnik A, Menaker RM (2015) Polymeric micelles in mucosal drug delivery: challenges towards clinical translation. Biotechnol Adv, in press
87. Talelli M, Barz M, Rijcken CJF, Kiessling F, Hennink WE, Lammers T (2015) Core-crosslinked polymeric micelles: principles, preparation, biomedical applications and clinical translation. Nano Today
88. Deng C, Jiang Y, Cheng R, Meng F, Zhong Z (2012) Biodegradable polymeric micelles for targeted and controlled anticancer drug delivery: promises, progress and prospects. Nano Today 7(5):467–480
89. Owen SC, Chan DPY, Shoichet MS (2012) Polymeric micelle stability. Nano Today 7(1):53–65
90. Aliabadi HM, Lavasanifar A (2006) Polymeric micelles for drug delivery. Expert Opin Drug Deliv 3(1):139–162

91. Torchilin VP (2004) Targeted polymeric micelles for delivery of poorly soluble drugs. Cell Mol Life Sci 61(19–20):2549–2559
92. Zhang J, Chen XG, Li YY, Liu CS (2007) Self-assembled nanoparticles based on hydrophobically modified chitosan as carriers for doxorubicin. Nanomedicine 3(4):258–265
93. Hu FQ, Wu XL, Du YZ, You J, Yuan H (2008) Cellular uptake and cytotoxicity of shell crosslinked stearic acid-grafted chitosan oligosaccharide micelles encapsulating doxorubicin. Eur J Pharm Biopharm 69(1):117–125
94. Xie YT, Du YZ, Yuan H, Hu FQ (2012) Brain-targeting study of stearic acid-grafted chitosan micelle drug-delivery system. Int J Nanomedicine 7:3235–3244
95. You J, Hu FQ, Du YZ, Yuan H (2007) Polymeric micelles with glycolipid-like structure and multiple hydrophobic domains for mediating molecular target delivery of paclitaxel. Biomacromolecules 8(8):2450–2456
96. Su Y, Hu Y, Du Y, Huang X, He J, You J, Yuan H, Hu F (2015) Redox-responsive polymer-drug conjugates based on doxorubicin and chitosan oligosaccharide-g-stearic acid for cancer therapy. Mol Pharm 12:1193–1202
97. Yang Y, Yuan SX, Zhao LH, Wang C, Ni JS, Wang ZG, Lin C, Wu MC, Zhou WP (2015) Ligand-directed stearic acid grafted chitosan micelles to increase therapeutic efficacy in hepatic cancer. Mol Pharm 12(2):644–652
98. Wang F, Chen Y, Zhang D, Zhang Q, Zheng D, Hao L, Liu Y, Duan C, Jia L, Liu G (2012) Folate-mediated targeted and intracellular delivery of paclitaxel using a novel deoxycholic acid-O-carboxymethylated chitosan-folic acid micelles. Int J Nanomedicine 7:325–337
99. Zhao Z, He M, Yin L, Bao J, Shi L, Wang B, Tang C, Yin C (2009) Biodegradable nanoparticles based on linoleic acid and poly(beta-malic acid) double grafted chitosan derivatives as carriers of anticancer drugs. Biomacromolecules 10(3):565–572
100. Yu J, Xie X, Zheng M, Yu L, Zhang L, Zhao J, Jiang D, Che X (2012) Fabrication and characterization of nuclear localization signal-conjugated glycol chitosan micelles for improving the nuclear delivery of doxorubicin. Int J Nanomed 7:5079
101. Chen Q, Sun Y, Wang J, Yan G, Cui Z, Yin H, Wei H (2013) Preparation and characterization of glycyrrhetinic acid-modified stearic acid-grafted chitosan micelles. Artif Cells Nanomed Biotechnol 1–7
102. Guerry A, Cottaz S, Fleury E, Bernard J, Halila S (2014) Redox-stimuli responsive micelles from DOX-encapsulating polycaprolactone-g-chitosan oligosaccharide. Carbohydr Polym 112:746–752
103. Jin X, Mo R, Ding Y, Zheng W, Zhang C (2014) Paclitaxel-loaded N-octyl-O-sulfate chitosan micelles for superior cancer therapeutic efficacy and overcoming drug resistance. Mol Pharm 11(1):145–157
104. Bhattarai N, Gunn J, Zhang M (2010) Chitosan-based hydrogels for controlled, localized drug delivery. Adv Drug Deliv Rev 62(1):83–99
105. Ladet S, David L, Domard A (2008) Multi-membrane hydrogels. Nature 452(7183):76–79
106. Shukla SK, Mishra AK, Arotiba OA, Mamba BB (2013) Chitosan-based nanomaterials: a state-of-the-art review. Int J Biol Macromol 59:46–58
107. Tian R, Chen J, Niu R (2014) The development of low-molecular weight hydrogels for applications in cancer therapy. Nanoscale 6(7):3474–3482
108. Li C, Ren S, Dai Y, Tian F, Wang X, Zhou S, Deng S, Liu Q, Zhao J, Chen X (2014) Efficacy, pharmacokinetics, and biodistribution of thermosensitive chitosan/beta-glycerophosphate hydrogel loaded with docetaxel. AAPS Pharm Sci Technol 15(2):417–424
109. Alexander A, Ajazuddin Khan J, Saraf S (2014) Formulation and evaluation of chitosan-based long-acting injectable hydrogel for PEGylated melphalan conjugate. J Pharm Pharmacol 66(9):1240–1250
110. Taleb MFA, Alkahtani A, Mohamed SK (2015) Radiation synthesis and characterization of sodium alginate/chitosan/hydroxyapatite nanocomposite hydrogels: a drug delivery system for liver cancer. Polym Bull 72:1–18

111. Ju C, Sun J, Zi P, Jin X, Zhang C (2013) Thermosensitive micelles-hydrogel hybrid system based on poloxamer 407 for localized delivery of paclitaxel. J Pharm Sci 102(8):2707–2717
112. Zhang D, Sun P, Li P, Xue A, Zhang X, Zhang H, Jin X (2013) A magnetic chitosan hydrogel for sustained and prolonged delivery of Bacillus Calmette-Guerin in the treatment of bladder cancer. Biomaterials 34(38):10258–10266
113. Verma IM, Somia N (1997) Gene therapy—promises, problems and prospects. Nature 389(6648):239–242
114. Kim T-H, Jiang H-L, Jere D, Park I-K, Cho M-H, Nah J-W, Choi Y-J, Akaike T, Cho C-S (2007) Chemical modification of chitosan as a gene carrier in vitro and in vivo. Prog Polym Sci 32(7):726–753
115. Lim MJ, Min SH, Lee JJ, Kim IC, Kim JT, Lee DC, Kim NS, Jeong S, Kim MN, Kim KD, Lim JS, Han SB, Kim HM, Heo DS, Yeom YI (2006) Targeted therapy of DNA tumor virus-associated cancers using virus-activated transcription factors. Mol Ther 13(5):899–909
116. Lee MK, Chun SK, Choi WJ, Kim JK, Choi SH, Kim A, Oungbho K, Park JS, Ahn WS, Kim CK (2005) The use of chitosan as a condensing agent to enhance emulsion-mediated gene transfer. Biomaterials 26(14):2147–2156
117. Kumar S, Garg P, Pandey S, Kumari M, Hoon S, Choung P-H, Kapavarapu R, Sobral A, Chung JH, Jang K-J (2015) Enhanced chitosan-DNA interaction by 2-Acrylamido-2-methylpropane coupling for efficient transfection in cancer cells. J Mater Chem B (in-press)
118. Ki MH, Kim JE, Lee YN, Noh SM, An SW, Cho HJ, Kim DD (2014) Chitosan-based hybrid nanocomplex for siRNA delivery and its application for cancer therapy. Pharm Res 31(12):3323–3334
119. Vegran F, Boidot R, Bonnetain F, Cadouot M, Chevrier S, Lizard-Nacol S (2011) Apoptosis gene signature of Survivin and its splice variant expression in breast carcinoma. Endocr Relat Cancer 18(6):783–792
120. Hu CM, Aryal S, Zhang L (2010) Nanoparticle-assisted combination therapies for effective cancer treatment. Ther Deliv 1(2):323–334
121. Parhi P, Mohanty C, Sahoo SK (2012) Nanotechnology-based combinational drug delivery: an emerging approach for cancer therapy. Drug Discov Today 17(17–18):1044–1052
122. Shen J-M, Gao F-Y, Yin T, Zhang H-X, Ma M, Yang Y-J, Yue F (2013) cRGD-functionalized polymeric magnetic nanoparticles as a dual-drug delivery system for safe targeted cancer therapy. Pharmacol Res 70(1):102–115
123. Anitha A, Deepa N, Chennazhi KP, Lakshmanan VK, Jayakumar R (2014) Combinatorial anticancer effects of curcumin and 5-fluorouracil loaded thiolated chitosan nanoparticles towards colon cancer treatment. Biochim Biophys Acta 1840(9):2730–2743
124. Jia M, Li Y, Yang X, Huang Y, Wu H, Huang Y, Lin J, Li Y, Hou Z, Zhang Q (2014) Development of both methotrexate and mitomycin C loaded PEGylated chitosan nanoparticles for targeted drug codelivery and synergistic anticancer effect. ACS Appl Mater Interfaces 6(14):11413–11423
125. Taratula O, Garbuzenko OB, Chen AM, Minko T (2011) Innovative strategy for treatment of lung cancer: targeted nanotechnology-based inhalation co-delivery of anticancer drugs and siRNA. J Drug Target 19(10):900–914
126. Wei W, Lv PP, Chen XM, Yue ZG, Fu Q, Liu SY, Yue H, Ma GH (2013) Codelivery of mTERT siRNA and paclitaxel by chitosan-based nanoparticles promoted synergistic tumor suppression. Biomaterials 34(15):3912–3923
127. Deng X, Cao M, Zhang J, Hu K, Yin Z, Zhou Z, Xiao X, Yang Y, Sheng W, Wu Y, Zeng Y (2014) Hyaluronic acid-chitosan nanoparticles for co-delivery of MiR-34a and doxorubicin in therapy against triple negative breast cancer. Biomaterials 35(14):4333–4344
128. Ragelle H, Riva R, Vandermeulen G, Naeye B, Pourcelle V, Le Duff CS, D'Haese C, Nysten B, Braeckmans K, De Smedt SC, Jerome C, Preat V (2014) Chitosan nanoparticles for siRNA delivery: optimizing formulation to increase stability and efficiency. J Controlled Release 176:54–63

Chitosan: A Promising Substrate for Regenerative Medicine in Drug Formulation

Madhu Kashyap, D. Archana, Alok Semwal, Joydeep Dutta and Pradip Kumar Dutta

Abstract Chitosan plays a most important role in the regenerative medication for wound healing. The adhesive nature of chitosan, with their antifungal and bactericidal character, and their permeability to oxygen, is a very important property associated with the treatment of wounds. Different derivatives and combination of chitosan have been reported for this purpose in the form of hydrogels, fibers, membranes, scaffolds and sponges. The purpose of the chapter is to have a closer look in the work done directly by different researchers on the chitosan formulation with potential medicinal applications to provide a better understanding of its usability in regenerative medicine.

Keywords Chitosan · Regenerative medicine · Wound healing · Tissue engineering · Drug delivery

M. Kashyap
Department of Chemistry, Nehru Gram Bharti University,
Allahabad 211002, U.P, India

D. Archana · P.K. Dutta
Department of Chemistry, MN National Institute of Technology,
Allahabad 211004, U.P, India

A. Semwal
Department of Pharmacy, Himachal Institute of Pharmacy,
Paonta Sahib 173025, H.P, India

J. Dutta
Department of Chemistry, Amity School of Applied Sciences, Amity University Haryana,
Pachgaon, Near IMT, Manesar 122413, Haryana, India

P.K. Dutta (✉)
Member, Centre for Medical Diagnostic & Research, MN National Institute of Technology, Allahabad 211004, U.P, India
e-mail: pkd_437@yahoo.com

P.K. Dutta (ed.), *Chitin and Chitosan for Regenerative Medicine*, Springer Series on Polymer and Composite Materials, DOI 10.1007/978-81-322-2511-9_10

Abbreviations

PEC	Polyelectrolyte complex
PVA	Poly(vinyl alcohol)
AgNPs	Silver nanoparticles
NPs	Nanoparticles
MCNs	Magnetic-Chitosan Nanogels
LPO	Lactoperoxidase
E. coli	Escherichia coli
P. aeruginosa	Pseudomonas aeruginosa
Cu	Copper

1 Introduction

Regenerative medicine is a "process for replacing or regenerating cells, tissues or organs, to restore or establish normal function". Regenerative medicine is generally connects the fields of tissue engineering, stem cell research, gene therapy and therapeutic cloning [1].

Although many polymers have been investigated for regenerative medicine, chitosan drawn more attention towards this field. Chitosan is the principle derivative of chitin and it is obtained from crustacean shells and also from fungi [2]. Chitosan is polycationic in nature and is composed of glucosamine and *N*-acetylglucosamine units linked with each other by β-1,4 linkage.

Chitosan possesses some special characters such as biodegradability, biocompatibility, hemostatic and antimicrobial activity [3–6]. Another important feature is that it can be easily fabricated into gel, film, microspheres, scaffolds and fibers [7–9].

To enhance biological properties such as hemostatic and wound healing of chitosan, it can be modified by incorporating antibiotics, anti-inflammatories, antiseptics, antibodies and other bioactives. The authors' research laboratory has established milestone in the field of functionalization of chitosan by chemicals, nanoparticles, quantum dots and biomolecule [4, 10–13]. In this book chapter we have discussed the importance of chitosan based drug formulation in wound healing, tissue engineering and drug delivery towards regenerative medicine.

2 Chitosan with Drug Formulations in Haemostatic and Wound Healing

Opening or breaking of skin results as wound. Proper healing with minimum time is important for repairing the damaged skin. The wound dressing plays important role by providing protection to wounded skin against microbes and dust particles.

Wound healing process through three overlapped phases such as inflammation, proliferation and maturation. Chitosan helps the process of heling by playing active role in every phase of wound healing.

The chitosan controls hemorrhage by forming clot. Chitosan elevates the agglutination of erythrocytes and bounds with it to create artificial clot [14]. Chitosan induces platelet adhesion, aggregation and activates intrinsic blood coagulation [15]. It simulates the migration of mononuclear and polymorph nuclear (PMN) cells. These cells accelerates the process of the reepithelialization and regeneration of normal skin [16]. Chitosan is known to have the ability to deliver drugs. Chitosan has the capability to encapsulate, stabilize, and deliver drugs that enhance connective tissue matrix growth (i.e., granulation tissue formation) during the proliferative phase of healing [17].

Chitosan will gradually depolymerize to release *N*-acetyl-β-D-glucosamine, which initiates fibroblast proliferation, helps in ordered collagen deposition and stimulates increased level of natural hyaluronic acid synthesis at the wound site. It helps in faster wound healing and scar prevention [18]. Chitosan possesses antibacterial activity against positive as well as negative bacteria which prevents delay in healing and scar formation. It provides moist environment and helps in absorbing exudates from wound. The controlled drug delivery to wound site from dressings plays important role because it provides constant delivery for long term so that dressing changing frequency reduced.

Chitosan and alginate bilayer composite membrane containing ciprofloxacin hydrochloride was developed. They observed that the membrane plays important role as reservoir because the amount of drug added was directly proportional to antibacterial activity. In drug release study they observed that (i) increasing crosslinking time slows drug release and (ii) lower concentration of calcium chloride prevents erosion of alginate from film and delays kinetics of drug release. So they selected 10 % calcium chloride for further studies. While New Zealand rabbits were used to study primary skin irritation test, in vivo pharmacodynamics efficacy and histopathological studies. The drug loaded composite membrane was compared with saline and Vaseline gauze. The healing was more in drug loaded composite membrane compared to other groups. Newly formed skin show bilayer epithelium tissue and no rotten flesh outside the skin [19].

Sung et al. [20] were prepared chitosan hydrogel films by using Poly (vinyl alcohol) (PVA) and chitosan loaded with minocycline by freeze thawing method for wound healing applications. Results indicate that amount of chitosan affects crosslinking strength of hydrogel fraction and also flexibility of prepared gel. On increasing chitosan concentration, gel fraction and cross linking ability decreases. It also decreases thermal stability of PVA hydrogel, increases swelling ability up to 400 %, water vapour transmission rate and porosity of hydrogel. Data analysis reveals that loading of minocycline does not affect physical property of hydrogel. Minocycline loaded PVA hydrogel shows faster rate of healing in rat dorsum than the control (sterile gauze) in the wound healing test. Hydrogel loaded with drugs were found to be more favourable for reepithelization micro vessel formation, collagen tissue and reconstruction of skin tissue than conventional product.

A PEC made up of chitosan and sodium alginate prepared for wound dressing applications. First of all, they prepared freeze-dried chitosan sponge and kept in alginate solution to form PEC complex and freeze dried and then silver sulfadiazine and dehydroepiandrosterone. The efficacy of prepared dressing was compared by an in vivo study on mice with vaseline gauze, chitosan sponge, silver sulfadiazine-impregnated PEC, and silver sulfadiazine and dehydroepiandrosterone-impregnated PEC wound dressing and results showed that complete wound contraction in silver sulfadiazine-impregnated PEC, and silver sulfadiazine and dehydroepiandrosterone-impregnated PEC wound dressing whereas, irregularly thickened epidermis observed in vaseline gauze treated wounds and infiltration of leukocytes and proliferation of fibroblasts are more in the upper dermis. Well organised epidermis, and the thickness is thinner in silver sulfadiazine-impregnated PEC than vaseline treated groups. In silver sulfadiazine and dehydroepiandrosterone-impregnated PEC wound dressing almost normal thickness of epidermis was observed. Kim et al. [21] shows that dehydroepiandrosterone can prevent the systemic change of the immune function after burn, as well as progressive destruction of the thermally injured skin tissues which makes silver sulfadiazine and dehydroepiandrosterone-impregnated PEC wound dressing as plays effective role in wound healing than other treatment groups.

Composite films prepared with chitosan, sago starch, silver nanoparticles and gentamycin. To see the effect of gentamycin some films were without it. They found that addition of 2 mL of ethylene glycol as plasticizer increased tensile strength. Male Albino Wister rats were used to compare activity between cotton gauze dipped with gentamicin (control), chitosan-sago-silver nanoparticles and chitosan-sago-silver nanoparticles with gentamicin. Complete healing time for control, chitosan-sago-silver nanoparticles and chitosan-sago-silver nanoparticles with gentamicin were 24, 18 and 16 days respectively. Same proved by histopathological study. Increased tensile strength thus increase in collagen matrix in chitosan-sago-silver nanoparticles with gentamicin treated wounds reflects faster healing. They concluded that this prepared dressing may be used as wound healing material [22].

Chitosan and dipotassium hydrogen orthophosphate were used to prepare hydrogel. Prepared hydrogels are non-toxic against human SaOS-2 osteoblast-like cell line. The release of FITC-dextran from hydrogel showed controlled release which quickly reached plateau so they concluded that it could be used for drug delivery applications [23].

Thiolated chitosan with poly(*N*-isopropyl acrylamide) loaded with ciprofloxacin formulations were prepared. They found that scratching of dressing from wound is easy without causing any pain or discomfort. The thermo responsive material was cytocompatible. The drug release was in controlled and continuous so it protects wound for a long time. They exhibited antibacterial activity against *E. coli* [24].

Chitosan film prepared by using drug Streptomycin and diclofenac with Polyox®, hydroxypropylmethylcellulose, carrageenan, sodium alginate and chitosan. These dressings showed good antibacterial activity, swelling, mechanical properties with controlled drug release. They concluded that streptomycin acts as anti-infectional and diclofenac helps by reliving the swelling and pain in inflammatory stage [25].

In a very recent study chitosan based derivatives 5-methyl pyrrolidinone chitosan, *N*-carboxymethyl chitosan, and *N*-succinyl chitosan, for the sustained delivery of neurotensin (NT) were prepared with the aim of wound healing applications. The order of the hydrophilicity changes is SC > 5-methyl pyrrolidinone chitosan > *N*-carboxymethyl chitosan and reduced glutathione release per mpolymer (g) is found the order of *N*-carboxymethyl chitosan (32.33 $\pm$ 0.72 %) < 5-methyl pyrrolidinone chitosan (67.65 $\pm$ 6.77 %) < *N*-succinyl chitosan (287.18 $\pm$ 14.92 %). C57BL/6 mice were used and neurotensin alone, 5-methyl pyrrolidinone chitosan foam alone and neurotensin-loaded 5-methyl pyrrolidinone chitosan foam in both control and diabetic mice. neurotensin—loaded 5-methyl pyrrolidinone chitosan treatment was more effective than 5-methyl pyrrolidinone chitosan alone treated groups. neurotensin-loaded 5-methyl pyrrolidinone chitosan can help healing in three phases by reducing the inflammatory response, higher collagen formation and stimulating re-epithelialization [26].

Recently, the authors' laboratory is involved on chitosan based formulation for wound healing by using three solvents, namely, glacial acetic acid, glycerol and propylene glycol [27]. The current study enumerates the positive characteristics of chitosan hydrogels for wound healing. The rheological properties were determined by advance rheometer. The results showed that viscosity decreases with increasing shear rate indicating that gel is formed. Elastic modulus has been found to be higher than loss modulus shows the formation of gel has taken. The results also confirmed that in certain range shear stress does not depend upon shear rate. The study shows that chitosan hydrogels are suitable in terms of gel strength for wound healing applications. Further study of chitosan hydrogels formulation for wound healing applications is under progress.

3 Chitosan with Drug Formulations in Tissue Engineering Applications

Tissue engineering is an interdisciplinary field between engineering and the life sciences and its main objective is to create therapies either by repairing or replacing the tissues and organs to restore their original functions which resulted due to congenital defects, disease, trauma and aging [28]. It is evident in Veda's that peoples were using different types of dressing materials from ancient time but from past few decades the field of biomaterials has evolved greatly [29, 30]. An ideal scaffolding material should be biocompatible, porous with inter connected pores, desired mechanical strength, degradation rate as per required function and ability to support cells in three dimensional way [31, 32]. The process of regeneration involves implantation of cells onto the designed scaffolding materials for regenerating tissue and subsequent deposition of extracellular matrix. These artificial scaffolds plays very important role because they act as an artificial ECM and provide a temporary environment for cells to infiltrate, adhere, proliferate and differentiate [33–35]. They provide an initial structural support and retain cells in the defective area for cell growth, metabolism and matrix production, and hence, their role during the development of engineered tissues is of prime importance [36].

Chitosan is a versatile biopolymer and its derivatives offer a wide range of biomedical applications due to their biodegradability, biocompatibility and nontoxicity. They can also provide controlled release of growth factors and extracellular matrix components. To improve the adherent ability for seeding cells, the chitosan allow for a wide range of molecules to be modified. The incorporation of collagen to chitosan as a chitosan-collagen scaffold will enhance its cell attachment ability [37]. Recent studies further show that chitosan has good characteristics for the attachment, proliferation and viability of mesenchymal stem cells [38, 39]. Chitosan and their derivatives are mostly used for biomedical applications. Chitosan coated microparticles have many advantages such as improvement of drug payloads, bioadhesive property and prolonged drug release properties over the uncoated particles [40]. With these promising features, they are considered as a very interesting biomaterial for use in cell transplantation and tissue regeneration. Tissue engineering has been used to create various tissue analogs including skin, cartilage, bone, liver, and nerve in the past decades.

The unique characteristics of chitosan like acts as replacement for missing or damaged tissue [41] and supporting cell attachment and proliferation will make suitable for tissue engineering applications [42]. Chitosan has favourable conditions like low immuno-genic activity, controlled biodegradability and porous structure for tissue engineering applications.

Tı and lı et al. in his study found high deacetylation degrees (>85 %) showed higher mechanical strength. From cell adhesion studies, cell attachment favoured high in deacetylation degrees [43]. Insulin is essentially the only effective drug for the treatment of diabetes as well as known as a key adipogenic factor for adipose regeneration. Tan et al. prepared hydrogel with succinyl-chitosan, aldehyde hyaluronic acid and insulin for adipose tissue regeneration. They found that controlled release of insulin from swelled hydrogel and suitable for adipose tissue engineering applications [44].

4 Chitosan with Drug Formulations in Drug Delivery Applications

Biodegradable and biocompatible are the prerequisite condition for polymers for used in drug delivery applications. So generally natural polymers such as polysaccharides, polypeptides, or phospholipids used as building blocks for the formulations. The drug delivery systems and relevant release materials is gaining attention of many researchers because they could enhance the effectivity of some available medicines in market. The delivery of drugs takes few days to years. Because of controlled release hydrogels frequently used for drug delivery applications. Hydrogels are polymeric networks with three-dimensional structure with capacity to absorb high amounts of water or biological fluids and this is due to hydrophilic groups such as –OH, –CONH–, $-CONH_2-$, and $-SO_3H$ in polymers forming hydrogel structures [45, 46].

Drug rizatriptan benzoate loaded chitosan microparticles were prepared. The dissolution of drug was considerably slow (approximately less than 40 % of the drug dissolved in 3 h) where as it is 90–100 % of released drug was achieved in less than 2 h. Chitosan decreases mucocilliary clearance residence time of drug in nasal cavity [47].

Ampicillin trihydrate-loaded chitosan nanoparticles were prepared. The drug loaded nanoparticles showed higher antibacterial activity than unloaded nanoparticles [48]. A chitosan based dressing consist of two layers. First layer is carboxymethyl-chitin hydrogel and second layer consist of chitosan acetate foam and chlorherxidine gluconate. The prepared dressing shown antibacterial activity against *Pseudomonas aeruginosa* and Staphlycoccus aureus which were main cause of infection in burn wounds [49].

Chitosan was mixed with dexamethasone for dexamethasone delivery. Dexamethasone loaded chitosan films were potential sustained-release of drug and release time was longer than convensional ocular topical delivery dosage form. They concluded that it could be used as ocular delivery carrier [50].

Remuñán-López et al. [51] proposed two types of microparticulate systems using chitosan as a carrier for site-specific oral delivery. One of the delivery systems included amoxycillin-loaded chitosan microspheres for targeted delivery to the gastric cavity. The authors proposed that chitosan microspheres adhered to the stomach mucosa to release the drug for the treatment of gastric ulcer caused by Helicobacter pylori. The other delivery system was based on pH sensitive microparticles for intestinal delivery.

Chitosan and silver nanoparticles were used for the preparation of hydrogel beads. Ibuprophen was incorporated to test the drug delivery activity. They found that silver nanoparticles prolonged the release of drug from chitosan beads. The drug release time from hydrogel beads prolonged because of longer path for drugs to migrate from nanocomposite bead to the media. So they concluded that it can be used in controlled delivery of drugs [52].

Chitosan and hyaluronic acid to used for preparation of mucoadhesive nanocomplexes. Heparin was loaded to this complex. High drug loading efficiency (approximately 70 %) and high stability in PBS showed by nanomaterial. From ex vivo experiments they concluded that it can be used for effective pulmonary delivery systems [53].

To chitosan nanoparticles, cyclosporine A was added. These loaded nanoparticles showed high encapsulation and loading efficacy efficiency of 73 and 9 % facilitated the transport of cyclosporin A to the inner eye, and significantly increased the bioavailability [54]. The oral bioavailability of paclitaxel loaded in *N*-octyl-O-sulfate chitosan micelles was improved sixfold in comparison with that

Table 1 Chitosan based formulations for controlled drug delivery

Composition of matrix	Drugs used	Positive points	Reference
Chitosan micro/ nanoparticles	Ampicillin	High antibacterial activity, initially bust after that constant release of drug	[57]
Chitosan film	Ibuprofen	Deliver drug across the rabbit buccal mucosa in controlled manner, oral mucosal drug delivery	[58]
Poly(methyl methacrylate)-chitosan microspheres	Ampicillin	Drug release and bio activity high in Poly(methyl methacrylate)-chitosan microspheres compared to virgin chitosan microspheres, oral drug delivery	[59]
Nanoparticulate hydroxyapatite/chitosan composites	Fluorescein	Excellent cell spreading and proliferation of osteoblastic MC3T3-E1 cells, controlled release of drug	[60]
Chitosan micro and nanoparticles	Vancomycin	At pH 2.0 partial release (resembling the release in the stomach), at pH 7.4 approx. tenfold increased drug release, colon-targeted delivery	[61]
Chitosan, poly lactic acid, polyethylene glycol, gelatin nanoparticles	Rifampicin	Potential carrier for controlled delivery of drug	[62]
N-gly-cyrrhetinic acid-polyethylene glycol (PEG)-chitosan, quaternaryammonium-chitosan	Brucine	Inhibit the growth of tumor, mitochondria targeting anticancer drug delivery	[63]
Chitosan, sodium starch gly, pectin, lactose monohydrate (tablet)	Theophylline	PEC (1.1 %) and coat-ing (3 %) showed highest swelling, 98.58 ± 1.92 % in presence of rat caecal content, buccal drug delivery	[64]
Chitosan coated magnetic iron oxide nanoparticles	Doxorubicin	Targeted drug delivery to tumor cells	[65]
Chitosan-egg albumin-PEG nanoparticles	Alprazolam	Sustained drug release over a period of 24 h	[66]
Chitosan/o-carboxymethyl chitosan nanoparticles	Doxorubicin hydrochloride	Oral delivery system, prolong retention of drug in liver, spleen, lung and decrease level of drug in heart and kidney	[67]

(continued)

Table 1 (continued)

Composition of matrix	Drugs used	Positive points	Reference
Chitosan-Alginate	verapamil hydrochloride	Drug to polymer ratio 1:1 and cross linking agent 4 % showed highest swelling ratio, drug release and in vivo bioavailability, oral drug delivery	[68]
Chitosan/PEG blended PLGA nanoparticle	paclitaxel	Superior anti-proliferative effect and cell cycle inhibition, tumor drug delivery	[69]
Chitosan, graphene oxide	FLuorescein sodium	controlled transdermal drug delivery	[70]

of free paclitaxel, which was resulted from the interference of P-gp ATPase rather than down-regulation of P-gp expression by *N*-octyl-O-sulfate chitosan [55].

Jain et al. [56] revealed that glutamate chitosan can enhance the transport of insulin across the nasal mucosa of sheep and rats. Only slight structural changes were observed in rat nasal membrane treated with 0.5 % (w/v) chitosan glutamate for 60 min. They concluded that chitosan microspheres can be used as carrier system for vaccine delivery. In fact, chitosan microspheres have been studied as a promising carrier system for mucosal vaccination, especially via the oral and nasal route to induce enhanced immune responses. Some of the chitosan based formulations for the release of drugs were discussed in Table 1.

5 Chitosan as Drug Targeting Applications

5.1 Chitosan Magnetic Nanoparticles

Magnetic NPs consisting of magnetite (Fe_3O_4) possess unique characteristics that make them promising agents for antibacterial applications [71] because the Food and Drug Administration approved that superparamagnetic iron oxide nanoparticles are biocompatible with human body [72]. It is interesting to mention that only magnetite particles with size of less than 30 nm have a large surface area and exhibits superparamagnetic properties that make them prone to magnetic fields and they do not become permanently magnetized without an external magnetic field to support them.

In a study conducted by Shrifian-Esfahni et al. [73] Superparamagnetic iron oxide nanoparticles/Citosan biocompatible nanocomposites were formulated by dissolving $FeCl_2{\cdot}4H_2O$ and $FeCl_3{\cdot}6H_2O$ in acidic solution of chitosan by using N_2 and optimum pH of 6.9 was maintained by using sodium hydroxide solution. Slow rate addition is an important and critical factor, leading to a uniform and black stable aqueous colloidal suspension. The solid forms of nanoparticles were obtained centrifuging the black stable aqueous colloidal suspension.

The antibacterial activity of magnetic nanoparticles was tested against two pathogenic strains of gram-negative bacteria *P. aeruginosa* (ATCC 27853) and *E. coli* (ATCC 25922) and ciprofloxacin was used as a standard drug. NPs showed enhancement in antibacterial activity due to the greater stability and arrest of bacterial growth. Chitosan interacts very easily with bacterium and binds to DNA, glycosaminoglycans and most of the proteins enhancing the antimicrobial effect of NPs. The minimum inhibitory concentration of NPs was found to be much lower than reported value.

5.2 Chitosan-Hybrid Hydrogel

Critical observation and applications of Cox-2 and NSAIDs has influenced the need for alternative non-medicinal products for osteo-arthritis. Cross-linked Hyaluronate has been approved as medical device for several applications in orthopedics such as visco-supplement for knee OA and others synovial joints pain. Zeitoun et al. [74] successfully formulated chitosan and hyaluronate hydrogels for innovative biocompatible implants and visco-supplements in painful orthopedic pathologie. Hydrogels were formulated and various studies were conducted to evaluate their stability and usefulness.

The study concluded that while no effective method for supplementing the intervertebral disc gel does exists, chitosan and Hyaluronate hydrogel should act as partial replenishment of disc gel with minimum or no damage to the disc structure. Formulated hydrogels restored the mechanical properties of the disc in a biomechanical model of animal spine and allowed restoration in part of the disc height lost during aging after injection (tested in animal spine).

5.3 Copper-Chitosan Nanoparticles

Chitosan is naturally occurring bio-polymer having strong affinity towards variety of transition metal ions. In various research studies chitosan is complexed with Cu, Fe, La, Mo, Ni and Zr etc. The prepared complexes can be used for the remediation and preconcentration of heavy metals in the wastewater [75, 76].

In a research conducted by Manikandan et al. [77] chitosan-metal ions (Cu ions) nanoparticles has been synthesized in order to generate an green (non-hazardous) alternative for antimicrobials currently in used. Chitosan can form co-ordinate bonding with variety of metal ions. When Cu salts dissolve in acidified chitosan solution, Cu ion binds to the polymer chains via amino groups. The reduction of these ions takes place further, coupled with the oxidation of the hydroxyl groups [78, 79].

For the determination of antimicrobial activity gram negative (*E. coli*, *Salmonella paratyphi*) and gram positive bacterial (*Bacillus*) stains were used. Antimicrobial activity of nanoparticles was compare with standard drugs ampicillin and streptomycin and it was found that chitosan stabilized Cu-NPs exhibited inhibitory activity towards gram negative, positive bacteria. In fact, one of the reasons for the antimicrobial character of chitosan it's positively charged amino group which interacts with negatively charged microbial cell membranes, leading to the leakage of proteinaceous and other intracellular constituents of the microorganisms [80, 81]. Furthermore greater surface area to volume ratio of chitosan-Cu nanoparticles allows NPs to be effective in very small amounts.

5.4 PH-Triggered Magnetic-Chitosan Nanogels (MCNs) for Doxorubicin Delivery

Magnetic nanoparticles (MNPs) are a rapidly evolving materials contributing to their multifunctional properties including small size, superparamagnetism and low toxicity [82]. Particularly in the pharmaceutical field, they have now attracted a lot of interests owing to their intrinsic magnetic properties for targeted drug delivery [83]. Doxorubicin is among the most widely used anticancer drugs in chemotherapy treatment; however, it suffers from various side effects. Sadighian et al. [84] formulated (Sodium tripolyphosphate) TPP cross linked magnetic chitosan nanogels (TPP/MCNs) and Glutaraldehyde cross-linked magnetic chitosan-nanogels (GA/MCNs) which were then followed by doxorubicin loading. Magnetized nanogels were formulated by using Ferric chloride hexahydrate ($FeCl_3{\cdot}6H_2O$) and ferrous chloride tetrahydrate ($FeCl_2{\cdot}4H_2O$). The drug release behavior of MCNs was studied in physiological pH of 7.4. To evaluate the pH dependency of drug release from MCNs, the release study was also conducted in acidic media with pH value of 5.3.

The finding revealed that both MCNs provide a sustained release pattern. It was found that the maximum drug release attainable for TPP/MCNs was higher than the one for GA/MCNs, probably due to the high stability of chemically cross linking compared to the physically cross linking method. The release study of doxorubicin also revealed that the extent of drug release at pH = 5.3 was promisingly more than drug release at pH = 7.4 for both MCNs. Generally, it can be concluded that type of drug cross linking provide an opportunity to tune drug release pattern of MCNs according to the specific drug delivery purpose [85].

5.5 Immobilization of Lactoperoxidase by Tragacanth-Chitosan Nano-Biopolymers

Lactoperoxidase (LPO) is a glycoprotein enzyme with a wide range of antimicrobial activity. The only problem with LPO is its stabilization. Gum tragacanth is a biopolymer, which is used for encapsulation and chitosan is used as a matrix for protein immobilization. In a study conducted by Nayeri et al. [86]

tragacanth-chitosan-LPO nanoparticles were formulated. The existence of negatively charged carboxyl and hydroxyl groups on tragacanth [87] and positively charged amino groups on chitosan [88] facilitated the ionic interaction between tragacanth and chitosan.

Considering the cationic properties of chitosan, LPO encapsulated in chitosan molecule and the COO– groups on tragacanth and NH3 + on chitosan electrostatically bond together, to produce the tragacanth-chitosan-LPO complexes. The results of the study showed that formulated tragacanth-chitosan nanocomposites are more stable at higher temperature.

5.6 Starch-Chitosan Nanocomposites for Control Drug Release of Curcumin

Since the discovery of chitosan and its derivatives, pharmaceutical field have witnessed a revolutionary development of novel drug delivery systems. Carrier materials such as chitosan have been employed to protect the pharmaceutical agent so that it may be released under optimal absorption conditions, or to adjust the timing of release of different pharmaceutical agents administered simultaneously [88].

Mohanty et al. [89] prepared novel nanocomposites of Starch, which is one of the most abundant polysaccharides polymer, which blended with the chitosan and Cloisite 30B. Starch/chitosan/MMT blend films were prepared by varying the concentration of Cloisite 30B. Cloisite 30B is a montmorillonite modified with a quaternary ammonium salt. Cloisite 30B was incorporated in the formulation as a matrix material component which also plays the role of a co-emulsifier in the nanocomposite preparation. Furthermore the kinetics of the drug release was investigated and it was found that drug was released in a controlled manner. The drug release studies indicated that the release was much more pronounced in the basic medium than the acidic medium.

6 Conclusion

Chitosan is a unique polymer and can be fabricated into gel, membranes, nanofibers, beads, microparticles, nanoparticles, scaffolds and sponges. It is used in regenerative medicine due their structural similarities with native extracellular matrix, chemical versatility and typically good biological performance such as better biocompatibility and low immunogenicity. But due to its weak mechanical modulus, it is used as blends, polyelectrolyte complexes or hybrids with different biodegradable synthetic/biopolymers. Continuous supply of growth factor is one of the most challenging task because seeded cells need slow and continuous release to proliferate and differentiate into identical tissue. It is also important to give attention that during regeneration process in vivo, specific drugs are needed to avoid infections. To overcome this problem, growth factors/drugs were encapsulated into biodegradable polymeric materials and prepared in different shape and size such as micro/nano beads, nano capsules, micelles, micro/nano fibers, etc. But at the same time it is also important to note down that the structural integrity of scaffolds should be retained. Moreover, it was realised that the incorporated growth factors/drugs were abundantly present at the surface of the scaffold, hence cells will likely prefer to stay on the surface without migrating into the bulk which represents that modification or optimization of different parameters of regenerative medicine is yet to be solved.

Acknowledgments One of the authors (Madhu Kashyap) gratefully acknowledged the spontaneous support and cooperation from Garm Bharti University during her research work under Ph.D.programme.

References

1. Shi C, Zhu Y, Ran X, Wang M, Su Y, Cheng T (2006) Therapeutic potential of chitosan and its derivatives in regenerative medicine. J Surg Res 133:185–192
2. Dutta PK, Dutta J, Tripathi VS (2004) Chitin chitosan: chem properties & applications. J Sci Indu Res 63:20–31
3. Jayakumar R, Menon D, Manzoor K, Nair SV, Tamura H (2010) Biomedical applications of chitin and chitosan based nanomaterials—A short review. Carbohydr Polym 82(2010):227–232
4. Singh J, Dutta PK (2009) Preparation, circular dichroism induced helical conformation and optical property of chitosan acid salt complexes for biomedical applications. Int J Biol Macromol 45:384–392
5. Mehta AS, Singh BK, Singh N, Archana D, Snigdha K, Harniman R, Rahatekar SS, Tewari RP, Dutta PK (2015) Chitosan silk-based three-dimensional scaffolds containing gentamicin encapsulated calcium alginate beads for drug administration and blood compatibility. J Biomater Appl 29:1314–1325
6. Singh BK, Archana D, Sirohi R, Jain A, Dutta PK (2015) Porous chitosan scaffolds: A systematic study for choice of crosslinker and growth factor incorporation. Int J Polymer Mater 64:242–252
7. Archana D, Upadhyay L, Tewari RP, Dutta J, Huang YB, Dutta PK (2013) Chitosan-pectin-alginate as a novel scaffold for tissue engineering applications. Indian J Biotechnol 12:475–4828
8. Muzzarelli RAA (2009) Chitins and chitosans for the repair of wounded skin, nerve, cartilage and bone. Carbohydr Polym 76:167–182
9. Muzzarelli RAA, Morganti P, Morganti G, Palombo P, Palombo M, Biagini G, Belmontec MM, Giantomassic F, Orlandif F, Muzzarelli C (2007) Chitin nanofibrils/chitosan glycolate composites as wound medicaments. Carbohydr Polym 70:274–284
10. Tripathi S, Mehrotra GK, Dutta PK (2009) Preparation and physicochemical evaluation of chitosan/poly(vinyl alcohol)/pectin ternary film for food-packaging applications. Carbohyd Polym 79:711–716
11. Singh J, Dutta PK (2010) Preparation, antibacterial & physicochemical behavior of chitosan/ofloxacin complexes. Int J Polym Mater 59:793–807
12. Archana D, Dutta J, Dutta PK (2013) Evaluation of chitosan nano dressing for wound healing: Characterization, *in vitro* and in vivo studies. Int J Biol Macromol 57:193–203
13. Archana D, Singh BK, Dutta J, Dutta PK (2013) In vivo evaluation of chitosan–PVP–titanium dioxide nanocomposite as wound dressing material. Carbohydr Polym 95:530–539

14. Arand AG, Sawaya R (1986) Intraoperative Chem hemost Neurosurg. Neurosurg 18:223–233
15. Wang XH, Li DP, Wang WJ, Feng QL, Cui FZ, Xu YX, Song XH, van der Werf M (2003) Crosslinked collagen/chitosan matrices for artificial livers. Biomaterials 24:3213–3220
16. Kojima K, Okamoto Y, Miyatake K, Kitamura Y, Minami S (1989) Collagen typing of granulation tissue induced by chitin and chitosan. Carbohydr Polym 37:109–113
17. Harkins AL, Duri S, Kloth LC, Tran CD (2014) Chitosan-cellulose composite for wound dressing material. Part Antimicrobial activity, blood absorption ability, and biocompatibility: Chitosan-cellulose composite for wound dressing material. J Biomed Mater Res B Appl Biomater 102:1199–1206
18. Paul W, Sharma CP (2004) Chitin and alginates wound dressings: a short review. Trends Biomater Artif Organs 18:18–23
19. Han F, Dong Y, Song A, Yin R, Li S (2014) Alginate/chitosan based bi-layer composite membrane as potential sustained-release wound dressing containing ciprofloxacin hydrochloride. Appl Surf Sci 311:626–634
20. Sung JH, Hwang MR, Kim JO, Lee JH, Kim YI, Kim JH, Chang SW, Jin SG, Kim JA, Lyoo WS, Han SS, Ku SK, Yong CS, Choi HG (2010) Gel characterisation and *in vivo* evaluation of minocycline-loaded wound dressing with enhanced wound healing using polyvinyl alcohol and chitosan. Int J Pharm 39:2232–2240
21. Kim HJ, Lee HC, Oh JS, Shin BA, Oh CS, Park RD, Yang KS, Cho CS (1999) Polyelectrolyte complex composed of chitosan and sodium alginate for wound dressing application. J Biomater Sci Polym Ed 10:543–556
22. Alemdaroğlu C, Değim Z, Celebi N, Zor F, Oztürk S, Erdogan D (2006) An investigation on burn wound healing in rats with chitosan gel formulation containing epidermal growth factor. Burns 32:319–327
23. Arockianathan PM, Sekar S, Kumaran B, Sastry TP (2012) Preparation, characterization and evaluation of biocomposite films containing chitosan and sago starch impregnated with silver nanoparticles. Int J Biol Macromol 50:939–946
24. Radhakumarya C, Antontyb M et al (2011) Drug loaded thermoresponsive and cytocompatible chitosan basedhydrogel as a potential wound dressing. Carbohydr Polym 83:705–713
25. Pawar HV, Tetteh J, Boatang JS (2013) Preparation, optimisation and characterisation of novel wound healing film dressings loaded with streptomycin and diclofenac. Colloid Surf B Biointerface 102:102–110
26. Moura LIF, Dias AMA, Leal EC, Carvalho L, De Sousa HC, Carvalho E (2014) Chitosan-based dressings loaded with neurotensin an efficient strategy to improve early diabetic wound healing. Acta Biomater 10:843–857
27. Kashyap M, Dutta J, Dutta PK (2011) Rheological characteristics of chitosan hydrogels in different solvent systems for wound healing applications. Asian Chitin J 7:15–20
28. Chien KR (2008) Regenerative medicine and human models of human disease. Nature 453:302–305
29. Khademhosseini A, Vacanti JP, Langer R (2009) Progress in tissue engineering. Sci Am 300:64–71
30. Place ES, Evans ND, Stevens MM (2009) Complexity in biomaterials for tissue engineering. Nat Mater 8:457–470
31. Senni K, Pereira J, Gueniche F, Delbarre-Ladrat C, Sinquin C, Ratiskol J, Godeau G, Fischer AM, Helley D, Colliec-Jouault S (2011) Marine polysaccharides: a source of bioactive molecules for cell therapy and tissue engineering Mar. Drugs 9:1664–1681
32. Wu J, Tan H, Li L, Gao C (2009) Covalently immobilized gelatin gradients within three-dimensional porous scaffolds. Chin Sci Bull 54:3174–3180
33. Lee KY, Mooney DJ (2012) Alginate: properties and biomedical applications. Prog Polym Sci 37:106–126
34. Lee KY, Yuk SH (2007) Polymeric protein delivery systems. Progr Polym Sci 32:669–697
35. Pawar SN, Edgar KJ (2012) Alginate derivatization: a review of chemistry, properties and applications. Biomaterials 33:3279–3305

36. Tan H, Gong Y, Lao L, Mao Z, Gao C (2007) Gelatin/chitosan/hyaluronan ternary complex scaffold containing basic fibroblast growth factor for cartilage tissue engineering. J Mater Sci Mater Med 18:1961–1968
37. Cuy JL, Beckstead BL, Brown CD, Hoffman AS, Giachelli CM (2003) Adhesive protein interactions with chitosan: Consequences for valve endothelial cell growth on tissue-engineering materials. J Biomed Mater Res A 67:538–547
38. Dang JM, Sun DD, Shin-Ya Y, Sieber AN, Kostuik JP, Leong KW (2006) Temperature-responsive hydroxybutyl chitosan for the culture of mesenchymal stem cells and intervertebral disk cells. Biomaterials 27:406–418
39. Cho JH, Kim SH, Park KD, Jung MC, Yang WI, Han SW, Noh JY, Lee JW (2004) Chondrogenic differentiation of human mesenchymal stem cells using a thermo sensitive poly(N-isopropylacrylamide) and water-soluble chitosan copolymer. Biomaterials 25:5743–5751
40. Agnihotri SA, Mallikarjuna NN, Aminabhavi TM (2004) Recent advances on chitosan-based micro- and nanoparticles in drug delivery. J Control Release 100:5–28
41. Dutta P, Rinki K, Dutta J (2011) Chitosan: a promising biomaterial for tissue engineering scaffolds. Chitosan for biomaterials II. In: Jayakumar R, Prabaharan M, Muzzarelli RAA (eds) Advances in polymer science. Heidelberg: Springer Berlin, p 45–79
42. Jagur-Grodzinski J (2003) Biomedical applications of polymers 2001–2002. e-Polym Paper No 12
43. Ma Z, Lim LY (2003) Uptake of chitosan and associated insulin in Caco-2 cell monolayers: a comparison between chitosan molecules and chitosan nanoparticles. Pharm Res 20:1812–1819
44. Tan H, Rubin JP, Marra KG (2010) Injectable in situ forming biodegradable chitosan-hyaluronic acid based hydrogels for adipose tissue regeneration. Organogenesis 6:173–180
45. Brannon-Peppas L (1990) Preparation and characterization of crosslinked hydrophilic networks. In: Brannon-Peppas L, Harland RS (eds) Absorbent Polymer Technology. Elsevier, Amsterdam, pp 45–66
46. Peppas NA, Khare AR (1993) Preparation, structure and diffusional behavior of hydrogels in controlled release. Adv Drug Del Rev 11:1–35
47. Jyotsna CD, Rajendra DC (2010) Formulation and evaluation of chitosan based microparticulate nasal drug delivery system of rizatriptan benzoate. Int J Pharm Tech Res 2:2391–2402
48. Saha P, Goyal AK, Rath G (2010) Formulation and evaluation of chitosan-based ampicillin trihydrate nanoparticles. Trop J Pharm Res 9:483–488
49. Loke WK, Lau SK, Yong LL, Khor E, Sum CK (2000) Wound dressing with sustained antimicrobial capability. J Biomed Mater Res 53:8–17
50. Rodrigues LB, Leite HF, Yoshida MI, Saliba JB, Cunha AS Jr, Faraco AA (2009) In vitro release and characterization of chitosan films as dexamethasone carrier. Int J Pharm 368:1–6
51. Remuñán-Lopez C, Bodmeier RJ (1997) Mechanical water uptake and permeability properties of crosslinked chitosan glutamate and alginate films. Control J Control Release 44:215–225
52. Yadollahia M, Farhoudiana S, Namazi H (2015) One-pot synthesis of antibacterial chitosan/silver bio-nanocomposite hydrogel beads as drug delivery systems. Int J Biol Macromol. doi:10.1016/j.ijbiomac.2015.04.032
53. Oyarzun-Ampuero F, Brea J, Loza M, Torres D, Alonso M (2009) Chitosan–hyaluronic acid nanoparticles loaded with heparin for the treatment of asthma. Int J Pharm 381:122–129
54. De Campos AM, Sanchez A, Alonso MJ (2001) Chitosan nanoparticles: a new vehicle for the improvement of the delivery of drugs to the ocular surface. application to cyclosporin a. Int J Pharm 224:159–168
55. Mo R, Jin X, Li N, Ju C, Sun M, Zhang C, Ping Q (2011) The mechanism of enhancement on oral absorption of paclitaxel by N-octyl-O-sulfate chitosan micelles. Biomaterials 32:4609–4620

56. Jain SK, Jain NK, Gupta Y, Jain A, Jain D, Chaurasia M (2007) Mucoadhesive chitosan microspheres for non-invasive and improved nasal delivery of insulin. Indian J Pharm Sci 69:498–504
57. Arya N, Chakraborty S, Dube N, Katti DS (2009) Electrospraying: a facile technique for synthesis of chitosan-based micro/nanospheres for drug delivery applications. J Biomed Mater Res Part B: Appl Biomater 88B:17–31
58. Tang C, Guan YX, Yao SJ, Zhu ZQ (2014) Preparation of ibuprofen-loaded chitosan films for oral mucosal drug delivery using supercritical solution impregnation. Int J Pharm 473:434–441
59. Changerath R, Nair PD, Mathew S, Nair CP (2009) Poly(methyl methacrylate) grafted chitosan microspheres for controlled release of ampicillin. J Biomed Mater Res B Appl Biomater 89:65–76
60. Uskoković V, Desai TA (2014) In vitro analysis of nanoparticulate hydroxyapatite/chitosan composites as potential drug delivery platforms for the sustained release of antibiotics in the treatment of osteomyelitis. J Pharm Sci 103:567–579
61. Cerchiara T, Abruzzo A, di Cagno M, Bigucci F, Bauer-Brandl A, Parolin C, Vitali B, Gallucci MC, Luppi B (2015) Chitosan based micro- and nanoparticles for colon-targeted delivery of vancomycin prepared by alternative processing methods. Eur J Pharm Biopharm 92:112–119
62. Rajan M, Raj V (2013) Formation and characterization of chitosan-polylacticacid-polyethylene glycol-gelatin nanoparticles: a novel biosystem for controlled drug delivery. Carbohydr Polym 98:951–958
63. Chen Z, Zhang L, Song Y, He J, Wu L, Zhao C, Xiao Y, Li W, Cai B, Cheng H, Li W (2015) Hierarchical targeted hepatocyte mitochondrial multifunctional chitosan nanoparticles for anticancer drug delivery. Biomaterials 52:240–250
64. Pandey S, Mishra A, Raval P, Patel H, Gupta A, Shah D (2013) Chitosan-pectin polyelectrolyte complex as a carrier for colon targeted drug delivery. J Young Pharm. 5:160–166
65. Unsoy G, Khodadust R, Yalcin S, Mutlu P, Gunduz U (2014) Synthesis of Doxorubicin loaded magnetic chitosan nanoparticles for pH responsive targeted drug delivery. Eur J Pharm Sci 62:243–250
66. Jana S, Maji N, Nayak AK, Sen KK, Basu SK (2013) Development of chitosan-based nanoparticles through inter-polymeric complexation for oral drug delivery. Carbohydr Polym 98:870–876
67. Feng C, Wang Z, Jiang C, Kong M, Zhou X, Li Y, Cheng X, Chen X (2013) Chitosan/o-carboxymethyl chitosan nanoparticles for efficient and safe oral anticancer drug delivery: in vitro and in vivo evaluation. Int J Pharm 457:158–167
68. Zhang XZ, Tian FJ, Hou YM, Ou ZH (2015) Preparation and in vitro in vivo characterization of polyelectrolyte alginate–chitosan complex based microspheres loaded with verapamil hydrochloride for improved oral drug delivery. J Incl Phenom Macrocycl Chem 81:429–440
69. Parveen S, Long Sahoo SK (2011) circulating chitosan/PEG blended PLGA nanoparticle for tumor drug delivery. Eur J Pharmacol 670:372–383
70. Justin R (2014) Chen B Strong and conductive chitosan-Reduced graphene oxide nanocomposites for transdermal drug delivery. J Mater Chem B 2:3759–3770
71. Niemirowicz K, Markiewicz KH, Wilczewska AZ, Car H (2012) Magnetic nanoparticles as new diagnostic tools in medicine. Adv Med Sci 57:196–207
72. Lattuada M, Hatton TA (2007) Functionalization of monodisperse magnetic nanoparticles. Langmuir 23:2158–2168
73. Shrifian-Esfahni A, Salehi MT, Nasr-esfahni M, Ekramian E (2015) Chitosan-modified superparamgnetic iron oxide nanoparticles: design, fabrication, characterization and antibacterial activity. CHEMIK 69:19–32
74. Zeitoun A, Robinson D, Nevo Z, Ben-Shalom N, Patchornik S (2010) Chitosan-Hybrid hydrogel technology for innovative biocompatible implants and visco-supplements in painful orthopedic pathologies, Founded 2008, at OHV Incubator, Israel

75. Shinde RN, Pandey AK, Acharya R, Guin R, Das SK, Rajurkarb NS, Pujari PK (2013) Chitosan-transition metal ions complexes for selective arsenic(V) preconcentration. Water Res 47:3497–3506
76. Shanmugapriya A, Hemalatha M, Scholastica B, Prasad AAT (2013) Adsorption studies of lead (II) and nickel (II) ions on chitosan-G-polyacrylonitrile. Der Pharma Chemica 5:141–155
77. Manikandan A, Sathiyabama M (2015) Green synthesis of copper-chitosan nanoparticles and study of its antibacterial activity. J Nanomed Nanotechnol 6:157–7439
78. Tran HV, Tran LD, Ba CD, Vu HD, Nguyen TN, Phamc DG, Nguyen PX (2010) Synthesis, characterization, antibacterial and antiproliferative activities of monodisperse chitosan-based silver nanoparticles. Colloids Surf A Physicochem Eng Asp 360:32–40
79. Wei D, Qian W (2008) Facile synthesis of Ag and Au nanoparticles utilizing chitosan as a mediator agent. Colloids Surf B Biointerfaces 62:136–142
80. Shahidi F, Arachchi JKV, Jeon YJ (1999) Food applications of chitin and chitosans. Trends Food Sci Tech 10:37–51
81. Semwal A, Singh R, Dutta PK (2013) Chitosan: a promising substrate for pharmaceuticals. J Chitin Chitosan Sci 1:1–16
82. Ko JA, Park HJ, Hwang SJ, Park JB, Lee JS (2002) Preparation and characterization of chitosan microparticles intended for controlled drug delivery. Int J Pharm 249:165–174
83. Sahu SK, Maiti S, Pramanik A, Ghosh SK, Pramanik P (2012) Controlling the thickness of polymeric shell on magnetic nanoparticles loaded with doxorubicin for targeted delivery and MRI contrast agent. Carbohydr Polym 87:2593–2604
84. Sadighian S, Hosseini-Monfared H, Rostamizadeh K, Hamidi M (2015) pH-triggered magnetic-chitosan nanogels (MCNs) For doxorubicin delivery: physically vs. chemically cross linking approach. Adv Pharm Bull 5:115–120
85. Islam MA, Firdous J, Choi YJ, Yun CH, Cho CS (2011) Design and application of chitosan microspheres as oral and nasal vaccine carriers: an updated review. Int J Nanomed 7:6077–6093
86. Nayeri H, Fattahi A, Iranpoor-mobarakeh M, Nori P (2015) Stabilization of lactoperoxidase by tragacanth-chitosan nano biopolymer. Int J Biosci 6:418–426
87. Singh B, Sharma V (2014) Influence of polymer network parameters of tragacanth gum-based pH responsive hydrogels on drug delivery. Carbohydr Polym 101:928–940
88. Jayakumar R, Chennazhi KP, Muzzarelli RAA, Tamura H, Nair SV, Selvamurugan N (2010) Chitosan conjugated DNA nanoparticles in gene therapy. Carbohydr Polym 79:1–8
89. Mohanty DP, Biswal SK, Nayak L (2015) Preparation of starch-chitosan nanocomposites for control drug release of curcumin. Inter J Curr eng Technol 5:336–31

D-Glucosamine and *N*-Acetyl D-Glucosamine: Their Potential Use as Regenerative Medicine

Tanvi Jain, Hridyesh Kumar and Pradip Kumar Dutta

Abstract Glucosamine (GlcN), an amino sugar, is a compound derived from substitution of a hydroxyl group of a glucose molecule with an amino group. GlcN and its acetylated derivative, *N*-acetylglucosamine (GlcNAc), have been widely used in food, cosmetics, and pharmaceutical industries and are currently produced by acid hydrolysis of chitin (a linear polymer of GlcNAc) extracted from crab and shrimp shells. In this review, distribution and production of GlcN and GlcNAc, their chemistry and determination in the complex samples will be treated first. This review will describe the procedure to identify a high-quality glucosamine product for Glucosamine/chondroitin Arthritis Intervention Trial (GAIT) and to clarify confusing product information and nomenclature. GlcN is a precursor of the glycosaminoglycans and proteoglycans that make up articular cartilage. Glucosamine sulfate and glucosamine hydrochloride have used for the treatment of osteoarthritis for more than 30 years, with no major known side effects. The notion that augmenting the intake of the precursor molecule, glucosamine, may directly stimulate articular proteoglycan synthesis to modulate osteoarthritis has provided the rationale for its widespread use. Theoretically, exogenous glucosamine may augment glycosaminoglycan synthesis in cartilage. There is a simultaneous theoretical concern that it might also induce insulin resistance in insulin-sensitive tissues. While the efficacy of glucosamine was published in the definitive medical journals, there were views against it. This concern will be also discussed. While glucosamine was not effective

T. Jain
Department of Chemical Engineering, MN National Institute of Technology, Allahabad 211004, U.P, India

T. Jain · H. Kumar · P.K. Dutta
Department of Chemistry, MN National Institute of Technology, Allahabad 211004, U.P, India

P.K. Dutta (✉)
Member, Centre for Medical Diagnostic & Research, MN National Institute of Technology, Allahabad 211004, U.P, India
e-mail: pkd_437@yahoo.com

P.K. Dutta (ed.), *Chitin and Chitosan for Regenerative Medicine*, Springer Series on Polymer and Composite Materials, DOI 10.1007/978-81-322-2511-9_11

without combination with chondroitin sulfate in the some trial, glucosamine alone was effective in the other trial. Some concerns about these trials will be discussed together with the mechanism of action of glucosamine and chondroitin for antiarthritic potential. Finally, the review will focus on the biomedical and other application of the glucosamine and chitosan oligosaccharide. Such biomedical applications include wound healing, bone regeneration, antibacterial effect, and oral hygiene. It also discusses the role of chitosan oligosaccharide as a drug carrier for molecular therapies, such as the drug and the gene delivery systems and the role in imaging for tumor and cancer detection.

Keywords Chitin · Glucosamine · D-Glucosamine · *N*-Acetyl D-glucosamine · Osteoarthritis

Abbreviations

GlcN	Glucosamine
GlcNAc	*N*-acetylglucosamine
NAG	*N*-acetyl-D-glucosamine
GS	Glucosamine Sulphate
OA	Osteoarthritis
GAGs	Glycosaminoglycans
GlcNHCl	Glucosamine hydrochloride
RP-HPLC	Reversed phase high-performance liquid chromatography
HCl	Hydrochloric acid
NaOH	Sodium hydroxide
rRNA	Ribosomal ribonucleic acid
GEO	Genetically engineered organisms
IGF-I	Growth factor-I
bFGF	Basic fibroblast growth factor
TGF-b1	Transforming growth factor bea1
TMJ	Temporomandibular joint
HA	Hyaluronic acid
QD	Quantum Dots
GAIT	Glucosamine/chondroitin Arthritis Intervention Trial

1 Introduction

Chitin is the most abundantly available polysaccharide on earth after cellulose, and chitin is a main constituent of cell walls of fungus, insect exoskeletons, and crustaceans. It is one of the most ample renewable biopolymer on earth [1]. Accordingly, a very huge number of applications of D-glucosamine and *N*-acetyl

D-glucosamine like food technology, biomaterial science, microbiology, agriculture, techniques for wastewater treatment, drug delivery, tissue engineering, bionanotechnology had been earlier described. The structure of chitin is a biopolymer of *N*-acetyl-D-glucosamine (NAG) allied by beta-glycosidic bonding. Polysaccharides, oligosaccharides, and monosaccharides are the main derivatives of chitin, which also includes several therapeutic activities such as immunomodulation [2, 3], antitumor [4] osteoarthritis treatment [5], wound healing, tissue engineering, drug delivery, and for regenerative medicines [6–10]. Also, in the recent years, NAG, the end hydrolytic product of chitin, has become an attractive biomaterial as food supplements and cosmetics [11–13]. Also, *N*-acetyl glucosamine has a pure sweet taste that has also been projected to be utilized as a substitute for similar applications [14].

Researchers has investigated about the glucosamine and analyzed that it has two major beneficial properties. First, Glucosamine declines the level of an enzyme that degrades articular cartilages. Secondly, it has also helped in repairing the damaged articular tissues [15].

In recent years Glucosamine sulphate (GS) has been widely studied as a treatment selection for osteoarthritis (OA) which is a great boon in the field of tissue engineering and regenerative medicines. Glucosamine helps efficiently in regenerating the damaged cartilages. Glucosamine is also a natural building block of any of the articular cartilage [16].

1.1 Chemical Structure of D-Glucosamine and N-Acetyl D-Glucosamine

Chitin polysaccharides, which are commonly known as amino sugars with the fundamental structural entity of *N*-acetyl-β-D-glucosamine are identified for enormous biological functions such as antitumor activity, elicitor action, and anti-inflammatory activity. Chitin is fragmented by endo and exo enzymes called as chitinases (EC3.2.1.14) and β-*N* acetyl hexosaminidase (EC3.2.1.52), respectively [17]. The degradation of chitin is generally initiated by sudden breakage within the chain of chitin by chitinases to liberate the oligomers of N-acetyl glucosamine chains. The molecular formula of this monosaccharide is C8H15NO6. The chemical structure of glucosamine and N-acetyl glucosamine is shown in Fig. 1.

Fig. 1 Chemical structure of glucosamine and *N*-acetylglucosamine

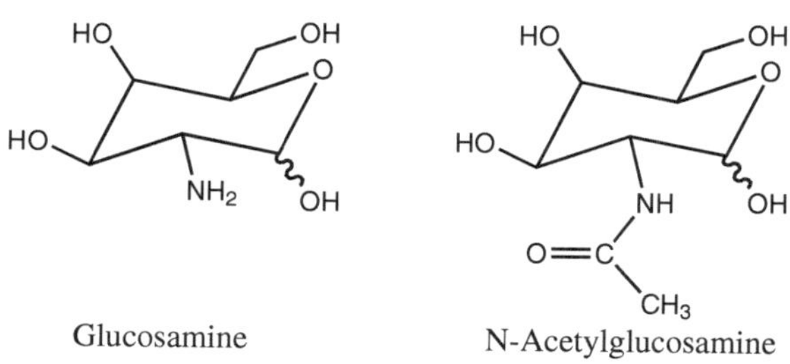

Generally, it's a whitish and sweetish powder that decays at 221 °C. The solubility of GlcNAc is about 25 % in water and 1 % aqueous solutions are colorless, odorless, and transparent [18]. Apart from the structural constituent of homogeneous polysaccharide like chitin, *N*-acetyl-D-glucosamine is also a component of heterogeneous oligosaccharides like murein [19], hyaluronic acid (HA) [20, 21].

2 Production of D-Glucosamine and *N*-Acetyl-D-Glucosamine

GlcNAc and GlcN are the basic constituents for glycoproteins, proteoglycans, glycosaminoglycans (GAGs) and other building units used in connective tissue. [22–27]. Apart from GlcNAc existed in free form in nature, except in human milk [28, 29]. The methods of GlcNAc production summarizes in Fig. 2.

About 150 billion tons of chitin/D-glucosamine is crafted in nature [30–32], which makes chitin a appropriate biomass source for the synthesis of D-glucosamine and GlcNAc. GlcNAc can be synthesized through the procedure that was established on chitin hydrolysis utilizing chitin as feed-stock.

2.1 Preparation of D-Glucosamine and N-Acetyl D-Glucosamine Using Chemical Method

Hydrolysis of chitin using acids such as hydrochloric acid, sulphuric acid etc., which is type of an industrial process for the synthesis of glucosamine hydrochloride salt (GlcNHCl), one of the most popular food supplements, drugs utilized for osteoarthritis and other biomedical applications [33]. The main focus is to understand that the treating temperature and the concentration of the acid should be

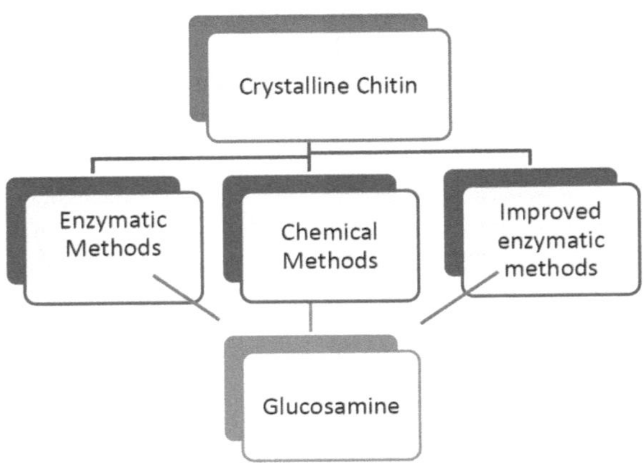

Fig. 2 Methods for preparation of glucosamine

chosen wisely. All such factors must be highly adequate to degrade only chitin not so high temperature that glucosamine or GlcNAc products are demolished.

Falk et al. discovered that the rates of development of acetic acid and glucosamine were the same but at starting it was less than the rate of *N*-acetylglucosamine synthesis. From this, it was presumed that the vast majority of the acetic acid was produced by the hydrolysis of *N*-acetylglucosamine, instead of by hydrolysis of the polymers' *N*-acetyl groups. Falk also establish a maximum concentration of *N*-acetylglucosamine when 10 M HCl was used for hydrolysis of 50 % of the chitin. The authors concluded that controlled acid hydrolysis of chitin did not result best yield of *N*-acetylglucosamine [34]. There are a few instances which show chemically synthesized D-glucosamine and GlcNAc. Like Kuyama et al. [35] synthesized totally deacetylated chitosan dodecamers started with glucosamine monomer using amino protective group. Aly et al. [36] explained a technique to synthesize fully *N*-acetylated D-Glucosamine from GlcN monomers using dimethyl maleoyl as an amino protective group for production of two enzymes, i.e.; chitotetraose and chitohexaose. Removal of the amino protective group and *N*-acetylation was also achieved in a one reaction to provide the required product [37].

The typical procedure for the preparation of glucosamine by direct acid hydrolysis [38]. But this process has a disadvantage of chitin, which includes high cost, lower yield (below 65 %), and acidic leftover which was formed by HCl. The estimated production is 6.42 g/L GlcNAc can be produced in 1 h [39].

Also Varum et al. studied the effect of concentration of hydrochloric acid on hydrolysis of the glycosidic linkages and the *N*-acetyl linkage of moderately N-acetylated. Hydrolysis of *N*-acetyl linkage was achieved by SN_2 reaction mechanism where water addition in carbonium ion was rate determining stage. It was concluded due to SN_2 that in concentrated acid depolymerization in tenfolds higher than de-*N*-acetylation and in dilute acid it was equal. [40].

Likewise Zhan synthesized an improved method for *N*-acetylation of Glucoasmine in which he has used pyridine as a solvent in the presence of tributylamine and acetic anhydride. Approximate 99 % pure GlcNAc can be attained after a series of purification methods. The estimated yield of this method had been reported to be higher than 70 % [41].

In another related study, Trombotto et al. [42] prepared homogeneous series of chito-oligomers having 2–12 degrees of polymerization (DP) and different (0–90 %) degrees of *N*-acetylation (DA) was also obtained via two-step chemical modification. First, they hydrolyzed completely *N*-deacetylated chitosan followed by selective precipitations to produce glucosamine oligomers. In second step hydro-alcoholic solution of acetic anhydride was used in controlled way for partial *N*-acetylation glucosamine unit of oligomers. The drawback of this method, as for the enzymatic preparation of chitooligosaccharides,is the heterogeneity of the prepared chitooligosaccharides. Goyal et al. synthesized and screened an innovative class of *N*-acetyl glucosamine with methoxybenzylidene acetal protective group. This modification did not exert a negative influence on the gelation. On the contrary, it actually enhanced the gelation tendency for many derivatives. The introduction of the additional methoxy group on the phenyl ring led to low molecular weight gelators with a higher pH responsiveness. The resulting gels were stable at neutral pH values but degraded in an acidic environment [43]. Also, Trombotto et al. [44] prepared glucosamine hydrochloride (GluNHCl) from various crustacean shells like *Penaeus monodon* (Indian shrimp), *Portunus pelagicus* (blue crab) and *Portunus sanguinolentus* (three spot crab) of acids through the hydrolysis and its evaluation was analyzed by reversed phase high performance liquid chromatography (RP-HPLC). The results showed that the yield of chitin after demineralization with 0.5 M HCl was 87.83, 89.18 and 51.11 % and deacetylation of chitin with 2 N NaOH resulted in the yield of 68.91, 75.67, and 30 % for *P. sanguinolentus*, *P. pelagicus,* and *P. monodon,* respectively. The HPLC analysis also gives that the obtained glucosamine hydrochloride revealed species of *Portunus* (21.64 mg g^{-1} and 21.83 mg $g-^{1}$) were better source of Glu-HCl than *P. monodon* (3.32 mg $g-^{1}$). Although the manufacture of GlcNAc by chemical methods were assessed to be adequately profitable, thus, the product was not reflected as a usual natural polysaccharide due to the chemical alteration of GlcNAc. That is why, GlcNAc prepared using chemicals typically flavors bitter due to these alterations. In addition, the usage of ozone has also been developed [45].

Thus, the use chemically modified GlcNAc and glucosamine covers a wide range of applications like in food additives, cosmetics and pharmaceuticals, and medicines.

2.2 *Preparation of D-Glucosamine and N-Acetyl D-Glucosamine Using Enzymatic Method*

The drawback of chemical method not only due to practical causes but also because of the other environment interests of producing D-glucosamine and *N*-acetyl D-glucosamine, this is why synthesis via chemical method is not widely commercialized.

Kuk et al. investigated how a bacteria, GJ18, which are having robust chitinolytic activity which was isolated from soil of coastal areas. Then isolated strain was recognized as *A. hydrophila* by its structural and biochemical characteristics along with 16S rRNA gene sequences. The results showed up to 45 °C, it was effectually hydrolyzed to GlcNAc by *A. hydrophila* GJ18 crude enzyme; however the hydrolysis process started decreasing when the temperature is above 50 °C. However, the enzymatic hydrolysis of chitin can synthesize the same under trivial surroundings [46].

They both not only performs physiochemical functions but also utilized in the field of pathological aspects [47–52].

The whole collection of chitinolytic enzymes which consists of endochitinases, exochitinases, chitobiosidases, and acetylglucosaminidases (NAGases) that are synthesized and secreted through a prokaryotic or a eukaryotic entities [53].

Earlier, Muraki et al. separated D-glucosamine oligosaccharides from a chitosan hydrolyzate prepared with the cellulose enzyme derived from *Trichoderma viride.* The separation was done by using weak ion exchange chromatography with the elution by 0.01 N Hcl. At last titration procedure was used to find the glucosamine content. The results showed that this technique is efficient for the synthesis of highly pure oligosaccharides [7]. In contrasting study, few microorganisms, like *Trichoderma hamatum* strain or *Trichoderma harzianum* strain, produced *N*-acetyl glucosamine in the medium in the absence of chitin or its derivatives [54].

A huge variety of enzymes is sterilized from major production of microorganisms, genetically engineered organisms (GEO) for the purpose of the industrial production of D-glucosamine and *N*-acetyl glucosamine [55–57].

Setthakaset et al. prepared *N*-acetyl glucosamine from squid pen β-chitin utilizing *Aspergillus species* chitinase using hydrolysis process through enzymes. The results showed that the growth of the fungus in the media at pH 3.5 and 40 °C for 5 days which provides the maximum chitinolytic activity of 3.1 U/mL. The crude enzyme achieved from the growth of fungus is used in the hydrolysis process of milled squid pen β-chitin at acidic pH 4 and 45 °C using 22 U of enzyme per gram of chitin [58]. Chen et al. proved to synthesize N-acetyl glucosamine *Chitinibacter tainanensis* which was isolated from Southern Taiwan soil. The results showed that when *Chitinibacter tainanensis* was incubated with chitin. Chitin degrading factors was successfully converted to *N*-acetyl-glucosamine [59].

Recently, Sun et al. produced D-glucosamine by hydrolyzing chitosan with the use of two crude enzymes chitosanase and β-*d*-glucosaminidase showed that the optimal conditions were temperature 50 °C, substrate conc is 20 mg/mL, the optimum ratio of enzyme to chitosan 1.5 U/60 mg, pH 6.8. Under above conditions, it was completely hydrolyzed in 5 h and thus this method can be used efficiently used for producing D-glucosamine which can be further utilized for other applications [60].

In one study, around 100 % pure *N*-Acetylglucosamine was synthesized by utilizing enzymes from *Aeromonas hydrophila* [61], while another enzyme chitobiose was also synthesized using *B. chitinases* [62]. The study clearly showed that the endochitinases, exochitinases, and *N*-acetylglucosaminidases are essential for the ample absorption of chitin, and more *N*-acetylglucosaminidases activity gives a higher and efficient purity of GlcNAc. Thus, chitin crystalline structure and the enzymatic constituents of these reactions are the vital characteristics in the synthesis of GlcNAc through enzymatic reactions.

3 Application of D-Glucosamine and *N*-Acetyl D-Glucosamine

Glucosamine belongs to a huge section of amino groups of sugars that perform various numbers of vital roles in human system and are also located in the humans. It has become of great interest not only as an under utilized resource, but also as a novel efficient substance with extraordinary ability in enormous fields, and latest progress in chitin or chitosan chemistry is quite remarkable. All the above mentioned properties, together with the very safe toxicity, make them an

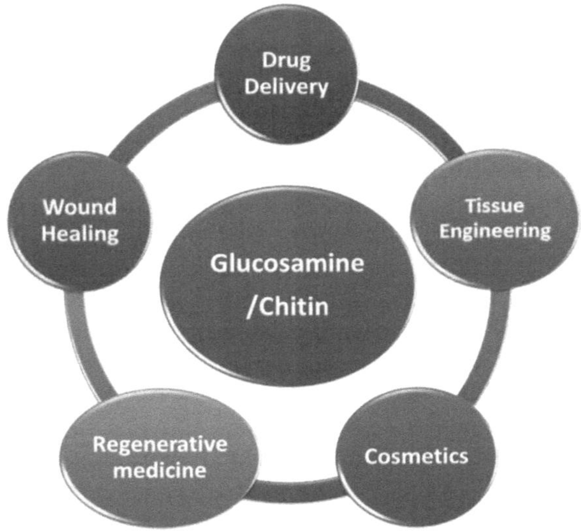

Fig. 3 Applications of glucosamine and chitin

exciting and promising product to get utilized for various applications such as biomedical applications, tissue engineering, wound healing, cosmetics, etc. The various application of glucosamine is shown in Fig. 3.

3.1 Application in Tissue Engineering

Glucosamine likewise assumes a part during the course of tissue engineering. Regardless of numerous studies in glucosamine and the more endeavors on supramolecular gelators or self-collected nanofibers and other tissue designing applications [15, 35, 63, 64].

Shaunak et al. utilized used anionic, polyamidoamine dendrimers to prepare innovative water-soluble conjugates of D-glucosamine and D-glucosamine 6-sulfate for effective immunogenic and antiangiogenic characteristics, respectively. After various evaluation and clinical trials conducted on rabbits the study showed that specially designed macromolecules such as the dendrimers depicted in this study might be altered to have specific immunogenic and antiangiogenic characteristics, and they utilized collectively to check wound tissue formation which can be majorly utilized in the field for regenerative medicine [65].

Likewise, Wang et al. studied the comparative study between the various outcomes of insulin-like growth factor-I, basic fibroblast growth factor, transforming growth factor beta1, and glucosamine sulphate on temporomandibular joint condylar cartilage and ankle cartilage cells in monolayered culture. The results showed that GS enhanced the cell propagation and synthesis at explicit concentrations and exceed growth elements in definite cases [66].

3.2 Application in Cosmetics

Human skin comprises of stratum corneum and dermis to protect the skin from different extreme natural circumstances like aridness and UV illumination. Stratum corneum assumes an indispensable part in maintaining the solidness of the skin. It additionally comprises of smoothed dead cells, corneocytes and a complex of lipid network, and characteristic dampness properties [67].

Chitin has a high review of UV radiation resistance, purification resistance, and oxidation resistance. Henceforth, it turns into the ideal common beautifiers crude materials. Chitin with saturating, film, and dynamic cell capacity can keep the skin damp, glossy, and adaptable in the readiness of a mixture of healthy skin restorative. It can likewise upgrade the metabolic capacity of epidermal cells, repress free radical oxidation, and wipe out lipofuscinosis quality, age spots, repair skin harm, concealment of bugs, and other harm to the skin. HA is primarily delivered from fibroblasts and keratinocytes and *N*-acetyl glucosamine are the primary building unit of HA so the last is chiefly included in skin change and assists in treating with skin maturing and detachment. Fitting organization of GlcNAc is utilized to enhance skin quality.

Bisset et al. [68] mulled over that GlcNAc helps in lessening the presence of facial hyperpigmentation in 8 week, which was demonstrated by the twofold visually impaired clinical trial [69]. Utilizing as a part of in vitro genomic trials, the mechanism by which GlcNAc decreases melanin creation has been found to include the up-regulation of a few gene, for example, epidermal turnover qualities and cell reinforcement-related qualities, and the down-regulation of the cytoskeleton qualities included in melanosome transport [70]. It is likewise viewed as that an important fixing in beautifiers for enhancing skin wrinkles and shading due to its flexible properties of GlcNAc [71]. As examined in past segment, glucosamine advances the proliferation of keratinocytes and fibroblasts and builds the creation of HA in the skin, it has likewise been effectively used to recuperate wounds in skin [72, 73].

3.3 Chitin as Drug Delivery Carriers

As chitin is the main constituent of Glucosamine, its important function is to act as a carrier for drug delivery. But the main feature of chitin is that it must not accrue inside the human system nor must it be toxic [74].

Chitin formulations like carboxymethyl chitin, succinyl chitosan [75], chitosan derived hydrogel [76], hydroxyethyl chitin [77] had been clearly mentioned to contain such properties. The colloidal structure of chitosan can easily entrap macromolecules by enormous methods. All such associated molecules have been revealed to transport via mucosa membrane and epithelial cells more effectively. Cationic chitosan when combined with other natural biopolymers have also clearly been indicated to increase the drug encapsulation efficiency of liposomes through layer by layer self-assembly method [78]. Nanoparticles prepared from chitosan/chitin with polyethylene oxide which has widely utilized as an effective carrier for proteins as well. Furthermore, an oral route for delivering drugs has also been established by using chitosan and tripolyphosphate, alginate, etc., In such system, microparticles and nanoparticles were get entrapped inside the beads which are synthesized from chitosan in solution of tripolyphosphate, etc.,

3.4 Application in Regenerative Medicine- Treatment of Osteoarthritis

Osteoarthritis (OA) is the most widely recognized manifestation of joint inflammation, is a general wellbeing issue all through the world. A few elements have been painstakingly examined for the symptomatic and auxiliary administration of OA. This review provides published studies regarding the effect of GS and chondroitin sulfate synthesis on the development of knee or hip OA. Since 1969, when first clinical trial of glucosamine conducted in Germany, various studies have contrasted glucosamine sulfate and nonsteroidal anti-inflammatory agents or with placebo. In the early 1980s, a little number of controlled human trials was directed in Europe and in Asia to evaluate the clinical utilization of glucosamine sulfate in the treatment of OA [79]. Glucosamine along with chondroitin sulfate are successfully utilizing in treat of osteoarthritis. Glucosamine and chondroitin sulfate, when combined, are found efficient in the subclass of patients with moderate-to-severe knee pain. In North America, glucosamine hydrochloride or sulfate along with chondroitin sulfate are measured as nutraceuticals (nutrition + pharmaceutical), although in maximum European nations they are promoted as pharmaceuticals and hence preparation, manufacture and marketing of glucosamine are watched more precisely in Europe [80]. Laverty et al. investigated that in maximum clinical related outcomes in Glucosamine/chondroitin Arthritis Intervention Trial (GATI) a perfect blend of sodium chondroitin sulfate and glucosamine hydrochloride were taken due to enhance the bioavailability of sulfates [81]. Clegg et al. showed the Glucosamine/chondroitin Arthritis Intervention Trial (GAIT) for 24 week, randomized, double blind, placebo and celecoxib controlled, multicenter trial to estimate severely the worth and protection of glucosamine-chondroitin sulfate, and the two in combination in the handling of pain because of the knee osteoarthritis [82].

Alindon et al. conducted a 12 week study and the consequences of this study recommended that glucosamine was no more viable than placebo in treating the indications of knee OA. Be that as it may, it ought to be observed that this study was intended for an alternate reason than evaluation of glucosamine [83, 84].

Hochberg et al. conducted a compare study of the effectiveness and safety of chondroitin sulfate and glucosamine hydrochloride against celecoxib over 606 patients in France, Germany, Poland, and Spain having knee osteoarthritis and severe pain. It was concluded that in comparison with glucosamine + chondroitin sulfate was reducing pain, stiffness, functional limitation, and joint swelling/effusion after 6 months in patients and having a decent safety outline [85].

3.5 Application in Cancer/Tumor Imaging

Tada et al. prepared 2-deoxy-2-[^{18}F] fluoroacetamido-*d*-glucopyranose via amide bond formation with ^{18}F labeled carboxylic acid [Fig. 4]. Rediochemical yield and purity was 9.1 and 98 %, respectively. The compound utilized in diagnosing tumor through imaging [86]. After an hour of injection of 2-deoxy-2-[^{18}F] fluoroacetamido-d-glucopyranose, a high uptake were measured in tumor cells which were used to diagnosed and visualized by positron emission tomography (PET) [87].

Qi et al. designed and synthesized 99 % pure *N*-(2-[^{18}F] fluoro-4-nitrobenzoyl) glucosamine, a glucosamine derivative, via many step reaction and their performance were checked by in vivo and in vitro study. Results showed that tumor/blood ratio of glucosamine derivative was 4.40 and tumor/muscle ratio of glucosamine derivative was found 4.84. The prepared ^{18}F derivative can showed it importance in tumor imaging using by positron emission tomography (PET) technique [88]. Carroll et al. designed ^{18}F-labeled glucosamine analogs because glucosamine had a role in the biochemical production of glycosylated proteins and lipids and this property was used for imagining of cancer cells. They design three ^{18}F labeled glucosamine by exploring ^{18}F radiochemistry imine formation via aldehyde condensation, alkylation, and copper mediated click chemistry reaction. The three synthesized ^{18}F labeled glucosamine were verified in vivo in mice by using HEY Rab25 tumors (human ovarian carcinoma cell-line. Figure 5c displayed favorable in vivo characteristics, with no bone uptake, quick clearance and most significantly noticeable tumor uptake [89].

Yang et al. [90] labeled glucosamine derivative with gallium 68 (^{68}Ga) to design a radiotracer for diagnosing tumor cells using micro-positron emission tomography (μPET). Glucosamine was first modified with 1,4,7,10-tetraazacyclododecane-1,4,7,10-tetraacetic acid (DOTA) by reaction of D-glucosamine hydrochloride and 2-(4-isothiocyanatobenzyl)-DOTA. Microwave irradiation of 25 W for 5 min and 50 W for 5 min were used for labeling of ^{68}Ga on

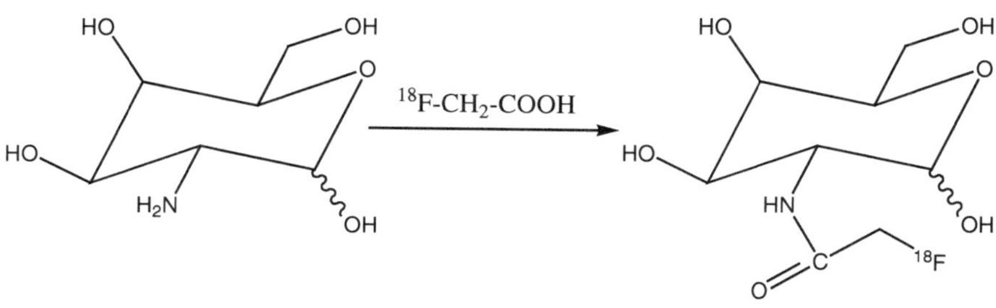

Fig. 4 Preparation of 2-deoxy-2-[^{18}F] fluoroacetamido-d-glucopyranose

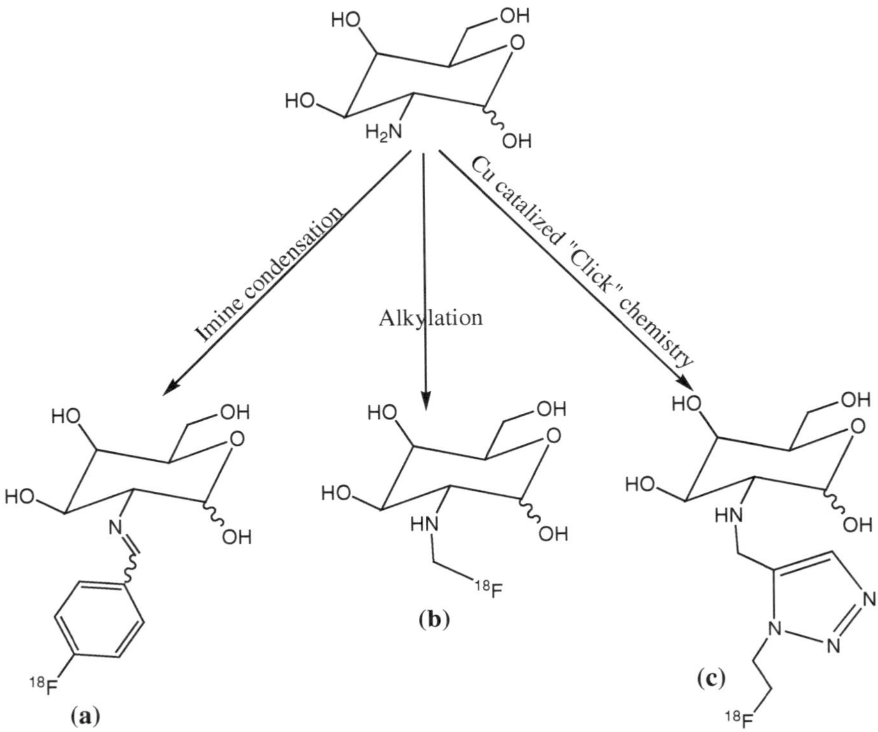

Fig. 5 ^{18}F labeled glucosamine **a**, **b**, and **c** are prepared by Carroll et al. for imaging of cancer cells [89]

DOTA-D-glucosamine derivative. After an hour of incubation of ^{68}Ga labeled glucosamine treated human epithelial carcinoma cells (A431), 15.7 % uptake was observed. μPET images of A431 tumors bearing mice displayed acceptance of ^{68}Ga labeled glucosamine in the tumors [91]. Igawa et al. [92] studied uptake of D-glucosamine conjugate carboxylated QDs into cultured osteoblastic cells. Size of the core carboxylated QDs and D-glucosamine conjugated QDs were calculated around 3 nm and 20–30 nm, respectively. Glucosamine modified QDs adhere on cell surface and conveyed into the cytoplasm in three hours of incubation and the process was confirmed by using specific fluorescent staining [93].

4 Concluding Remarks

Glucosamine has always shown low cytotoxicity, biodegradable, and biocompatible in nature which makes them utilized for various biomedical applications. It also consists of chitinases which is widely available in nature and found almost in all the bacteria, fungi as well as in plants, and in the digestive systems of various animals. D-glucosamine and *N*-acetyl D-glucosamine is synthesized using substrate, i.e.; chitin by various chemical, enzymatic methods. New methods using glucose as a substrate can also be applied to obtain n acetyl glucosamine using genetically engineered microorganisms. D-Glucosamine and *N*-acetyl D-glucosamine has a wide application in regenerative medicine like for osteoarthritis through which the damaged cartilages can be replaced by using Glucosamine. At last, the versatile characteristics of both enhance its usage as a novel candidate for utilization not only in regenerative medicine but also for drug development and other biomedical applications.

Acknowledgments Tanvi Jain and Hridyesh Kumar gratefully acknowledged the research support both from Institute and TEQIP-II, MNNIT Allahabad.

References

1. Flash J, Pilet PE, Jolles P (1992) What's new in chitin research? Experientia 48:701–716
2. Shibata Y, Foster LA, Metzger WJ, Myrvik QN (1997) Alveolar macrophage priming by intravenous administration of chitin particles, polymers of *N*-acetyl-D-glucosamine, in mice. Infect Immun 65:1734–1741
3. Shibata Y, Honda I, Justice JO, Van Scott MR, Nakamura RM, Myrvik QN (2001) The adjuvant *N*-acetyl-D-glucosamine polymer up-regulates Th1 immunity but down-regulates Th2 immunity against a mycobacterial protein (MPB-59) in interleukin-10-knockout and wild-type mice. Infect Immun 69:6123–6130
4. Suzuki K, Mikami T, Okawa Y, Tokoro A, Suzuki S, Suzuki M (1986) Antitumor effect of hexa-*N*-acetylchitohexaose and chitohexaose. Carbohydr Res 151:403–408
5. Creamer P (2000) Osteoarthritis pain and its treatment. Curr Opin Rheumatol 12:450–455
6. Aam BB, Heggset EB, Norberg AL, Sorlie M, Varum KM, Eijsink VG (2010) Production of chitooligosaccharides and their potential applications in medicine. Mar Drugs 8:1482–1517
7. Yang CJ, Liu YK, Liu CL, Shen CN, Kuo ML, Su CC, Tseng CP, Yen TC, Shen CR (2009) Inhibition of acidic mammalian chitinase by RNA interference suppresses ovalbumin-sensitized allergic asthma. Hum Gene Ther 20:1597–1606
8. Shen CR, Juang JH, Tsai ZT, Wu ST, Tsai FY, Wang JJ, Liu CL, Yen TC (2011) Preparation, characterization and application of superparamagnetic iron oxide encapsulated with *N*-[(2-hydroxy-3-trimethylammonium) propyl] chitosan chloride. Carbohydr Polym 84:781–784
9. Chen JK, Shen CR, Liu CL (2010) N-acetylglucosamine: production and applications. Mar. Drugs 8:2493–2516
10. Jain T, Dutta PK (2011) Chitin nanoparticles for drug delivery: preparation, characterization and evaluation. Asian Chitin J 7:13–18
11. Salvatore S, Heuschkel R, Tomlin S, Davies SE, Edwards S, Walker-Smith JA, French I, Murch SH (2011) A pilot study of *N*-acetyl glucosamine, a nutritional substrate for glycosaminoglycan synthesis, in pediatric chronic inflammatory bowel disease. Aliment Pharmacol Ther 14:1567–1579
12. Kumar H, Srivastava R, Dutta PK (2013) Highly luminescent chitosan-l-cysteine functionalized CdTe quantum dots film: synthesis and characterization. Carbohydr Polym 97:327–334
13. Kimball AB, Kaczvinsky JR, Li J, Robinson LR, Matts PJ, BergeCA Miyamoto K, Bissett DL (2011) Reduction in the appearance of facial hyperpigmentation after use of moisturizers with a combination of topical niacinamide and *N*-acetyl glucosamine. Int J Mol Sci 12:1194–1198

14. Dutta PK, Dutta J, Tripathi VS (2004) Chitin and chitosan: Chemistry, properties and applications. J Sci Ind Res 63:20–31
15. Rath P, Sanya K, Kamontip K, Mongkol S, Seiichi A (2002) Quantitative production of 2-acetamido-2-deoxy-D-glucose from crystalline chitin by bacterial chitinase. Carbohydr Res 337:557–559
16. Dutta PK, Tripathi S, Mehrotra GK, Dutta J (2009) Perspectives for chitosan based antimicrobial films in food applications. Food Chem 114:1173–1182
17. Dutta J, Chattopadhyaya MC, Tripathi VS, Dutta PK (2004) Chitin & Chitosan: novel biomaterials waiting for future developments. J Polym Mater 21:321–332
18. Fasbender MH (1994) Glucosamine sulfate compared to ibuprofen in osteoarthritis of the knee. Osteoart Cart 2:61–69
19. Hackman RH, Goldberg M (1974) Light-scattering and infrared-spectrophotometric studies of chitin and chitin derivatives. Carbohydr Res 38:35–45
20. Jollès P, Muzzarelli RAA (eds) (1999) Chitin and Chitinases. Birkhauser Verlag publishers, Basel Switzerland
21. Tanaka T, Fujiwara S, Nishikori S, Fukui T, Takagi M, Imanaka TA (1999) Unique chitinase with dual active sites and triple substrate binding sites from the hyperthermophilic archaeon pyrococcus kodakaraensis KOD1. Appl Environ Microbiol 65:5338–5344
22. Mao JS, Liu HF, Yin YJ, Yao KD (2003) The properties of chitosan-gelatin membrane and scaffolds modified with hyaluronic acid by different methods. Biomaterials 24:1621–1629
23. Talent JM, Gracy RW (1996) Pilot study of oral polymeric *N*-acetyl-D-glucosamine as a potential treatment for patients with osteoarthritis. Clin Ther 18:1184–1190
24. Angelis PL (1999) Molecular directionality of polysaccharide polymerization by the pasteurella multocida hyaluronan synthase. J Biol Chem 274:26557–26562
25. Turely EA, Bowman P, Kytryk MA (1985) Effects of hyaluronate and hyaluronate binding proteins on cell motile and contact behaviour. J Cell Sci 78:133–145
26. Alho AM, Underhill CB (1989) The hyaluronate receptor is preferentially expressed on proliferating epithelial cells. J Cell Biol 108:1557–1565
27. Toole BP, Jackson G, Gross J (1972) Hyaluronate in morphogenesis: inhibition of chondrogenesis in vitro. Proc Natl Acad Sci 69:1384–1386
28. Brecht M, Mayer U, Schlosser E, Prehm P (1986) Increased hyaluronate synthesis is required for fibroblast detachment and mitosis. J Biochem 239:445–450
29. Dube B, Luke HJ, Aumaille M, Prehm P (2001) Hyaluronan reduces migration and proliferation in CHO cells. Biochim Biophys Acta 1538:283–289
30. Kobata A, Ginsburg V (1969) Oligosaccharides of human milk: isolation and characterization. Arch Biochem Biophys 130:509–513
31. Miller JB, Bull S, Miller J, McVeagh P (1994) The oligosaccharide composition of human milk: temporal and individual variations in monosaccharide components. J Pediatri. Gastroenterol Nutr 19:371–376
32. Tharanathan RN, Kittur FS (2003) Chitin—the undisputed biomolecule of great potential. Crit Rev Food Sci Nutr 43:61–87
33. Howard MB, Ekborg NA, Weiner RM, Hutcheson SW (2003) Detection and characterization of chitinases and other chitin-modifying enzymes. J Ind Microbiol Biotechnol 30:627–635
34. Kurita K (2006) Chitin and chitosan: functional biopolymers from marine crustaceans. Mar Biotechnol 8:203–226
35. Hitoshi S, Shizu F, Naoko Y, Norioki K, Atsuyoshi N (2002) Production of N-acetyl-D-glucosamine from α-chitin by crude enzymes from aeromonas hydrophila H-2330. Carbohydr Res 337:761–763
36. Falk M, Smith DG, McLachla J, McInnes AG (1966) Studies on chitin (Beta-(1-]4)-Linked 2-Acetamido-2-Deoxy-D-Glucan) fibers of diatom thalassiosira fluviatilis hustedt. J Canad Chem 44:2269–2281
37. Kuyama H, Nakahara Y, Nukada T, Ito Y, Ogawa T (1993) Stereo controlled synthesis of chitosan dodecamer. Carbohydr Res 243:1–7

38. Aly MRE, Ibrahim E, Ashry ESH, Schmidt RR (2001) Synthesis of chitotetraose and chitohexaose based on dimethylmaleoyl protection. Carbohydr Res 331:129–142
39. Aly MRE, Castro-Palomino JC, Ibrahim ESI, El-Ashry FSH, Schmidt RR (1998) The dimethylmaleoyl group as amino protective group-application to the synthesis of glucosamine-containing oligosaccharides. Eur J Org Chem 8:2305–2316
40. Bohlman JA, Schisler DO, Hwang KO, Hennling JP, Trinkle JR, Anderson TB, Steinke JD, Vanderhoff A, (2010) N-Acetyl-D-glucosamine and Process for Producing N-Acetyl-D-glucosamine. US Patent NO. 6693188B2
41. Sashiwa H, Fujishima S, Yamano N, Kawasaki N, Nakayama A, Muraki E, Aiba S (2001) Production of N-Acetyl-D-glucosamine from β-Chitin by enzymatic hydrolysis. Chem Lett 31:308–309
42. Vårum KM, Ottoy MH, Smidsrod O (2001) Acid hydrolysis of chitosans. Carbohydr Polym 46:89–98
43. Zhan WS (2007) Process for Preparing Refined N-Acetyl-D-aminoglucose. CN Patent NO 1907993
44. Trombotto S, Ladavière C, Delolme F, Domard A (2008) Chemical preparation and structural characterization of a homogeneous series of chitin/chitosan oligomers. Biomacromolecules 9:1731–1738
45. Goyal N, Mangunuru HPR, Parikh B, Shrestha S, Wang S (2014) Beilstein synthesis and characterization of pH responsive D-glucosamine based molecular gelators. J Org Chem 10:3111–3121
46. Sibi G, Dhananjaya K, Ravikumar KR, Mallesha H, Venkatesha RT, Trivedi D, Bhusal PK, Gowda K (2013) Preparation of glucosamine hydrochloride from crustacean shell waste and It's quantitation by RP-HPLC. American-Eurasian J Sci Res 8:63–67
47. Seo S, King JM, Prinyawiwatkul W (2007) Simultaneous depolymerization and decolorization of Chitosan by ozone treatment. J Food Sci 72:522–526
48. Kuk JH, Jung WJ, Jo GH, Kim YC, Kim KY, Park RD (2005) Production of N-acetyl-β-D-glucosamine from chitin by Aeromonas sp. GJ-18 crude enzyme. Appl Microbiol Biotechnol 68:384–389
49. Gooday BW (1977) Biosynthesis of the fungal wall—mechanisms and implications. the first fleming lecture. J Gen Microbiol 99:1–11
50. Fuchs RL, McPherson SA, Drahos DJ (1986) Cloning of a serratia marcescens gene encoding chitinase. Appl Environ Microbiol 51:504–509
51. Abeles FB, Bosshart RP, Forrence LE, Habig WH (1971) Preparation and purification of glucanase and chitinase from bean leaves. Plant Physiol 47:129–134
52. Cohen-Kupiec R, Chet I (1998) The molecular biology of chitin digestion. Curr Opin Biotechnol 9:270–277
53. Zhu Z, Zheng T, Homer RJ, Kim YK, Chen NY, Cohn L, Hamid Q, Elias JA (2004) Acidic mammalian chitinase in asthmatic Th2 inflammation and IL-13 pathway activation. Science 204:1678–1682
54. Gooday GW (1990) The ecology of chitin degradation. Adv Microb Ecol 11:387–430
55. Muraki E, Yaku F, Kojima H (1993) Preparation and crystallization of D-glucosarnine oligosaccharides with dp 6-8. Carbohydr Res 239:227–237
56. Mori T, Ichikawa W, KitaY, Tetsuka Y (2010) Method for fermentative production of N-Acetyl-D-glucosamine by microorganism. US Patent NO. 20100055746
57. Roberts RL, Cabib E (1982) Serratia marcescens chitinase: one-step purification and Use for the determination of chitin. Anal Biochem 127:402–412
58. Oppenheim AB, Chet I (1992) Cloned chitinase in fungal plant—pathogen control strategies. Trends Biotechnol 10:392–394
59. Pan CH, Rim SL, Kim SI (1996) Expression of two cDNAs encoding class I chitinases of rice in Escherichia coli. Biosci Biotech Biochem 60:1346–1348
60. Setthakaset P, Pichyangkura R, Ajavakom A, Sukwattanasinitt M (2008) Preparation of N-acetyl-D-Glucosamine using Enzyme from Aspergillus sp. J Metals, Mater Minerals 18:53–57

61. Chen JK, Shen CR, Yeh CH, Fang BS, Huang TL, Liu CL (2011) N-Acetyl Glucosamine obtained from chitin by chitin degrading factors in chitinbacter tainanesis. Int J Mol Sci 12:1187–1195
62. Sun Y, Zhang J, Wua S, Wang S (2013) Preparation of d-glucosamine by hydrolysis of chitosan with chitosanase and β-d-glucosaminidase. Int J Biol Macromol 61:160–163
63. Estroff LA, Hamilton AD (2004) Water gelation by small organic molecules. Chem Rev 104:1201–1218
64. Kiyonaka S, Sada K, Yoshimura I, Shinkai S, Kato N, Hamachi I (2004) Semi-wet peptide/protein array using supramolecular hydrogel. Nat Mater 3:58–64
65. Silva GA, Czeisler C, Niece KL, Beniash E, Harrington D, Kessler JA (2004) Stupp SI (2004) selective differentiation of neural progenitor cells by high-epitope density nanofibers. Science 303:1352–1355
66. Zhang SG, Holmes TC, Dipersio CM, Hynes RO, Su X, Rich A (1995) Self-complementary oligopeptide matrices support mammalian cell attachment. Biomaterials 16:1385–1393
67. Shaunak S, Thomas S, Gianasi E, Godwin A, Jones E, Teo I, Mireskandari K, Luthert P, Duncan R, Patterson S, Khaw P, Brocchini S (2004) Polyvalent dendrimer glucosamine conjugates prevent scar tissue formation. Nat Biotechnol 22:977–984
68. Wang L, Detamore MS (2009) Effects of growth factors and glucosamine on porcine mandibular condylar cartilage cells and hyaline cartilage cells for tissue engineering applications. Arch Oral Biol 54:1–5
69. Wertz PW, Bergh VB (1998) The physical, chemical and functional properties of lipid in the skin and other biological barrier. Chem Phys Lipids 91:85–96
70. Bissett D, Robinson LR, Raleigh PS, Miyamoto K, Hakozaki T, Li J, Klem GR (2007) Reduction in the appearance of facial hyperpigmentation by topical N-Acetyl Glucosamine. J Cosmet Dermatol 6:20–26
71. Bissett D, Farmer T, McPhail S, Reichling T, Tiesman JP, Juhlim KD, Hurley GJ, Robinson MK (2007) Genomic expression changes induced by topical N-acetylglucosamine in skin equivalent cultures in vitro. J Cosmet Dermatol 6:232–238
72. Riordan NH (1999) Skin Treatment System. US Patent NO 5866142
73. Hwang JI, Kim KS (2005) Cosmetics Composition Comprising Extract of Natural Materials for Improving Acne and Skin Wrinkles and Whitening Skin. KR Patent NO 20050004355 A
74. Minami S, Okamoto Y (2007) Drug for Remedy or Treatment of Wound. EP Patent NO 1749532
75. Flessner MF, Lofthouse J, Williams A (2002) Chronic alteration of sub-peritoneal tissue and peritoneal transport. Adv Perit Dial 18:12–14
76. Dev A, Mohan JC, Sreeja V, Tamura H, Patzke GR, Hussain F, Weyenet S, Nair SV, Jayakumar R (2010) Novel carboxymethyl chitin nanoparticles for cancer drug delivery applications. Carbohydr Polym 79:1073–1079
77. Kamiyama K, Onishi H, Machida Y (1999) Biodisposition characteristics of N-succinyl-chitosan and glycol- chitosan in normal and tumor-bearing mice. Biol Pharm Bull 22:179–186
78. Ishihara M, Obara K, Nakamura S, Fujita M, Masuoka K, Kanatani Y, Takase B, Hattori H, Morimoto Maehara T, Kikuchi M (2006) Chitosan hydrogel as a drug delivery carrier to control angiogenesis. J Artif Organs 9:8–16
79. Zhao Y, Chen G, Sun M, Jin Z, Gao C (2006) Study on preparation of the pH sensitive hydroxyethyl chitin/poly (acrylic acid) hydrogel and its drug release property. Sheng Wu Yi Xue Gong Cheng Xue Za Zhi 23:338–341
80. Haidar ZS, Hamdy RC, Tabrizian M (2008) Protein release kinetics for core-shell hybrid nanoparticles based on the layer-by-layer assembly of alginate and chitosan on liposomes. Biomaterials 29:1207–1215
81. Drovanti A, Bignamini AA, Rovati AL (1980) Therapeutic activity of oral glucosamine sulfate in osteoarthritis: a placebo-controlled double-blind investigation. Clin Ther 3:260
82. Russell AS, Aghazadeh-Habashi A, Jamali F (2002) Active ingredient consistency of commercially available glucosamine sulfate products. J Rheumatol 29:2407–2409

83. Laverty S, Sandy JD, Celeste C, Vachon P, Marier JF, Plaas AH (2005) Synovial fluid levels and serum pharmacokinetics in a large animal model following treatment with oral glucosamine at clinically relevant doses. Arthritis Rheum 52:181–191
84. Clegg DO, Reda DJ, Harris CL, Klein MA, O'Dell JR, Hooper MM, Bradley JD, Bingham CO 3rd, Weisman MH, Jackson CG, Lane NE, Cush JJ, Moreland LW, Schumacher HR Jr, Oddis CV, Wolfe F, Molitor JA, Yocum DE, Schnitzer TJ, Furst DE, Sawitzke AD, Shi H, Brandt KD, Moskowitz RW, Williams HJ (2006) Glucosamine, chondroitin sulfate, and the two in combination for painful knee osteoarthritis. N Engl J Med 354:795–808
85. McAlindon T, Formica M, LaValley M, Lehmer M, Kabbara K (2004) Effectiveness of glucosamine for symptoms of knee osteoarthritis: results from an internet-based randomized double-blind controlled trial. Am J Med 117:643–649
86. McAlindon T, Formica M, Kabbara K, LaValley M, Lehmer M (2003) Conducting clinical trials over the internet: feasibility study. BMJ 327:484–487
87. Hochberg MC, Martel-Pelletier J, Monfort J, Möller I, Castillo JR, Arden N, Berenbaum F, Blanco FJ, Conaghan PG, Doménech G, Henrotin Y, Pap T, Richette P, Sawitzke A, du Souich P, Pelletier JP (2015) Combined chondroitin sulfate and glucosamine for painful knee osteoarthritis: a multicentre, randomised, double-blind, non-inferiority trial versus celecoxib. Ann Rheum Dis 2014:206792
88. Tadal M, Oikawal A, Iwata R, Fujiwara T, Kubota K, Matsuzawa T, Sugiyama H, Ido T, Ishiwata K, Sato T (1989) An efficient, one-pot synthesis of 2-deoxy-2-[18F]fluoroacetamido-D-glucopyranose (N-[18F]fluoroacetyl-D-glucosamine), potential diagnostic imaging agent. J Labelled Compd Radiopharm 27:1317–1324
89. Fujiwara T, Kubota K, Sato T, Matsuzawa T, Tada M, Iwata R, Itoh M, Hatazawa J, Sato K, Fukuda H, Ido T (1990) N-[18F] fluoroacetyl-D-glucosamine: a potential agent for cancer diagnosis. J Nucl Med 31:1654–1658
90. Yang Z, Xiong C, Zhang R, Zhu H, Li C (2012) Synthesis and evaluation of 68 Ga-labeled DOTA-2-deoxy-D-glucosamine as a potential radiotracer in μPET imaging. Am J Nucl Med Mol Imaging 2:499–507
91. Qi CM, He Y, Wang X, Feng M, Xu JL, Ding R, Liu H, Chen YR, Li F, Zhu ZH, Dang YH, Zhang ST, Xie Y (2011) Synthesis and evaluation of N-(2-[18F]fluoro-4-nitrobenzoyl)glucosamine: a preliminary report. J Radioanal Nucl Chem 287:913–920
92. Igawa K, Xie MF, Ohba H, Yamada S, Hayashi Y (2014) D-glucosamine conjugation accelerates the labeling efficiency of quantum dots in osteoblastic cells. Biomed Res Int 2014:821607
93. Carroll L, Witney TH, Aboagye EO (2013) Design and synthesis of novel 18F-radiolabelled glucosamine derivatives for cancer imaging. Med Chem Commun 4:653

Functionalized Chitosan: A Quantum Dot-Based Approach for Regenerative Medicine

Hridyesh Kumar and Pradip Kumar Dutta

Abstract Quantum dots (QDs) are the semiconducting inorganic substances that form luminescent nanocrystals with unique optical properties. The formation of shell and or functionalization of it may be utilized as probes or carriers for target-specific cells or tissues for proper utilization in the field of regenerative medicine. Thus, the association of chitosan makes the entire body as biocompatible and suitable for optical stability in physiological environment. QDs-bound hybridization probe design reported for detection of intracellular pre-miRNA using chitosan poly(γ-glutamic acid) complex as a gene vector toward the progress and prognosis of cancer. It is also demonstrated that chitosan-based QD hybrid nanospheres can be internalized by tumor cells and hence act as labeling agent in cell imaging by optical microscopy. The challenge of such cell imaging in the field of molecular imaging is also being discussed. Overall, the interest in using chitosan–QDs in regenerative medicine and the current barriers to moving the technique into the clinic as great challenges will also be discussed.

Keywords Chitosan · Functionalization · Quantum dot · Regenerative medicine

Abbreviations

AFt	Apoferritin
ATP	Adenosine triphosphate
CMC	Carboxymethyl chitosan
CNT	Carbon nanotubes
CD	Cyclodextrin

H. Kumar · P.K. Dutta (✉)
Department of Chemistry, MN National Institute of Technology, Allahabad 211004, U.P, India
e-mail: pkd_437@yahoo.com

P.K. Dutta (ed.), *Chitin and Chitosan for Regenerative Medicine*, Springer Series on Polymer and Composite Materials, DOI 10.1007/978-81-322-2511-9_12

DTG	Dithioglycerol
DCC	*N,N′*:dicyclohexylcarbodiimide
DNA	Deoxyribose nucleic acid
ECM	Extracellular network
EDAC	1-Ethyl-3-(-3-dimethylaminopropyl)carbodiimide
GNP	Gold nanoparticle
GSH	Glutathione
LA	Lipoic acid
MPA	Mercaptopropionic acid
MPS	Mercaptopropyltris(methyloxy)silane
MSA	Mercaptosuccinic acid
MUA	Mercaptoundecanoic acid
NPs	Nanoparticles
NHS	*N*-Hydroxysuccinimide
PEG	Polyethylene glycol
PVA	Polyvinyl alcohol
PVK	Polyvinylcarbazole
QDs	Quantum Dots
RNA	Ribose nucleic acid
SA	Sialic acid
TEM	Transmission electron microscope
TGA	Thioglycolic acid
TGL	Thioglycerol
TOP	Trioctylphosphine
TOPO	Trioctylphosphine oxide

1 Introduction

Nanoscale biological systems such as proteins, lipids, nucleic acids, antibodies, antigens, and enzymes which work throughout the life without any functional problems were synthesized and developed by a chief architect, "Nature." These nanodimension components are the building block components of our life. Nanomaterials together with nanotechnology have fascinated science world for to develop and deliver new tools in the area of therapeutics, molecular imaging, regenerative medicine, and labeling applications. Growth of nanosized probe for understanding of molecular and cell biology results in a new approach for regenerative medicine. Nanosized probe like NPs [1] and QDs [2] has propelled high expectation in modern molecular [3], genetic [4] and biochemical field [5]. It can be assumed that these probes can explain many fundamental questions concerning molecular level at biological system. To explore these fundamental processes in biological sciences entails straightforward tools and technique for easy, fast, accurate, trustworthy, and reproducible detection of targeted molecule in a complicated molecular and genetic reaction.

From the last two decades, QDs emerges as a new light emitting particle, a major breakthrough in the vast and deep area of fluorescent labeling for analysis [6], imaging [7, 8] and therapeutic studies [9]. By their unique electronic and optical properties like size-tunable light emission, narrow and symmetric emission spectra, broad absorption spectra, and simultaneous excitation of multiple fluorescence colors [10], QDs stop domination of organic fluorophores in the field of imaging and labeling [11]. Electronic property of QDs play key role in QDs widespread application. Once the dimension of the QDs goes below than the size of the Bohr radius, quantum confinement effects correlate between band gap and quantization of the energy levels [12].

Regenerative medicine, significant area of research in life science [13, 14], has appealed a countless attention of investigators from several fields such as stem cell research [15–17], tissue engineering [18], therapeutic cloning [19], and gene therapy [20] aiming to develop artificial tissues for repair or replacement of missing or injured tissue, body part and organs [21, 22]. New regenerate tissues are transplanted with the support of scaffolds and biomolecules [23]. Regenerative medicine and tissue engineering are somehow closely related term that can substitute each other. Embryonic stem cells or adult stem cells [24, 25], biomaterials [26] and growth factors [27] are three central constituents' of regenerative medicine and tissue engineering too. So recently, biomaterial is the main focus of research to develop new tissue-specific scaffolds as improved drug delivery systems for regenerative medicine. Three-dimensional permeable frameworks for the most part serves three capacities one is to give basic upkeep of the state of a deformity and void, second is to help as a starting provisional extracellular network (ECM) for cell bond, expansion, separation and development, and third one is capacity as a bioactive atoms, development components, and cells stacked conveyance vehicle for tissue morphogenesis and imperfection mending [28].

Chitosan, a natural cationic polymer, display unique physicochemical properties and their unique ability that allows further adaptation renders them appealing for various therapeutic, regenerative, and biomedical applications [29]. In addition their inherent bioactive properties such as biocompatibility, biodegradability to harmless products, nontoxicity, physiological inertness, antibacterial properties, heavy metal ions chelation, gel forming properties, hydrophilicity, and remarkable affinity to proteins make chitosan more promising for further enhanced regenerative medicines, therapeutic potential and tissue engineering applications, and future demands on bioproducts [30]. Due to their potential to form polyelectrolyte complexes with nucleic acids (DNA, RNA, and PNA) chitosan seeking interest of scientist working in the field of therapeutic applications and regenerative medicines [31]. Chitosan is becoming excellent candidates for gene delivery by providing shield from enzymatic degradation of DNA, RNA, and facilitate cellular uptake and endolysosomal escape. In the arena of regenerative medicine, chitosan is the most extensively investigated because of their suitable significant potential like it can used as an artificial extracellular matrix (ECM) material to study cell–biomaterial interactions, development of tissue-specific chitosan-based scaffolds with or without cells and growth factors, repair and regenerate a variety of human tissues.

2 Functionalized Chitosan

The utmost importance of carbohydrate science using nanotechnology is to advance the current knowledge of existing therapeutic strategies like cell-based transplantation, tissue engineering, and gene therapy. Conjugated saccharide moieties with proteins or lipids form glycoproteins and glycolipids, commonly known as Glycoconjugates, are critical components of biological organisms, including cells, tissues of animals and plants due to its structural maintaining properties.

Historically, in 1811, Professor Henri Braconnot, isolated fibrous substances from mushroom and found them to be insoluble in aqueous acidic solution. A decade later in 1823, Ojer named it "chitin" from Greek "khiton" meaning "envelope" present in certain insects. In 1894, Hope Seyle named it as "chitosan." From 1930 to 1940, this biopolymer of glucosamine gained much interest in the field of medicine [32]. Chitosan, a natural biopolymer, is a polysaccharide composed of repeating units of glucosamine and *N*-acetyl-glucosamine which is produced from chitin, the second most abundant polysaccharide in nature. Chitin is isolated from shells of several crustaceans such as krill and squid pen. Other sources are also there like exoskeleton of insects and cell walls of fungi. Chitin is a high-molecular weight polymer which is then further processed by base hydrolysis under forcing conditions which removes the acetyl groups and at this point the material is generally described as chitosan. *N*-deacetylation of chitin produces chitosan. One can draw a border between chitin and chitosan only after at the point when chitin is sufficiently deacetylated and only after a high degree of deacetylation, the resultant polymer is soluble in dilute acid. Along with molecular weight, degree of deacetylation (DD) plays a main role for deciding the physicochemical properties of chitosan.

The authors' research laboratory has established milestone in the field of functionalization of chitosan by chemicals [33–36], nanoparticles [37–41], QDs [42, 43], and biomolecule [44]. The preparation of chitosan-based QDs, chitosan-GNP derivatives, and fullerene-based chitosan derivatives, and the immobilization of gold nanoparticles onto thiol-functionalized chitosan were reported by author [45].

2.1 Nanoparticles

Inspite of enormous biomedical and theronostics application of nanoparticles, bare nanoparticles still have restricted in many fields. Making these bare nanoparticles functionalized is the new strategy that has been adopted. In this strategy, according to the need ligands have been chosen. Organic ligands, inorganic molecules and ligands, biomolecules like lipid, DNA, RNA, enzymes, polymers, biopolymers like chitin, chitosan, cellulose, etc., have been used it make bare nanoparticles functionalized and quality of their application upsurges rapidly. Biocompatibility and stability are the factors that decide the mode of application. Functionalization is a process which not only enhances its biocompatibility but also it stabilizes nanomaterials against agglomeration to render them compatible with another phase, to use modified inorganic nanofillers in organic polymers, and to enable their self-organization. Any inorganic, organic molecule, ligand or biopolymer material whose incorporation or coating will enhance the functional properties of potential interest like nanoparticles biocompatibility and stability can be used as functionalizing agent. Mostly insertion of well-designed and functional organic groups (–COOH, NH_2, –SH, etc.) or biomolecule (DNA, RNA, antigen, antibody, proteins, etc.) or polymer (chitosan, chitin, PEG, PVA, etc.) on the surface of nanoparticles gives us another efficient tool in the field of biotechnology, biomedical, classical reaction chemistry, solar cells, biosensing, bioimaging, etc. [46]. Preparation and properties of various nanoparticles functionalized chitosan are given in Table 1.

2.2 *Organic Compound/Complex*

A lot of work has been published and still published in the field of functionalization of chitosan. The reason behind functionalization of chitosan is to alter its solubility and other physiochemical properties [61–64]. Introduction of a group not only modify its chemical and physical property but also transform it biological properties. Majority of functionalization reaction of chitosan depend on nucleophilicity of the $–NH_2$ group present on chitosan. Schiff base formation, reductive amination, and amide bond formation cover chitosan chemistry significantly. Biological activity is enhanced by modifying its solubility which is one of the important factor while consider its use in biomedical. Chemically modified chitosan has been used in tissue engineering [65], regenerative medicine [66], wound healing [67], efficient gene delivery [68], etc. Maximum number of chitosan-based reaction follow three type of reaction, i.e., Schiff base formation, reductive amination, and amide bond formation (Fig. 2). Thiolated chitosan is a thiol-functionalized chitosan which show better chelating property due to addition of another group having sufficient electron. Basically, three type of synthesis strategy was followed for preparation of thiolated chitosan which shown in Fig. 3. Reaction of chitosan with different organic reactants are summarized in Table 2.

2.3 *Biomolecule*

Biomolecules like DNA, RNA, Antigen, Antibody, ATP, liposomes, and growth factors are conjugated directly to chitosan. Widely, conjugation strategy is based on either amide bond formation or electrostatic interaction. Positively charged chitosan easily form complex with negatively charged biomolecule. Chitosan is used to form various shape and size of chitosan–biomolecule complex. Chitosan-encapsulated nanoparticle, spherical micro- and macrobeads, nanofiber, gels, and spray dried molecule are successfully synthesized. Some of the chitosan functionalized biomolecules are listed in Table 3.

Table 1 Preparation and properties of nanoparticles functionalized chitosan

Chitosan	Nanoparticles	Synthesis procedure and applications	References
Chitosan	Au and Fe_3O_4	Au NPs/chitosan and magnetic Au NPs/chitosan/Fe_3O_4 composites were synthesized without any toxic chemicals and were used efficiently as biochemical catalysis or therapy	[47]
Chitosan	Au NPs	Au NPs were synthesized on chitosan that further transform into self-collapsing gels. The gels show good biocompatibility that can be used to encapsulate drug for controlled release in drug delivery	[48]
Chitosan	Fe_3O_4	Fe_3O_4–chitosan nanocatalyst was used as a reusable organocatalyst. Reaction between substituted benzils and urea/thiourea derivatives were catalyzed by nanocatalyst for synthesis of 5,5-diphenyl-hydantoins and 5,5-diphenyl-2-thiohydantoins effectively	[49]
Chitosan	Silica coated Fe_3O_4	Silica coated Fe_3O_4 NPs were functionalized by biopolymer, chitosan. So formed Fe_3O_4-silica-chitosan NPs showed outstanding adsorbent for DNA with efficiency of ~88 % and a high elution efficiency of ~98 %	[50]
Chitosan	Copper-silver NPs	In the synthesis of core-shell copper-silver NPs, chitosan used as a template. Reduction of $CuSO_4$ was done by hydrazine hydrate in the presence of chitosan as stabilizer then $AgNO_3$ was added to shell Cu NPs. Synthesized CS–Cu–Ag NP composite was assessed on *Escherichia coli* and *Bacillus cereus* bacteria by turbidity measurement and flowcytometry. Result revealed that the CS-Cu@Ag NP composite was highly effective in reducing bacterial growth due to synergistic effect of CS and Cu@Ag NPs	[51]
Chitosan oligosaccharide	Au NPs	In this study Au NPs was stabilized by multidentate zwitterionic chitosan oligosaccharide. Lipoic acid and zwitterionic acryloyloxyethyl phosphorylcholine were used to obtain zwitterionic chitosan. Dithiolane end have high affinity to Au metals. Zwitterionic modifcation of chitosan remarkably boosted the stability and biocompatibility of Au NPs. The Zwitterionic chitosan-modified Au NPs might moderately escape from endosomes/lysosomes to cytosol and accrue around the nucleus; which can be used as intracellular delivery efficacy of NPs for biomedical applications	[52]
Chitosan	Fe_3O_4, Au NPs	In this paper, chitosan was modified to carboxymethyl chitosan and fabricate with Fe_3O_4 NPs. Then this chitosan–Fe_3O_4 nanoparticles were thiolated by 2-iminothiolane which were used to conjugate with Au NPs. To develop the biorecognition toward J5 cancer target cells, Au–chitosan-Fe_3O_4 nanoparticles were further modified to Transferrin	[53]

(continued)

Table 1 (continued)

Chitosan	Nanoparticles	Synthesis procedure and applications	References
Chitosan	Silver-ruthenium hybrid nanoparticles	MPA-stabilized Ag NPs were synthesized by in situ reduction reaction with $NaBH_4$. Thiol group the Ag surface atoms. Electrostatic interaction chemistry were used to form stable bond between COO^- groups of AgNPs stabilized with 3-MPA and aqueous $[Ru(bpy_3)]^{2+}$ complex . Abundant amine groups of chitosan helped to form a hybrid having coordination bond in between AgNPs@ $[Ru(bpy)_3]^{2+}$. Chitosan-modified core-shell NPs exhibiting a selective fluorescence quenching effect. Aqueous solutions of sialic acid (SA) with interferents exhibited significant fluorescence quenching of chitosan–hybrid NPs in comparision with SA free interferent solutions	[54]
Chitosan	Cu NPs	Different chemically modified chitosan–carbon (carbon nanotubes or graphite) were prepared to create new environment for Cu NPs growth. chitosan was first react with graphite then this graphite-modified chitosan was transform into *N*-carboxyethyl chitosan methyl ester–graphite, amine group grafted chitosan–graphite and thiol group grafted chitosan–graphite by different chemical reactions as shown in Fig. 1. These different group grafted chitosan-graphite was decorated with Cu NPs by reduction of copper (II) sulfate by sodium borohydride. This modified chitosan/CNT/CuNPs utilize to prepare a sensitive sensor application for glucose oxidation and hydrogen peroxide reduction	[55]
Chitosan	ZnO NPs	Xanthine oxidase (XOD), a Mo, Fe, and S containing flavoprotein was immobilized on ZnO NPs/chitosan/carboxylated MWCNT/polyaniline modified Pt electrode. This electrode was working electrode and Ag/AgCl as reference electrode. This amperometric xanthine biosensor successfully detects xanthine level in fish meat during its 16 days of storage at room temperature. The ZnO/MWCNT decorated Pt electrode was checked up to 80 uses over 1 month and 30 % activity lost was measured	[56]
Chitosan	Cu NPs	Iodine-stabilized Cu NPs were prepared in chitosan solution and antibacterial activity was checked against Gram-negative green fluorescent expressing *E. coli* and Gram-positive *B. cereus* bacteria. $CuSO_4$ were reduced in situ in chitosan environment by hydrazine to synthesize Cu NPs. Molecular iodine was used to stabilize Cu NPs and also enhance efficacy as a potent antibacterial agent	[57]

(continued)

Table 1 (continued)

Chitosan	Nanoparticles	Synthesis procedure and applications	References
Chitosan	Ag NPs	Ag NPs were grown in situ by UV irradiation on the cellulose microfibril bundles coated with titania/chitosan composite films. Titania/chitosan film were deposited layer by layer on common filter paper then chelating power of chitosan is used to form chelate with silver ion which was tuned to silver nanoparticles by UV irradiation (365 nm, 16 W) for 2 h. Evidently, 4–20 nm diameters polydisperse Ag-NPs on cellulose/titania/chitosan composite microfibril were display impressive antibacterial activities against *E. coli* and *Staphylococcus aureus*	[58]
Chitosan	$CoFe_2O_4$	In this paper, simple and general protocol was described to produce well disperse and stable in solution $CCoFe_2O_4$ NPs in chitosan polymer. Chitosan were also functionalized $CoFe_2O_4$ NPs to enhance radiation released through the chemiluminescence oxidation of luminol by H_2O_2. Prepared NPs show its catalytic performance to H_2O_2 decomposition into $^{\cdot}OH$ radicals	[59]
Chitosan	ZnO	Sonochemical method was used to prepare hybrid antimicrobial coatings of ZnO NPs presence of chitosan on cotton fabrics. Chitosan-enhanced antibacterial efficiency of ZnO NPs against *S. aureus and E. coli* and chitosan significantly upgraded the biocompatibility of the ZnO coatings without any adverse effects on human health. ZnO–chitosan treated fabrics showed potential resistant to *S. aureus and E. coli* even after multiple washing cycles at 75 °C	[60]

Table 2 Preparation and properties of organic compound functionalized chitosan

Chitosan	Organic reactant	Reaction	Final product	Application/property	References
Chitosan	2-Pyridine carboxaldehyde	Schiff's base formation	Chitosan-pyridyl-imine complex	Adsorbents and catalysts	[69]
Chitosan	2-*O*-(Formylmethylated)-α-cyclodextrin	Reductive amination $NaCNBH_3$	α-CD–Chitosan complex	α-CD–Chitosan show better inclusion ability with *p*-nitro phenolate which is useful in cosmetic industry	[70]
Chitosan	β-Cyclodextrin		β-Cyclodextrin-2-chitosan	Absorption of iodine	[71]
Chitosan	Salicylaldehyde derivative	Schiff's Base Formation	Chitosan-salicyldehyde Schiff base	Heterogeneous catalysts	[72]
Chitosan	4-Mercaptobenzoic acid	Amide bond formation	Chitosan-4-Mercaptobenzoic acid	Oral or in situ gelling formulations	[73]
Chitosan	Aldehyde-ending sialodendrimers	Reductive Amination $NaCNBH_3$	Chitosan sialic acid dendrimer hybrid	Synthesis and property	[74]
Chitosan	Acrylic acid esters	Michael Reaction	*N*-Alkylated chitosan product having PEG, quaternary ammonium, and an amide group	Modified chitosan derivatives are water-soluble and biodegradable	[75]
Chitosan	Acyl chloride, $MeSO_3H$	Acetylation	*N*,*O*-acetylated chitosan derivative	Water soluble, stable in water, important role on the biological activity	[76]
Chitosan	*p*-Formylphenyl β-melibioside, $NaCNBH_3$	Reductive amination	α-Galactosyl-chitosan conjugates	α-Galactosyl-chitosan are water soluble and show robust affinity toward human anti α-galactoside antibodies	[77]

(continued)

Table 2 (continued)

Chitosan	Organic reactant	Reaction	Final product	Application/property	References
Chitosan	Lauryl aldehyde	Schiff's base formation	*N,N*-Dilauryl chitosan pentamer	Stable and monolayer of Dilauryl chitosan pentamer were formed on water and solid surface with 1.7 nm thickness	[78]
Chitosan	Phosphorous acid, formaldehyde	Reductive amination	*N*-Methylene phosphonic chitosan	Powerful chelating agent	[79]
Chitosan	Phosphorous acid, formaldehyde, poly(ethyleneglycol)-aldehyde	Reductive amination	Poly(ethylene glycol)-crosslinked *N*-methylene phosphonic chitosan	Prepared film leads to cell attachment and proliferation	[80]
Chitosan	Lactobionic acid	Amide formation (EDC and NHS)	Galactosylated chitosan	Galactosylated chitosan coated polystyrene disk has been used for hepatocytes adhesion	[81]
Chitosan	Deoxycholic acid	Amide formation (EDC)	Deoxycholic acid-modified chitosan	Modified chitosan was used for physical entrapment and carrier of adriamycin drug	[82]
Glycol chitosan	Palmitic acid *N*-hydroxysuccinimide (PNS)	Amide formation	Chitosan laurylamide	Synthesized hydrogel was used as erodible controlled release system for the delivery of hydrophilic macromolecules	[83]
Chitosan	Nicotinic acid, L-Alanine	Amide bond formation	Chitosan dihydronicotinamide conjugates	Ethyl benzoylformate was reduced by the conjugate. Reducing performance, asymmetric, selectivity and chemical yield were increased by introduction of L-alanine	[84]

(continued)

Table 2 (continued)

Chitosan	Organic reactant	Reaction	Final product	Application/property	References
Chitosan	Nicotinic acid, L- and D-phenylalanine	Amide bond formation	Chitosan dihydronicotinamide conjugates	Effect of L- and D-phenylalanine spacer arms on asymmetric reducing performance of chitosan dihydronicotinamide	[85]
Chitosan	Laminin-related peptide	Amide bond formation	Chitosan hybrid of a laminin-related peptide (Ac-Tyr-Ile-Gly-Ser-Arg(HCl)-βAla-Gly-chitosan)	Synthesized conjugate enhanced the antimetastatic effect in mice and can be used as drug carrier	[86]
Chitosan	Dodecanoyl chloride	Acylation	Poly(acyl)chitosans	Synthesis of poly(acyl)chitosan soluble in organic solvent	[87]
Chitosan	Maltodextrin	Reductive Amination $NaCNBH_3$	Chitosan-*graft*-maltodextrin	The copolymer was used as DNA vector. maltodextrin enhances water solubility, stability and self-assembling ability into spherical nanoparticles	[88]

Table 3 Preparation and properties of biomolecule functionalized chitosan

Chitosan	Biomolecule	Application	References
Chitosan oligosaccharide	Adenosine triphosphate (ATP)	ATP was encapsulated in chitosan oligosaccharide (CSO) via W/O miniemulsion, chemical cross-linking and ionic complexation. CSO NPs with 40.6–69.5 % ATP loading efficiencies formed ATP/CSO ionic complex having 110 nm diameter and 20 mV surface zeta potential. ATP loading efficiency and ATP release rate were affected by amount and the molecular weight of chitosan oligosaccharide and molar ratio of glutaraldehyde to chitosan oligosaccharide	[89]
Chitosan	DNA	Coacervation process was used to prepare chitosan-DNA NPs by varying important parameters. pH value regulate surface charge of these particles, i.e., zeta potential. The transfection efficiency of chitosan-DNA NPs was 3–4 orders of degree than background level in HEK293 cells and two to ten times lesser than that attained by LipofectAMINE-DNA	[90]
Chitosan	siRNA	Chitosan-siRNA nanoparticles were formulated for RNA-mediated therapy of systemic and mucosal disease. Formulated nanoparticles were also introduced as novel delivery system for RNA interference in vitro (Peritoneal macrophages from GFP transgenic mice) and in vivo (H1299 GFP)	[91]
Chitosan	Liposome	Chitosan were used to coat liposomal to be tested as the new siRNA delivery system. Chitosan coated liposome loaded with siHIF1-α and siVEGF showed good stability and guard siRNA from serum deprivation after 1 day incubation and prominently quash expression level of VEGF mRNA in MCF-7 and MDA-MB435 cells	[92]

2.4 Inorganic Compound

Chitosan and its derivatives are projected as effective selective adsorbents and in high demand due to recovery, regeneration, and separation of precious and other industrially important metals. Chelating property of chitosan is used to adsorb an inorganic salt or complex on chitosan surface. When these chitosan–inorganic complexes are using as a catalyst, it showed its excellent efficiency in synthetic organic chemistry as compared to bare inorganic complex. Also chitosan–inorganic complex were reported effective as potential antitumor agents, antimicrobial, and antibacterial activities. Some of them are summarized in Table 4.

3 Quantum Dots

Semiconductor nanocrystals materials with characteristic size in the range between 1 and 20 nm are popularly known as QDs. Nanosized luminescent semiconductor crystal from group II–VI elements (e.g., CdS, CdSe, CdTe, ZnS, ZnSe, ZnTe) and III–V elements (e.g., GaN, GaP, GaAs, GaSb, InN, InP, InSb, InAs) are most studied and applied. As the structure is concerned in Table 5, different QDs show different crystal structure.

The unique property of QDs arises from its size and electronic property. Quantum confinement effect works at the place where size of the matter is below Bohr radius as a result of it energy bandgap between the valence and conduction bands increases. Quantum confinement is resulted by opposite charge carrier interaction or coulomb interaction. QDs are photoelectrochemically active. When QDs were excited by a photon, energy was absorbed by valence band electron. This high energy electron jump to conduction band leaving a hole behind. QDs most prominent quality, i.e., luminescence is a phenomenon that originates from the recombination route of the photogenerated holes and electrons from the valence and conduction bands across the bandgap. Size of the QDs regulates the emission wavelength directly. Figure 1 shows tuning in color of L-cysteine-capped CdTe QDs by collecting the samples from reaction solution at different time [42]. As the reaction proceed with time, size of the QDs increases resulting emission to higher wavelengths. The small-sized semiconductor crystal has a large band gap which means it needs more energy to excite its valance band electron to jump into conduction band electron leaving a hole behind. When this electron returns for recombination, it releases more radiative energy which we see in the form of light. This is the reason why when we move from smaller (2 nm) QDs to larger (6 nm), we see tuning over the entire visible part of the electromagnetic spectrum, from blue to red as shown in Fig. 4.

Brus and Henglein set a milestone by successfully synthesizing colloidal QDs in 1980s [101, 103]. The synthesis procedure uses basic principles of aqueous and ionic chemistry which lead to increase our knowledge about QDs's electronic

Table 4 Preparation and properties of inorganic compound/complex functionalized chitosan

Chitosan	Inorganic compound	Applications	References
Chitosan nanofiber	Paladium (II)	Palladium-catalyzed Mizoroki–Heck cross-coupling reaction is used for formation of carbon–carbon bonds. Glutaraldehyde crosslinked chitosan nanofiber were used as a platform to support Pd. chitosan nanofibres-Pd complex with high surface to volume ratio was formed soaking of chitosan naofiber in aqueous Na_2PdCl_4. Nanofiber-Pd composite materials were found effective without loss of activity and ultrafine catalytic loading of the metal at 0.17 mol%	[93]
Chitosan	Neodymium	Neodymium modified chitosan were prepared to for the removal of excess fluoride ions from water. Effect of temperature, pH, adsorbent dose, particle size and the presence of co-anions on elimination of fluoride ions were reported	[94]
Chitosan	Copper (II)	Copper (II) chitosan had been produced by simple mixing of copper sulfate in chitosan solution at different molar ratios. Different Chitosan-Cu complexes were tested in vitro (293 cells and HeLa cells) and founde effective as potential antitumor agents	[95]
Chitosan	Ruthenium (II)	Bis(2,2′-bipyridine)[4-methyl-4′-(6-bromohexyl)-2,2′-bipyridine] ruthenium (II) perchlorate were mixed with chitosan solution to form chitosan-Ru (II) complex. Chitosan-Ru (II) coated platinum electrode was used in electrochemiluminescence detection of oxalic acid	[96]
Chitosan	Nickel (II)	Immobilization of Ni (II) ion along with antibiotic cefixime using chitosan matrix were used sensor for methanol sensor. Chitosan-Ni modified glass electrode showed a vast electrocatalytic movement to the oxidation of methanol oxidation intermediates to the last product	[97]
Thiourea Chitosan	Silver	Thiurea chitosan were prepared in ethanol by mixing of ammonium thiocyanate with chitosan. Silver ion (Ag^+) form complex with thiourea chitosan by coordination with sulpher as major electron donor and oxygen as minor electron donor. Antimicrobial activities of the complex were checked against *S. aureus, E. coli, Bacillus subtilis, Aspergillus flavus, Mucor kacilliformis, Paccilomgces uarioti* and found MIC values much lower than chitosan	[98]
Carboxymethylchitosan (CMC)	Titanium (Ti)	Dopamine anchored Ti substrate was functionalized with CMC in presence of EDAC/NHS. Bone morphogenetic protein-2 (BMP-2) was then conjugated with CMC functionalized Ti substrate. Bacterial adhesion and cell functions were assayed with bacteria (*S. aureus* and *Staphylococcus epidermidis*) and osteoblasts and human bone marrow-derived mesenchymal stem cells respectively	[99]

Table 5 Bulk semiconductor parameter [100]

Type	Bulk semiconductor	Band gap E_{gap} (eV)	Crystal structure
II–VI	CdS,	2.49	Wurtzite
	CdSe	1.74	
	CdTe	1.43	Zinc Blend
	ZnS	3.61	Zinc Blend
	ZnSe	2.69	
	ZnTe	2.39	
III–V	GaN	3.44	Wurtzite
	GaP	2.27	Zinc Blend
	GaAs	1.42	
	GaSb	0.75	
	InN	0.8	Wurtzite
	InP	1.35	Zinc Blend
	InAs	0.35	
	InSb	0.23	
IV–VI	PbS	0.41	Sodium Chloride
	PbSe	0.28	
	PbTe	0.31	

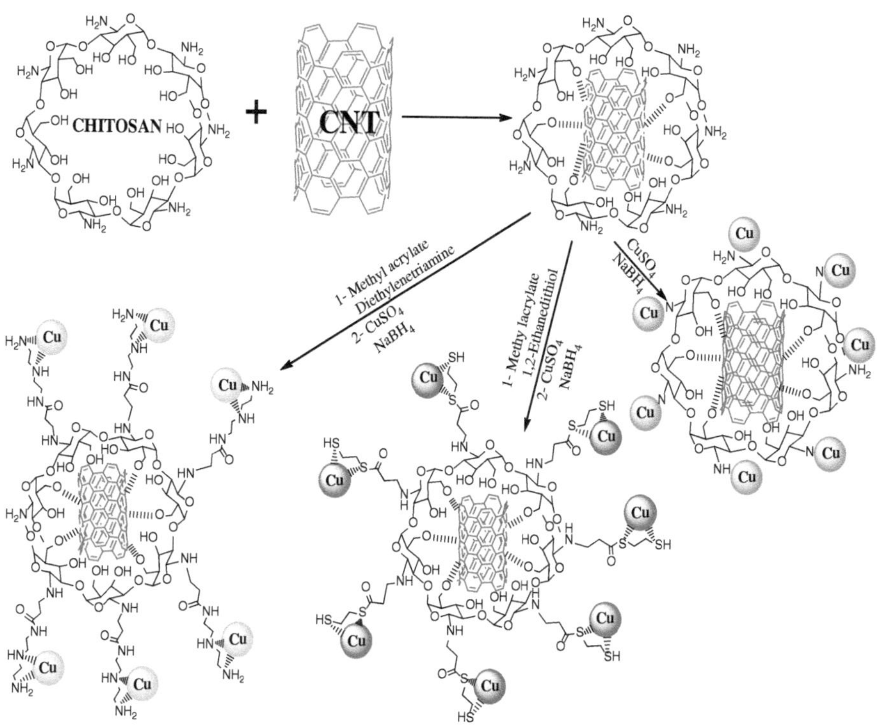

Fig. 1 Schematic presentation of synthesis of Cu NPs on various chitosan-CNT derivatives [55]

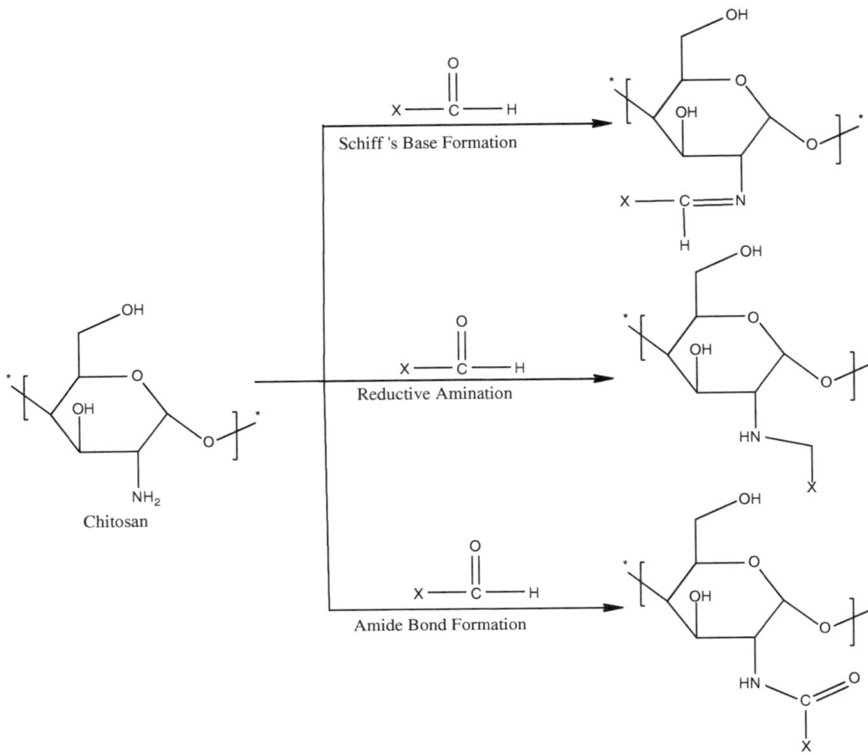

Fig. 2 Reaction on $-NH_2$ of chitosan through schiff base formation, reductive amination, and amide bond formation

structure. Synthesis till 1990s were based on either one phase or two phase synthesis where stabilizers (surfactant-type or polymer-type) were used for precipitation in homogeneous aqueous solutions in one phase synthesis and inverse micelles-mediated precipitation of QDs by amphiphilic surfactant was two phase synthesis [104–106].

First, revolutionary synthesis of high-quality optical colloidal QDs was reported by Murray, Norris and Bawendi in 1993 where organic solvent is used to produce QDs with narrow size distribution high quantum yield by pyrolysis of organometallic compound with hot coordinating ligands [107]. Dimethyl cadmium used as a cadmium precursor, phosphine selenide used as a selenium precursor, phosphine telluride used as a tellurium precursor, and bistrimethylsilyl sulfide used as a sulpher precursor in a mixture of coordinating solvent like trioctylphosphine (TOP) and trioctylphosphine oxide (TOPO). Discrete homogeneous nucleation, slow growth, and annealing were the key steps in their synthesis. The use of the coordinating solvent mixture is to provide high temperature ($\approx$300 °C) which

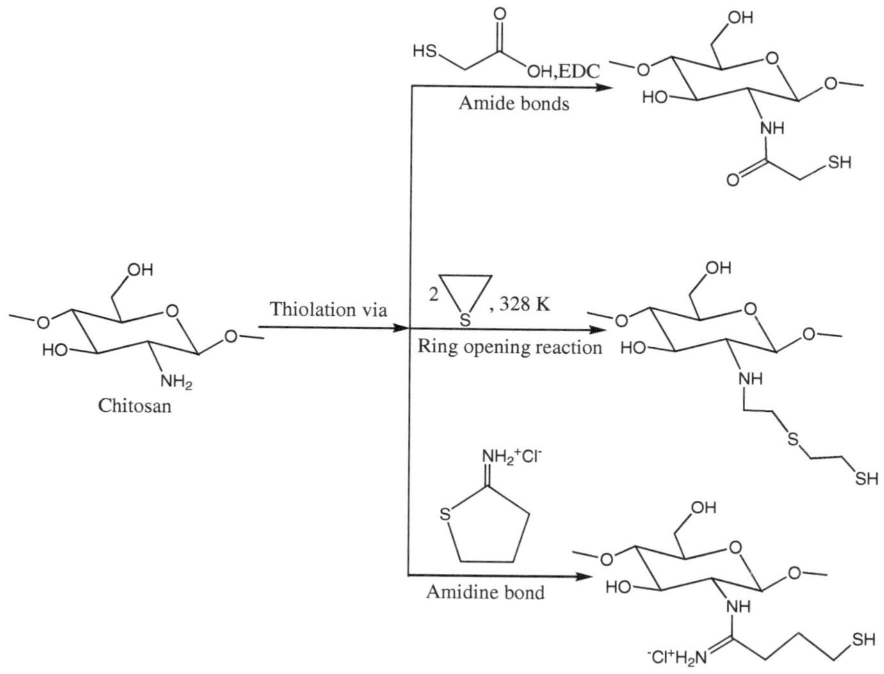

Fig. 3 Chemistry of functionalization of chitosan with thiol-containing organic compound via amide bond, amidine bond formation and ring opening reaction

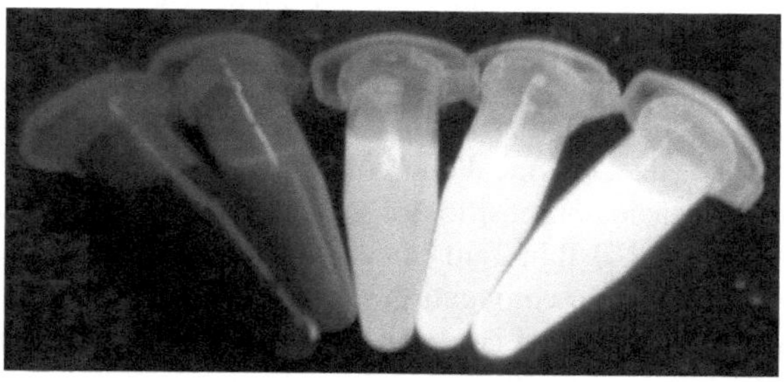

Fig. 4 Showing tuning in color of L-cysteine-capped CdTe QDs at different time intermittent fluorescence was a blinking phenomenon of QDs occurred when during crystal growth some surface defects developed in the structure act which trap the electron or hole preventing the radiative recombination and results alternation of trapping and untrapping events which cause blinking

potentiated crystallinity, annealed out defects, control size, and monodispersity. The hot-injection method is a quick injection of organometallic precursors into an organic solvent at high temperature.

Synthesis of QDs with high-quality optical and electronic properties got a new energy when Murray, Norris and Bawendi developed a new route by pyrolysis of organometallic reagents in hot coordinating solvent to synthesize CdE (E = S, Se, Te). For high quantum yield and narrow size distribution, the main steps in their synthesis were discrete homogeneous nucleation, slow growth, and annealing. Methods used in starting years to synthesize QDs utilize hydrophobic molecule like TOPO, TOP, and TBP as a ligand for surface passivation surface of such QDs is covered by hydrophobic molecules such as TOPO, TOP, and TBP. On the other hand, QDs with hydrophilic surface molecules and reactive functional groups are necessary for biological applications.

The challenges related to synthesis of QDs bioconjugate are not trifling. Due to fascinating optical and electronic properties, applications of QDs in the applied field biology, nanotechnology, and regenerative studies are significantly increasing. For practical application nanosized QDs must be modified chemically or fixed in a solid matrix. Regardless potential and demonstrated application of QDs ultimately depends, on a large extent, on the ability to bioconjugate regardless of whatever the formation chemistry utilized in a controlled and desirable manner. QDs bioconjugate is defined as an assembly where QDs pair with biomolecules such as protein (antibodies, enzymes, collagen, receptors, myoglobin, etc.), peptides (glutathione, cell penetrating peptides, enzyme inhibitors), nucleic acids (DNA,RNA, ribozymes, DNAzymes), fatty acids, lipids, carbohydrates, chitin, chitosan, chitosan derivatives, and all related members of each of these super-families of biomolecules [108]. Intrinsic functionalities of these QDs decorated biomolecules are not altered and they can explore in targeting and imaging purposes. The biomolecule act as an encapsulating agent or structural intermediary. All other those are actively participate in biological application (drugs, enzymatic cofactors) or that deliver a chosen activity in the living matrix (fluorophores, radiolabels, MRI probes) are also can be used to attach or encapsulate QDs to look after the change in process inside the living organism [109].

Surface modification strategies are mainly governed by QDs size, shape, structure and surface chemistry. Other factors that dictated bioconjugation strategies include nature, shape, size, chemical composition and functional group availability of surface ligand. Intrinsic composition of QDs also plays an important role while choosing a surface ligand. Aubin-Tam et al. showed how size, shape, structure, and morphology took attention in conjugation strategy. Surface modification is mainly achieved by ligand exchange process, surface silanization and amphiphilic combination. Ligand exchange is a process where existing nonfunctional ligands are replaced by mono or multifunctional ligand. Ligand exchange delivers QDs with bonus properties such as water solubility, flexibility, biocompatibility, and aiming [110, 111]. Hexadecylamine (HDA), trioctylphosphine (TOP), trioctylphosphine oxide (TOPO), oleic acid (OA), and tetradecyl phosphonic acid (TDPA) are the organic ligands having long chain to provide excellent stability and solubility in organic non-coordinating solvents, suppress continues crystal growth, and aggregation during synthesis and passivate surface defects to retain its fluorescence [112–115]. Ligands maintain dynamic equilibrium with surface ligands and ligand in solvent. They attach and detach continuously on the surface of QDs

[116, 117]. Dynamic equilibrium process shift toward forward side by adding new ligand. New ligand competes for the surface of QDs for passivation. Concentration of the replacing ligand should be higher than existing ligand for maximum ligand exchange on QDs surface [118]. If replacing ligand has low passivation affinity even then increasing concentration of replacing ligand push ligand exchange because the probability of attaching ligand was increased [119].

In a biological system, QDs could induce toxicity either by slow deliverance of cadmium ion by oxidation of QD on ultraviolet light introduction or it involves the generation of reactive oxygen species (ROS) like hydroxyl, superoxide free radicals, and singlet oxygen, therefore before biomedical applicationwide analysis is obviously desired. Strength and quality of capping agent regulate the appearance symptom of meddling with cell functioning by release of cadmium ion [120]. ROS is responsible for permanent damage to nucleic acids, enzymes, plasma and nuclear membranes, and cellular components [121].

3.1 Cadmium Chalcogenides (CdX)

A chalcogenides are a chemical compound containing of at least one chalcogen anion and at least one cation. Group 16 elements are demarcated as chalcogens but sulfides, selenides, and tellurides are commonly termed as chalcogenides. Cadmium chalcogenides (CdX) (X = S, Se, Te) were known as first synthesized chalcogenides. In early phase synthesis of CdX, organic solvent were used for synthesis of CdX by mixing a solution of trioctylphosphine (TOP) chalcogenide (either S, Se, or Te) and cadmium alkylphosphonate (a mixture of hexylphosphonate and octadecylphosphonate) in trioctylphosphine oxide (TOPO) at 280–380 °C, under nitrogen. Now many easy methods were developed using aqueous solution and at room temperature. In Table 6, many important cadmium chalcogenides are summarized along with their short synthesis procedure and application.

3.2 Zinc Chalcogenides

Non- or low-toxic nanosized semiconductor materials are always in demand and zinc chalcogenides appeal researcher's attention due to the environment benign properties. Most importantly, decreasing demand of heavy metal pollutants like Cd, Hg, Pb in preparation of semiconductor QDs which serve our society without much harmful effect. The most innovative materials developed via an aqueous approach among zinc chalcogenides are ZnSe NCs. Shavel et al. successfully developed aqueous colloidal synthesis of strongly emitting ZnSe NCs [135]. Table 7 summarizes synthesis of zinc chalcogenides core QDs with different shell composition and their applications.

Table 6 Core, shell, coating material, short synthesis procedure, and application of selected cadmium chalcogenides

Core QD	Shell/NP	Coating material	Synthesis	Application/conclusion	References
CdSe		Polyvinylcarbazole (PVK) and an oxadiazole derivative	Organic phase synthesis of QDs, TOPO passivated	Electroluminescence efficiency of the films rises when it cooled down to very low temperatures	[122]
CdSe	ZnS	Maltose Binding Protein-Basic Leucine Zipper Fusion Protein	TOP/TOPO passivated synthesis of QDs then ligand exchange with dihydrolipoic acid	Purified proteins were successfully labeled with QDs by facile electrostatic conjugation	[123]
CdSe	ZnS	Silica	Silica-capped CdSe/ZnS was prepared by silanization (using mercaptopropyltris (methyloxy)silane (MPS) of TOPO-capped CdSe/ZnS	Silanized CdSe/ZnS exhibit excellent optical properties, photochemical stability, water/PBS stability, and permit conjugation biological entities	[124]
CdSe	ZnS	Lipoic acid (LA) based ligands	TOP/TOPO-QDs were UV irradiated in presence different reactant having LA-PEG	Develop a strategy to transfer of QDs from nonpolar to polar solvents and buffers mediated by UV-irradiation	[125]
CdSe		Acid-functionalized poly(vinyl alcohol) (PVA-COOH)	Cadmium and selenium precursor was mixed in PVA-COOH solution and PVA-COOH-capped CdSe QDs were synthesized in situ	Results conclude that PVA–COOH at concentration of 1.0 $molL^{-1}$ was found effective to prepare ~2 nm size CdSe QDs	[126]
CdSe	CdS	Tri-sodium citrate	Fe_3O_4 NPs was prepare by coprecipitation method then Cd (II), Se (II),Fe_3O_4 NPs, and citrate were used as seeds for the preparation of the CdSe/CdS-Fe_3O_4 nanocomposites	Facile and green methods under mild conditions were used for preparing low cytotoxicity bifunctional nanocomposites that could be directed with a magnetic force for in vitro delivery into a human breast cancer cells (MCF-7)	[127]

(continued)

Table 6 (continued)

Core QD	Shell/ NP	Coating material	Synthesis	Application/conclusion	References
CdSe		Bacterial cellulose nanofibers (BC)	BC was used as a matrix to grow CdSe by in situ method	CdSe/BC nanocomposite shown decent optical properties, superb mechanical properties which could be used in security papers, sensors and flexible luminescent membranes	[128]
CdSe	CdS	Dendron ligand with two carboxylate groups	Dendron-coated CdSe/CdS were prepared by sonication of alkylamine coated CdSe/CdS and Dendron ligand in $CHCl_3$ for 4 h at pH = 10	Aliquots of Dendron-coated CdSe/CdS was highly luminescent, stable and soluble in various aqueous media, including all common biological buffer solutions tested, for at least 1.5 years	[129]
CdTe		MUA, MPA	Microwave-assisted MPA and MUA-capped CdTe QDs were prepared by using $CdCl_2$ as a Cd source and NaHTe as a Te source in a microwave-assisted heating system for different time	By using different capping agents, weak relationship was shown in between brightness per particle (BPP) and photoluminescence quantum yield (PLQY) and QDs were used as a probe for fluorescence imaging	[130]
CdTe	ZnTe	Cysteine	Aqueous synthesis of CdTe/ZnTe QDs involved mixing $NaBH_4$ with Te power to produce Te solution, which was then introduced into the Cd and Zn precursor solution under vigorous stirring, and heating at 100 °C	The fabricated CdTe/ZnTe QDs have an ultrasmall (3-5 nm) size, high QY (52 %), nontoxic to cells and IC_{50} value of these QDs is ≈ 160 μgmL^{-1}. Further specific targeting of pancreatic and esophageal cancer cells in vitro and time-dependent biodistribution, long-term fate, and tumor targeting ability in in vivo were the application	[131]

(continued)

Table 6 (continued)

Core QD	Shell/ NP	Coating material	Synthesis	Application/conclusion	References
CdTe	CdS	MPA, octadecylamine-modified poly(acrylicacid)	Amphiphilic polymer encapsulates the QDs when MPA-capped CdTe/CdS QDs were exchanged by L-dodecanethiol	Encapsulation by amphiphilic polymer significantly improves the optical properties (PLQY = 50 %) and also plays a crucial role to advances its photostability	[132]
CdTe		GSH, 3-mercaptopropyl trimethoxysilane	Displacement of the GSH molecules from CdTe with MPS and promotion of the condensation reaction of silane precursors along the QD surfaces	Silica coating rises the biocompatibility of CdTe by protecting biological milieu to direct contact of CdTe in vivo applications. Silica-coated CdTe exhibit a different biodistribution pattern compared to commercially available QDs	[133]
CdTe		Thioglycolic acid (TGA), rutin and caffeine	Aqueous solution of TGA-CdTe QDs were mixed with rutin and caffeine and kept in an ultrasonic bath for 15–20 min to form QDs–rutin and QDs–caffeine complex	Antibacterial activity was checked by disk diffusion method against *E. coli.* Rutin efficiently quench the CdTe-TGA QDs fluorescence emission compared to caffeine and QDs–rutin conjugate improved antibacterial activity against *E. coli*	[134]

Table 7 Core, shell, coating material, short synthesis procedure and application of selected zinc chalcogenides

Core QD	Shell	Coating material	Synthesis	Application/conclusion	References
ZnSe	ZnS, ZnSeS		By the use of ZnO as a Zn precursor, TOPS as a S precursor and TOPSe as Se precursor in lauric acid/hexadecylamine solution, colloidal ZnSe, ZnSe/ZnS, and ZnSe/ZnSeS QDs were synthesized	TEM images showed the average diameter of QDs in the range of 2.5–6 nm. The PL was found from 400 to 440 nm and PLQY was 6–10 % at room temperature but After coating with ZnS by two-step method, PLQY of ZnSe/ZnS QDs were amplified ~4.5 times (PLQY ~ 32 %). By coating with ZnSeS PLQY were improved ~3.8 times	[136]
ZnSe	ZnS	MPA	At basic pH zinc perchlorate was added in aqueous MPA solution. Then after adding selenium (Se + $NaBH_4$) precursor, solution transfer in Teflon-lined autoclave and place in oven at 70 °C	The substitution of ZnSe/ZnS for CdSe/ZnS nanocrystals as energy donors enabled the use of a green fluorescent protein, GFP5, as energy acceptor. Violet emitting ZnSe/ZnS nanocrystals and green GFP5 constitute a system with 43 % FRET efficiency and an unusually strong sensitized emission	[137]
ZnS		Graphene	ZnS QDs and ZnS quantum flakes were successfully prepared on graphene nanosheet used as a template. Thiourea, zinc chloride, and graphene ethanol solution was sealed in Teflon-lined steel autoclave at 180 °C for 12–16 h	ZnS/graphene quantum flakes-on-sheet heterostructures show higher photovoltage	[138]

(continued)

Table 7 (continued)

Core QD	Shell	Coating material	Synthesis	Application/conclusion	References
ZnSe	ZnS	MPA	ZnSe/ZnS core/shell was synthesized via a two-step method. In first step oleic acid-capped ZnSe was transferred into MPA solution in water for ligand exchange then shell ZnS was coated	UV-blue emitting ZnSe/ZnS core/shell QDs were prepared that showed water-solubility and highly. QDs give out tunable emissions ranging from 390 to 460 nm, high crystallinity, high PL quantum yield up to 45 %	[139]
ZnSe	InP, ZnS		ZnSe cores were synthesized with Et_2Zn and TOP-Se as the Zn and Se precursors. To form the InP shell, Tributylphosphine indium (TBP-In) and tris(trimethylsilyl) phosphine ($(TMS)_3P$) was added to the core solution at 210 °C and for ZnS shell, zinc acetate, and dodecanethiol were mixed in ZnSe/InP solution at 230 °C	QY for ZnSe/InP core/shell QDs was found between 5 and 13 %. When ZnS was added as another shell layer QY was increased to 60 %. by controlling InP shell thickness, emission color could be tuned from violet to red. The mean particle diameter of the ZnSe core was 3 nm, for the ZnSe/InP 4.4 nm and for the ZnSe/InP/ZnS it was 5.4 nm	[140]
ZnSe		MPA	Organic solvent soluble Mn^{2+}-doped ZnSe QDs were made to be water soluble by coating them with a hydrophilic thiol (mercaptopropionic acid)	Effect of Mn^{2+} and Cu^{2+} doping agent were studied in ZnSe QDs. It was concluded that position of doping center in QDs surface control the photoluminescence property. 7–8 nm sized QDs were found excellent permeability in biological issues. Doped QDs were conjugated with avidine and used for the recognition of a biotin pattern	[141]

3.3 Lead Chalcogenides

Lead chalcogenides are group III–V semiconductor materials which are most commonly used in modern infrared optoelectronics. It has rocksalt crystal lattice and face-centered cubic unit cell structure. After synthesis of first lead chalcogenides, first application was observed in infrared detector. Synthesis and application of lead chalcogenides core with dfferent shell are summarized in Table 8.

3.4 Carbon Dot

Buckminsterfullerene (C60), single-walled (SWCNT)/multiple-walled (MWCNT), nanodiamonds, carbon nanofibers, and graphene are the family members of carbon nanomaterials. These members are well known for their remarkable physicochemical properties. Carbon Dot is the new generation of its popular cousins since 2004. Discovery of carbonaceous QDs, so-called Carbon dots or C-dot, were serendipity. In 2004, during purification of arc-discharge shoot-derived single-walled carbon nanotubes electrophoratically, two new classes of nanomaterials were isolated one is tubular carbon and other was fluorescent nanomaterials which was later termed as Carbon dots [147].

Carbon dots display excellent water solubility due to their surface that typically covered with many carboxylic acid and hydroxyl groups. These groups enable carbon dots for their water solubility and surface functionalization with various species. The ultrafine dimensions of carbon dot, isotopic size along with tunable emission make carbon dot a strong competitor against semiconductor QDs. Most important is the toxicity issue where carbon dot leave semiconductor QDs far behind. Carbon dots are nearly nontoxic. They have the potential to replace semiconductor QDs in various biomedical fields including bioimaging and sensing. Nowadays carbon dot is at the focus of substantial research efforts for green, low cost, and less toxic agents that can be used for gene therapy, therapeutic, and regenerative medicines.

There are many synthetic routes developed to generate carbon dot. Top-down and bottom-up approaches are two main groups for synthesizing carbon dot. Top-down approaches are based on broken off a larger carbon structure and this approach is used by arc discharge [148], laser ablation [149–152], and electrochemical oxidation [153–156]. In bottom-up approaches carbon dots are synthesized from molecular precursor where surface of carbon material are oxidized by nitric acid (HNO_3) and further purified using centrifugation, dialysis, or another separation technique. Bottom-up approaches comprise combustion/thermal [157], supported synthetic [158, 159] or microwave methods [160].

Table 8 Core, shell, coating material, short synthesis procedure and application of selected lead chalcogenides

Core QD	Shell	Coating material	Synthesis	Application/conclusion	References
PbS	ZnS		PbS QDs-capped amorphous ZnS was produced by successive ionic layer adsorption and reaction	PbS QDs-embedded ZnS bulk heterojunctions displays advanced photovoltaic performance because of the small QD size and small distance between the QDs, and consequently more efficient charge injection and transport	[142]
PbS	CdS	Polystyrene	Different sized PbS was synthesized in oleylamine by different scheme mentioned. For CdS shell synthesis, cadmium oleate solution was mixed in PbS core QDs in toluene solvent	PbS/CdS QDs in hexane or implanted in polystyrene films displayed advanced free-carrier absorption (FCA) as compared to bare PbS QDs. It was detected that due to equal distribution of PbS/CdS QDs on the polystyrene film, Not only FCA cross-section but nonlinear refractive index also showed probable independence of the depth of PbS/CdS-embedded polymer film	[143]
PbS		Thioglycerol (TGL) dithioglycerol (DTG) Apoferritin (AFt)	TGL/DTG-capped PbS QDs were synthesized by colloidal chemistry in deionized water. AFt-encapsulated PbS was prepared by two routes. In first one, PbS was encapsulated in AFt cavity and in second one PbS was prepared in situ in AFt cavity	PbS QDs display valuable NIR photoluminescence. AFt-encapsulated PbS QDs were selectively chock to proliferative capacity of colorectal carcinoma cell as a result cells undergo apoptosis by reactive oxygen species (ROS). AFt-PbS QDs did not show any negative effect on health and behavioral change in in vivo study	[144]

(continued)

Table 8 (continued)

Core QD	Shell	Coating material	Synthesis	Application/conclusion	References
PbS	CdS	Amphiphilic polymer	By varying cation exchange condition, different CdS shell thickness were explored for same core (~4.5 nm) PbS QDs. Aqueous solution of PbS QDs were prepared via PEG-based amphiphilic polymer	It was observed that by increment of shell thickness ~0.7 nm, the quantum yield (QY) of PbS cores in water reached 33 % due to better surface passivation and then reduces to 1.7 % when the shell thickness reaches 2.3 nm	[145]
PbS		Glutathione (GSH)	GSH-capped PbS QDs by reacting lead acetate and sodium sulfide in the presence of GSH. Higher pH and high temperature increased synthesis of GSH-PbS QDs	Amounts of Na_2S, pH, and temperature of the solution of reactants regulate the emission wavelength of GSH-coated PbS QDs and tuned in between second NIR window (1000–1400 nm). Due to low cytotoxicity NIR emitting GSH-coated PbS QDs used in fluorescence imaging of noninvasive tissue	[146]

4 Functionalization of Chitosan by QDs

Biopolymers have gained focus in biomedical field where abundant studies examine resources such as alginate, chitosan, and gelatin as regenerative medicine. In recent years, carbohydrate polymers have been surprisingly mark their impact in thereputics biomedical and pharmaceutical applications because of their biocompatibility and biodegradability. Polysaccharides represent one of the most abundant industrial raw materials and have been the subject of intensive research owing to their sustainability, biodegradability, and biosafety. Currently, the uses of chitin and chitosan have been postulated in numerous areas of biopharmaceutical research, such as mucoadhesion, permeation enhancement, vaccine technology, gene therapy, and wound healing.

It is an accepted fact that bare QDs are not feasible for applications related to biological environment for numerous explanations. First is related to their synthesis. Many of the synthesis strategies prepare water insoluble QDs using organic solvent like TOP & TOPO. Heavy metal composition is the second reason that discourages their use typically in biological application due to their toxicity and biocompatibility issue. Third, high reactivity and large surface area-volume ratio of QDs make vague contacts with biological macromolecules that leading to the increase in particle size by aggregation and this affect their fluorescence intensity.

Chitosan provides a suitable matrix for the synthesis of QDs and also biofabrication of QDs with chitosan makes it suitable for other application [161]. There are few important reasons why chitosan is suitable for QDs:

1. QDs are formed by crystal growth and its main property is linked with its size, so its size should be small to produce different emissions used in biosensing and bioimaging. Well-established chains of chitosan hole the QDs well and prevent agglomeration during crystal growth which maintains its size.
2. Chitosan provides active sites for the formation of QDs. Abundant hydroxyl and amine group initiate the synthesis of QDs by chelation of metal (Cd, Zn, Ag, Pb, and Cu).
3. Surface functionalization of QDs passivates the surface of QDs and enhances its properties.
4. Encapsulation of QDs by chitosan can reduce toxic nature of QDs.
5. Cationic nature of chitosan is used to encapsulate anionic QDs by electrostatic attraction.
6. Using film-forming ability of chitosan it is easy to impart favorable bulk mechanical properties to QDs.
7. Active amine group of chitosan provides a site for attachment of different amino acid-functionalized QDs.
8. Chitosan is deeply studied in the field of drug delivery. We can easily track our drug using combined effect of fluorescence property of QDs and drug delivery ability of chitosan.

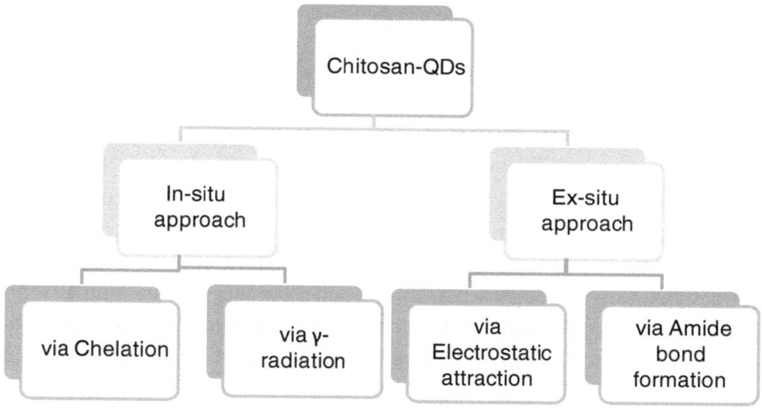

Fig. 5 Stepwise presentation of synthesis strategy

4.1 Synthesis Strategy

In today's world, research has been more focused to develop a method to accomplish our necessity and most importantly that method does not affect our environment. This type of synthesis is known as green chemistry. For the fabrication of the QDs, both academia and industry are constantly working for an environmentally benign method. Chitosan is a biopolymer that provides a suitable matrix which can be used for both synthesis of QDs and postsynthesis application of QDs. Among the 12 principles of green chemistry, synthesis of QDs using chitosan fulfills two key principles of relevance, i.e., "safer solvents" and to "design for energy efficiency" [162]. So preparation of QDs using chitosan as a solvent can be claimed as green synthesis. On the basis of use of chitosan in QDs synthesis strategy, it can be divided into in situ and ex situ synthesis. Different approaches of chitosan-QDs synthesis are shown in Fig. 5.

4.1.1 In Situ Synthesis

In situ is a Latin phrase typically means "in the reaction mixture." In situ preparation of QDs any polymer has a better compatibility with living cells. In situ synthesis of QDs does not require any chemical that harm our environment. Due to ease of synthesis and biocompatibility research, in situ synthesis of QDs has been increased in the last 3 years tremendously. Here we use this term in the context of synthesized QDs in chitosan matrix and take this QDs–chitosan composite or nanocomposite for different purposes.

By Chelation

Chelating capacity of chitosan is the main basic property that prepares a platform on which different QDs synthesis takes place. Having much electron-rich oxygen as hydroxyl group and nitrogen as primary amine group makes chitosan a valuable choice for synthesis of QDs under ordinary environment. Chitosan has the maximum chelating capability when compared with seafood waste-based natural polymers [163] and natural substances like bark, activated sludge, etc. Chitosan has binding abilities of more than 1 mmol/g for heavy and toxic metals other than chromium. Chelating capacity of a compound depends on the number of functional group containing donor atoms such as O, N, S, and P. Chitosan has a better binding ability for transition metals even in presence of alkali and alkaline earth metals [164].

Adsorption capacity of chitosan had been used for removal of harmful metal ion. Chitosan have been modified with numerous agents like sand [165], tripolyphosphate beads [166], bentonite [167], PVC beads [168], epichlorohydrin-triphosphate [169], hydroxyapatite [170], glutaraldehyde [171], magnetic phenylthiourea resin [172], magnetic 2-aminopyridine glyoxal [173], etc., for removal of lead, copper, nickel, cadmium, mercury, zinc, cobalt which are important environmental concern. Several studies have been carried out to check adsorption capacity of chitosan and modified chitosan. Isotherm, kinetic, and thermodynamic studies of acid-modified chitosan have been studied to establish a connection between adsorption capacity of chitosan and metal [223].

In situ syntheses of various QDs were reported time to time. CdS QDs [174], ZnS QDs [175], Mn-doped ZnS QDs [176], Cu-doped ZnS NPs [177], CdSe NPs [178], CdSe QDs [179], CdS-Ag core-shell QDs [180], etc. It is the safest way to synthesize QDs with a capping agent. QDs synthesized on chitosan templet were monodispersed and well passivated. Explicit polymer chains of chitosan could check QDs from agglomeration during the crystal formation. Highly regular structure and orderly sited $-NH_2$ and –OH groups of chitosan favors formation of chelation complex with cadmium and zinc metal which makes it possible for its metal ion complexes to be used as precursors to synthesize QDs. Figure 6 displays that how chitosan forms chelate with cadmium and zinc. After forming metal–chitosan complex, counterpart (S or Se) is added by simply adding S or Se precursor. Before adding sulfur precursor, pH of chitosan–metal complex solution was maintained at 5.5 [181]—6.0 [175]. For sulfide addition, mostly sodium sulfide (Na_2S) solution was used under ordinary condition. For selenide addition, NaHSe was used as a selenium precursor [179]. There is also many reports where modified chitosan was used to fabricate various QDs like anti-CD20 polyclonal antibody (pAbCD20)-modified chitosan for ZnS [181], poly(ethylene glycol)-*graft*-chitosan for CdSe [182], chitosane-poly(methacrylic acid) networks for CdSe [183].

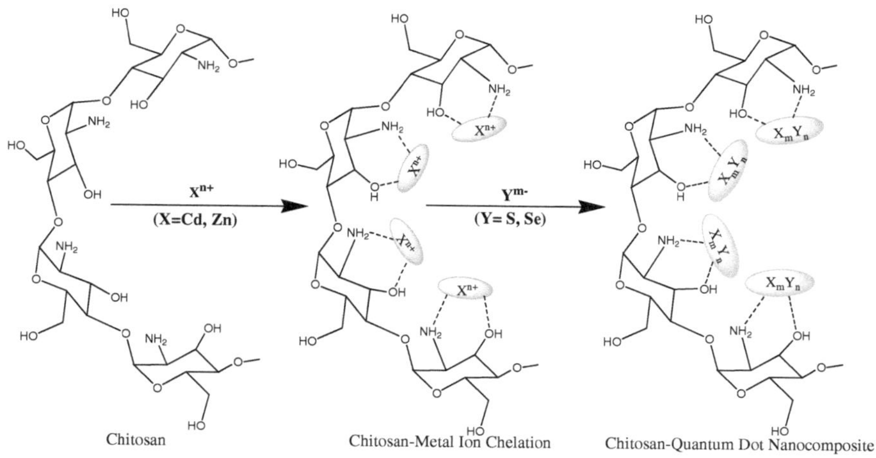

Fig. 6 Synthesis of QDs in chitosan by chelation

By γ-Radiation

Irradiation by γ ray is another simple, green, and environment friendly route to synthesize QDs in a polymer matrix. This route is also an example of in situ synthesis of QDs in chitosan matrix. Kang et al. [179] had prepared CdSe QDs first time by γ-radiation route using chitosan as a molecular template. Results indicated that size of CdSe QDs were about 4 nm with cubic zinc blende structure, narrow size distribution, and good solubility in water. By TEM images, it was clearly explained that chitosan glaze the surface of CdSe QDs core and exhibits light-green fluorescence. Later, Chang et al. [176] fabricated 5 nm size chitosan-coated ZnS QDs and chitosan-coated manganese-doped ZnS QDs through γ-radiation method. Figure 7 shows schematic presentation of synthesis of QDs in chitosan solution.

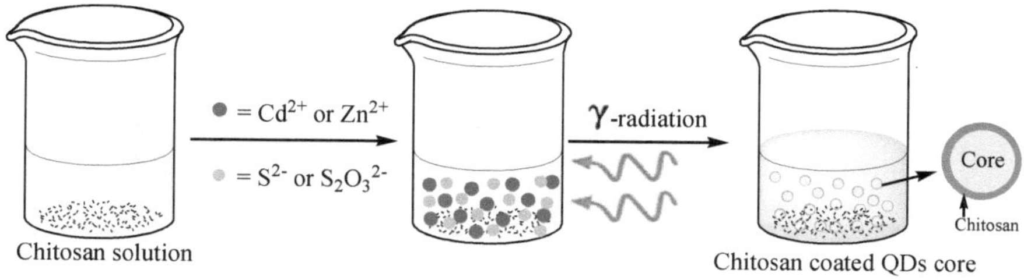

Fig. 7 Schematic presentation of synthesis of QDs in chitosan by γ-radiation

4.1.2 Ex Situ Synthesis

Ex situ is a Latin phrase classically used in chemistry at place to describe "outside the reaction mixture". Here we use this term in the perspective of synthesis of different QDs outside the chitosan matrix and after synthesis it transfer in chitosan solution for desired application. As discussed earlier, there are many ways to prepare various QDs in organic solvent and water both. QDs synthesized in organic solvent need to go through a ligand exchange process to make it biocompatible and used in biomedical and regenerative purposes. QDs like L-cysteine-capped CdTe [42], MPA-capped CdTe [184], MPA-capped ZnS [185].

Electrostatic Attraction

Two strategies have been used mostly to interact QDs with chitosan. First, strategy utilizes cationic property of chitosan. As discussed earlier in this chapter, chitosan is soluble in slightly acidic solution. Chitosan is a weak polybase with a pKa around 6.5, implying that its charge density varies in the pH range of 6–6.5. By accepting one proton from the acidic solution, primary amine group of chitosan have been changed into tertiary amine ($-NH_3^+$) and that makes chitosan a cationic polymer. Cationic property of chitosan attracts intense research interest due to their potential to form polyelectrolyte complexes with nucleic acids (DNA, RNA, and PNA). Figure 8 displays how cationic chitosan electrostatically attract anionically charged QDs. Nowadays maximum approach in synthesis of QDs is based on aqueous route with thiol capping agent. While choosing thiol-containing

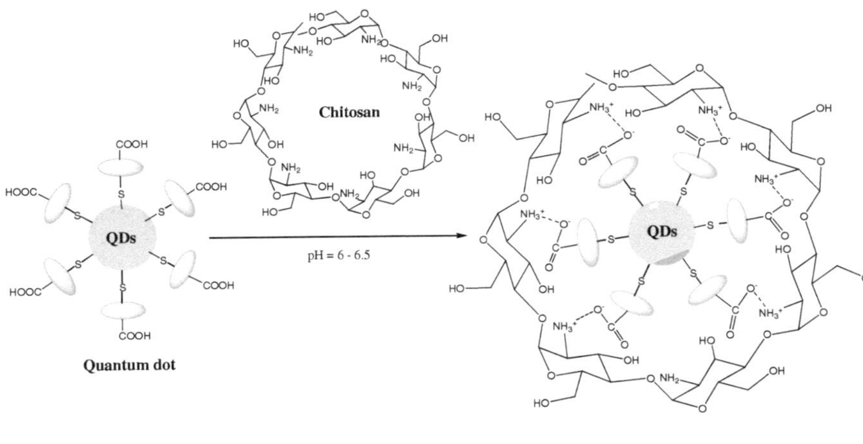

Fig. 8 Electrostatic attraction between QDs and chitosan

ligands, other functional groups (–COOH, –NH_2) are also being considered for bioconjugation. Carboxylic acid is an interesting group that is mostly used in bioconjugation.

When carboxylic acid containing ligand on QDs is incorporated in acidic solution, it deprotonate a proton and become anionic. These anionic QDs electrostatically trapped in chitosan cationic structure and form a uniform coating on QDs. carboxymethyl chitosan-coated CdTe QDs [186], gelatin-modified chitosan-CdS [187], carboxymethyl chitosan-coated CdTe/CdS [188].

Amide Bond Formation

Amide bond is formed between carbon of carbonyl group (–C=O) and nitrogen of amine group (–NH_2). Generally, amide bond is formed by removal of a water molecule between carboxyl group (–COOH) and amine group (–NH_2). Suitable reagent is used to assist amide bond formation. 1-Ethyl-3-(-3-dimethylaminopropyl) carbodiimide (EDAC) along with *N*-hydroxysuccinimide (NHS) is a reagent in trend for amide bond synthesis (Fig. 9). *N*,*N*′-dicyclohexylcarbodiimide (DCC) is also used a lot in developing a bond between carboxyl group (–COOH) and amine group (–NH_2). L-cysteine, MPA, MSA, TGA, GSH are thiol ligands having multiple functional groups (–COOH, –NH_2), has been used tremendously over years for biocompatible QDs synthesis. Addition of chitosan by amide bond formation is adding one more protective and reactive functional layer that not only enhances their biocompatibility but also alter their optical properties which have been used for fluorescence resonance energy transfer (FRET) [189], cell imaging, antibacterial agent [18], DNA trafficking [189], and drug delivery.

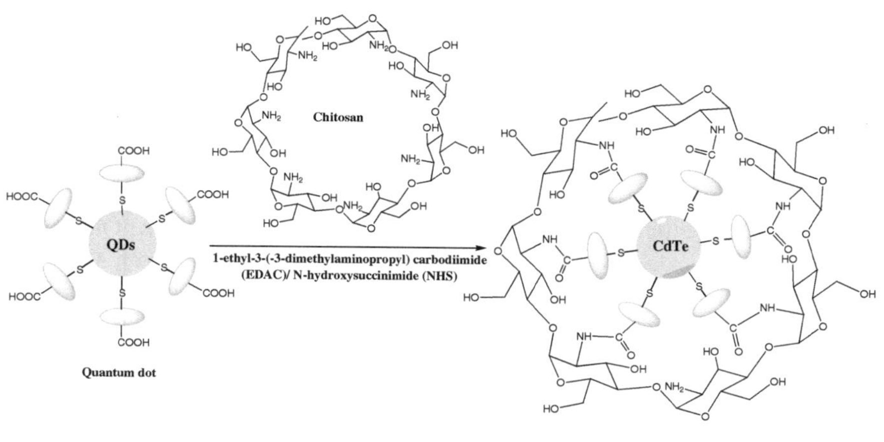

Fig. 9 Amide bond formation between QDs and chitosan

5 QDs-Functionalized Chitosan: Applications in Regenerative Medicine, Biomedical and Industrial Field

Regenerative medicine, an emergent multidisciplinary area, aimed to refurbish, preserve, and improve tissues/organ roles. With the combination of tissue and material, aim of regenerative medicine can be achieved. Specific cell deliver functionality and material used as base for cell proliferation. In experiments, human cells behave in vivo in nature to the biological signals attained from the neighboring environment. The methodology of regenerative medicine mainly includes two steps, first, to find the targeted area, and second use a 3D porous biomaterial that can launch tissue formation and mimic the exact signals that can direct the cells toward a required performance. Scaffolds are suitable 3D structure and fulfill the criteria of regenerative medicine. Chitosan, unlike the expensive natural polymers, induces minimal foreign body response [190, 191] and freeze-dried chitosan solution produce 3D scaffold. Not only Chitosan 3D scaffold [192] but also the liquid nitrogen-dried chitosan beads [193] are found suitable for cell delivery. Cell delivery is second area of interest in regenerative medicine because after getting a suitable base (scaffold), cells are directly seeded on it. Mainly, two approaches can be used to apply chitosan scaffold support in regenerative medicine. One is to implant scaffold at target site and permit cells to transfer in scaffold and multiply. In second approach, scaffold with seeded cells are graft at the targeted site and allow them to grow [194]. Stem cells, gifted with a property to differentiate, are reported in regenerative medicine [195]. Chitosan fibers in association with mesenchymal stem cells are successfully used in bone regeneration [196]. Monitoring cell occupation and its communication in the cellular atmosphere unlocks enormous area of insinuations regenerative medicine. Applications of QDs-functionalized chitosan are shown in Fig. 10.

To understand the whole process, starting from implantation to regenerative outcome of stem cells, it is required to tag the stem cells with some imaging agent which can be monitored throughout the process [197, 198]. Superparamagnetic iron oxide (SPIO) [199], Eu^{3+}-doped Gd_2O_3 hybrid NPs [200], gadolinium hexanedione NPs [201] are magnetic resonance imaging (MRI)-based contrasting agents, used to track transplanted stem cells. For long-term tagging of stem cells throughout its self-replication and differentiation process, various QDs are reported working efficiently in this area like RGD peptide conjugated QDs [202], NIR QDs [203], QDs [204–207], graphene QDs [208, 209]. Chitosan–QDs bioconjugate is making its deep impact mainly in bioimaging which unties new opportunities in regenerative medicine. Here we discuss chitosan–QDs conjugate used in different areas such as regenerative medicine, biomedical, and industrial applications. Cell imaging, cancer diagnosis, controlled drug delivery, gene delivery, pathogen detection, DNA detection; insulin delivery, and pollutant dye degradation, etc., are major areas where chitosan–QDs give their strong impression.

Fig. 10 Application of QDs-functionalized chitosan

5.1 Detection of Pathogen

Detection of *Escherichia coli* by fluorescence Imaging of Microbial Cells with chitosan-capped Mn-doped ZnS QDs was established by Dutta et al. [210]. They tried a strategy to dope manganese in ZnS QDs by coprecipitation reaction in acidic solution of chitosan. Well-distributed 4–5 nm size Mn-doped ZnS QDs on chitosan matrix with 3.2 eV band gap can be easily seen in TEM images.

Wu et al. [211] grew CdS QDs in chitosan to develop a fast, delicate, and biocompatible biosensor for *Pseudomonas aeruginosa* and *Staphylococcus aureus.* Bacteria membranes have a comparable chemical structure like chitosan and that similarity helps to interact with bacterial membrane. Bioaffinity of chitosan was used to prepare chitosan-coated CdS probe for pathogenic *S. aureus* and *P. aeruginosa* at detection rang of 1.5×10^2–18.0×10^2 cfu mL^{-1} within 1 min. They also conclude negative enthalpy (ΔH) while positive entropy (ΔS) is responsible for hydrophobic and electrostatic interaction between chitosan-modified CdS and bacterial membrane whereas negative free energy (ΔG) was accountable for spontaneous progressions.

Kumar et al. [42] conjugated L-cysteine-capped CdTe QDs with chitosan via amide bond formation by EDAC. Mechanism of amide bond formation was proposed. Novel CdTe QDs–chitosan film was screened out for their antibacterial activity against Gram-positive (*S. aureus*) as well as Gram-negative (*Pseudomonas aurigionasa* and *E. coli*) bacteria by disk diffusion method. Result concluded that prepared functionalized film shows greater zone of inhibition when comapred to chitosan and bare L-cysteine CdTe. Chitosan–CdTe fluorescent film can be used as a good antibacterial agent.

5.2 DNA Detection

Ying Lia [187] developed a tactic for facile preparation and highly stable chitosan–gelatin-modified electrode and fabricated with CdS electrostatically. Modified electrode used as an electrochemical biosensors for the recognition of real sample DNA. Over 100 continuous scan cycles of successful measurement of guanine (G) and adenine (A) by electrode, the electrode loses its peak heights of the cyclic voltammograms less than 5 % and loses current response 8 % after 20 days.

5.3 Gene Delivery

Tan et al. [212] encapsulate QDs with biopolymer chitosan to fulfill their objective, i.e., delivery of siRNA, tracking their delivery, and monitoring their transfection efficiency. Targeted delivery of HER2 siRNA to HER2-overexpressing SKBR3 breast cancer cells was shown to be specific with chitosan/QD NP surface labeled with HER2 antibody targeting the HER2 receptors on SKBR3 cells. HER2 ELISA assay confirmed that chitosan-encapsulated QDs NPs efficiently used to convey siRNA into cells with a great precision. This experiment concludes that with the help of chitosan-encapsulated QDs labeled with suitable ligand, we can successfully track gene delivery with high precision.

Ho et al. [213] established a very delicate technique to illustrate the structural composition and intracellular fate of polymeric DNA nanocomposite, designed by reducing plasmid DNA with chitosan biopolymers via electrostatic connections. Sensible strategy of additional capable polymeric gene carriers was designed for

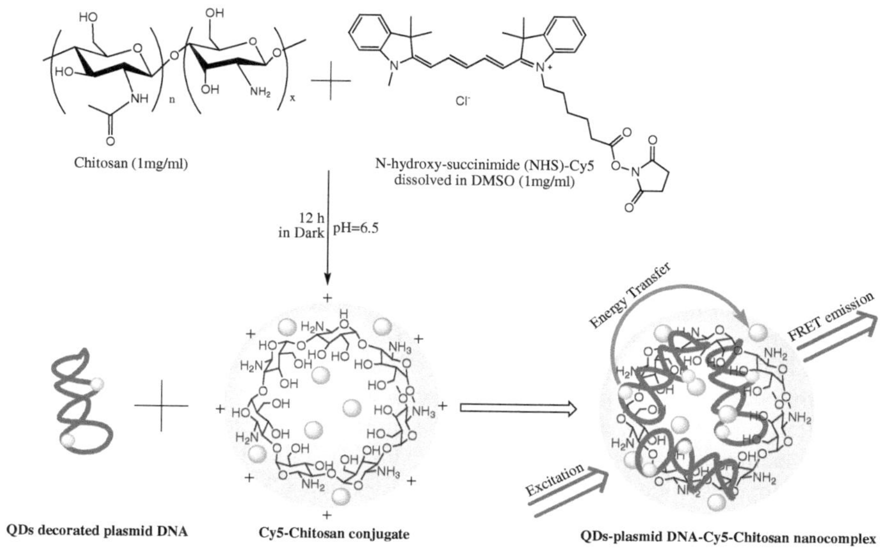

Fig. 11 Illustration of structural composition and intracellular fate of polymeric DNA nanocomposite by FRET

possible mechanistic insights of non-viral gene transfer process. Arrangement and binding dynamics chitosan–plasmid DNA nanocomposite were characterized by fluorescence resonance energy transfer (FRET). Plasmid, labeled with QDs works as donor and chitosan, a polymer carrier behave as acceptor, was labeled with *N*-hydroxy-succinimide (NHS)-functionalized Cy5. Figure 11 represents modification of chitosan with fluorescent organic dyes and condensing pDNA through complex coacervation method. Fluorescence microscopy used to localize red emission of pDNA-QDs and green emission of Cy5-chitosan. Colocalization of red and green emissions displayed their close association and effective energy transfer from donor to acceptor.

5.4 Sensing Application

Fang et al. [178] prepared CdSe–chitosan hybrid fluorescent films through in situ synthesis of CdSe QDs on chitosan film. QDs were fabricated by dipping the cadmium containing chitosan film in selenium precursor (NaHSe) under inert atmosphere. Sensing performance of the prepared hybrid film was tested by quenching mechanism and it was observed that tertiary amines display a slight quenching to the emission. Out of other polyamine, ethylenediamine quench the emission maximum probably due to chelating of EDA to Cd^{2+} of the hybrid film.

Chitosan was started using in preparation of QDs since 2003. Pang el al synthesized CdS QDs using chitosan matrix under mild conditions. They use chitosan to synthesize CdS QDs and also to prevent agglomeration of QDs. They adopt a strategy to chelate cadmium bivalent ion with acidic solution of chitosan. Fresh Na_2S aqueous solution was used to add counterpart, i.e., sulfur.

Lai et al. [214] prepared CdS QDs modified with chitosan and developed fluorescence probe for copper (II) ion determination based on the quenching of fluorescence signals of the functionalized CdS QDs. Quenching in fluorescence signal was due to chemical displacement of Cd^{2+} of CdS by Cu^{2+} and form CuS.

Sharma et al. [215] report a DNA biosensor for detection of chronic myelogenous leukemia (CML). Electrochemical sensor was assembled by electrophoretic deposition of chitosan-encapsulated cadmium–telluride QDs nanostructured onto indium tin oxide (ITO)-coated glass substrate. CML-specific probe DNA (pDNA) sequence identified from the BCR–ABL gene was immobilizing on chitosan–CdTe/ITO electrode. This nucleic acid sensor detected as low as 2.56 pM concentration of complementary target DNA within 60 s with a shelf life of about 6 weeks and can be used for about 5–6 times. This leukemia biosensor responded excellent when experiments conducted using clinical patient samples.

Chen et al. [216] described a direct gas–liquid microfluidic method to produce uniform-sized chitosan microcapsules covering CdS QDs. These microcapsules showed fluorescent reactions to chemical environmental stimuli. α-cyclodextrin (α-CD) brings clear deviations on the fluorescent color of the microcapsules whether β-cyclodextrin (β-CD) had a slight impact on the outline and fluorescent color of the microcapsules. So these stimuli-responsive microcapsules can be useful as a possible microdetector to chemicals, such as CDs.

5.5 Cancer Diagnosis, Cell Imaging, and Controlled Drug Delivery

The combination of strong fluorescent QDs and chitosan is an advanced emerging research area. Specifically, when high-fluorescence quantum yield property of QDs engages unique physicochemical and intrinsic bioactive properties of chitosan, it unlocked new prospects for in vivo and in vitro imaging at exceptional sensitivity and three-dimensional resolution cancer diagnosis, controlled drug delivery, and theranostics. QDs-functionalized chitosan bioconjugate plays an active role in diagnosing cancer cells. Drug delivery with the introduction of chitosan passes many hurdles that suppress its effectiveness. Specific targeting, intercellular transport, and biocompatibility are main issues that have been resolved before picking up a drug delivery agent. Cationic nature, high cellular uptake efficiency, and biocompatibility make chitosan a good candidate in drug delivery. Chitosan itself and along with its chemically modified derivatives efficiently used in sustained release of proteins/peptides [217], growth factors (rhBMP-2) [218], drugs [219–222], antibiotics via gelation, hydrogel, drug conjugate, and biodegradable release system.

Chang et al. [176] fabricated chitosan-capped ZnS QDs and chitosan-coated Mn^{2+}–ZnS QDs and successfully employed as fluorescent labels for Human pancreatic carcinoma cells (PANC-1). As compared to PANC-1 cells containing TGA–CdSe QDs where metabolic activity decreased to 32 % at the consumption of QDs 10 $\mu g\ mL^{-1}$, chitosan-fabricated QDs maintain 70 % metabolic activity at 1,000 $\mu g\ mL^{-1}$ concentration. Chitosan-coated ZnS QDs and chitosan-coated Mn^{2+}–ZnS QDs accumulate around nuclei and stain cytoplasm bright blue light and orange light, respectively.

Wu et al. [183] reported a covalently crosslinked chitosan–poly(methacrylic acid) (chitosan–PMAA) hybrid nanogels to create a base on which CdSe QDs immobilize in situ. Chitosan–PMAA hybrid nanogels reveal excellent colloidal and structural stability and pH-dependent reversible physical property. Low cytotoxicity of in situ synthesized CdSe QDs helped to culture nutrient medium with mouse melanoma B16F10 cells. B16F10 cells show the luminescence after 2 h of chitosan–PMAA-CdSe hybrid nanogels treatment. Due to reversible pH-dependent property of the drug delivery by chitosan–PMAA–CdSe nanogel had been checked. Anticancer drug temozolomide (TMZ) trapped in hybrid nanogel to control and discharge drug at around extracellular tissues of tumors where pH was found in between 5 and 7.4. So chitosan–PMAA–CdSe nanogel is an example of theronostics where CdSe QDs is used for diagnosis and pH-dependent chitosan–PMAA network is used for drug delivery or therapy. Carboxymethyl chitosan-coated CdTe/CdS QDs were developed as luminescent fluorescent probe for imaging live Madin–Darby canine kidney (MDCK) cells by He et al. [188]. Microwave-assisted aqueous synthesis of TGA-capped CdTe/CdS was adopted. Chitosan was chemically transformed into carboxymethylchitosan(CMC) to modify its low solubility in water to water-soluble material. TGA-capped CdTe/CdS QDs chelated in water soluble CMC via amino and carboxyl groups. CMC incorporation modified QDs stability up to 10 months at room temperature and photoluminescence quantum yield enhanced significantly up to 75 %. CMC-coated CdTe/CdS QDs labeled MDCK cells considerable fluorescent cytoplasm without any inhibited growth and damage of cells.

Mansur et al. [181] designed a novel approach for the synthesis of ZnS quantum dot–chitosan nano-immunoconjugates for cancer diagnosis. Passivation of ZnS QDs and detection of cancer cells strictly via an environment friendly green route. Chitosan was used as a ligand for capping of ZnS QDs in situ and an antibody-specific cancer biomarker (abCD20) for non-Hodgkin lymphoma were chemically conjugated to chitosan–ZnS QDs via amide bond formation by EDAC/

sulfo-NHS. CD20 is a specific transmembrane antigen receptor that is overexpressed in non-Hodgkin lymphoma and chitosan-ZnS conjugated with anti-CD20 antibodies specifically labelled tumor cell (a lymphoma B-cell line). Blue luminescent chitosan-ZnS-antibody (abCD20) immunoconjugate were active for specific targeting, labeling, and bioimaging in oncology diagnosis.

Ma et al. [186] had established an innovative intracellular nanoprobe built on Zn^{2+} sensing capability CMC–CdTe QDs in prostate cancer cells (PC-3 M). He utilized amino groups of carboxymethyl chitosan polymeric chains and carboxyl groups of the CdTe QDs for electrostatic interaction. CMC show its excellent biocompatibility and strong binding ability with Zn^{2+}. On the other hand, the CMC coating on the QDs surface effectively reduced the potential toxicity CdTe QDs, enhanced their biocompatibility and stability, and stimulated the endocytosis of the nanoprobe.

6 Concluding Remarks

Chitosan–QDs-based approach in the field of regenerative research is now focused on active research where cationic property of chitosan is well utilized in the light of fluorescent multifunctional QDs. Findings of previous reports direct us to achieve our future goal where we can not only design regenerative approach but also visualize whole process from cell proliferation to organ transplantation by the use of chitosan–QDs bioconjugate. Chitosan–QDs-based approach was used in the field to therapeutics. Figure 12 is a schematic presentation of how lipid bilayer encapsulated drug is conjugated with chitosan–QDs and cancer/tumor targeting antibody for fluorescent diagnosing of cancer/tumor cell and Fig. 12 showing after diagnosing cancer/tumor cell lipid bilayer dissolve to release drug on cancer/tumor cells. In future, functionalize chitosan-QDs deliver an active role in cell imaging based regenerative medicine.

Acknowledgments One of the authors (HK) gratefully acknowledged the financial support from UGC, New Delhi and TEQIP-II, MNNIT Allahabad during his Ph.D.Thesis work.

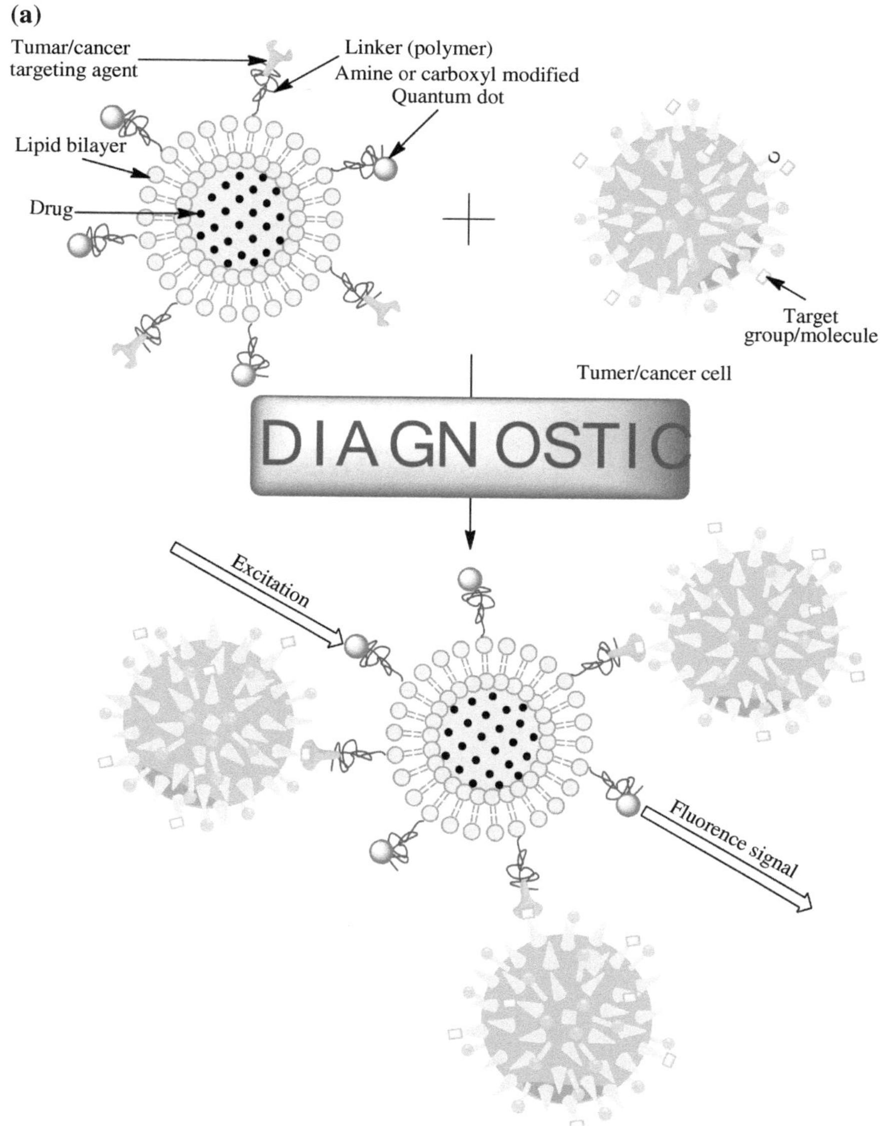

Fig. 12 Pictorial representation theronostics where **a** antibody–QDs–chitosan-modified drug encapsulated liposome that first detect and attach to antigen of tumor cell (diagnostic) and in second phase **b** dissolution of liposome occur resulting death of cancer cells (therapeutic)

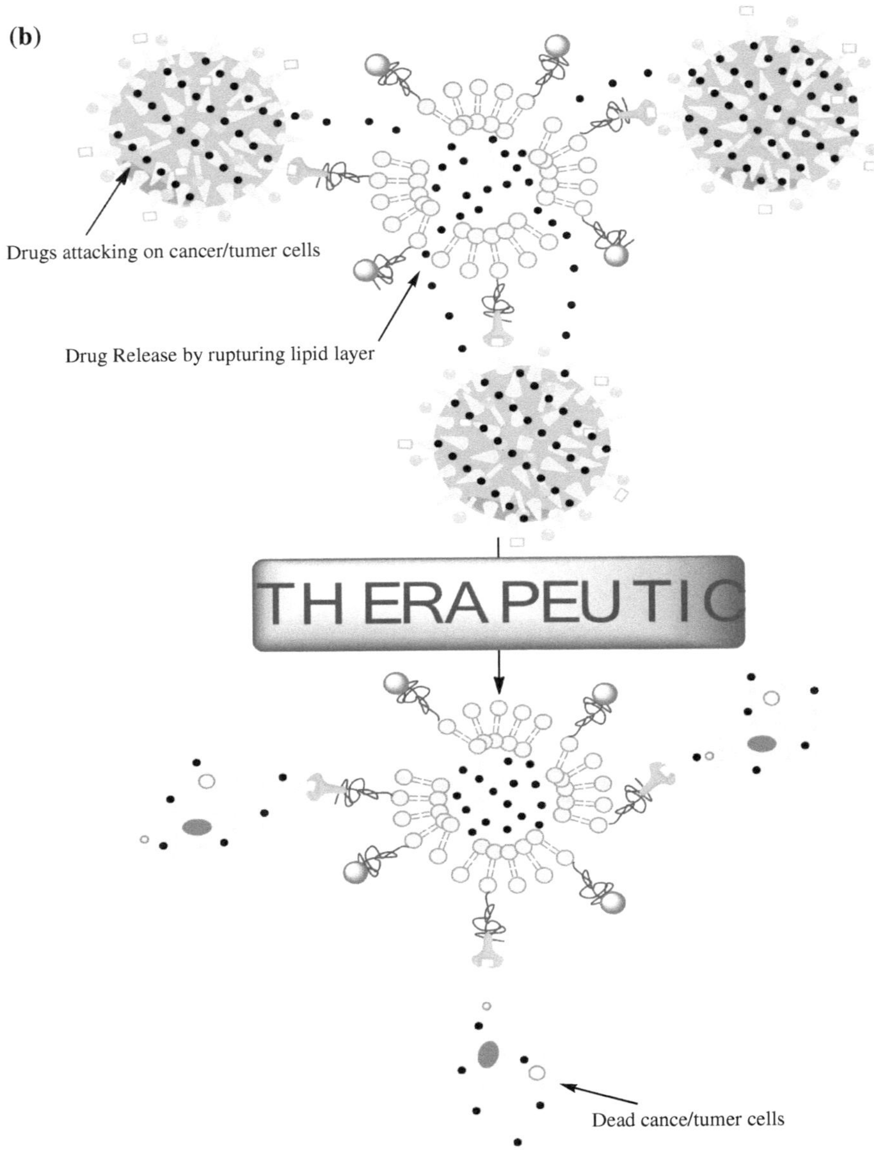

Fig. 12 (continued)

References

1. Bäuerlein E (ed) (2004) Biomineralization: progress in biology, molecular biology and application. Wiley-VCH Verlag GmbH & Co. KGaA, Weinheim
2. Medintz IL, Uyeda HT, Goldman ER, Mattoussi H (2005) Quantum dot bioconjugates for imaging, labelling and sensing. Nat Mater 4:435–446
3. Arai Y, Nagai T (2014) Real-time chemiluminescence imaging using nano-lantern probes. Curr Protoc Chem Biol 6:221–236
4. Sun NF, Liu ZA, Huang WB, Tian AL, Hu SY (2014) The research of nanoparticles as gene vector for tumor gene therapy. Crit Rev Oncol Hematol 89:352–357
5. Lima AC, Mano JF (2015) Micro/nano-structured superhydrophobic surfaces in the biomedical field: part II: applications overview. Nanomedicine 10:271–297
6. Frigerio C, Ribeiro DSM, Rodrigues SSM, Abreu VLRG, Barbosa JAC, Prior JAV, Marques KL, Santos JLM (2012) Application of quantum dots as analytical tools in automated chemical analysis: a review. Anal Chim Acta 735:9–22
7. Bruchez JM, Moronne M, Gin P, Weiss S, Alivisatos AP (1998) Semiconductor nanocrystals as fluorescent biological labels. Science 281:2013–2016
8. Chan WCW, Nie SM (1998) Quantum dot bioconjugates for ultrasensitive nonisotopic detection. Science 281:2016–2018
9. Ho YP, Leong KW (2010) Quantum dot-based theranostics. Nanoscale 2:60–68
10. Chan WCW, Maxwell DJ, Gao XH, Bailey RE, Han MY, Nie SM (2002) Luminescent quantum dots for multiplexed biological detection and imaging. Curr Opin Biotechnol 13:40–46
11. Resch GU, Grabolle M, Cavaliere JS, Nitschke R, Nann T (2008) Quantum dots versus organic dyes as fluorescent labels. Nat Methods 5:763–775
12. Kagan CR, Murray CB, Nirmal M, Bawendi MG (1996) Electronic energy transfer in CdSe quantum dot solids. Phys Rev Lett 76:1517–1520
13. Xu Y, Shi Y, Ding SA (2008) A chemical approach to stem-cell biology and regenerative medicine. Nature 453:338–344
14. Levin M (2011) The wisdom of the body: future techniques and approaches to morphogenetic fields in regenerative medicine, developmental biology and cancer. Regen Med 6:667–673
15. Garate Z, Davis BR, Quintana BO, Segovia JC (2013) New frontier in regenerative medicine: site-specific gene correction in patient-specific induced pluripotent stem cells. Hum Gene Ther 24:571–583
16. Zack-Williams SD, Butler PE, Kalaskar DM (2015) Current progress in use of adipose derived stem cells in peripheral nerve regeneration. World J Stem Cells 7:51–64
17. Senarath YK, McArdle A, Renda A, Longaker MT, Quarto N (2014) Adipose-derived stem cells: a review of signaling networks governing cell fate and regenerative potential in the context of craniofacial and long bone skeletal repair. Int J Mol Sci 15:9314–9330
18. Griffith LG, Naughton G (2002) Tissue engineering–current challenges and expanding opportunities. Science 295:1009–1014
19. Kfoury C (2007) Therapeutic cloning: promises and issues. Mcgill J Med 10:112–120
20. Niidome T, Huang L (2002) Gene therapy progress and prospects: nonviral vectors. Gene Ther 9:1647–1652
21. Atta HM (2010) Gene therapy for liver regeneration: experimental studies and prospects for clinical trials. World J Gastroenterol 16:4019–4030
22. Kimelman BN, Kallai I, Lieberman JR, Schwarz EM, Pelled G, Gazit D (2012) Gene therapy approaches to regenerating bone. Adv Drug Deliv Rev 64:1320–1330
23. Mekhail M, Tabrizian MA (2014) Injectable chitosan-based scaffolds in regenerative medicine and their clinical translatability. Adv Healthc Mater 3:1529–1545
24. Zhao XJ, Zhang SG (2007) Designer self-assembling peptide materials. Macromol Biosci 7:13–22

25. Mata A, Hsu L, Capito R, Aparicio C, Henrikson K, Stupp SI (2009) Micropatterning of bioactive self-assembling gels. Soft Matt 5:1228–1236
26. Pastist CM, Mulder MB, Gautier SE, Maquet V, Jérôme R, Oudega M (2004) Freeze-dried poly(D, L-lactic acid) macroporous guidance scaffolds impregnated with brain-derived neurotrophic factor in the transected adult rat thoracic spinal cord. Biomaterials 25:1569
27. Niklason LE, Gao J, Abbott WM, Hirschi KK, Houser S, Marini R, Langer R (1999) Functional arteries grown in vitro. Science 284:489–493
28. Ma PX, Langer R (1999) Morphology and mechanical function of long-term in vitro engineered cartilage. J Biomed Mater Res 44:217–221
29. Dutta J, Dutta PK (eds) (2005) Chitosan a material for 21st century. Publication Contai, SSM Intl
30. Dutta J, Dutta PK, Rinki K (2008) Current research on chitin and chitosan for tissue engineering applications and future demands on bioproducts. In: Jayakumar R, Prabaharan M (eds) Current research and developments on chitin and chitosan in biomaterials science. Research Signpost, Trivandrum
31. Jiang T, Kumbar SG, Nair LS, Laurencin CT (2008) Biologically active chitosan systems for tissue engineering and regenerative medicine. Curr Top Med Chem 8:354–364
32. Jana S, Gandhi A, Sen KK, Basu SK (eds) (2014) Biomedical applications of chitin and chitosan derivatives. CRC Press Taylor & Francis Group, London
33. Singh J, Dutta PK, Dutta J, Hunt AJ, Macquarrie DJ, Clark JH (2009) Preparation and properties of highly soluble chitosan-L-glutamic acid aerogel derivative. Carbohyd Polym 76:188–195
34. Singh J, Dutta PK (2009) Preparation, circular dichroism induced helical conformation and optical property of chitosan acid salt complexes for biomedical application. Int J Biol Macromol 45:384–392
35. Tripathi S, Mehrotra GK, Dutta PK (2009) Preparation and physicochemical evaluation of chitosan/poly(vinyl alcohol)/pectin ternary film for food-packaging applications. Carbohyd Polym 79:711–716
36. Singh J, Dutta PK (2010) Preparation, antibacterial & physicochemical behavior of chitosan/ofloxacin complexes. Int J Polym Mater 59:793
37. Archana D, Dutta J, Dutta PK (2013) Evaluation of chitosan nano dressing for wound healing: characterization, in vitro and in vivo studies. Int J Biol Macromol 57:193–203
38. Archana D, Singh BK, Dutta J, Dutta PK (2013) In vivo evaluation of chitosan–PVP–titanium dioxide nanocomposite as wound dressing material. Carbohydr Polym 95:530–539
39. Srivastava R, Tiwari DK, Dutta PK (2011) 4-(Ethoxycarbonyl) phenyl-L-amino-oxobutanoic acid–chitosan complex as a new matrix for silver nanocomposite film: preparation, characterization and antibacterial activity. Int J Biol Macromol 49:863–870
40. Yadav SK, Mahapatra SS, Yadav MK, Dutta PK (2013) Mechanically robust biocomposite films of chitosan grafted carbon nanotubes via the [2+1] cycloaddition of nitrenes. RSC Adv 3:23631–23637
41. Singh J, Srivastava M, Dutta J, Dutta PK (2010) Preparation and properties of hybrid monodispersed magnetic -Fe_2O_3 based chitosan nanocomposite film for industrial and biomedical applications. Int J Biol Macromol 48:170–176
42. Kumar H, Srivastava R, Dutta PK (2013) Highly luminescent chitosan-L-cysteine functionalized CdTe quantum dots film: synthesis and characterization. Carbohydr Polym 12(97):327–334
43. Kumar H, Srivastava R, Singh BK, Dutta PK (2012) Synthesis and antibacterial activity of CdTe quantam dots (QD)-chitosan nanocomposite film. Asian Chitin J 8:27–30
44. Singh BK, Sirohi R, Archana D, Jain A, Dutta PK (2015) Porous chitosan scaffolds: a systematic study for choice of crosslinker and growth factor incorporation. Int J Polym Mater 64:242–252
45. Dutta PK, Kumar H, Tiwari DK, Archana D, Rizvi KS, Kumar A, Singh BK, Srivastava R (2011) The glimpses of chitosan nanoparticles. Asian Chitin J 7:103–106

46. Dutta PK, Srivastava R, Dutta J (eds) (2013) Functionalized nanoparticles and chitosan-based functional nanomaterials. In: Dutta PK, Dutta J (ed) Multifaceted development and application of biopolymers for biology, biomedicine and nanotechnology. Adv Polym Sci 254:1–50
47. Qiu Y, Ma Zand HuPA (2014) Environmentally benign magnetic chitosan/Fe3O4 composites as reductant and stabilizer for anchoring Au NPs and their catalytic reduction of 4-nitrophenol. J Mater Chem A 2:1347
48. Ramasamy RP, Maliyekkal MS (2014) Formation of gold nanoparticles upon chitosan leading to formation and collapse of gels. New J Chem 38:63–69
49. Safari J, Javadian L (2014) Chitosan decorated Fe_3O_4 nanoparticles as a magnetic catalyst in the synthesis of phenytoin derivatives. RSC Adv 4:48973–48979
50. Tiwari AP, Satvekar RK, Rohiwal SS, Karande VA, Raut AV, Patil PG, Shete PG, Ghosh SJ, Pawar SH (2015) Magneto-separation of genomic deoxyribose nucleic acid using pH responsive Fe3O4@silica@chitosan nanoparticles in biological samples. RSC Adv 5:8463–8470
51. Mallick S, Sanpui P, Ghosh SS, Chattopadhyay A, Paul A (2015) Synthesis, characterization and enhanced bactericidal action of a chitosan supported core–shell copper–silver nanoparticle composite. RSC Adv 5:12268–12276
52. Liu X, Huang H, Liu G, Zhou W, Chen Y, Jin Q, Ji J (2013) Multidentate zwitterionic chitosan oligosaccharide modified gold nanoparticles: stability, biocompatibility and cell interactions. Nanoscale 5:3982–3991
53. Huang JH, Lai TC, Cheng LC, Liu RH, Lee CH, Hsiao M, Chen CH, Her LJ, Tsai DP (2011) Modulating cell-uptake behavior of Au-based nanomaterials via quantitative biomolecule modification. J Mater Chem 21:14821–14829
54. Veerapandian M, Zhu XX, Giasson S (2015) Chitosan-modified silver@ruthenium hybrid nanoparticles: evaluation of physico-chemical properties and bio-affinity with sialic acid. J Mater Chem B 3:665–672
55. Ensafi AA, Jafari AM, Dorostkar N, Ghiaci M, Martıınez-Huerta MV, Fierro JLG (2014) The fabrication and characterization of Cu-nanoparticle immobilization on a hybrid chitosan derivative-carbon support as a novel electrochemical sensor: application for the sensitive enzymeless oxidation of glucose and reduction of hydrogen peroxide. J Mater Chem B 2:706–717
56. Devi R, Yadav S, Pundir CS (2012) Amperometric determination of xanthine in fish meat by zinc oxide nanoparticle/chitosan/multiwalled carbon nanotube/polyaniline composite film bound xanthine oxidase. Analyst 137:754–759
57. Mallick S, Sharma S, Banerjee M, Ghosh SS, Chattopadhyay A, Paul A (2012) Iodine-stabilized Cu nanoparticle chitosan composite for antibacterial applications. ACS Appl Mater Interfaces 4:1313–1323
58. Xiao W, Xu J, Liu X, Hu Q, Huang J (2013) Antibacterial hybrid materials fabricated by nanocoating of microfibril bundles of cellulose substance with titania/chitosan/silver-nanoparticle composite films. J Mater Chem B 1:3477–3485
59. Fan Y, Huang Y (2012) The effective peroxidase-like activity of chitosan-functionalized $CoFe_2O_4$ nanoparticles for chemiluminescence sensing of hydrogen peroxide and glucose. Analyst 137:1225–1231
60. Petkova P, Francesko A, Fernandes MM, Mendoza E, Perelshtein I, Gedanken A, Tzanov T (2014) Sonochemical coating of textiles with hybrid ZnO/chitosan antimicrobial nanoparticles. ACS Appl Mater Interfaces 6:1164–1172
61. Sashiwa H, Aiba S (2004) Chemically modified chitin and chitosan as biomaterials. Prog Polym Sci 29:887–908
62. Kurita K (2001) Controlled functionalisation of the polysaccharide chitin. Prog Polym Sci 26:1921–1971
63. Morimoto M, Saimoto H, Usui H, Okamoto Y, Minami S, Shigemasa Y (2002) Control of functions of chitin and chitosan by chemical modification. Trends Glycosci Glycotech 14:205–222

64. Macquarrie DJ, Hardy JEE (2005) Applications of functionalized chitosan in catalysis. Ind Eng Chem Res 44:8499–8520
65. Kim IY, Seo SJ, Moon HS, Yoo MK, Park IY, Kim BC, Cho CS (2008) Chitosan and its derivatives for tissue engineering applications. Biotechnol Adv 26:1–21
66. Shi C, Zhu Y, Ran X, Wang M, Su Y, Cheng T (2006) Therapeutic potential of chitosan and its derivatives in regenerative. J Surg Res 133:185–192
67. Felice F, Zambito Y, Belardinelli E, Fabiano A, Santoni T, Stefano RD (2015) Effect of different chitosan derivatives on in vitro scratch wound assay: a comparative study. Int J Biol Macromol 76:236–241
68. Saranya N, Moorthi A, Saravanan S, Pandima Devi M, Selvamurugan N (2011) Chitosan and its derivatives for gene delivery. Int J Biol Macromol 48:234–238
69. Hardy JF, Hubert S, Macquarrie DJ, Wilson AJ (2004) Chitosan-based heterogeneous catalysts for Suzuki and Heck reactions. Green Chem 6:53–56
70. Tojima T, Katsura H, Han S, Tanida F, Nishi N, Tokura S, Sakaira N (1998) Preparation of an β cyclodextrin linked chitosan derivative via reductive amination strategy. J Polym Sci Polym Chem 36:1965–1968
71. Chen S, Wang Y (2001) Study on β-cyclodextrin grafting with chitosan and slow release of its inclusion complex with radioactive iodine. J Appl Polym Sci 82:2414–2421
72. Sun W, Xia CG, Wang HW (2002) Efficient heterogeneous catalysts for the cyclopropanation of olefins. New J Chem 26:755–758
73. Millotti G, Samberger C, Fröhlich E, Sakloetsakun D, Bernkop SA (2010) Synthesis and characterization of a novel thiolated chitosan. J Mater Chem 20:2432–2440
74. Sashiwa H, Shigemasa Y, Roy R (2000) Chemical modification of chitosan 3 hyperbranched chitosan-sialic acid dendrimer hybrid with tetraethylene glycol spacer. Macromolecules 33:6913–6915
75. Sashiwa H, Yamamori N, Ichinose Y, Sunamoto J, Aiba SI (2003) Michael reaction of chitosan with various acryl reagents in water. Biomacromolecules 4:1250–1254
76. Sashiwa H, Kawasaki N, Nakayama A, Muraki E, Yamamoto N, Aiba SI (2002) Chemical modification of chitosan 14 synthesis of water-soluble chitosan derivatives by simple acetylation. Biomacromolecules 3:1126–1128
77. Sashiwa H, Thompson JM, Das SK, Shigemasa Y, Tripathy S, Roy R (2000) Chemical modification of chitosan: preparation and lectin binding properties of α-galactosyl-chitosan conjugates potential inhibitors in acute rejection following xenotransplantation. Biomacromolecules 1:303–305
78. Li M, Xin M, Miyashita T (2002) Preparation of N, N dilauryl chitosan langmuir–blodgett film. Polym Int 51:889–891
79. Heras A, Rodríguez NM, Ramos VM, Agulló E (2003) N-methylene phosphonic chitosan: a novel soluble derivative. Carbohydr Polym 52:39–46
80. Ramos VM, Rodríguez NM, Henning I, Díaz MF, Monachesi MP, Rodríguez MS, Abarrategi A, Correas MV, López JL, Agulló E (2006) Poly(ethylene glycol)-crosslinked N-methylene phosphonic chitosan-preparation and characterization. Carbohydr Polym 64:328–336
81. Park IK, Yang J, Jeong HJ, Bom HS, Harada I, Akaike T, Kim SI, Cho CS (2003) Galactosylated chitosan as a synthetic extracellular matrix for hepatocytes attachment. Biomaterials 24:2331–2337
82. Lee KY, Kim JH, Kwon IC, Jeong SY (2000) Self-aggregates of deoxycholic acid-modified chitosan as a novel carrier of Adriamycin. Colloid Polym Sci 278:1216
83. Martin L, Wilson CG, Koosha F, Tetley L, Gral AI, Senel S, Uchegbu IF (2002) The release of model macromolecules may be controlled by the hydrophobicity of palmitoyl glycol chitosan hydrogels. J Control Rel 80:87–100
84. Nishiyama Y, Yoshida T, Mori T, Ishii S, Kurita K (1998) Asymmetric reduction with chitos an/dihydronicotinamide conjugates: influence of L-alanine spacer arms on reducing performance. React Funct Polym 37:83–91

85. Kurita K, Hiyakama M, Nishiyama Y, Harata M (2002) Polymeric asymmetric reducing agents: preparation and reducing performance of chitosan/dihydronicotinamide conjugates having L- and D-phenylalanine spacer arms. Carbohydr Polym 47:7–14
86. Hojo K, Maeda M, Mu Y, Kamada H, Tsutsumi Y, Nishiyama Y, Yoshikawa T, Kurita K, Block LH, Mayumi T, Kawasaki K (2000) Facile synthesis of a chitosan hybrid of a lamininrelated peptide and its antimetastatic effect in mice. J Pharm Pharmacol 52:67–73
87. Fujii S, Kumugai H, Noda M (1980) Preparation of poly(acyl) chitosans. Carbohydr Res 83:389–393
88. Paiva D, Ivanova G, Pereira MC, Rocha S (2013) Chitosan conjugates for DNA delivery. Phys Chem Chem Phys 15:11893–11899
89. Du YZ, Ying XY, Wang L, Zhai Y, Yuan H, Yu RS, Hu FQ (2010) Sustained release of ATP encapsulated in chitosan oligosaccharide nanoparticles. Int J Pharm 392:164–169
90. Mao HQ, Roy K, Troung-Le VL, Janes KA, Lin KY, Wang Y, August JT, Leong KW (2001) Chitosan-DNA nanoparticles as gene carriers: synthesis, characterization and transfection efficiency. J Control Rel 70:399–421
91. Howard KA, Rahbek UL, Liu X, Damgaard CK, Glud SZ, Andersen MØ, Hovgaard MB, Schmitz A, Nyengaard JR, Besenbacher F, Kjems J (2006) RNA interference in vitro and in vivo using a novel chitosan/siRNA nanoparticle system. Mol Ther 14:476–484
92. Salva E, Turan SO, Eren F, Akbuğa J (2014) The enhancement of gene silencing efficiency with chitosan-coated liposome formulations of siRNAs targeting HIF-1α and VEGF. Int J Pharm 478:147–154
93. Bradshaw M, Zou J, Byrne L, Iyer KS, Stewart SG, Raston CL (2011) Pd(II) conjugated chitosan nanofibre mats for application in Heck cross-coupling reactions. Chem Comm 47:12292–12294
94. Yao R, Meng F, Zhang L, Ma D, Wang M (2009) Defluoridation of water using neodymium-modified chitosan. J Hazard Mater 165:454–460
95. Zheng Y, Yi Y, Qi Y, Wang Y, Zhang W, Du M (2006) Preparation of chitosan–copper complexes and their antitumor activity. Bioorg Med Chem Lett 16:4127–4129
96. Zhao CZ, Egashira N, Kurauchi Y, Ohga K (1998) Electrochemiluminescence oxalic acid sensor having a platinum electrode coated with chitosan modified with a ruthenium (II) complex. Electrochim Acta 43:2167–2173
97. Liu Y, Luo S, Wei W, Liu X, Zeng X (2009) Methanol sensor based on the combined electrocatalytic oxidative effect of chitosan-immobilized nickel(II) and the antibiotic cefixime on the oxidation of methanol in alkaline medium. Microchim Acta 164:351–355
98. Chena S, Wua G, Zeng H (2005) Preparation of high antimicrobial activity thiourea chitosan–Ag^+ complex. Carbohydr Polym 60:33–38
99. Shi Z, Neoh KG, Kang ET, Poh CK, Wang W (2009) Surface functionalization of titanium with carboxymethyl chitosan and immobilized bone morphogenetic protein-2 for enhanced osseointegration. Biomacromolecules 10:1603–1611
100. Madelung O, Schulz M, Weiss H (eds) (1982) Landolt-Bornstein: numerical data and functional relationships in science and technology. In: Crystal and solid state physics. New Series, Group III, vol III/17b. Springer, Berlin
101. Weller H, Schmidt H, Koch U, Fojtik A, Baral S, Henglein A, Kunath W, Weiss K, Dieman E (1986) Photochemistry of colloidal semiconductors. Onset of light absorption as a function of size of extremely small CdS particles. Chem Phys Lett 124:557
102. Spanhel L, Haase M, Weller H, Henglein A (1987) Photochemistry of colloidal semiconductors. 20. Surface modification and stability of strong luminescing CdS particles. J Am Chem Soc 109:5649–5655
103. Henglein A (1989) Small-particle research: physicochemical properties of extremely small colloidal metal and semiconductor particles. Chem Rev 89:1861–1873
104. Steigerwald ML, Brus LE (1990) Semiconductor crystallites: a class of large molecules. Acc Chem Res 23:183–188

105. Brus L (1986) Electronic wave functions in semiconductor clusters: experiment and theory. J Phys Chem 90:2555–2560
106. Brus LE (1984) Electron-electron and electron-hole interactions in small semiconductor crystallites: the size dependence of the lowest excited electronic state. J Chem Phys 80:4403–4409
107. Murray CB, Norris DJ, Bawendi MG (1993) Synthesis and characterization of nearly monodisperse CdE (E = sulfur, selenium, tellurium) semiconductor nanocrystallites. J Am Chem Soc 115:8706–8715
108. Sapsford KE, Algar WR, Berti L, Gemmill KB, Casey BJ, Oh E, Stewart MH, Medintz IL (2013) Developing chemistries that facilitate nanotechnology. Chem Rev 113:1904–2074
109. Algar WR, Prasuhn DE, Stewart MH, Jennings TL, Blanco-Canosa JB, Dawson PE, Medintz IL (2011) The controlled display of biomolecules on nanoparticles: a challenge suited to bioorthogonal chemistry. Bioconjug Chem 22:825–858
110. Parak WJ, Boudreau R, Le Gros M, Gerion D, Zanchet D (2002) Cell motility and metastatic potential studies based on quantum dot imaging of phagokinetic tracks. Adv Mater 14:882–885
111. Mattoussi H, Mauro J, Goldman ER, Anderson GP, Sundar VC, Mikulec FV (2000) Selfassembly of CdSe–ZnS quantum dot bioconjugates using an engineered recombinant protein. J Am Chem Soc 122:12142–12150
112. Ong WL, Rupich SM, Talapin DV, McGaughey AJ, Malen JA (2013) Surface chemistry mediates thermal transport in three-dimensional nanocrystal arrays. Nat Mater 12:410–415
113. Bera D, Lei Qian L, Tseng TK, Holloway PH (2010) Quantum dots and their multimodal applications: a review. Mater 3:2260–2345
114. Talapin DV, Haubold S, Rogach AL, Kornowski A, Haase M, Weller H (2001) A novel organometallic synthesis of highly luminescent CdTe nanocrystals. J Phys Chem B 105:2260–2263
115. Hammer NI, Emrick T, Barnes MD (2007) Quantum dots coordinated with conjugated organic ligands: new nanomaterials with novel photophysics. Nanoscale Res Lett 2:282–290
116. Mei BC, Susumu K, Medintz IL, Delehanty JB, Mountziaris TJ, Mattoussi H (2008) Modular poly(ethylene glycol) ligands for biocompatible semiconductor and gold nanocrystals with extended pH and ionic stability. J Mater Chem 18:4949–4958
117. Bloemen M, Stappen TV, Willot P, Lammertyn J, Koeckelberghs G, Geukens N, Gils A, Verbiest T (2014) Heterobifunctional PEG ligands for bioconjugation reactions on iron oxide nanoparticles. PLoS ONE 9:e109475
118. Kim S, Bawendi MG (2003) Oligomeric ligands for luminescent and stable nanocrystal quantum dots. J Am Chem Soc 125:14652–14653
119. Zhang Y, Clapp A (2011) Overview of stabilizing ligands for biocompatible quantum dot nanocrystals. Sensors 11:11036–11055
120. Kloepfer JA, Mielke RE, Wong MS, Nealson KH, Stucky G, Nadeau JL (2003) Quantum dots as strain- and metabolism-specific microbiological labels. Appl Environ Microbiol 69:4205–4213
121. Derfus AM, Chan WC, Bhatia SN (2004) Probing the cytotoxicity of semiconductor quantum dots. Nano Lett 4:11–18
122. Dabbousi BO, Bawendi MG, Onitsuka O, Rubner MF (1995) Electroluminescence from CdSe quantumdot/polymer composites. Appl Phys Lett 66:1316–1320
123. Mattoussi H, Mauro JM, Goldman ER, Anderson GP, Sundar VC, Mikulec FV, Bawendi MG (2000) Self-assembly of CdSe-ZnS quantum dot bioconjugates using an engineered recombinant protein. J Am Chem Soc 122:12142–12150
124. Gerion D, Pinaud F, Williams SC, Parak WJ, Zanchet D, Weiss Alivisatos AP (2001) Synthesis and properties of biocompatible water-soluble silica-coated CdSe/ZnS semiconductor quantum dots. J Phys Chem B 105:8861–8871
125. Palui G, Avellini T, Zhan N, Pan F, Gray D, Alabugin I, Mattoussi H (2012) Photoinduced phase transfer of luminescent quantum dots to polar and aqueous media. J Am Chem Soc 134:16370–16378

126. Mansur HS, Mansur AP (2011) CdSe quantum dots stabilized by carboxylic-functionalized PVA: synthesis and UV–vis spectroscopy characterization. Mater Chem Phys 125:709–717
127. Zhou S, Chen Q, Hu X, Zhao T (2012) Bifunctional luminescent superparamagnetic nanocomposites of CdSe/CdS-Fe3O4 synthesized via a facile method. J Mater Chem 22:8263–8270
128. Yang Z, Chen S, Hu W, Yin N, Zhang W, Xiang C, Wang H (2012) Flexible luminescent CdSe/bacterial cellulose nanocomoposite membranes. Carbohydr Polym 88:173–178
129. Liu Y, Kim M, Wang Y, Wang YA, Peng X (2006) Highly luminescent, stable, and water-soluble CdSe/CdS core-shell dendron nanocrystals with carboxylate anchoring groups. Langmuir 22:6341–6634
130. Dong C, Ren J (2011) Water-soluble mercaptoundecanoic acid (MUA)-coated CdTe quantum dots: one-step microwave synthesis, characterization and cancer cell imaging. Luminescence 27:199–203
131. Law WC, Yong KT, Roy I, Ding H, Hu R, Zhao W, Prasad PN (2009) Aqueous-phase synthesis of highly luminescent CdTe/ZnTe core/shell quantum dots optimized for targeted bio-imaging. Small 5:1302–1310
132. Yuwen L, Bao B, Liu G, Tian J, Lu H, Luo Z, Zhu X, Boey F, Zhang H, Wang L (2011) One-pot encapsulation of luminescent quantum dots synthesized in aqueous solution by amphiphilic polymers. Small 7:1456–1463
133. Ma N, Marshall AF, Gambhir SS, Rao J (2010) Facile synthesis, silanization, and biodistribution of biocompatible quantum dots. Small 6:1520–1528
134. Ananth DA, Rameshkumar A, Jeyadevi R, Jagadeeswari S, Nagarajan N, Renganathan R, Sivasudha T (2015) Antibacterial potential of rutin conjugated with thioglycolic acid capped cadmium telluride quantum dots (TGA-CdTe QDs). Spectrochim Acta 138:684–692
135. Shavel A, Gaponik N, Eychmuller A (2004) Efficient UV-blue photoluminescing thiol-stabilized water-soluble alloyed ZnSe(S) nanocrystals. Phys Chem B 108:5905–5908
136. Chen HS, Lo B, Hwang JY, Chang GY, Chen CM, Tasi SJ, Wang SJ (2004) Colloidal ZnSe, ZnSe/ZnS, and ZnSe/ZnSeS quantum dots synthesized from ZnO. J Phys Chem B 108:17119–17123
137. Hering VR, Faulin TS, Triboni ER, Rodriguez SD, Bernik DL, Schumacher RI, Mammana VI, Alario AF, Abdalla DP, Gibson G, Politi JM (2004) Violet ZnSe/ZnS as an alternative to green CdSe/ZnS in nanocrystal-fluorescent protein FRET systems. Bioconjugate Chem 20:1237–124
138. Li C, Jiang D, Zhang L, Xia J, Li Q (2012) Controlled synthesis of ZnS quantum dots and ZnS quantum flakes with graphene as a template. Langmuir 28:9729–9734
139. Dong B, Cao L, Sua G, Liu W (2010) Facile synthesis of highly luminescent UV-blue emitting ZnSe/ZnS core/shell quantum dots by a two-step method. Chem Commun 46:7331–7333
140. Kim S, Park J, Kim T, Jang E, Jun S, Jang H, Kim B, Kim SK (2011) Reverse type-I ZnSe/InP/ZnS core/shell/shell nanocrystals: cadmium-free quantum dots for visible luminescence. Small 7:70–73
141. Pradhan N, Battaglia DM, Liu Y, Peng X (2007) Efficient, stable, small, and water-soluble doped ZnSe nanocrystal emitters as non-cadmium biomedical labels. Nano Lett 7:312–317
142. Sun L, Wang Q (2014) PbS quantum dots capped with amorphous zns for bulk heterojunction solar cells: the solvent effect. ACS Appl Mater Interfaces 6:14239–14246
143. Neo MS, Venkatram N, Li GS, Chin WS, Ji W (2010) Synthesis of PbS/CdS Core-Shell QDs and their nonlinear optical properties. J Phys Chem C 114:18037–18044
144. Bradshaw TD, Junor M, Patane A, Clarke P, Thomas NR, Li M, Mann S, Turyanska L (2013) Apoferritin encapsulated PbS quantum dots significantly inhibit growth of colorectal carcinoma cells. J Mater Chem 1:6254–6260
145. Zhao H, Chaker M, Ma D (2011) Effect of CdS shell thickness on the optical properties of water-soluble, amphiphilic polymer-encapsulated PbS/CdS core/shell quantum dots. J Mater Chem 21:17483–17491

146. Nakane Y, Tsukasaki Y, Sakata T, Yasuda H, Jin T (2013) Aqueous synthesis of glutathione-coated PbS quantum dots with tunable emission for non-invasive fluorescence imaging in the second near-infrared biological window (1000–1400 nm). Chem Commun 9:7584–7586
147. Xu X, Ray R, Gu Y, Ploehn HJ, Gearheart L, Raker K, Scrivens WA (2004) Electrophoretic analysis and purification of fluorescent single-walled carbon nanotube fragments. J Am Chem Soc 126:12736
148. Shen JH, Zhu YH, Yang XL, Li CZ (2012) Graphene quantum dots: emergent nanolights for bioimaging, sensors, catalysis and photovoltaic devices. Chem Commun 48:3686
149. Cao L, Wang X, Meziani MJ, Lu F, Wang H, Luo PG, Lin Y, Harruff BA, Veca LM, Murray D, Xie SY, Sun YP (2007) Carbon dots for multiphoton bioimaging. J Am Chem Soc 129:11318–11319
150. Hu SL, Niu KY, Sun J, Yang J, Zhao NQ, Du XW (2009) One-step synthesis of fluorescent carbon nanoparticles by laser irradiation. J Mater Chem 19:484–488
151. Sun YP, Wang X, Lu F, Cao L, Meziani MJ, Luo PG, Gu L, Veca LM (2008) Doped carbon nanoparticles as a new platform for highly photoluminescent dots. J Phys Chem C 112:18295–18298
152. Sun YP, Zhou B, Lin Y, Wang W, Fernando KA, Pathak P, Meziani MJ, Harruff BA, Wang X, Wang H, Luo PG, Yang H, Kose ME, Chen B, Veca LM, Xie SY (2006) Quantum-sized carbon dots for bright and colorful photoluminescence. J Am Chem Soc 128:7756–7757
153. Lu J, Yang JX, Wang J, Lim A, Wang S, Loh KP (2009) One-pot synthesis of fluorescent carbon nanoribbons, nanoparticles, and graphene by the exfoliation of graphite in ionic liquids. ACS Nano 3:2367–2375
154. Zhao QL, Zhang ZL, Huang BH, Peng J, Zhang M, Pang DW (2008) Facile preparation of low cytotoxicity fluorescent carbon nanocrystals by electrooxidation of graphite. Chem Commun 41:5116–5118
155. Zheng L, Chi Y, Dong Y, Lin J, Wang B (2009) Electrochemiluminescence of water-soluble carbon nanocrystals released electrochemically from graphite. J Am Chem Soc 131:4564–4565
156. Zhou J, Booker C, Li R, Zhou X, Sham TK, Sun X, Ding Z (2007) An electrochemical avenue to blue luminescent nanocrystals from multiwalled carbon nanotubes (MWCNTs). J Am Chem Soc 129:744–745
157. Bourlinos AB, Stassinopoulos A, Anglos D, Zboril R, Karakassides M, Giannelis EP (2008) Surface functionalized carbogenic quantum dots. Small 4:455–458
158. Bourlinos AB, Stassinopoulos A, Anglos D, Zboril R, Georgakilas V, Giannelis EP (2008) Photoluminescent carbogenic dots. Chem Mater 20:4539–4541
159. Dutta PK, Ghosh T, Kumar H, Jain T, Singh Y (2015) Hydrothermal and solvothermal synthesis of carbon dots from chitosan-ethanol system. Asian Chitin J 11:1-4
160. Zhu H, Wang XL, Li YL, Wang ZJ, Yang F, Yang XR (2009) Microwave synthesis of fluorescent carbon nanoparticles with electrochemiluminescence properties. Chem Commun 34:5118–5120
161. Kumar H, Dutta PK (2014) Chitosan-QD nanocomposite: synthetic strategy and application. Asian Chitin J 10:19–24
162. Dallinger D, Kappe CO (2007) Microwave-assisted synthesis in water as solvent. Chem Rev 107:2563–2591
163. Maeng YJ, Choi SW, Kim HO, Kim JH (2010) Culture of human mesenchymal stem cells using electrosprayed porous chitosan microbeads. J Biomed Mater Res, Part A 92:869–876
164. Jayakumar R, Prabaharan M, Nair SV, Tamura H (2010) Novel chitin and chitosan nanofibers in biomedical applications. Biotechnol Adv 28:142–150
165. Wan MW, Petrisor IG, Lai HT, Kim D, Yen TF (2004) Copper adsorption through chitosan immobilized on sand to demonstrate the feasibility for in situ decontamination. Carbohydr Polym 55:249–254
166. Ngah WS, Fatinathan S (2010) Adsorption characterization of Pb(II) and Cu(II) by chitosan-tripolyphosphate beads. J Environ Manage 91:958–969

167. Futalan CM, Kan CC, Dalida ML, Hsien KJ, Pascua C, Wan MW (2011) Comparative and competitive adsorption of copper, lead and nickel using chitosan immobilized on bentonite. Carbohydr Polym 83:528–536
168. Popuri SR, Vijaya Y, Boddu VM, Krishnaiah A (2009) Adsorptive removal of copper and nickel ions from water using chitosan coated PVC beads. Bioresour Technol 100:194–199
169. Laus R, de Favere VD (2011) Competitive adsorption of Cu(II) and Cd(II) ions by chitosan crosslinked with epichlorohydrin-triphosphate. Bioresour Technol 02:8769–8776
170. Gupta N, Kushwaha AK, Chattopadhyaya MC (2012) Adsorptive removal of Pb(II), Co(II) and Ni(II) by hydroxyapatite/chitosan composite from aqueous solution. J Taiwan Inst Chem Eng 43:125–131
171. Suguna M, Siva Kumar N, Subba Reddy A, Boddu VM, Krishnaiah A (2011) Biosorption of Lead(II) and from aqueous solutions on glutaraldehyde cross-linked chitosan beads. Can J Che Eng 89:833–843
172. Monier M, Abdel-Latif DA (2012) Preparation of cross-linked magnetic chitosan-phenylthiourea resin for adsorption of Hg(II), Cd(II) and Zn(II) ions from aqueous solutions. J Hazard Mater 209–210:240–249
173. Monier M, Ayad DM, Abdel-Latif DA (2012) Adsorption of Cu(II), Cd(II) and Ni(II) ions by cross-linked magnetic chitosan-2-aminopyridine glyoxal Schiff's base. Colloid Surf B: Biointerf 94:250–258
174. Li Z, Du Y, Zhang Z, Pang D (2003) Preparation and characterization of CdS quantum dots chitosan biocomposite. React Funct Polym 55:35–43
175. Mansur AAP, Mansur HS, Ramanery FP, Oliveira LC, Souza PP (2014) "Green" colloidal ZnS quantum dots/chitosan nano-photocatalysts for advanced oxidation processes: Study of the photodegradation oforganic dye pollutants. Appl Catal B 158–159:269–279
176. Chang SQ, Kang B, Dai YD, Zhang HX, Chen D (2011) One-step fabrication of biocompatible chitosan coated ZnS and ZnS:Mn2+quantum dots via γ gradiation route. Nanoscale Res Lett 6:591
177. Wang L, Sun Y, Xie X (2014) Structural and optical properties of Cu-doped ZnS nanoparticles formed in chitosan/sodium alginate multilayer films. Luminescence 29:288–292
178. Xia H, He G, Peng J, Li W, Fang Y (2010) Preparation and fluorescent sensing applications of novel CdSe–chitosan hybrid films. Appl Surf Sci 256:7270–7275
179. Kang B, Chang SQ, Dai YD, Chen D (2008) Synthesis of green CdSe/chitosan quantum dots using a polymer-assisted γ-radiation route. Radiat Phys Chem 77:859–863
180. Gong HM, Wang XH, Du YM, Wang QQ (2006) Optical nonlinear absorption and refraction of CdS and CdS-Ag core-shell quantum dots. J Chem Phys 125:024707
181. Mansur AAP, Mansur HS, Soriano-Araújo A, Lobato ZIP (2014) Fluorescent nanohybrids based on quantum dot−chitosan−antibody as potential cancer biomarkers. ACS Appl Mater Interfaces 6:11403–11412
182. Jiang Z, Zhao C, Liu X (2014) Synthesis of poly(ethylene glycol)-graft-chitosan and using as ligandfor fabrication of water-soluble quantum dots. Colloids Surf, B 115:260–266
183. Wu W, Shen J, Banerjee P, Zhou S (2010) Chitosan-based responsive hybrid nanogels for integration of optical pH-sensing, tumor cell imaging and controlled drug delivery. Biomaterials 31:8371–8381
184. Shen M, Jia W, Lin C, Fan G, Jin Y, Chen X, Chen G (2014) Facile synthesis of folate-conjugated magnetic/fluorescent bifunctional microspheres. Nanoscale Res Lett 9:558
185. Wang Y, Geng Z, Guo M, Chen Y, Guo X, Wang X (2014) Electroaddressing of ZnS quantum dots by codeposition with chitosan to construct fluorescent and patterned device surface. ACS Appl Mater Interf 6:15510–15515
186. Ma Q, Lin ZH, Yang N, Li Y, Su XG (2014) A novel carboxymethyl chitosan–quantum dot-based intracellular probe for Zn^{2+} ion sensing in prostate cancer cells. Acta Biomater 10:868–874
187. Li Y, Chen WC, Chen SM, Lou BS, Ali MA, Al-Hemaid FMA (2014) Detection of real sample DNA at a cadmium sulfide—chitosan/gelatin modified electrode. Colloids Surf, B 113:85–91

188. He Z, Zhu H, Zhou P (2012) Microwave-assisted aqueous synthesis of highly luminescent carboxymethyl chitosan-coated CdTe/CdS quantum dots as fluorescent probe for live cell imaging. J Fluoresc 22:193–199
189. Lee JI, Ha KS, Yoo HS (2008) Quantum-dot-assisted fluorescence resonance energy transfer approach for intracellular trafficking of chitosan/DNA complex. Acta Biomater 4:791–798
190. Udaybhaskar P, Iyengar L, Rao AVSP (1990) Cr(VI) interaction with chitosan. J Appl Polym Sci 39:739–747
191. Bhattarai N, Edmondson D, Veiseh O, Matsen FA, Zhang M (2005) Electrospun chitosan-based nanofibers and their cellular compatibility. Biomaterials 26:6176–6184
192. Yang TC, Zall RR (1984) Absorption of metals by natural polymers generated from seafood processing wastes. Ind Eng Chem Prod Res Dev 23:168–172
193. Maeng YJ, Choi SW, Kim HO, Kim JH (2010) Culture of human mesenchymal stem cells using electrosprayed porous chitosan microbeads. J Biomed Mater Res, Part A 92:869–876
194. Choi SW, Zhang Y, Yeh YC, Wooten AL, Xia Y (2012) Biodegradable porous beads and their potential applications in regenerative medicine. J Mater Chem 22:11442
195. Bhumiratana S, Vunjak-Novakovic G (2012) Concise review: personalized human bone grafts for reconstructing head and face. Stem Cells Transl Med 1:64–69
196. Ghosh P, Rameshbabu AP, Dogra N, Dhara S (2014) 2,5-Dimethoxy 2,5-dihydrofuran crosslinked chitosan fibers enhance bone regeneration in rabbit femur defects. RSC Adv 4:19516
197. Wang Z, Ruan J, Cui DX (2009) Advances and prospect of nanotechnology in stem cells. Nanoscale Res Lett 4:593–605
198. Ruan J, Ji JJ, Song H, Qian QR, Wang K, Wang C, Cui DX (2012) Fluorescent magnetic nanoparticle-labeled mesenchymal stem cells for targeted imaging and hyperthermia therapy of in vivo gastric cancer. Nanoscale Res Lett 7:309
199. Patel D, Kell A, Simard B, Deng J, Xiang B, Lin HY, Gruwel M, Tian G (2010) Cu^{2+}-labeled, SPION loaded porous silica nanoparticles for cell labeling and multifunctional imaging probes. Biomaterials 31:2866–2873
200. Shi Z, Neoh KG, Kang ET, Shuter B, Wang SC (2010) Bifunctional Eu^{3+}-doped Gd2O3 nanoparticles as a luminescent and T1 contrast agent for stem cell labeling. Contrast Media Mol Imaging 5:105–111
201. Tseng CL, Shih IL, Stobinski L, Lin FH (2010) Gadolinium hexanedione nanoparticles for stem cell labeling and tracking via magnetic resonance imaging. Biomaterials 31:5427–5435
202. Shah BS, Clark PA, Moioli EK, Stroscio MA, Mao JJ (2007) Labeling of mesenchymal stem cells by bioconjugated quantum dots. Nano Lett 7:3071–3079
203. Yang K, Li Z, Cao Y, Yu X, Mei J (2009) Effect of peptide-conjugated near-infrared fluorescent quantum dots (NIRF-QDs) on the invasion and metastasis of human tongue squamous cell carcinoma cell line Tca8113 in vitro. Int J Mol Sci 10:4418–4427
204. Seleverstov O, Zabirnyk O, Zscharnack M, Bulavina L, Nowicki M, Heinrich JM, Yezhelyev M, Emmrich F, O'Regan R, Bader A (2006) Quantum dots for human mesenchymal stem cells labelling. A size-dependent autophagy activation. Nano Lett 6:2826–2832
205. Solanki A, Kim JD, Lee KB (2008) Nanotechnology for regenerative medicine: nanomaterials for stem cell imaging. Nanomedicine 3:567–578
206. Lin S, Xie X, Patel MR, Yang YH, Li Z, Cao F, Gheysens O, Zhang Y, Gambhir SS, Rao JH (2007) Quantum dot imaging for embryonic stem cells. BMC Biotechnol 7:67–76
207. Muller-Borer BJ, Collins MC, Gunst PR, Cascio WE, Kypson AP (2007) Quantum dot labelling of mesenchymal stem cells. J Nanobiotechnol 5:9
208. Shang W, Zhang X, Zhang M, Fan Z, Sun Y, Han M, Fan L (2014) The uptake mechanism and biocompatibility of graphene quantum dots with human neural stem cells. Nanoscale 6:5799–5806
209. Zhang M, Bai L, Shang W, Xie W, Ma H, Fu Y, Fang D, Sun H, Fan L, Han M, Liub C, Yang S Facile synthesis of water-soluble, highly fluorescent graphene quantum dots as a robust biological label for stem cells. J Mater Chem 22:7461

210. Baruah S, Ortinero C, Shipin OV, Dutta J (2012) Manganese doped zinc sulfide quantum dots for detection of Escherichia coli. J Fluoresc 22:403–408
211. Abdelhamid HN, Wu HF (2013) Probing the interactions of chitosan capped CdS quantum dots with pathogenic bacteria and their biosensing application. J Mater Chem B 1:6094
212. Tan WB, Jiang S, Zhang Y (2007) Quantum-dot based nanoparticles for targeted silencing of HER2/neu gene via RNA interference. Biomaterials 28:1565–1571
213. Ho YP, Chen HH, Leong KW, Wang TH (2006) Evaluating the intracellular stability and unpacking of DNA nanocomplexes by quantum dots-FRET. J Control Release 116:83–89
214. Lai S, Chang X, Fu C (2009) Cadmium sulfide quantum dots modified by chitosan as fluorescence probe for copper (II) ion determination. Microchim Acta 165:39–44
215. Sharma A, Pandey CM, Sumana G, Soni U, Sapra S, Srivastava AK, Chatterjee T, Malhotra BD (2012) Chitosan encapsulated quantum dots platform for leukemia detection. Biosens Bioelectron 38:107–113
216. Chen Y, Yao R, Wang Y, Chen M, Qiu T, Zhang C (2015) CdS QDs-chitosan microcapsules with stimuli-responsive propertygenerated by gas–liquid microfluidic technique. Colloids Surf, B 125:21–27
217. Ma L, Liu C (2010) Preparation of chitosan microspheres by ionotropic gelation under a high voltage electrostatic field for protein delivery. Colloids Surf B Biointerfaces 75:448–453
218. Chang CH, Tsao CT, Chang KY, Young TH, Han JL, Hsieh KH (2010) Chitosan membrane with surface-bonded growth factor in guided tissue regeneration applications. J Bioact Compat Polym 25:465–482
219. Saber A, Strand SP, Ulfendahl M (2010) Use of the biodegradable polymer chitosan as a vehicle for applying drugs to the inner ear. Eur J Pharm Sci 39:110–115
220. Ghendon Y, Markushin S, Vasiliev Y, Akopova I, Koptiaeva I, Krivtsov G, Borisova O, Ahmatova N, Kurbatova E, Mazurina S, Gervazieva V Evaluation of properties of chitosan as an adjuvant for inactivated influenza vaccines administered parenterally. J Med Virol 81:494–506
221. Mahkam M (2010) Modified chitosan cross-linked starch polymers for oral insulin delivery. J Bioact Compat Polym 25:406–418
222. Zhang N, Li J, Jiang W, Ren C, Li J, Xin J, Li K (2010) Effective protection and controlled release of insulin by cationic β-cyclodextrin polymers from alginate/chitosan nanoparticles. Int J Pharm 393:213–219
223. Kamari A, Ngah WSW (2009) Isotherm, kinetic and thermodynamic studies of lead and copper uptake by H_2SO_4 modified chitosan. Colloid Surf B: Biointerf 73:257–266

Development and Selection of Porous Scaffolds Using Computer-Aided Tissue Engineering

Nitin Sahai, Tanvi Jain, Sushil Kumar and Pradip Kumar Dutta

Abstract Tissue engineering is considered a multidisciplinary field where the involvement of many course of studies as well as utilization of knowledge of various researchers/scientists and medical practitioners provided good health care and esthetics concepts. Porous scaffolds play very important role in tissue engineering and many of the relevant disciplines. Henceforth, the development and selection of proper scaffolds, particularly in the form of porous due to multifaceted microstructure-like veins and capillaries is essential. Computer-Aided Tissue Engineering (CATE) is an important tool to categorize the porous scaffolds in terms of design, modeling as well as experimental validation. The present article describes the selection of materials, facilitated properties, experimental methods, knowledge of computer to fabricate scaffolds.

Keywords Porous scaffold · Computer-aided tissue engineering · Chitosan · Design · Characterization

N. Sahai
Department of Biomedical Engineering, North-Eastern Hill University, Shillong, Meghalaya 793022, India

T. Jain · S. Kumar
Department of Chemical Engineering, M.N. National Institute of Technology, Allahabad, U.P. 211004, India

T. Jain · P.K. Dutta
Department of Chemistry, M.N. National Institute of Technology, Allahabad, U.P. 211004, India
e-mail: pkd_437@yahoo.com

P.K. Dutta (✉)
Center for Medical Diagnostic and Research, M.N. National Institute of Technology, Allahabad, U.P. 211004, India
e-mail: pkd_437@yahoo.com

P.K. Dutta (ed.), *Chitin and Chitosan for Regenerative Medicine*, Springer Series on Polymer and Composite Materials, DOI 10.1007/978-81-322-2511-9_13

1 Introduction

Tissue Engineering is a field of basic sciences that employs the fundamentals of engineering and the life sciences for the designing and fabrication of biological substitutes that repair, maintain, or enhance the biological actions at tissue or organ level by designing the scaffolds. The methodology adopted is by transplantation of cells in a suitable polymeric matrix. A biopolymer-based scaffold should have the following features: (i) three-dimensional and highly porous with an interconnected pore network for cell growth and flow transport of nutrients and metabolic waste; (ii) biocompatible and bioresorbable with a well-regulated degradation and resorption rate to balance cell/tissue growth in vitro and/or in vivo; (iii) appropriate surface chemistry for cell attachment, proliferation, and differentiation; and (iv) mechanical characteristics to match those of the tissues at the site of implantation [1]. There are various kinds of tissue-engineered implants used in tissue engineering such as physical, physiological, chemical, and combination of all the above varieties. The 3D tissue fabrication can also be attained by applying CAD-based manufacturing technologies. Some of the major commercial ventures based on tissue engineering are Apligraf (Organogenesis canton, MA), Dermagraft (Smith and Nephew, London, UK), and Orcel (Ortec international, New York). Tissue engineering is appearing as a fascination tool or an alternative solution, by which tissue and organ damage is inscribed by implanting natural, man-made, or semisynthetic tissue and organ mimics that are completed functional from the beginning, or that grow into the important performance. Primarily, the efforts have only been mainly concentrated toward the skin treatments for managing burns and wounds, but huge number of different types of tissues are now being evaluated and then engineered, also various types of biomaterials (synthetic or natural) and scaffolds are used as delivery entities [2]. Various kinds of biomaterials derived from the synthetic as well as natural polymers have been investigated as scaffolds for tissue engineering applications [3].

Significant analysis involves tissue-engineered bone using different modulations, blood vessels, muscle, and even nerve ducts. As a consequence, there is momentous academic and corporate attraction in this newly emerging technology which is of great market potential.

A computer-aided characterization modality for the interpretation of mechanical characteristics and structural arrays of porous tissue scaffolds has been a promising technique in modern day. The central idea for utilizing Computer-Aided Tissue Engineering (CATE) is that it enables the characterization of the design and preparation of a porous tissue scaffold of accurate pore size and also of appropriate mechanical strength where this porous scaffold is implemented.

It is always necessary to understand the effect of the design and process factors on the structural characteristics of the scaffolds, the effective mechanical characteristics of porous tissue scaffolds of chitosan fabricated through a precision extruding process for its envision. The experimental evaluation of the characterization demonstrates the effective mechanical characteristics of the chitosan scaffold which is due to the materials usually biopolymers, the direction of the deposition layout pattern, and the total porosity of the scaffold structure. Normally, all the scaffold formulations perform with anisotropic mechanical characteristics and the deposition layout pattern is responsible for its various levels of anisotropy.

The dynamic mechanical characteristics decline with the increase in porosity for the scaffolding biomaterials like, hydroxyapatite (HA), polycaprolactone (PCL), poly(DL-lactic acid-coglycolic acid) (PLGA). Chitosan scaffolds also show the same kind of variation in their mechanical characteristics, as an increase in porosity results in a decrease of mechanical properties. The mechanical strength of chitosan is very low, so it can be concluded that chitosan has a potential to be utilized as a biomaterial for healing for nerve tissue. Chitosan has been enormously studied for biomedical applications because of its excellent biodegradable and biocompatible properties.

Chitosan being nontoxic, biodegradable, and biocompatible, has drawn much attention from chemists, biologists, and materials scientists. It can easily be fabricated into films, fibers, foams, etc. Chitosan scaffolds and its hybrid scaffolds are extensively used for tissue engineering. Chitosan scaffolds are synthesized by freeze–drying of chitosan gels and also by electrospinning technique using chitosan solutions. Chitosan scaffolds are used for skin tissue regeneration, bone and cartilage tissue regeneration, nervous tissue regeneration, etc.

Chitosan is produced commercially by deacetylation of chitin, which is the structural element in the exoskeleton of crustaceans (crabs, shrimp, etc.) and cell walls of fungi. It is a linear copolymer of β-(1–4)-linked-2-acetamido-2-deoxy-β-*d*-glucopyranose and 2-amino-2-deoxy-β-*d*-glycopyranose which is the second most abundant natural biopolymer (polysaccharides) on earth after cellulose. Chitosan may be retrieved from alkalescent deacetylation of chitin which is originated from exoskeletons of crustaceans and cell walls of fungi or creatures [4]. The advantages of chitosan involve excellent biodegradability and biocompatibility and low toxicity with numerous biological activities, such as antimicrobial activity and low immunogenicity, which provides bountiful opportunities for numerous applications in the field of tissue engineering, drug delivery, wound healing, quantum dots, and other uses [5–11]. The degree of deacetylation (DD) and the molecular weight (MW) of chitosan often dominate its properties, like solubility in water, mechanical behavior, chemical properties and stability, biodegradability, etc. Normally, pure chitosan displays good biological properties, but low solubility in aqueous medium at pH, dominating the creation of fragile films with ineligible mechanical characteristics [12, 13]. The amino group in chitosan has a pKa value of ~6.5, thus, chitosan is positively charged and soluble in acidic to neutral solutions with a charge density dependent on the pH and %DA value.

Chitosan and its water soluble derivatives such as trimethylchitosan, glycolchitosan, carboxymethylchitosan, etc., have been continuously used in drug delivery and other biomedical applications [14, 15].

Like, Trimethylchitosan, has shown effective results to transfect breast cancer cells; with an increased degree of trimethylation increases the cytotoxicity, and at approximately 50 % trimethylation, the derivative is the most efficient at gene

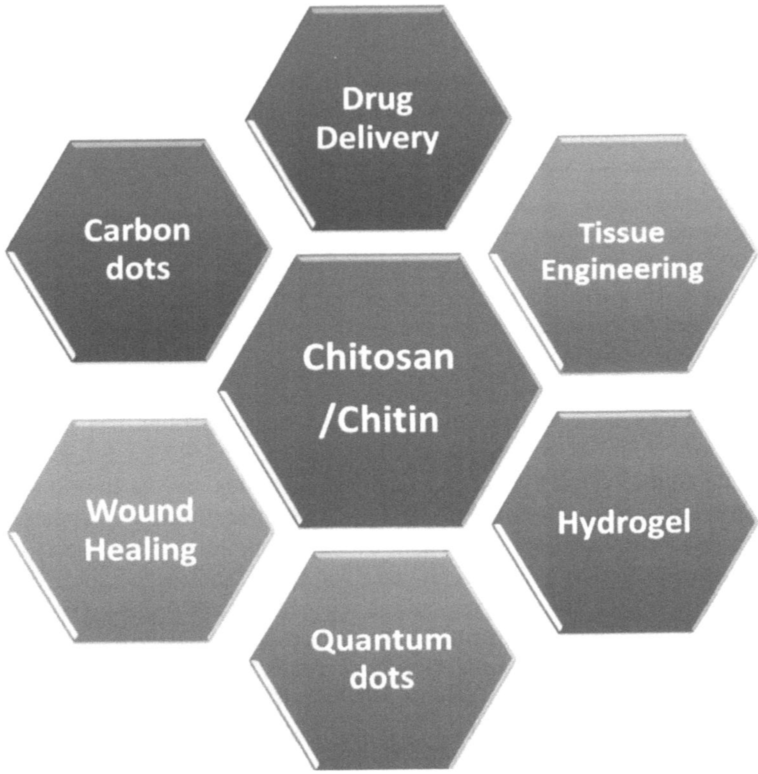

Fig. 1 Biomedical application of chitosan

delivery. Oligomeric derivatives (3–6 kDa) are relatively nontoxic and have good gene delivery properties [16]. Various applications of chitosan in biomedical area are shown in Fig. 1.

1.1 In Vitro Synthesis of Chitosan Scaffolds

Fabrication of a chitosan scaffold is by formation of a composition of chitosan with two other biopolymers. Aliginate and Pectin are used in different proportion as it is known that chitosan alone does not have good biocompatible and biodegradable properties. Both a lyophilization technique and a freeze–drying technique are utilized to help in giving porous scaffolds, e.g., porous chitosan alginate gelatin scaffolds were synthesized using freeze–drying technique. Here, the solutions were frozen to preserve ice crystals and freezing solutions were lyophilized

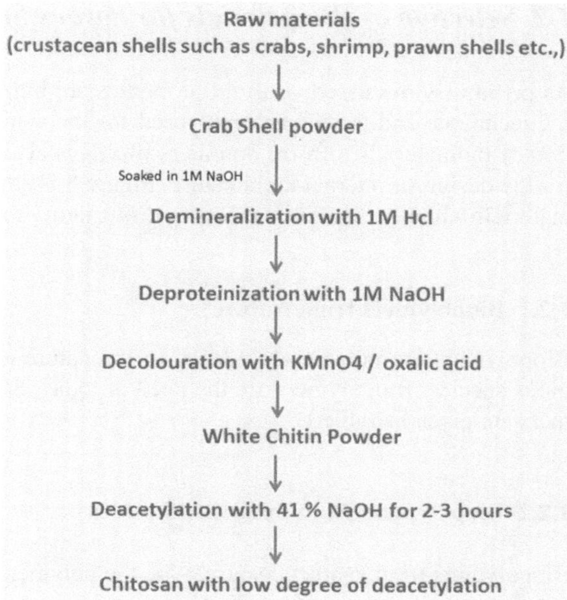

Fig. 2 Schematic representation for the preparation of chitosan from crab shells

in order to maintain porous and interconnected structure by discharging the ice crystals from the earlier frozen solutions. The three different pore size samples were fabricated with Scanning Microscopy done for pore size determination. The preparation method for chitosan from crab shell is shown in Fig. 2.

Porous chitosan scaffolds were synthesized by lyophization of chitosan solution or gels, and by the elecrospinning. Pure chitosan scaffolds give deficient mechanical properties. Some researchers have prepared hybrid chitosan scaffolds to improve its load-bearing properties. Phongying et al. [17] reported that chitosan scaffolds can be directly fabricated from the chitin whiskers. Wang et al. [18] reported on the preparation of porous chitosan scaffolds by thermally induced phase separation. Wang et al. [19] also reported the synthesis of porous chitosan tubular scaffolds by a novel molding and casting/lyophilizing method. Zheng et al. [20] prepared a new Gelatin/Montmorillonite–Chitosan (Gel/MMT-CS) composite scaffold through the interpolation process and the lyophilization technique, using the ice aggregates as the porogen materials. The Gel/MMT-CS scaffold has great mechanical properties and manageable degradation rate. Hydroxyapatite and alginate hybridized chitosan scaffolds were also reported.

Lyophilization and freeze–drying are the techniques used to generate the porous chitosan tissue scaffold through which the size of porous tissue scaffold is controlled. These processes will thus be helpful in fabrication of an appropriate mechanical strength tissue scaffold.

1.2 Selection of Biomaterials for Porous Scaffolds

As previously discussed, scaffold properties are highly influenced by the processing technique and the biomaterials used for the synthesis of scaffolds. The selection of biomaterials to form porous scaffolds is also one of the important aspects for the design of internal architecture. Figure 3 is showing various number of biomaterials which are used to fabricate porous tissue scaffolds.

1.2.1 Biopolymers from Nature

Biopolymers are polymers produced in the nature during the life cycles of enormous species. Biopolymers are the need of today. Some of the important biopolymers are given in Table 1.

1.2.2 Biopolymers from Synthetic Sources

Biopolymers from synthetic sources are the polymers which satisfy the shortfall of natural as well as hour of needs. Nowadays many synthetic biopolymers are used. Some of the important synthetic biopolymers which are used for the synthesis of porous tissue scaffolds are given in (Table 2).

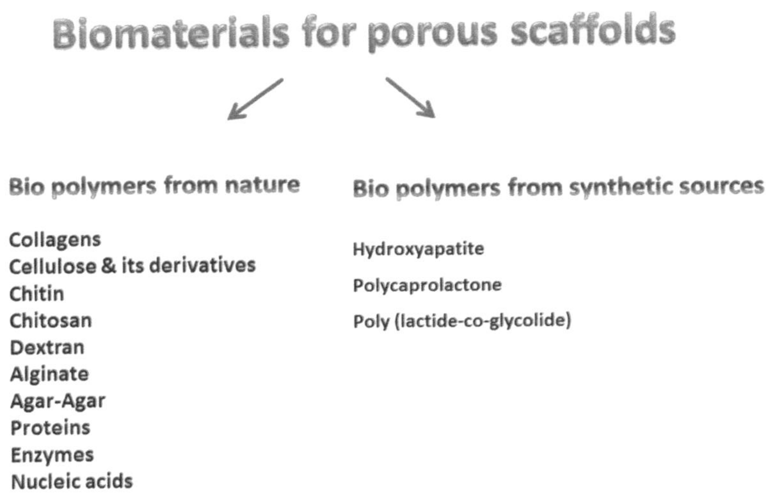

Fig. 3 Biomaterials for porous scaffolds

Table 1 Natural biopolymers with its applications

Biopolymer	Applications	References
Collagens	They are used as "glue" since ancient times. More recently, it has been widely used in soft tissue augmentation	[21]
	Employed in the management of burns and act as artificial skin substitutes	
	Collagen–GAG scaffolds loaded with basic fibroblast growth factor have been reported to significantly enhance tissue remodeling in animal models	
Cellulose and its derivatives	Carboxymethyl cellulose is used in coatings, detergents, food, toothpaste, adhesives, and cosmetics. Like wise, Methyl and ethyl cellulose are used as emulsifier, stabilizer, thickening agent, foaming agent. Also, cellulose nitrate (guncotton) used as low-order explosives	[21, 22]
Chitin	Chitin and its derivatives have been used for a variety of applications such as water treatment, textile and paper, cosmetic, food and health supplements, agriculture, and biotechnology	[21, 22]
	In biomedical area, it is useful for drug delivery and wound healing applications	
	Commercially used for making edible plastic food wrap and cleaning up of industrial wastewater and also acts as a thickening agent	
Chitosan	Chitosan and its chemically modified derivatives act as a penetration enhancer by opening epithelial tight junctions. In medical it forms a tough, water-absorbent, oxygen-permeable, wound-healing and has bacteriostatic effects, biodegradable and biocompatible films used as bandages and sutures, cosmetics and also for drug delivery in cancer chemotherapy. Chitosan carries a positive charge (cationic) in aqueous solution and is used as a flocculating agent to purify drinking water	[23–30]
	Various other uses in different areas are mentioned below	
	Skin: Many studies have been reported on the use of chitosan as a skin substitute material in skin tissue engineering due to its many advantages for wound healing such as hemostasis, accelerating the tissue regeneration and stimulating the fibroblast synthesis of collagen	
	Bone: Chitosan has been also extensively used in bone tissue engineering since exploration of its capacity to promote growth and mineral rich matrix deposition by osteoblasts in culture. Also, chitosan is biocompatible (additional minimizes local inflammation), biodegradable, and can be molded into porous structures (allows osteoconduction)	
	Cartilage: Chitosan and some of its degraded products could be involved in the synthesis of the articular matrix component such as chondroitin, chondroitin sulfate, dermatane sulfate, keratan sulfate, and hyaluronic acid, as their synergy was examined in vivo. Lu et al. has demonstrated that chitosan solution injected into the knee articular cavity of rats led to a significant increase in the density of chondrocytes in the knee articular cartilage, indicating that chitosan could be potentially beneficial to wound healing of articular cartilage	

(continued)

Table 1 (continued)

Biopolymer	Applications	References
	Nerve cells: Chitosan has been studied as a candidate material for nerve regeneration due to its properties such as antitumor, antibacterial activity, biodegradability, and biocompatibility. Haipeng et al. reported that neurons cultured on the chitosan membrane can grow well and that chitosan tube can greatly promote the repair of the peripheral nervous system	
Dextran	It is useful substitute for blood plasma, antithrombotic (anti-platelet) agents to reduce blood viscosity, as a volume expander in anemia, as a lubricant in some eye drops	[25]
Alginate	Alginate used as thickening or gelling agents, as colloidal stabilizers, as blood expanders and used to take impressions of the oral tissue due to their ability to undergo a transformation from the sol state to a gel state through ionotropic gelation in the presence of many multivalent ions such as Ca_2^+, for cartilage and bone regeneration, either as a scaffold or as carriers for biologically active molecules and drug delivery	[31]
Agar-agar	Nutrient agar is used throughout the world to provide a solid surface containing medium for the growth of bacteria and fungi. Agar-Agar is a natural vegetable gelatin counterpart originally eaten in Japan. Agar is used as an impression material in dentistry	[32]
Proteins	Proteins are condensation product of α-amino acid which are linked through peptide (–CO–NH–) bonds. A large number of proteins exist in nature. All type of proteins are derived from 20 amino acids. The structure of proteins is very complex	[33]
	Primary structure of proteins shows the chemical formulae of the amino acids arranged in a definite sequence through peptide linkage	
	Secondary structure of proteins describes the molecular shape (or conformation) of a protein molecule	
	Tertiary structure designates the shape of folding which results due to the sulfur–sulfur cross-links between macromolecular chains. The specificity of enzyme action in fact depends on tertiary structure	
Nucleic acids	Macromolecules of utmost biological importance. Two kinds of nucleic acids exist in the living cell (a) Deoxyribonucleic acid (DNA) (b) Ribonucleic acid (RNA). Nucleic acids are usually built up of bases (purine and pyrimidine bases), pentose sugar, and phosphoric acid	[34]

1.3 Tissue Engineering

Tissue damages and organ failures imitate life-threatening situations due to accidents and diseases. Repairing of tissue by analogous cell/tissue transplantation is one of the most encouraging modality for the regeneration of tissue. Nowadays, autografting and allografting are the two main approaches which are usually used to replace or repair the damaged tissue or organs. However, autografts are associated

Table 2 Synthetic biopolymers with its applications

Synthetic biopolymer	Applications	References
Hydroxylapatite	It is a naturally occurring mineral form of calcium apatite with the formula $Ca_5(PO_4)_3(OH)$, but is usually written $Ca_{10}(PO_4)_6(OH)_2$ to denote that the crystal unit cell comprises two entities. Hydroxylapatite is the hydroxyl end member of the complex apatite group. The OH–ion can be replaced by fluoride, chloride, or carbonate, producing fluorapatite or chlorapatite	[35]
Polycaprolactone	It is a biodegradable polyester with a low melting point of around 60 °C and a glass transition temperature of about −60 °C	[36]
	The most common use of polycaprolactone is in the manufacture of specialty polyurethanes and also imparts good water, oil, solvent, and chlorine resistance to the polyurethane	
	PCL is degraded by hydrolysis of its ester linkages in physiological conditions (such as in the human body) and has therefore received a great deal of attention for use as an implantable biomaterial	
Poly(lactide-*co*-glycolide)	A copolymer of polylactic acid and polyglycolic acid Poly(lactide-*co*-glycolide) is a very popular and an important biodegradable polyester due to its better degradation rates than other polymer rates, good mechanical properties, etc. It is widely used as a material for the synthesis of absorbable sutures and in scaffolds	[37]

with some demerits such as donor site melancholy and less availability. Whereas, allografts are not limited; however, they have promising feature to cause an immune response and also have the risk of transfer of disease. Tissue engineering has emerged as an attractive approach for the repair/regeneration of damaged tissues, with the capability to overcome all the disadvantages of autologous and allogenic tissue repair. Tissue engineering is a new approach for resolving the problems like missing tissue or any organ. Hence, tissue engineering is an emerging multidisciplinary area which includes the methods of engineering and life sciences which help in understanding the basic knowledge of structure–function relationships in normal as well as in therapeutic mammalian tissues, and also the development of biological entities that are able to restore, maintain, or improve tissue function.

There are three scenarios in tissue engineering:

(1) Replacement of those cells that provides genetic or other manipulations before the cell infusion by the use of isolated cells or its cell substitutes.
(2) The delivery of tissue-activating materials, like growth and differentiation factors, to specific locations.
(3) Growth of the cells in three-dimensional (3-D) matrices (scaffolds) or devices, where the cells can be either inducted from the host tissues in vivo nor seeded (encapsulated) in vitro.

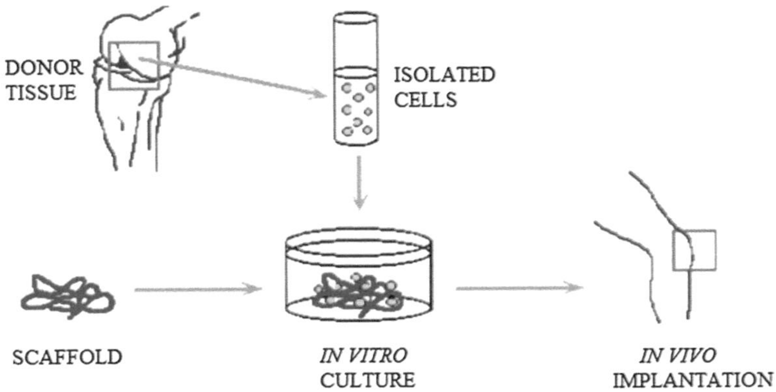

Fig. 4 Illustrative diagram of various phases utilized in tissue engineering

Figure 4 is showing the various stages used in tissue engineering. Biomaterials play a vital role in tissue engineering by serving as 3D synthetic foundation for cellular attainment, procreation, and in growth finally lead to the formation of a new tissue. Both synthetic and natural polymers have been enormously reviewed as biodegradable polymeric biomaterials. Synthetic polymers have unique adaptability in contrast because the composition and the structure of the biopolymeric scaffolds can be adjusted to the desired needs. Various novel approaches have been designed for the fabrication of biomaterial-based 3D scaffolds.

1.4 Tissue Scaffolds

From human health care point of views, the loss or failure of organs or tissue by any means is concerned as a costly and devastating problem and thus it leads to the developments of biological substitute organs and tissues either to replace or repair the failing organs and tissues and this process is called tissue engineering (TE). The most important aspect of TE is to seed cells on porous scaffold to grow them and form new tissue that can replace the damaged tissue.

Tissue scaffolds are highly devised structures that can behave as an artificial substrate for the cells to expedite the regeneration of the target tissue. The challenge in such scaffolding is to design and synthesize the commercial biodegradable scaffolds with appropriate properties that can supply nutrient to transplant and regenerate the cells and also endorse cell adhesion and support cell growth, proliferation, and differentiation, and thus formation of the extracellular matrix (ECM) for the tissues. Porous scaffolds are varied for different cell types on the basis of the characteristics like its porosity, size, and pore structure. A large void volume and a large surface area to volume ratio are desirable properties that can be used for cell growth, maximum cell seeding, formation of ECM, and vascularization.

Scaffold pores of the similar diameter are high surface area per volume yield provided the pore size is larger than the diameter of the cells [38, 39].

Besides the design of internal architecture, scaffold properties are highly influenced by the processing technique and the type of biomaterial used for the synthesis of scaffolds. There are a variety of processing methods to fabricate TE scaffolds, namely, lyophilization, fiber bonding, phase separation are few of them. Unfortunately, most of these processes have limitations in controlling the pore size, porosity, and pore interconnectivity. It is important to have a control over the scaffold microstructures, such as pore size, porosity, and pore geometry, since these properties directly affect cells ingrowth. For example, in the case of adult mammalian skin regeneration an optimal pore size is 20 and 200–400 μm for bone ingrowth [16, 33]. The pore size and crystalline nature of the polymer are the main reasons for the tissue invasion into porous scaffolds [40–44].

Recently, finite element method (FEM), computer-aided design (CAD), and computer-aided manufacturing (CAM) softwares have been found to be promising for the architecture and fabrication of tissue scaffolds of exact pore size and porosity as required for the growth of cell tissue in the scaffolds.

1.5 Computer-Aided Tissue Engineering

The implementation of computer-aided tools in tissue engineering has resulted in the development of a newly emerging discipline of CATE. CATE comprises of computer imaging technology CAD/CAM, linked with modern design and industrialized technology. By utilizing CATE, the design of an intricate three-dimensional structure of a scaffold can be analyzed, and these scaffolds can be fabricated with appropriate efficiency to assist the researchers in understanding and studying the multifaceted tissue engineering glitches. The classification of CATE is separated into three main categories; (1) computer-aided tissue anatomical modeling, (2) computer-aided tissue classification, and (3) computer-aided tissue implantation [5, 7]. An outline of CATE [8] is shown in Fig. 5. CATE empowers various innovative approaches in designing, modeling, and synthesis of complex tissue alternatives with enhanced functionality for research in patient-specific implant analysis and simulation, image-guided surgical planning, and scaffold-guided tissue engineering.

Computer-aided design and FEM can be merged with building technologies such as solid free-form fabrication (SFF). This is helpful in allowing the simulated designing, characterization, and synthesis of a scaffold that can be optimized for the replacement of tissues. Thus, it makes possible to design and manufacture very intricate tissue scaffold structure with functional modules that are complex to fabricate. The research will also enable cell biologists and engineers to expand their scope of research and study in the area of tissue engineering and regenerative medicine. If the mechanical properties of material of tissue scaffolds are known, then with the help of finite element modeling (FEM), it is possible to predict the

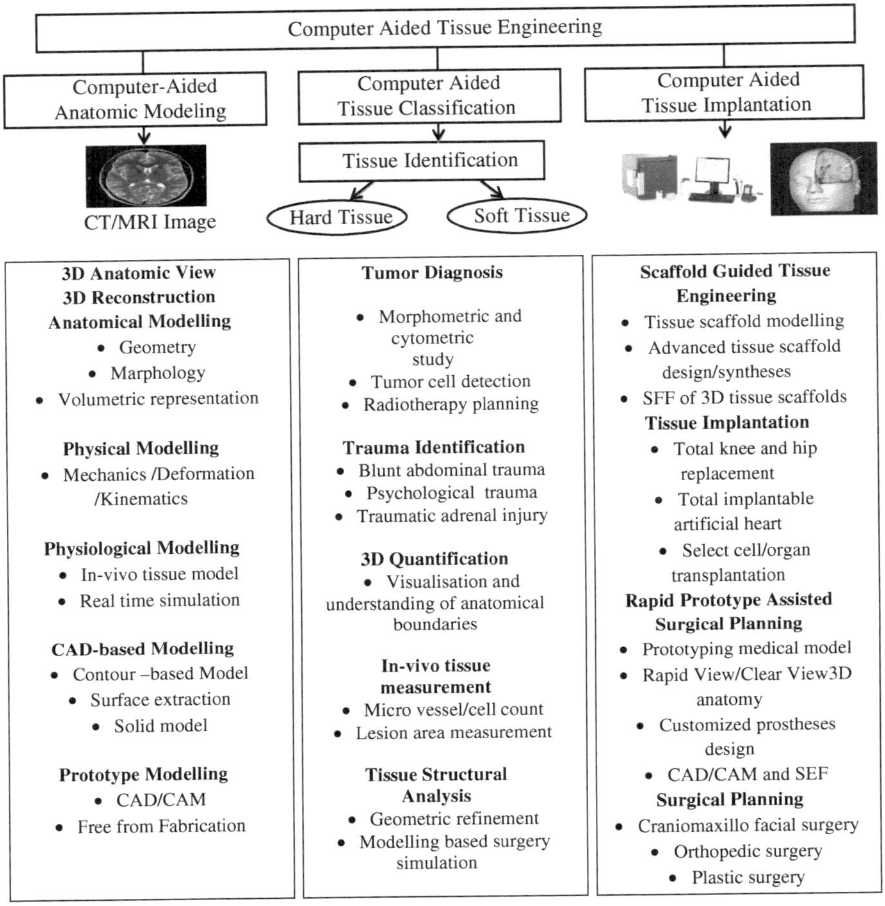

Fig. 5 Outline of computer-aided tissue engineering [8]

actions of the complex formations, such as multilayer systems [6–10, 12, 45, 46]. When tissue scaffold is in an in vivo condition, then there will be some microscopic loads (compressive loads and fluid flow). As the tissue differentiation proceeds, it is difficult to determine the local mechanical stimuli sensed by the cells at a microscopic level, hence the study of stress–strain relationship at microscopic level uses finite element analysis (FEA).

1.5.1 Finite Element Methods in Tissue Engineering

If the mechanical properties of the biomaterials are known, then FEM is capable of predicting the performance of multifaceted structures such as multilayer systems [47–51]. Miranda et al. [52] used FEM as a tool to compute the stress fields in complex tissue scaffold structures, and hence their predicted mechanical behavior (e.g., bone implants as load bearer) was calculated. This method was rationalized to identify the fracture modes and assessing the strength of robocast tissue scaffold. The calculation was performed for three testing configurations namely compression, tension, and shear. The tools applied for FEM are ANSYS, MSC Nastran, MSC Patran, etc. using FEM.

The computer-aided characterization approach used in estimating the efficient mechanical properties of a porous scaffold has also been reported in the literature [47, 48]. The computational algorithm that is applied in characterization is a finite element execution of asymptotic homogenization technique. The characterization results of PCL display that its efficient mechanical properties are the function of the scaffolding materials and the total porosity of the scaffold assemblies. Lacroix et al. [53] made a FEA of cylinders of CaP-based scaffolds to evaluate the stress–strain allocation of the larger scaffolds.

To calculate the mechanical properties of a microstructure, multiscale fine element models using the homogenization theory have been used [54]. Olivares et al. [55] modeled both gyroid and hexagonal scaffold have porosity of 55 and 77 % by using FEA. Submitting this to an inlet fluid flow and compressive strain gives results showing that gyroid architecture provides a better accessibility of the fluid than the hexagonal structure. This shows that the distribution of the shear stress induced by fluid perfusion is very much relied on the pore distribution within the scaffolds. Guo et al. [56] proposed a FEM-based method which is a direct method to analyze the material reconstruction inverse problem for soft tissue elastography. Using the method of direct material reconstruction inverse method, the results were obtained by minimizing the objective, defined as the sum of the squares of the residual norms, at all nodes where the nodal residual norm of the associated elements is defined as a linear function of elasticity factors. Satisfying the equilibrium at every node is utilized as the optimization objective, and the measured deformation is enforced directly as a result. The soft tissue elastography can be obtained directly by solving the resulting set of linear equations where no iterations are required.

Images of Human Tissue Using Computed Tomography and Magnetic Resonance Imaging

The anatomical model of human tissue is generated with the help of a noninvasive imaging modality, like CT or MRI technology. Each CT image is composed of tiny picture element (pixels) which turns into a small volume element (voxel) of patient tissue examined by the CT scanner. Both CT and MRI are also complementary to each other in that CT can produce a distinct contour of bones and MRI can show the shape of both tissues and bones 26. Table 3 gives a characteristics imaging of human tissue commonly used in production of a 3-D reconstruction CT and MRI [57, 58]. The novel use of CT and MRI data and image processing tools to segment and reconstruct tissue images in 3D is to determine characteristics of muscle, bone, and brain 28.

Table 3 Characteristics of CT and MRI

Characteristics	CT	MRI
Matrix size (Pixel)	512 × 512	256 × 256
Voxel size (mm)	0.5 × 0.5 × 2.0	0.5 × 0.5 × 1.5
Density resolution	4096 levels (12 bit)	128 levels (16 bit)
Transverse resolution	768 pixels × 768 pixels	512 pixels × 512 pixels
Segmentation	Threshold	Complex
Slice thickness	1.25 mm	1.20 mm
Field of view	230 × 230 mm	200 × 200 mm

Three-dimensional reconstruction techniques help in the development of 3-D anatomical models or images, and allow the third dimension to be precisely learned by revealing the 3D anatomical image [6, 7]. Voxel-based anatomical representation through CT/MRI images cannot be effectually utilized in various biomechanical engineering deeds such as anatomical structural development, modeling-based anatomical tissue biomechanical analysis and its simulation [59–63]. CAD systems and CAD-based solid modeling based on a vector modeling environment are used for structural modeling design, development analysis, and simulation.

1.6 CAD in Tissue Engineering

Computer-aided tissue engineering assimilates advanced technologies derived from biological sciences, biomedical engineering, computer science engineering, and modern designing and manufacturing utilized for the purpose of tissue engineering. In particular, computer-aided design (CAD), medicinal-based image processing, computer-aided manufacturing (CAM), and SFF are now actively used for the simulation, design, and manufacturing of tissue and organ surrogates. Therefore, CATE encompasses three main applications in tissue engineering: (1) computer-aided tissue modeling, including 3D anatomic visualization, 3D reconstruction, and CAD-based tissue modeling, and biophysical modeling for surgical planning and simulation; (2) computer-aided tissue scaffold informatics; and (3) CAD systems which are used as boundary representation (B-REP) in which a solid entity is explained by the surfaces which bind to it and these surfaces are selected with the support of nonuniform rational B-spline (NURBS) applications [64–66]. CAD contributes to shortening the design process of scaffolds and also contributes to the minimizing of experimental tests [8, 9]. Since the year 2000, chemical-based techniques for fabrication of scaffolds have made way for alternate methods of fabrication that have the capability to be integrated with CAD/CAM technologies [67–70]. This has made it possible for the scaffolds to be designed using

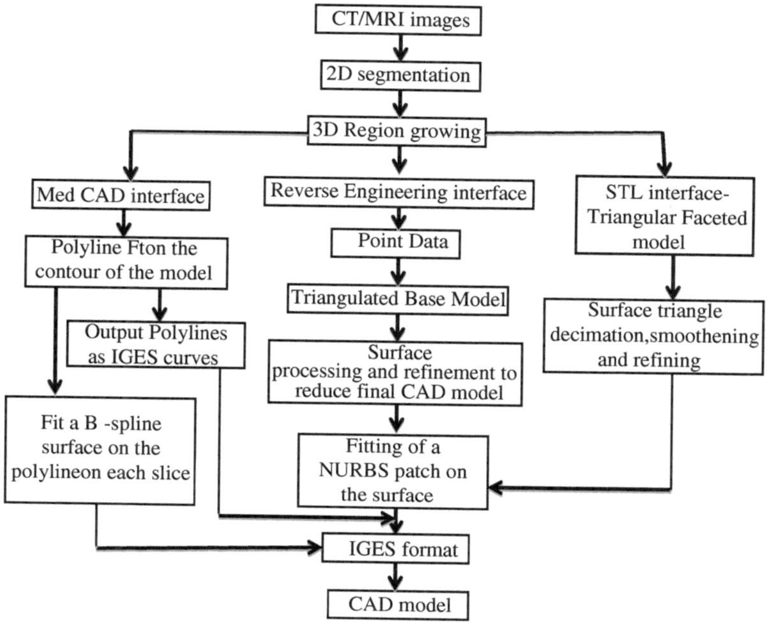

Fig. 6 Illustration of a CAD model from CT/MRI data [6]

CAD software and then transferred using rapid prototyping for their fabrication. A typical component Med CAD is used to mend the disparity between medical imaging techniques and CAD designing softwares. The Med CAD can integrate the export data from CT/MRI system to the CAD platform and vice versa by using various entities such as IGES (International Graphics Exchange Standard), STEPS (Standard for Exchange of Product), or STL (STereo Lithography) layout.

Sun et al. [5] has evaluated and compared the following three major methods for producing a CAD model from therapeutic imaging data: [38] Med CAD interface method [16], reverse engineering interface method, and [33] STL triangulated model converting method [5]. The outline of the processes is presented in Fig. 6.

The microarchitecture of tissue scaffold is believed to influence the biological function of tissue and the behavior of cells by providing spatial allocation for cell growth and specifying a nutritional pathways and proliferation to cells. A problem in the fabrication of scaffolds is often limitation of particle thickness, due to the difficulty in getting cells beneath into internal sections of the scaffold. This limitation can be eliminated by adopting an advance technology for the synthesis of scaffolds such as SFF [71–77]. As advancement took place in SFF, it becomes the most favorable technique in the synthesis of scaffolds. The internal architecture of these scaffolds were either made as a pattern of extrusions, cuts, or holes, across the structure surface of the tissue scaffolds in CAD [78–81].

1.6.1 Porous Tissue Scaffolds Using CAD

In scaffolds, the porosity and size of the pore will play a crucial function in tissue and formation of a cell in vitro and in vivo. The density and structure of many manufactured components could be designed to achieve optimum performance if the native porosity of formed material might be controlled. The designing of scaffolds for the therapies based on cell for its repairing of the impaired articular cartilage should provide the following features: (i) a three-dimensional (3D) and highly porous architecture to assist cell attachment, propagation, and extracellular matrix (ECM) production; (ii) an interconnected/permeable pore linkages to endorse nutrient and waste exchange; (iii) a biocompatible and bioresorbable substrate with sustainable rate of degradation; (iv) a appropriate structural chemistry for cell attachment, proliferation, and variation; (v) mechanical characteristics to support, or balance, those of the tissues at the targeted location of implantation, etc., [82–84]. Sun et al. [85] suggested a computer modeling method for composing a 3D microsphere-packed bone graft assembly, which consists of two extreme microsphere-packing paradigms to design tissue-engineering scaffolds. In tissue engineering bone scaffold modeling, to support the users for various CAD modeling, a standard parametric collection of scaffold structure was developed, from which a user could select cell unit and suitable size. A fiber deposition modality for fabricating three-dimensional poly(ethylene glycol)-terephthalate–poly(butylene terephthalate) (PEGT/PBT) block copolymer scaffolds with a 100 % interrelating pore linkage for engineering of articular cartilage is also prepared. Schroeder et al. [86] proposed the differences between the bone scaffolds synthesized by rapid prototyping technique and natural structures from SEM (Scanning Electron Microscope) images, by producing a stochastic paradigm for a berea sandstone and a fontainebleau sandstone, each with a recommended lineal-path utility, two-point probability utility, and "pore size" distribution tasks. Cai [87] has given a control modeling method for designing tissue engineering bone scaffold with identified pore size distribution based on the hexahedral mesh modification which adored controllability and higher precision in comparison to another methods such as varying managing parameter in supercritical fluid handling and multi-interior architecture designing.

The SEM images of 3D accumulated porous scaffolds with variable fiber accumulated geometries are presented in Fig. 7 [88].

2 History of Various Approaches in Modeling, Design, and Fabrication of Scaffolds

Computer-aided tissue engineering empowers various methodologies in modelling, design, and fabrication of intricate tissues with improved functionality and cell–matrix interfaces to fight against the diseases. This section elaborates the main application and advances in this field, particularly in computer-aided tissue

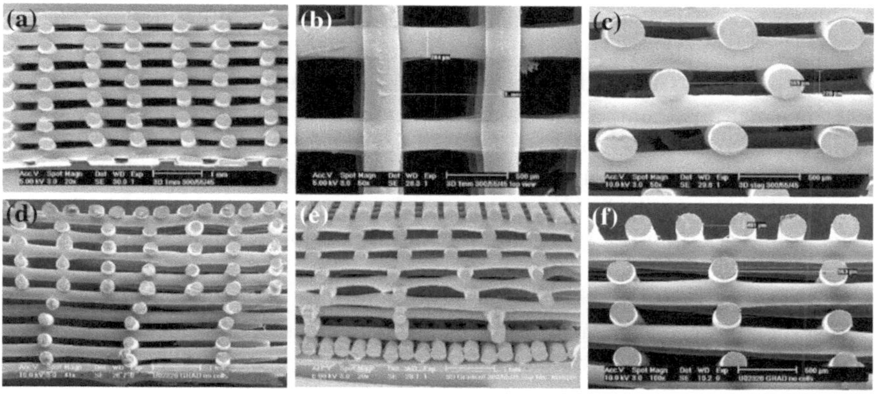

Fig. 7 SEM images of 3D credited porous scaffolds with varying fiber deposition geometries (Adapted from T.B.F. Woodfield et al. [88])

fabrication techniques for scaffolds and evaluation of their degradation properties and also some approaches used to generate bio-CAD models using high resolution noninvasive imaging and the medical imaging practices along with the 3D reconstruction method.

Sun et al. [85] reviewed recent development and application of enabling computer technology, imaging technology, computer-aided design and computer-aided manufacturing (CAD and CAM), and rapid prototyping (RP) technology, in tissue engineering, particularly in computer-aided tissue anatomical modeling, three-dimensional (3D) anatomy visualization and 3D reconstruction, CAD-based anatomical modeling, computer-aided tissue classification, computer-aided tissue implantation, and prototype modeling assisted surgical planning and reconstruction. And they also presented [86] some salient advances of bio-CAD modeling and application in CATE, including biomimetic design, analysis, simulation, and free-form fabrication of tissue-engineered substitutes. In another study, Qiang et al. [46] demonstrated that the chitosan conduits have appropriate mechanical strength, swelling, characteristics, and nerve cell similarity, and thus can be effectively used as a neural tissue engineering scaffold. Also, Li et al. [83] shows that chitosan–alginate scaffolds can be fabricated from the solutions of physiological pH, which may provide promising conditions for integrating proteins with fewer risk of denaturation. Chitosan and alginate solutions provide a promising scaffold with higher porosity, and better mechanical and biological characteristics which is much more suitable for rapid progression into clinical trials. In another method, Woodfield et al. [88] utilize rapid prototyping (RP) to generate novel scaffolds with honeycomb-like structures, complete interconnected channel networks, and manageable porosity and channel size. Variation of the parameters of PCL articular can give result to the scaffolds having standard geometrical honeycomb pores with a range of 160–700 mm of its channel size and 260–370 mm filament diameters.

The authors laboratory also fabricated double-hybrid silk fibroin, chitosan/gelatin, and double-hybrid silk fibroin/chitosan scaffolds and then gentamicin was encapsulated within calcium alginate beads for drug delivery application. The study revealed that hemocompatible nature of double-hybrid silk fibroin/chitosan and double-hybrid silk fibroin scaffolds of hemolysis was found to be % <5 and porosity >85 %. A comparative drug release profile of fabricated scaffolds showed that double-hybrid silk fibroin/chitosan scaffold is an appropriate model because of its prolonged drug release, optimal hemocompatability, and high compressive modulus [89].

Tai et al. [90] used carbon dioxide process to form foamed scaffolds in which the escape of CO_2 from a plasticized and melted polymer generated gas bubbles that finally shaped into pores. The effects of polymer chemical constituents, molecular weight, and processing factors on the concluding scaffold properties were studied which demonstrated that the pore size and design of the supercritical poly(DL-lactic acid) (PDLLA) and poly(DL-lactic acid-coglycolic acid) (PLGA) scaffolds can be altered by sensible control over processing environment. However, this alteration did not possess accurate controllability, specially for a targeted site in a bone scaffold. Also, Torres-Sanchezet et al. [91] inspected how ultrasound, applied during the synthesis of polyurethane foam, affected its cellular structure. It was described that factors related to ultrasound exposure could affect the volume of the pore and distribution within the final polyurethane matrix. The data demonstrated that porosity (i.e., volume fraction) varied in direct proportion to both the acoustic pressure and frequency of the ultrasound signal. The effect of ultrasound on porosity offered the outlook of a manufacturing method that can adjust functionally graded cellular microstructures of foam, and resulting properties matched the useful requirements. However, this method did not possess accurate level of controllability. Many researchers have already tried to design the bone scaffold from the geometric modeling. The other study done by Hollister [92] reviewed the incorporation of computational topology design (CTD) with SFF to design new tissue-engineering scaffolds. The mechanical characteristics as well as tissue regeneration attained using designed scaffolds were also explained in this study. In another approach, Lal et al. [85] suggested a new computer modeling approach for the fabrication of a three-dimensional microsphere-packed bone graft design. This geometric model utilized two utmost microsphere-packing models and a statistical packing model to calculate the total number of microspheres packed in a fabricated bone implant. Chua et al. [23] proposed a novel modeling method for a tissue engineering scaffold in which many polyhedral shapes which come out to be best suited for bone scaffold modeling were considered as a single cell unit. A paradigm library of various scaffold structures was developed and maintained, from which a researcher can choose a cell unit and size which suits the best in tissue engineering bone scaffold modeling to encourage researchers also in the area of CAD modeling. A programmed algorithm for producing complex polyhedral scaffolds was developed when an open polyhedral cell unit was selected and resized [93]; it helps in assembling the micro structure of the scaffold, while keeping the external geometry of the patient's anatomy created from medical imaging records.

In another study, Starly et al. [24] used the inner architecture design approach to create a layered scaffold from synthesis tool path without making complexed 3D CAD scaffold models. Here, the concept of layered manufacturing was employed to evaluate specific layered process planes and layered contours of the scaffold. Thereafter, 2D characteristic patterns of building blocks (cell units) were described to develop the interior pattern of the scaffold. At last, the tool path for free-form fabrication of the prepared scaffolds, along with the specific interior design was thus produced. Various hierarchies possessed different pore sizes developed on the basis of these above approaches. Schroeder et al. [86] observed the dissimilarities between natural structures and the bone scaffolds manufactured by rapid prototyping technology with the use of images of scanning electron microscope (SEM), and were elucidated by the stochastic reestablishment of sandstones [83]. Here, a simulated annealing approach was utilized to create a stochastic model for a berea sandstone and also for a fontainebleau sandstone, with each a proposed two-point probability function, lineal-path function, and "pore size" distribution function, respectively. They presented a representation of model density and porosity based on stochastic geometry. The pore size and its distribution were altered based on stochastic geometry theory and minimum distance between the pores. Here also, Li et al. [83] developed a process model for this manufacturing device using a swept volume representation of the capabilities of the tool. Specifically, a swept volume which depicted the tool path information and outlined the scaffold to be made, which could be used to simulate the scaffold synthesis process and thus created virtual scaffold prototyping. Wettergreen et al. [28] demonstrated the creation of a unit library of architectures that could be utilized to accumulate a complex scaffold of individual, satisfying characterized microstructure designs. Computer-aided design (CAD) and FEA combined with the manufacturing technology like SFF, was earlier theorized to permit for the virtually designed scaffolds and also fabrication of scaffolds optimized for tissue replacement studies. A library of microarchitectures were introduced for the scientists who got involved in the tissue engineering and other relevant areas, which allowed them to study and merge the ideas based on various characteristics like mechanical properties, flow perfusion characteristics, and porosity. However, the approach used here for porous bone scaffold modeling is based on the fundamental idea of finite element method. It is also possible to use the subdivided units to model porous bone scaffolding by using this approach in which first, a single pore based on the subdivided element is modeled and then a boolean expression union amid the pores is used to create the full model of the tissue engineering bone scaffold. At last, the bone scaffold which consists of numerous irregular pores can be modeled by the boolean expression by taking the difference between the solid model and the whole pore model. It is already studied that the subdivided units are not regular components like cube or regular tetrahedron. If reliable pores (e.g., spheres) under influence of the regular components (e.g., cubes) are modeled, then they would be deformed when main regular components were deformed. Porous bone scaffolds which contain various irregular pores can be modeled using this fact. This study also presented a control approach for the distribution pore size in a scaffold in tissue engineering relied on hexahedral mesh refinement.

3 Some Useful Facilitated Properties of Chitosan Scaffolds for Tissue Engineering with Special Reference to SEM

Chitosan scaffolds are used for tissue engineering purposes and also for biomedical applications like in drug delivery. Chitosan (CS) is widely used as scaffolds for the regeneration of bone tissue, skin tissue, nervous tissue, etc. Porous chitosan scaffolds were prepared by freeze–drying technique of chitosan solution or gels, and also by elecrospinning in Fig. 8. Pure chitosan scaffolds may appear as a poor biomaterial for mechanical properties.

EDS spectra display the chemical constituents of the cell surfaces and extracellular matrices. The spectra were procured from the regions highlighted with white square frames on the SEM micrograph images [94].

The SEM images of chitosan–alginate, namely, CAG1, 2, and 3 scaffolds are displayed in Fig. 9. All scaffolds demonstrate highly porous architecture with good interconnectivity. The size and the figure of pore would influence the capability of attaching the cell and thus growing, and also influence the efficacy of tissue reconstruction. The optimal size and geometry of scaffold pore are dependent on precise cell types. Porous chitosan structures are formed by freeze–drying and lyophilizing the solution. During the freezing process, ice structures are developed.

The type of porosity depends on the freeze–drying conditions to a larger extent. The thermal gradient and the rate of cooling have an outcome on the pore structure [95–97].

The SEM images of chitosan I scaffolds in Fig. 10 show the porous structure of chitosan I (high molecular weight) scaffolds at different degradation rates.

Figure 10a, b shows relatively uniform pore morphology in the original scaffold. Figure 10c, d represents SEM images of chitosan I after 14 days of degradation at flow rate of 0.5 ml/min. The figure shows matrix with nonuniform pores, some pores are very large as compared to Fig. 10a, b due to the fact that as degradation proceeds, the pore diameter increases. Figure 10e, f shows SEM images of chitosan I after 14 days of degradation at flow rate of 1 ml/min. The figure shows filamentous structures with decreased number of pores, and pores of very small size. Figure 10g, h shows SEM images of chitosan I scaffold after 14 days of degradation at 2 ml/min. The figure shows almost a plane surface with no pores as with increase in rate, pore decreases.

Likewise, in another case the morphology of chitosan–hydroxyapatite scaffolds was analyzed by Scanning Electron Microscopy. The SEM images of porous scaffolds with and without PVA–HA fusion coating are shown in Fig. 11.

Kim et al. [98] obtained bare HA scaffold by a polyurethane associated foam method which revealed a well-developed open pore structure.

Figure 11a shows the SEM image of bare chitosan–hydroxyapatite scaffold prepared; (B) shows the SEM image of PVA–HA coated chitosan–hydroxyapatite scaffold prepared using $CaCl_2$ and K_2HPO_4 without any cross-linker. It shows

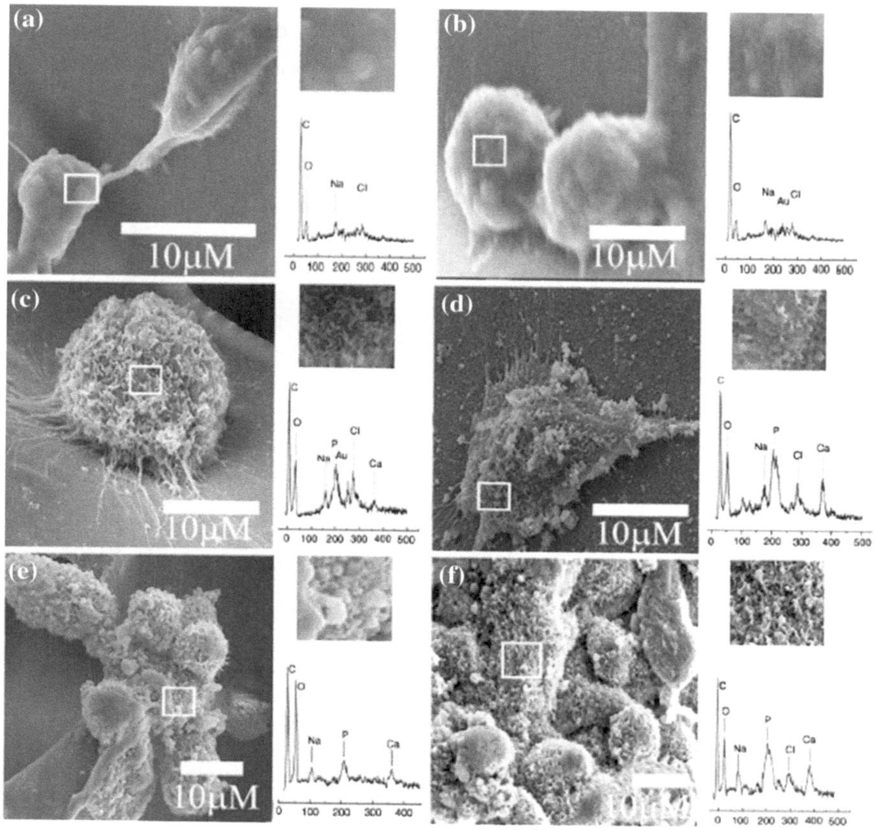

Fig. 8 SEM images of osteoblasts grown on chitosan and chitosan–alginate scaffolds. Osteoblast cells on chitosan after **a** 3 days and **b** 7 days of culture; specific osteoblast cells on chitosan–alginate after **c** 3 and **d** 7 days of cell culturing; bunch of cells on chitosan–alginate after **e** 3 and **f** 7 days of cell culture

nonuniform layered coating on the scaffold. Figure 15b shows the SEM image of PVA–HA coated chitosan–hydroxyapatite scaffold prepared using $CaCl_2$ and K_2HPO_4 without any cross-linker. It shows nonuniform layered coating on the scaffold. It shows bonded but less porous and cross-linked structure of the scaffold.

Figure 12a shows the SEM image of bare chitosan–hydroxyapatite nanopowder scaffold which interprets the morphology of directly nano-HA incorporated scaffold whereas Fig. 12b gives an idea about the morphology of PVA–HA coated chitosan–hydroxyapatite nanopowder scaffold. It shows that coating is not uniform on the scaffold and also the layer structure is not found. It may be due to the presence of projected structures in the scaffold.

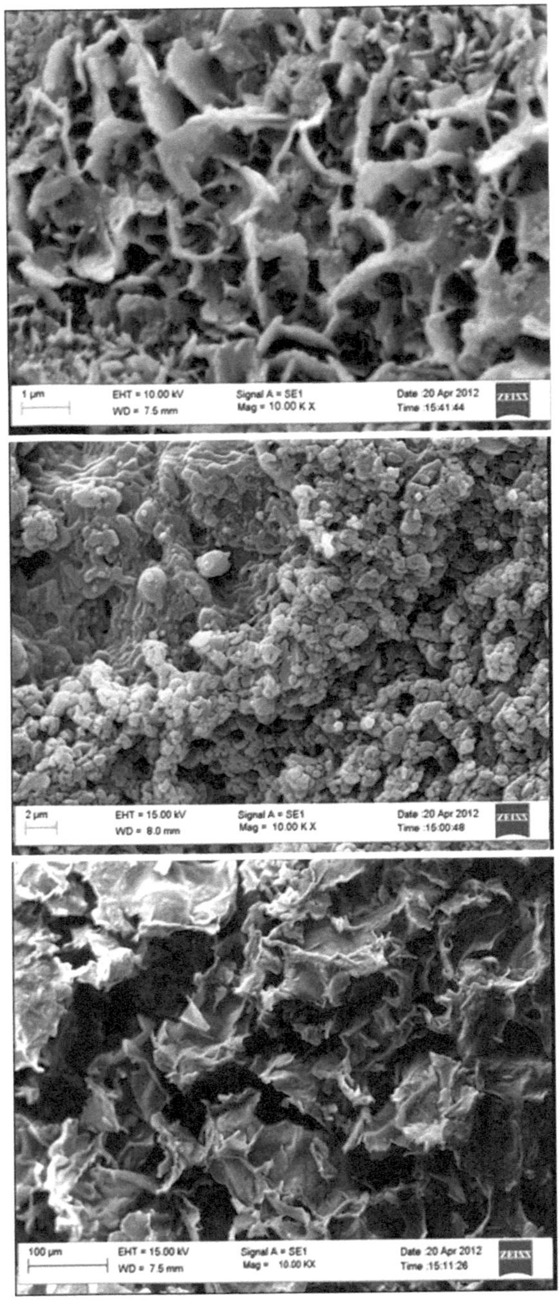

Fig. 9 SEM images of CAG1, CAG2, and CAG3 at 10K× maginification

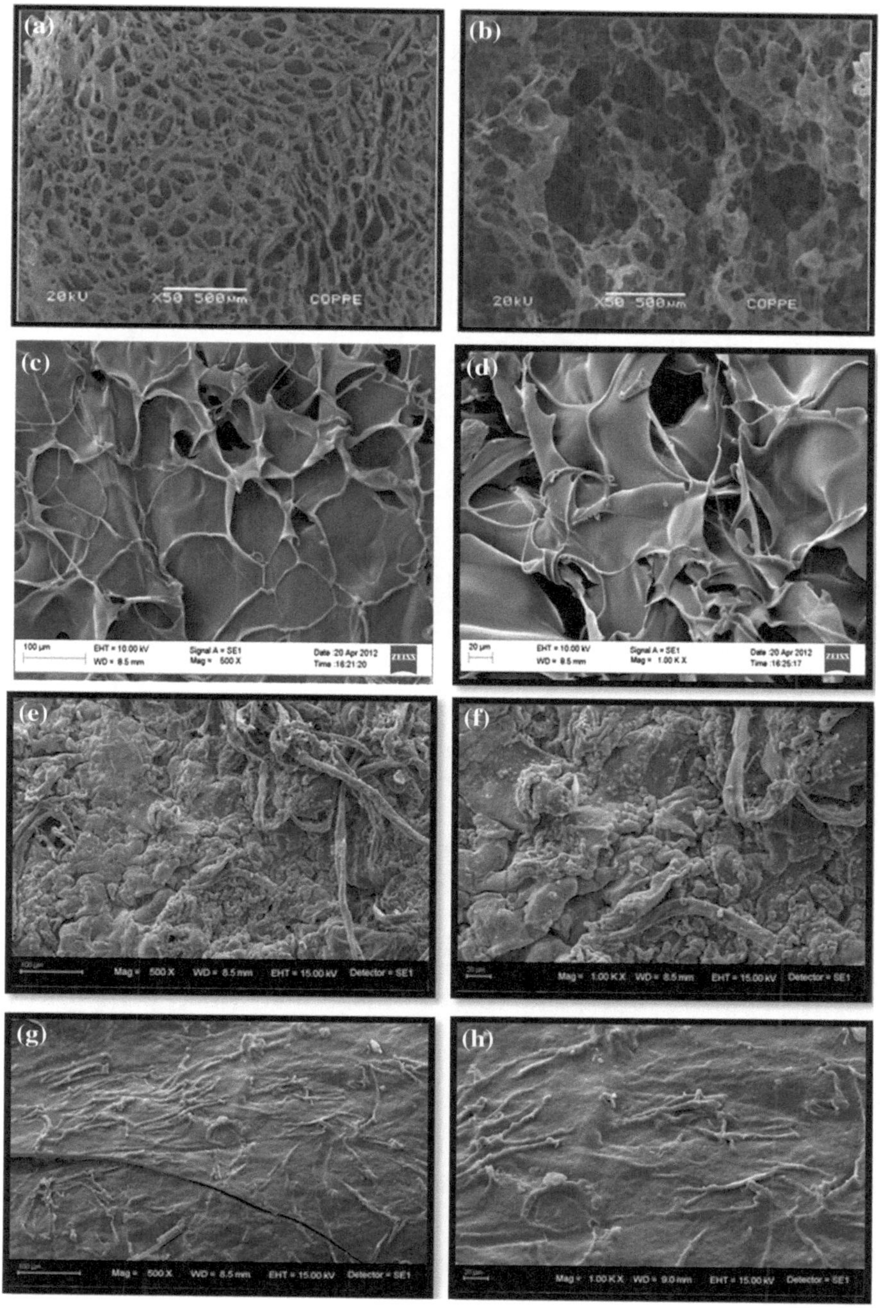

Fig. 10 SEM images of chitosan I: **a** and **b** Before degradation. **c** and **d** After 14 days degradation at 0.5 ml/min. **e** and **f** After 14 days degradation at 1 ml/min. **g** and **h** After 14 days degradation at 2 ml/min

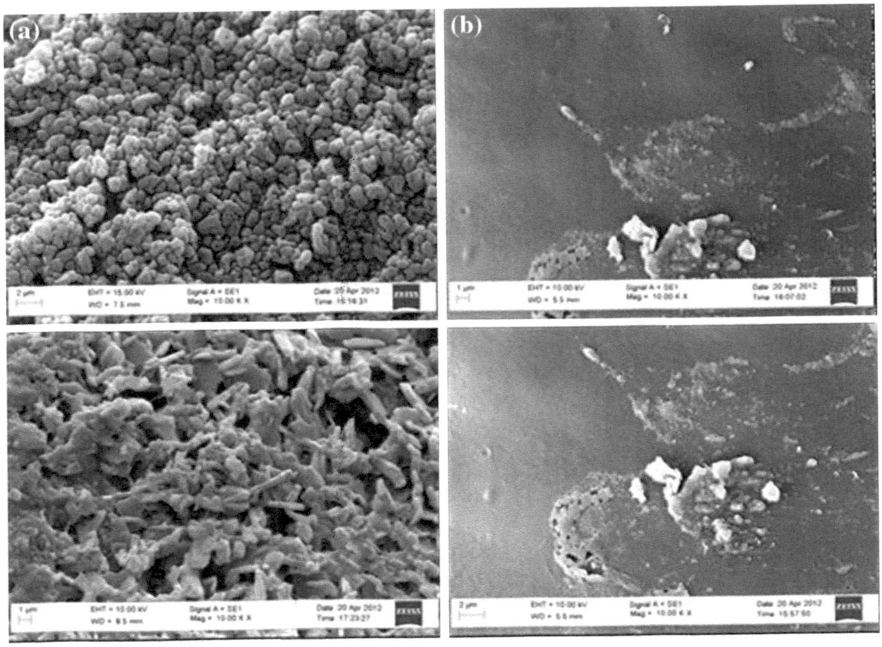

Fig. 11 **a** Cross-linked bare scaffold and **b** cross-linked coated scaffold

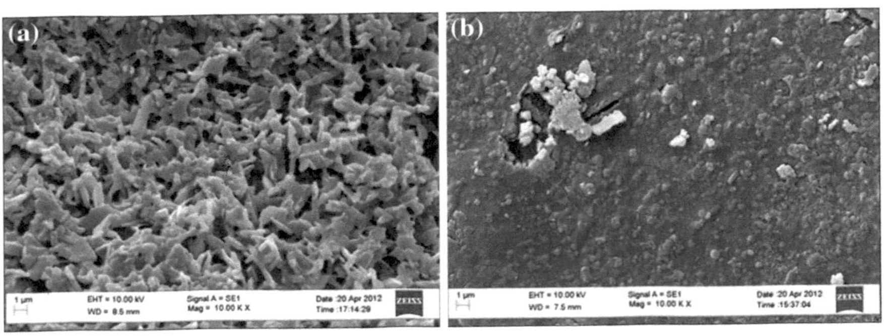

Fig. 12 **a** Bare CS-HA scaffold and **b** coated CS-HA scaffold

4 Mechanical Testing of Chitosan-Based Scaffolds

4.1 Porosity Testing

The sample of defined quantity is cut in the shape of a cylinder. Force was applied perpendicular to the longitudinal axis of the various channels at a crosshead speed of 1 mm/min, and the loads were measured at strains of 20, 40, and 60 %.

Fig. 13 **a** Dimension of porous chitosan tissue scaffolds. **b** Fabricated porous chitosan tissue scaffolds

The dimension of porous scaffolds is represented diagrammatically, shown in Fig. 13a and the fabricated porous chitosan tissue scaffolds are shown in Fig. 13b.

Cross Part area	153.938 mm^2
Gage length	6.00 mm
Full Scale	0.5 kN
Scale Factor	1.0

Briefly, the porosity is calculated by the approach of displacement of the liquid with absolute ethanol. Briefly, prepared scaffold sample was submerged in a graduated beaker containing a known volume (V_1) of ethanol. The sample was kept in the ethanol for 5 min permitting the ethanol in the pores of the porous scaffold.

The total volume of the ethanol and the scaffold was thus recorded as V_2.

At last, the prepared scaffold was removed from the beaker and the remaining ethanol volume was recorded as V_3.

The porosity of the scaffold (ε) is calculated by:

$$\frac{V_1 - V_3}{V_2 - V_3} \times 100$$

where, $V_1 =$ Initial volume of ethanol

$V_2 =$ volume of ethanol after scaffold is immersed

$V_3 =$ volume remained for ethanol after removal of the scaffold

4.1.1 Porosity Testing for Hybrid Scaffolds

The porosity of various hybrid scaffolds and control scaffolds is given in Fig. 14. The porosity of different scaffolds which includes C, CA, CAG1, CAG2, CAG3 is shown in Fig. 14a. The porosity testing results showed that porosity of ternary scaffold is near to 86–87 % which is comparable to that of pure chitosan and chitosan alginate and chitosan–gelatin scaffold. The possible reason for the increase in porosity of chitosan alginate gelatin scaffold as compared to the two hybrid

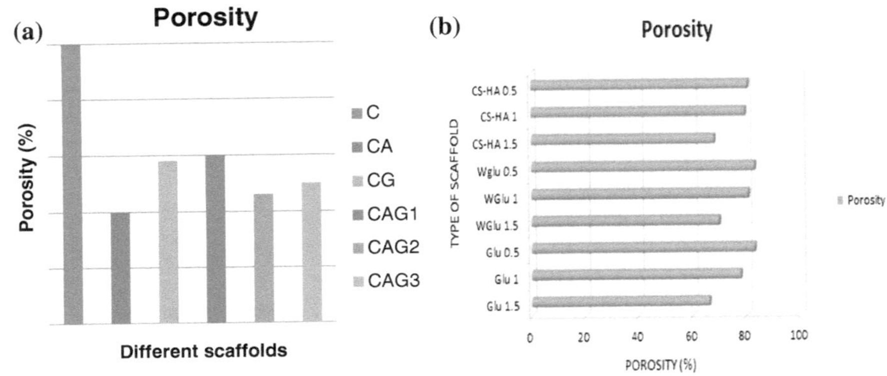

Fig. 14 **a** Graph showing porosity of different scaffolds C, CA, CAG1, CAG2, CAG3 (*C* chitosan, *CA* chitosan–alginate, *CG* chitosan–gelatin). **b** Graph showing porosity of different scaffolds Glu, WGlu, SC-HA (*Glu* With Glutaraldehyde; *WGlu* Without Glutaraldehyde; *CS–HA* Chitosan–Hydroxyapetite)

scaffolds may be due to the effect of different concentration of individual components and also the environmental factor that varies the porosity of each scaffold, and as seen not much difference is seen in the porosities of each scaffold. Preceding studies have shown that the scaffolds with porosity higher than 85 % can be suitable for tissue engineering applications [1, 99, 100]. Liuyun et al. [101] determines the porosity of CS-HA-Carboxy cellulose scaffold by the liquid displacement method which shows that porosity decreases with increasing HA concentration. Figure 14b shows that not only HA concentration but also cross-linking affects porosity of the scaffold. Scaffolds prepared using cross-linker are less porous in comparison to the scaffold prepared without using cross-linker and it also shows that in every scaffold porosity decreases with increasing HA concentration. This criterion arises from the demand for the cell ingrowth in sufficient space. Therefore, the prepared scaffold satisfied this criterion suitable for tissue engineering.

4.1.2 Porosity Testing for Chitosan Scaffolds

The porosity testing results revealed that chitosan I (low MW) scaffolds were more porous than chitosan II (high MW) scaffolds. Figure 15 shows that the porosity of scaffolds also decreases as the concentration of chitosan increases, i.e., 1 % is more porous than 1.5 % and 2 % is the least porous. This statement is true for both chitosan I and chitosan II. Chitosan I > chitosan II, and 1 % > 1.5 % > 2 %.

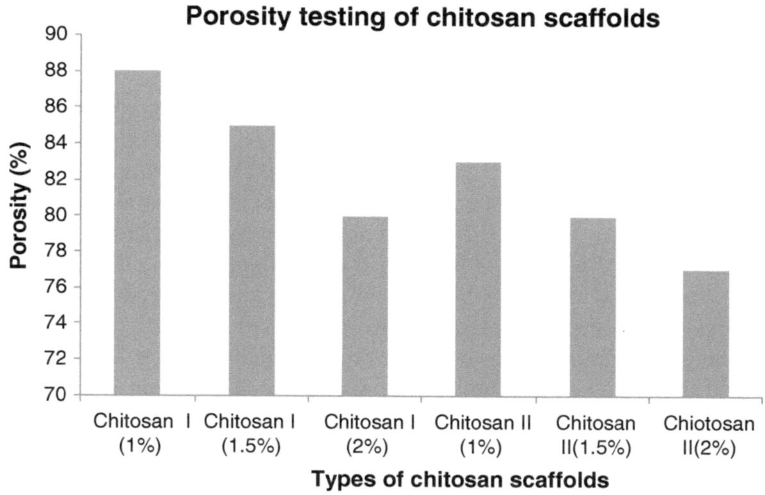

Fig. 15 Graph showing porosity of different scaffolds chitosan I and chitosan II (chitosan I-low Molecular weight chitosan) and (chitosan II-high Molecular weight chitosan)

4.2 Effective Young's Modulus

In the scaffolds, stress is marginally increased with the increase in the strain, which is in agreement with the performance of the polymer foam solids. Particularly, two different regions are followed; a plateau due to the plastic collapse, and an extra region where the stress increases hastily due to the efficient densification of the foam structure [102, 103].

Young's modulus, E, can be calculated after taking the ratio between the tensile stress by the tensile strain

$$E = \frac{\sigma}{\epsilon} = \frac{F/A_0}{\Delta L/L_0} = \frac{FL_0}{A_0 \Delta L} \quad (1)$$

Here,

E is the Young's modulus of elasticity, F is the applied force of the object, A_0 is the original cross-sectional area by which the force is applied, ΔL is the amount of change in the length of the object, and L_0 is the initial length of the object. Here, the average stress is used normally to calculate an Effective Young's Modulus.

This section will illustrate about the typical load versus compression and stress–strain features for various composite scaffolds. The scaffolds were analyzed until the samples were compressed to 90 % of its original height approximately.

Byrane et al. [101] states that the scaffold properties for different types of tissue range from 0.2 to 6000 MPa, i.e., 0.1, 0.2, 2, 10, 1000, 6000 MPa for soft tissue, granular tissue, fibrous tissue, cartilage, immature bone, and mature bone, respectively.

In case of chitosan–alginate scaffolds, the Young's modulus was found to be 0.148, 0.575, 0.309 MPa for CAG1, CAG2, CAG3 (CAG: Chitosan–Alginate with different composition), respectively. It can be noticed that the compressive strength of CAG2 was higher, which may be associated to the low porosity of scaffolds and the presence of more alginate content in CAG2 increases its mechanical strength from other ones. The obtained Young's modulus is in agreement with the data given by Byrane et al. [101] for different tissues, so it can be used for granular and epithelial tissues. The load versus compression graph is shown in Fig. 16.

Figure 17 shows the graph between load and displacement, and Fig. 18 shows graph between stress and strain of chitosan I (after and before degradation), chitosan II (Before and after degradation), bare and coated scaffolds.

The Young's Modulus of chitosan I before degradation was 0.23407 MPa from Fig. 17 and the Young's Modulus of chitosan I after degradation was 0.309 MPa. Young's modulus was obtained from stress–strain curve which finds out to be 0.1303693 MPa. On comparing the values of Young's Modulus for chitosan I before and after degradation, the values were higher after degradation. Thus results state that mechanical strength of scaffold of chitosan I increased after degradation.

The graph also shows that the Young's Modulus was 0.335 M Pa of chitosan II before degradation and the Young's Modulus obtained was 0.575 MPa of chitosan II after degradation. Similarly, on comparing the values of Young's Modulus for chitosan II before and after degradation we concluded that, the values were higher after degradation. Figures 17 and 18 also show the Load versus Displacement plot of bare and coated scaffolds and the stress–strain curve of bare scaffold and coated scaffolds, respectively. Young's modulus was calculated from stress–strain curve which finds out to be 0.1303693 MPa. The Young's modulus of coated scaffold was 0.274016 which is much higher in comparison to the bare scaffold. Hence it shows that mechanical strength improves after coating.

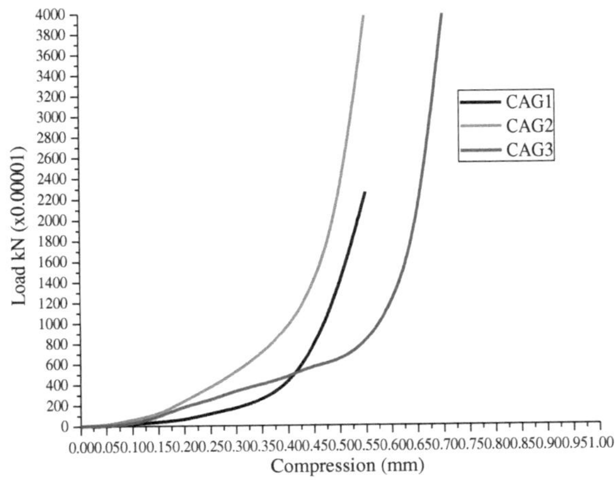

Fig. 16 a Load versus compression graph of CAG1, CAG2, and CAG3. **b** Graph of stress versus strain graph of CAG1, CAG2, and CAG3

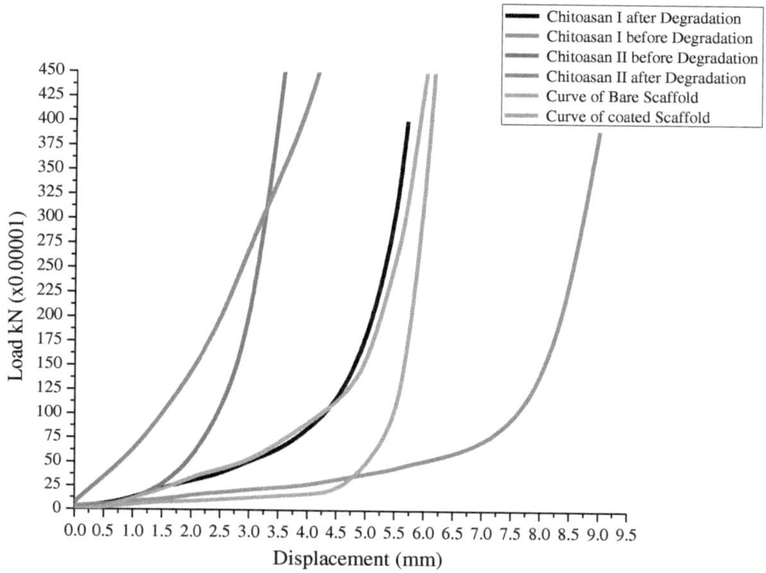

Fig. 17 Graph of stress versus strain of chitosan I (after and before degradation), chitosan II (before and after degradation), Bare and coated scaffolds

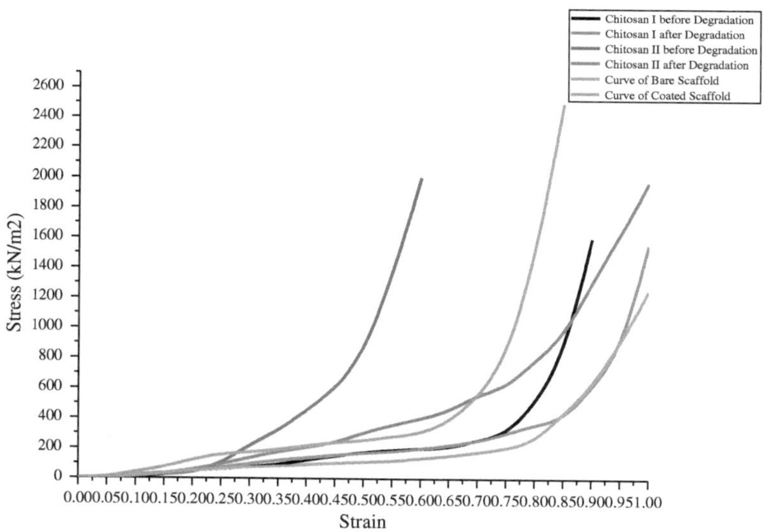

Fig. 18 On comparing chitosan I and II original samples, we concluded that, chitosan II is more mechanically stable than chitosan I. The results of the compressive tests were used to calculate the modulus of elasticity, i.e., the Young's Modulus of the different scaffolds

5 Modeling of Unit Cell of Scaffold

The design of the model unit cell is achieved with the Pro/Engineering CAD package which allows the use of three different biomaterials having different porosities. Like, Hydroxyapatite (HA, $E = 14{,}000$ MPa, $\nu = 0.25$), Polycaprolactone (PCL, $E = 400$ MPa, $\nu = 0.33$), and a copolymer of polylactic acid and polyglycolicacid (PLGA, $E = 1200$ MPa, $\nu = 0.33$) were taken [104]. If the edge length of the unit cell and the edge length of the square channel are L and A, respectively, the porosity of the scaffold can be determined by:

$$p = \frac{3LA^2 - 2LA^3}{L^3} A < L,$$

for square conduit as shown in Fig. 19.

In the design of the scaffold model using six different pore sizes (Fig. 20), the models were imported into the ANSYS software for stress analysis with different mechanical properties for all pore size models. The models were discredited into tetrahedral segments (Fig. 21) of size 0.5 mm through meshing. After discretization displacement of 0.1 mm was given to every pore size model in the positive direction of the X-axis. A total of 18 simulations were done with different porosities ranging between 0 and 95 %. An Effective Young's Modulus was then obtained which describes the mechanical behavior of the scaffold with an increase in its porosity.

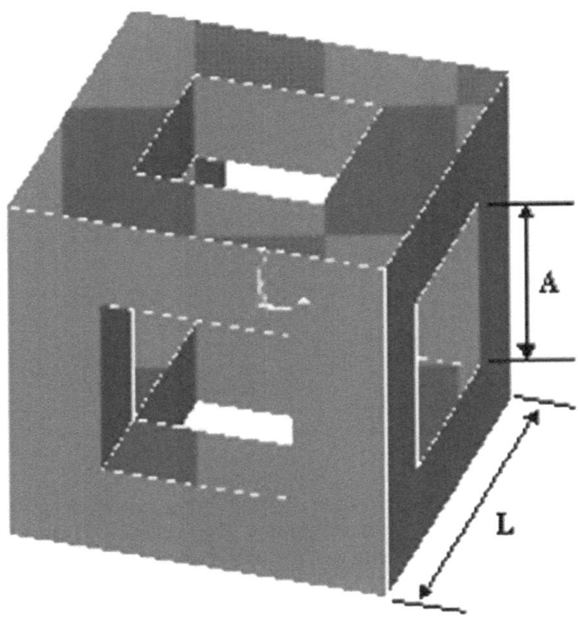

Fig. 19 Unit cell of a scaffold consists of open square pore

Fig. 20 Scaffold unit cells of six different pore sizes

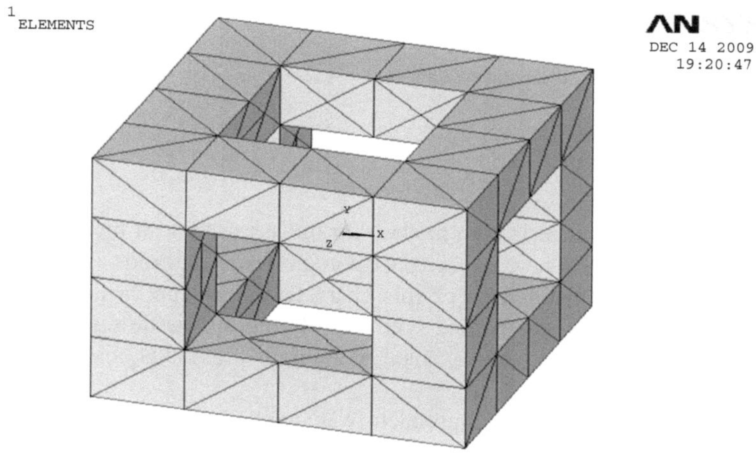

Fig. 21 Discretized unit cell of scaffold

Finally, the results were obtained after 18 simulations run on ANSYS for six different pore size models, and these results were compared with the results obtained by the Fang et al. [104] including the deviation in the effective Young's Modulus to approximate. Figure 22 shows the average stress in the *X* direction from which the Effective Young's Modulus is obtained for a sample porosity of 20 % [105, 106]. In this direction recently the heterogeneous minimal surface porous scaffold design using the distance field and radial basis functions for biomedical applications has also been studied [107–109].

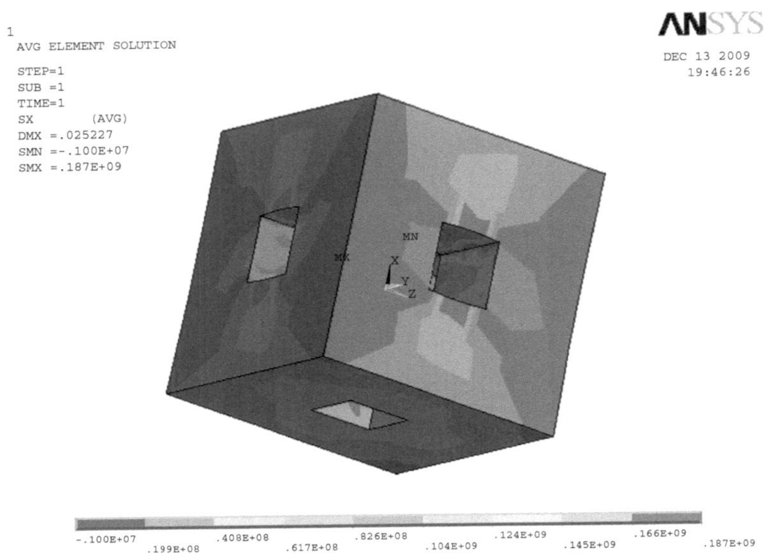

Fig. 22 Average stress in *X* direction

6 Concluding Remarks

A computer-aided characterization method for the assessment of mechanical characteristics and structural heterogeneity of porous tissue scaffolds is presented in this chapter. The significant idea of this characterization approach is the use of CATE which permits the design and synthesis of a porous tissue scaffold of accurate pore size and of appropriate mechanical strength where this porous scaffold is implemented.

Also, the above-mentioned approach was applied to predict the efficient mechanical characteristics of porous chitosan tissue scaffolds produced through a precision extruding method and to study the effect of the design and process factors on the structural characteristics of the scaffolds.

These characterization results showed that the effective mechanical properties of the chitosan scaffold are functions of the scaffolding materials, the orientation of the deposition layout pattern, and the overall porosity of the scaffold structure. In general, the scaffold structures behave with anisotropic mechanical properties and the degree of the anisotropy is dependent on the deposition layout pattern.

The chitosan scaffolds show the same type of deviation in their mechanical properties, since with the increase in porosity, the mechanical properties decrease. However, the mechanical strength of chitosan is very low, so it can be concluded that chitosan has a potential biomaterial and can be used in the repairing of the nerve tissues. Chitosan has been enormously studied in the biomedical field because of its tremendous biodegradable and biocompatible properties.

Lyophilization and freeze–drying are the techniques that are used to generate the porous chitosan tissue scaffold through which the size of porous tissue scaffold is controlled. This will be helpful in fabrication of an appropriate mechanical strength tissue scaffold. CATE will be helpful in the fabrication of chitosan tissue scaffold for bone and soft tissue like blood vessels, liver, skin, etc. with exact porosity and mechanical strength.

Acknowledgments One of the authors (Tanvi Jain) gratefully acknowledged the financial support from TEQIP-II, MNNIT Allahabad for her Ph.D. research work.

Appendix

Allograft: The transplant of an organ or tissue from one individual to another of the same species with a different genotype.

Computer-aided design: Computer-aided design (CAD) is the use of computer technology for the design of objects, real or virtual. CAD often involves more than just shapes. As in the manual drafting of technical and engineering drawings, the output of CAD must convey information, such as materials, processes, dimensions, and tolerances, according to application-specific conventions.

Freeze–drying: Freeze–drying or lyophilization is a dehydration process typically used to preserve a perishable material or make the material more convenient for transport. Freeze–drying works by freezing the material and then reducing the surrounding pressure and adding enough heat to allow the frozen water in the material to sublime directly from the solid phase to the gas phase.

Polysaccharide: Polysaccharides are polymeric carbohydrate structures, formed of repeating units (either mono- or disaccharides) joined together by glycosidic bonds.

Tissue: Tissue is a cellular organizational level intermediate between cells and a complete organism. Hence, a tissue is an ensemble of cells, not necessarily identical, but from the same origin, that together carry out a specific function. Organs are then formed by the functional grouping together of multiple tissues.

Tissue Scaffolds: Cells are often implanted or "seeded" into an artificial structure capable of supporting three-dimensional tissue formation.

References

1. Hutmacher DW (2000) Scaffolds in tissue engineering bone and cartilage. Biomaterials 21:2529–2543
2. Monique M, Dietmar WH (2007) Biodegradable polymers applied in tissue engineering research: a review. Polym Int 56:145–157
3. Singh BK, Sirohi R, Archana D, Jain A, Dutta PK (2015) Porous chitosan scaffolds: a systematic study for choice of crosslinker and growth factor incorporation. Int J Polym Mater 64:242–252

4. Sun W, Starly B, Nam J, Darling A (2005) Bio- CAD modeling and its application in computer-aided tissue engineering. Comput Aided Des 37:1097–1114
5. Sun W, Pallavi L (2002) Recent development on computer aided tissue engineering—a review. Comput Meth Prog Bio 67:85–103
6. Nguyen CM, Levy AJ (2011) Mechanics of interface failure in the trilayer elastic composite. Int J Solids Struct 48:2467–2484
7. Sornsuwan T, Swain MV (2012) The effect of margin thickness, degree of convergence and bonding interlayer on the marginal failure of glass-simulated all-ceramic crowns. Acta Biomater 8:4426–4437
8. Liu PF, Zheng JY (2010) Recent developments on damage modeling and finite element analysis for composite laminates. Mater Design 31:3825–3834
9. Hassania S, Bielawskic M, Beresc W, Balazinskib M, Martinua L, Klemberg-Sapieha JE (2010) Impact stress absorption and load spreading in multi-layered erosion-resistant coatings. Wear 268:770–776
10. Lee KS, Jang KS, Park JH, Kim TW, Han IS, Woo SK (2011) Designing the fiber volume ratio in SiC fiber-reinforced SiC ceramic composites under Hertzian stress. Mater Design 32:4394–4401
11. Kumar H, Srivastava R, Dutta PK (2013) Highly luminescent chitosan-l-cysteine functionalized CdTe quantum dots film: Synthesis and characterization. Carbohydr Polym 97:327–334
12. Hsueh CH, Kelly JR (2009) Simple solutions of multilayered discs subjected to biaxial moment loading. Dent Mater 25:506–513
13. Dutta J, Chattopadhyaya MC, Tripathi VS, Dutta PK (2004) Chitin & chitosan: novel biomaterials waiting for future developments. J Polym Mater 21:321–332
14. Jain T, Dutta PK (2011) Chitin nanoparticles for drug delivery: preparation, characterization and evaluation. Asian Chitin J 7:13–18
15. Dutta PK, Tripathi S, Mehrotra GK, Dutta J (2009) Perspectives for chitosan based antimicrobial films in food applications. Food Chem 114:1173–1182
16. Khang D, Choi J, Im YM, Kim YJ, Jang JH, Kang SS, Nama TH, Song J, Park JW (2012) Role of subnano-, nano- and submicron-surface features on osteoblast differentiation of bone marrow mesenchymal stem cells. Biomaterials 33:5997–6007
17. Phongying S, Aiba S, Chirachanchai S (2007) Direct chitosan nanoscaffold formation via chitin whiskers. Polymer 48:393–400
18. Wang AJ, Cao WL, Ao Q, Gong K, Kong LJ, Gong YD, Zhang XF (2006) Controlling morphology and porosity of 3-D chitosan scaffolds produced by thermally induced phase separation technique. Asian Chitin J 2:69–78
19. Wang AJ, Cao WL, Gong K, Ao Q, LiJun K, He CZ, Gong YD, Zhang XF (2006) Development of porous chitosan tubular scaffolds for tissue engineering applications. Asian Chitin J 2:53–60
20. Zheng JP, Wang CZ, Wang XX, Wang HY, Yao KD (2007) Preparation of biomimetic three-dimensional gelatin/montmorillonite-chitosan scaffold for tissue engineering. React Funct Polym 67:780–788
21. Dutta PK, Dutta J, Tripathi VS (2004) Chitin and chitosan: chemistry, properties and applications. J Sci Ind Res 63:20–31
22. Peter M, Binulal NS, Nair SV, Selvamurugan N, Tamura H, Jayakumar R (2010) Novel biodegradable chitosan–gelatin/nano-bioactive glass ceramic composite scaffolds for alveolar bone tissue engineering. Chem Eng J 158:353–361
23. Chua CK, Leong KF, Cheah CM, Chua SW (2003) Development of a tissue engineering scaffold structure library for rapid prototyping. Part 2: Parametric library and assembly program. Int J Adv Manuf Tech 21:302–312
24. Starly B, Lau W, Bradbury T, Sun W (2006) Internal architecture design and freeform fabrication of tissue replacement structures. Comput Aided Des 38:115–124
25. Cho TW, Cho YN, Chung SH, Yoo G, Ko SW (1999) Water-soluble chitin as a wound healing accelerator. Biomaterials 20:2139–2145

26. Ma J, Wang H, He B, Chen J (2001) A preliminary in vitro study on the fabrication and tissue engineering applications of a novel chitosan bilayer material as a scaffold of human fetal dermal fibroblasts. Biomaterials 22:331–336
27. Ueno H, Yamada H, Tanaka I, Kaba N, Matsuura M, Okumura M, Kadosawa T, Fujinaga T (1999) Accelerating effects of chitosan for healing at early phase of experimental open wound in dogs. Biomaterials 20:1407–1414
28. Haipeng G, Yinghui Z, Jianchun L, Yandao G, Nanming Z, Xiufang Z (2000) Studies on nerve cell affinity of chitosan-derived materials. J Biomed Mater Res 52:285–295
29. Taravel MN, Domard A (1995) Collagen and its interaction with chitosan II. Influence of the physicochemical characteristics of collagen. Biomaterials 16:865–871
30. Taravel MN, Domard A (1996) Collagen and its interaction with chitosan III. Some biological and mechanical properties. Biomaterials 17:451–455
31. Jong WR (2004) Physical and mechanical properties of water resistant sodium alginate films. LWT—Food Sci Technol 37:323–330
32. Rossi F, Chatzistavrou X, Perale G, Boccaccini AR (2012) Synthesis and degradation of agar-carbomer based hydrogels for tissue engineering applications. J Applied Polym Sci 123:398–408
33. Chien KB, Shah RN (2012) Novel soy protein scaffolds for tissue regeneration: material characterization and interaction with human mesenchymal stem cells. Acta Biomater 8:694–703
34. Cheatham TE, Young MA (2000) Molecular dynamics simulation of nucleic acids: successes, limitations and promise. Biopolymers 56:232–256
35. Koutsopoulos S (2002) Synthesis and characterization of hydroxyapatite crystals: a review study on the analytical methods. J Biomed Mater Res 62:600–612
36. Li Z, Tan BH (2014) Towards the development of polycaprolactone based amphiphilic block copolymers: molecular design, self-assembly and biomedical applications. Mater Sci Eng C Mater Biol Appl 45:620–634
37. Mehdikhani NM, Fathi MH, Mortazavi V, Mousavi SB, Akhavan A, Haghighat A, Hashemi-Beni B, Razavi SM, Mashhadiabbas F (2015) Biodegradable nanocomposite coatings accelerate bone healing: in vivo evaluation. Dent Res J 12:89–99
38. Ratner BD, Hoffman AS, Schoen FJ, Lemons JE (eds.) (2004) Biomaterials science: an introduction to materials in medicine, Academic Press
39. Ratner BD, Hoffman AS, Schoen FJ, Lemons JE (Third Edition) (2013) Introduction—biomaterials science: an evolving, multidisciplinary endeavour. Biomater Sci 25–39
40. Sondergaard CS, Witt R, Mathews G, Najibi S, Le L, Clift T, Si MS (2012) Prevascularization of self-organizing engineered heart tissue by human umbilical vein endothelial cells abrogates contractile performance. Cell Tissue Res 9:1492–1497
41. Cicuendez M, Izquierdo BI, Sanchez SS, Vila M, Vallet RM (2012) Biological performance of hydroxyapatite–biopolymer foams: in vitro cell response. Acta Biomater 8:802–810
42. Zimmermann M, Flechsig C, Monica NL, Tripodi M, Adler G, Dikopoulos N (2008) Hepatitis C virus core protein impairs in vitro priming of specific T cell responses by dendritic cells and hepatocytes. J Hepatol 48:51–60
43. Annabi N, Mithieux SM, Boughton EA, Ruys AJ, Weiss AS, Dehghani F (2009) Synthesis of highly porous crosslinked elastin hydrogels and their interaction with fibroblasts in vitro. Biomaterials 30:4550–4557
44. Jell G, Minelli C, Stevens MM, Meyer U, Handschel J, Wiesmann HP, Meyer T (eds) (2009) Fundamentals of tissue engineering and regenerative medicine. Springer publishers, Germany
45. Zhang Y (2012) Damage resistance of graded ceramic restorative materials. J Eur Ceram Soc 32:2623–2632
46. Qiang AO, Wang A, Cao W, Zhao W, Gong Y, Zhao N, Zhang X (2005) Fabrication and characterization of chitosan nerve conduits with microtubular architecture. Tsinghua Sci Technol 10:435–436

47. Sahai N, Tiwari RP (2012) Recent development in finite element methods and computer aided design in the development of porous scaffolds—a review. J Tissue Sci Eng 3:113
48. Sahai N (2012) Computer aided design in the development of porous scaffolds, M.Tech Thesis submitted to MNNIT Allahabad, India
49. Miranda P, Pajares A, Saiz E, Tomsia AP, Guiberteau F (2008) Mechanical properties of calcium phosphate scaffolds fabricated by robocasting. J Biomed Mater Res A 85:218–227
50. Adachi T, Osako Y, Tanaka M, Hojo M, Hollister SJ (2006) Framework for optimal design of porous scaffold microstructure by computational simulation of bone regeneration. Biomaterials 27:3964–3972
51. Miranda P, Pajares A, Guiberteau F, Deng Y, Lawn BR (2003) Designing damage-resistant brittle-coating structures. Bilayers Acta Mater 51:4347–4356
52. Miranda P, Pajares A, Guiberteau F (2008) Finite element modeling as a tool for predicting the fracture behavior of robocast scaffolds. Acta Biomater 4:1715–1724
53. Lacroix D, Chateau A, Ginebra MP, Planell A (2006) Micro-finite element methods of bone tissue-engineering scaffolds. Biomaterials 27:5326–5334
54. Hollister SJ, Maddox RD, Taboas JM (2002) Optimal design and fabrication of scaffolds to mimic tissue properties and satisfy biological constraints. Biomaterials 23:4095–4103
55. Olivares AL, Marsal E, Planel JA, Lacroix D (2009) Finite element study of scaffold architecture design and culture conditions for tissue engineering. Biomaterials 30:6142–6149
56. Guo Z, You S, Wang X, Bicanic N (2010) A FEM-based direct method for material reconstruction inverse problem in soft tissue elastography. Comput Struct 88:1459–1468
57. Chua CK, Leong KF, Cheah CM, Chua SW (2003) Development of a Tissue Engineering scaffold structure library for rapid prototyping. Part 1: investigation and classification. Int J Adv Manuf Tech 21:291–301
58. Sahai N, Jain T, Kumar S, Dutta PK (2015) Chitosan based porous scaffolds by Computer Aided Tissue Engineering. Asian Chitin J 10:11–18
59. Voor MJ, Yang S, Burden RL, Waddell SW (2008) In vitro micro-CT scanning of a rabbit distal femur: repeatability and reproducibility. J Biomech 41:186–193
60. Watling CP, Lago N, Benmerah S, FitzGerald JJ, Tarte E, McMahon S, Lacour SP, Cameron RE (2010) Novel use of X-ray micro computed tomography to image rat sciatic nerve and integration into scaffold. J Neurosci Methods 188:39–44
61. Giglio ED, Cometa S, Ricci MA, Zizzi A, Cafagna D, Manzotti S, Sabbatini L, Mattioli BM (2010) Development and characterization of rhVEGF-loaded poly(HEMA–MOEP) coatings electrosynthesized on titanium to enhance bone mineralization and angiogenesis. Acta Biomater 6:282–290
62. Tanigawa S, Aida Y, Kawato T, Honda K, Nakayama G, Motohashi M, Suzuki N, Ochiai K, Matsumura H, Maeno M (2011) Interleukin-17F affects cartilage matrix turnover by increasing the expression of collagenases and stromelysin-1 and by decreasing the expression of their inhibitors and extracellular matrix components in chondrocytes. Cytokine 56:376–386
63. Puppi D, Chiellini F, Piras AM, Chiellini E (2010) Polymeric materials for bone and cartilage repair. Prog Polym Sci 35:403–440
64. Wang J, Ye M, Liu Z, Wang C (2009) Precision of cortical bone reconstruction based on 3D CT scans. Comput Med Imag Grap 33:235–241
65. Galibarov PE, Prendergast PJ, Lennon AB (2010) A method to reconstruct patient-specific proximal femur surface models from planar pre-operative radiographs. Med Eng Phys 32:1180–1188
66. Rendina E, Lim YF, Marlow D, Wang Y, Clarke SL, Kuvibidila S, Lucas EA, Smith BJ (2012) Dietary supplementation with dried plum prevents ovariectomy-induced bone loss while modulating the immune response in C57BL/6 J mice. J Nutr Biochem 23:60–68
67. Li J, Zhang L, Lv S, Li S, Wang N, Zhang Z (2011) Fabrication of individual scaffolds based on a patient-specific alveolar bone defect model. J Biotech 151:87–93

68. Dash TK, Konkimalla BV (2012) Poly-ε-caprolactone based formulations for drug delivery and tissue engineering. J Control Release 158:15–33
69. Billiet T, Vandenhaute M, Schelfhout J, Vlierberghe SV, Dubruel P (2012) A review of trends and limitations in hydrogel-rapid prototyping for tissue engineering. Biomaterials 33:6020–6041
70. Baas Elbert, Jan HK (2008) A numerical model of heterogeneous surface strains in polymer scaffolds. J Biomech 41:1374–1378
71. Alves A, Duarte ARC, Mano JF, Sousa RA, Reis RL (2012) PDLLA enriched with ulvan particles as a novel 3D porous scaffold targeted for bone engineering. J Supercrit Fluid 65:32–38
72. Chen JSS, Feng HY (2011) Optimal layer setup generation in layered manufacturing with a given error constraint. J Manuf Syst 30:165–174
73. Li W, Lin D, Rungsiyakull C, Zhou S, Swain M, Li Q (2011) Finite element based bone remodeling and resonance frequency analysis for osseointegration assessment of dental implants. Finite Elem Anal Des 47:898–905
74. Vo TN, Kasper FK, Mikos AG (2012) Strategies for controlled delivery of growth factors and cells for bone regeneration. Adv Drug Deliver Rev 64:1292–1309
75. Marcos-Campos I, Marolt D, Petridis P, Bhumiratana S, Schmidt D, Vunjak NG (2012) Bone scaffold architecture modulates the development of mineralized bone matrix by human embryonic stem cells. Biomaterials 33:8329–8342
76. Davidenko N, Gibb T, Schuster C, Best SM, Campbell JJ, Watson CJ, Cameron RE (2012) Biomimetic ollagen scaffolds with anisotropic pore architecture. Acta Biomater 8:667–676
77. Sharifpoor S, Simmons CA, Labow RS, J. Santerre JP (2011) Functional characterization of human coronary artery smooth muscle cells under cyclic mechanical strain in a degradable polyurethane scaffold. Biomaterials 32:4816–4829
78. Lin S, Mequanint K (2012) The role of Ras-ERK-IL-1β signaling pathway in upregulation of elastin expression by human coronary artery smooth muscle cells cultured in 3D scaffolds. Biomaterials 33:7047–7056
79. Saito E, Liu Y, Migneco F, Hollister SJ (2012) Strut size and surface area effects on long-term in vivo degradation in computer designed poly(L-lactic acid) three-dimensional porous scaffolds. Acta Biomater 8:2568–2577
80. Fan X, Case ED, Ren F, Shu Y, Baumann MJ (2012) Part I: Porosity dependence of the Weibull modulus for hydroxyapatite and other brittle materials. J Mech Behav Biomed Mater 8:21–36
81. Vivanco J, Aiyangar A, Araneda A, Ploeg HL (2012) Mechanical characterization of injection-molded macro porous bioceramic bone scaffolds. J Mech Behav Biomed Mater 9:137–152
82. Manwart C, Torquato S, Hilfer R (2000) Stochastic reconstruction of sandstones. Phys Rev 62:893
83. Li J, Regli W, Sun W (2005) A case study of swept volume representation of tissue scaffolds. In: Bioengineering conference, proceedings of the IEEE 31st annual Northeast, pp 182–183
84. Wettergreen MA, Bucklen BS, Sun W et al (2005) Creation of a unit block library of architectures for use in assembled scaffold engineering. Comput Aided Des 37:1141–1149
85. Lal P, Sun W (2004) Computer modeling approach for microsphere-packed bone scaffold. Comput Aided Des 36:487–497
86. Schroeder C, Regli WC, Shokoufandeh A, Sun W (2005) Computer-aided design of porous artifacts. Comput Aided Des 37:339–353
87. Cai S, Xi J (2008) A control approach for pore size distribution in the bone scaffold based on the hexahedral mesh refinement. Comput Aided Des 40:1040–1050
88. Woodfield TBF, Malda J (2004) Design of porous scaffolds for cartilage tissue engineering using a three-dimensional fiber-deposition technique. Biomaterials 25:4149–4161

89. Mehta AS, Singh BK, Singh N, Archana D, Snigdha K, Harniman R, Rahatekar SS, Tewari RP, Dutta PK (2015) Chitosan silk-based three-dimensional scaffolds containing gentamicin encapsulated calcium alginate beads for drug administration and blood compatibility. J Biomater Appl 29:1314–1325
90. Tai H, Mather ML, Howard D (2007) Control of pore size and structure of tissue engineering scaffolds produced by supercritical fluids processing. Eur Cell Mater 14:64–77
91. Sanchez CT, Corney JR (2008) Effects of ultrasound on polymeric foam porosity. Ultrasonic Sonochem 15:408–415
92. Hollister Scott J (2005) Porous scaffold design for tissue engineering. Nat Mater 4:518–524
93. Tripathi G, Basu B (2012) A porous hydroxyapatite scaffold for bone tissue engineering: physico-mechanical and biological evaluations. Ceramic Int 38:341–349
94. Zhensheng L, Hassna RR, Kip DH, Demin X, Miqin Z (2005) Chitosan–alginate hybrid scaffolds for bone tissue engineering. Biomaterials 26:3919–3928
95. Hsieh CY, Tsai SP, Ho MH, Wang DM, Liu CE, Hsieh CH (2007) Analysis of freeze gelation and cross-linking processes for preparing porous chitosan scaffolds. Carbohydr Polym 67:124–132
96. Mao JS, Liu HF, Yin YJ, Yao KD (2003) The properties of chitosan-gelatin membranes and scaffolds modified with hyaluronic acid by different methods. Biomaterials 24:1621–1629
97. Kang Y, Xu X, Yin G, Chen A, Liao L, Yao Y, Huang Z, Liao X (2007) A comparative study of the *vitro* Degradation of Poly (l-lactic acid)/β-tricalcium phosphate scaffold in Static and dynamic stimulated body fluid. Eur Polym J 43:1768–1778
98. Kim Cui J, Goh J, Kim P, Choi S, Lee B (2000) Survival and stability of Bifidobacteria loaded in alginate poly-Llysine microparticles. Int J Pharm 210:51–59
99. Jin HH, Lee CH, Lee WK, Lee JK, Park HC, Yoon SY (2008) In-situ formation of the hydroxyapatite/chitosan-alginate composite scaffolds. Mat Lett 62:1630–1633
100. Byrane DP, Lacroix D, Planell JA, Kelly DJ, Prendergast PJ (2007) Simulation of tissue differentiation in a scaffold as a function of porosity, Young's modulus and dissolution rate: Application of mechanobiological models in tissue engineering. Biomaterials 28:5544–5554
101. Liuyun J, Yubao L, Chengdong X (2009) Preparation and biological properties of a novel composite scaffold of nano-hydroxyapatite/chitosan/carboxymethyl cellulose for bone tissue engineering. J Biomed Sci 16:65–75
102. Saey TH, Dietmar W, Hutmacher A (2006) comparison of micro CT with other techniques used in the characterization of scaffolds. Biomaterials 27:1362–1376
103. Ravi M, Warkhedkar Amba DB (2009) Material- solid modeling of human body: a heterogeneous B-spline based approach. Comput Aided Des 4:586–597
104. Fang Z, Starly B, Sun W (2005) Computer-aided characterization for effective mechanical properties of porous tissue scaffolds. Comput Aided Des 37:65–72
105. Yoo D (2012) Heterogeneous minimal surface porous scaffold design using the distance field and radial basis functions. Med Eng Phys 34:625–639
106. Landers R, Hübner U, Schmelzeisen R, Mülhaupt R (2002) Rapid prototyping of scaffolds derived from thermoreversible hydrogels and tailored for applications in tissue engineering. Biomaterials 23:4437–4447
107. Zeinalwan Hutmacher DW (2002) Fused deposition modeling of novel scaffold architectures for tissue engineering applications. Biomaterials 23:1169–1185
108. Sandino C, Planell JA, Lacroix D (2008) A finite element study of mechanical stimuli in scaffolds for bone tissue engineering. J Biomech 41:1005–1014
109. Martino AD, Sittinger M, Risbud MV (2005) Chitosan: a versatile biopolymer for orthopaedic tissue-engineering. Biomaterials 26:5983–5990

Author Index

P.K. Dutta (ed.), *Chitin and Chitosan for Regenerative Medicine*, Springer Series on Polymer and Composite Materials, DOI 10.1007/978-81-322-2511-9